Advances in Isotope Geochemistry

Series editor

Jochen Hoefs, Göttingen, Germany

More information about this series at http://www.springer.com/series/8152

Nikolaus Gussone
Anne-Desirée Schmitt
Alexander Heuser · Frank Wombacher
Martin Dietzel · Edward Tipper
Martin Schiller

Calcium Stable Isotope Geochemistry

Nikolaus Gussone
Institut für Mineralogie
Universität Münster
Münster
Germany

Anne-Desirée Schmitt
LHyGeS/EOST
Université de Strasbourg
Strasbourg
France

Alexander Heuser
Steinmann-Institut für Geologie,
  Mineralogie und Paläontologie
Universität Bonn
Bonn
Germany

Frank Wombacher
Institut für Geologie und Mineralogie
Universität zu Köln
Köln
Germany

Martin Dietzel
Institut für Angewandte
  Geowissenschaften
Technische Universität Graz
Graz
Austria

Edward Tipper
Department of Earth Sciences
University of Cambridge
Cambridge
UK

Martin Schiller
Natural History Museum of Denmark
University of Copenhagen
Copenhagen
Denmark

ISSN 2364-5105   ISSN 2364-5113   (electronic)
Advances in Isotope Geochemistry
ISBN 978-3-540-68948-5   ISBN 978-3-540-68953-9   (eBook)
DOI 10.1007/978-3-540-68953-9

Library of Congress Control Number: 2016934957

© Springer-Verlag Berlin Heidelberg 2016
This work is subject to copyright. All rights are reserved by the Publisher, whether the whole or
part of the material is concerned, specifically the rights of translation, reprinting, reuse of
illustrations, recitation, broadcasting, reproduction on microfilms or in any other physical way,
and transmission or information storage and retrieval, electronic adaptation, computer software,
or by similar or dissimilar methodology now known or hereafter developed.
The use of general descriptive names, registered names, trademarks, service marks, etc. in this
publication does not imply, even in the absence of a specific statement, that such names are
exempt from the relevant protective laws and regulations and therefore free for general use.
The publisher, the authors and the editors are safe to assume that the advice and information in
this book are believed to be true and accurate at the date of publication. Neither the publisher nor
the authors or the editors give a warranty, express or implied, with respect to the material
contained herein or for any errors or omissions that may have been made.

Printed on acid-free paper

This Springer imprint is published by Springer Nature
The registered company is Springer-Verlag GmbH Berlin Heidelberg

Preface

First studies on variations in Ca stable isotope compositions were carried out early on (Backus et al. 1964; Hirt and Epstein 1964; Artemov et al. 1966; Miller et al. 1966; Mesheryakov and Stolbov 1967). However, reliable results were not obtained before Russel et al. (1978) published their seminal paper entitled "Ca isotope fractionation on the Earth and other solar system materials". This work facilitates the reliable and accurate determination of Ca stable isotope compositions using double spike—thermal ionisation mass spectrometry. Albeit further improvements were made, their method is still the method of choice for most Ca stable isotope analysis. Only few studies focused on Ca stable isotopes until the late 1990s, since then the number of studies dealing with Ca stable isotope fractionation is steadily increasing. This reflects both analytical advances and promising results obtained in a large variety of applications in Earth and Life sciences, including medicine.

The aim of this book is to provide an overview of fundamentals and reference values for Ca stable isotope research ("Introduction"), current analytical methodologies including detailed protocols for sample preparation and isotope analysis ("Analytical Methods") and different fields of applications including low-temperature mineral precipitation and biomineralisation ("Calcium Isotope Fractionation During Mineral Precipitation from Aqueous Solution" and "Biominerals and Biomaterial"), Earth surface processes and global Ca cycling ("Earth-surface Ca Isotopic Fractionations" and "Global Ca Cycles: Coupling of Continental and Oceanic Processes"), high-temperature processes and cosmochemistry ("High Temperature Geochemistry and Cosmochemistry") and finally human studies and biomedical applications ("Biomedical Application of Ca Stable Isotopes"). These major areas of research are introduced and their current state of the art is discussed and open questions and possible future directions are identified.

We gratefully acknowledge Florian Böhm, Vasileios Mavromatis, Jianwu Tang, Peter Stille, François Chabaux, and Barbara Teichert for discussions and their constructive comments on pre-versions of selected chapters that helped to improve the present version. AH would like to thank Petra Frings-Meuthen, Jörn Rittweger, Steve Galer, Toni Eisenhauer, Katharina Scholz-Ahrens and Jürgen Schrezenmeir for numerous discussions regarding biomedical application of Ca isotopes. FW thanks Jan Fietzke, Michael Wieser and Victoria Kremser for data and figures. Finally, we are grateful to

Jochen Hoefs and Annett Büttner for encouragement, patience and support during the preparation of the manuscripts.

References

Artemov YM, Knorre KG, Strizkov VP et al (1966) $^{40}Ca/^{44}Ca$ and $^{18}O/^{16}O$ isotopic ratios in some calcareous rocks. Geochemistry (USSR) (English Transl.) 3:1082–1086

Backus MM, Pinson WH, Herzog LF et al (1964) Calcium isotope ratios in the Homestead and Pasamonte meteorites and a Devonian limestone. Geochim Cosmochim Acta 28:735–742

Hirt B, Epstein S (1964) A search for isotopic variations in some terrestrial and meteoritic calcium Trans Am Geophys Union 45:113

Meshcheryakov RP, Stolbov MY (1967) Measurement of the isotopic composition of calcium in natural materials. Geochemistry (USSR) (English Transl.) 4, 1001–1003

Miller YM, Ustinov VI, Artemov YM et al (1966) Mass spectrometric determination of calcium isotope variations. Geochem Intern 3, p 929

Russell WA, Papanastassiou DA, Tombrello TA (1978) Ca isotope fractionation on the Earth and other solar system materials. Geochim Cosmochim Acta 42:1075–1090

Contents

Introduction

Frank Wombacher, Anne-Desirée Schmitt,
Nikolaus Gussone and Alexander Heuser

Abstract

The first part of this chapter briefly introduces the application of Ca stable isotopes in Earth sciences and other branches of natural sciences and summarizes the early history of Ca isotope research. Furthermore, tables with fundamental reference data are provided. The second part introduces the principles of equilibrium and kinetic isotope fractionation theory and discusses isotope fractionation in open and closed systems. It briefly touches on the mass-dependence of equilibrium and kinetic isotope fractionation and approaches to obtain equilibrium isotope fractionation factors experimentally. The reference at the end of this chapter provides a list of general introductions to stable isotope geochemistry and to previous reviews that focused on Ca stable isotopes.

Keywords

Calcium stable isotope geochemistry · Equilibrium isotope fractionation · Kinetic isotope fractionation · Rayleigh fractionation · Mass fractionation laws · Calcium isotope fractionation factors

F. Wombacher (✉)
Institut für Geologie und Mineralogie, Universität zu Köln, Köln, Germany
e-mail: fwombach@uni-koeln.de

A.-D. Schmitt
LHyGeS/EOST, Université de Strasbourg, Strasbourg Cedex, France
e-mail: adschmitt@unistra.fr

A.-D. Schmitt
Laboratoire Chrono-Environnement, Université Bourgogne Franche-Comté, Besançon, France

N. Gussone
Institut für Mineralogie, Universität Münster, Münster, Germany
e-mail: nikolaus.gussone@uni-muenster.de

A. Heuser
Steinmann-Institut für Geologie, Mineralogie und Paläontologie, Universität Bonn, Bonn, Germany
e-mail: aheuser@uni-bonn.de

© Springer-Verlag Berlin Heidelberg 2016
N. Gussone et al., *Calcium Stable Isotope Geochemistry*,
Advances in Isotope Geochemistry, DOI 10.1007/978-3-540-68953-9_1

1 Introduction to Calcium Stable Isotope Geochemistry

Until the beginning of the 21th century, stable isotope effects for elements others than the traditional light elements C, O, H, N and S where little investigated mainly due to the restrictions of suitable mass spectrometers. In step with analytical advances, the application of Ca isotopes increased steadily (Fig. 1) and Ca isotopes are now applied in different areas as diverse as (bio-) mineralogy, Earth-surface processes, ecosystem science, global geochemical cycles, paleoceanography, cosmochemistry, archeology and biomedical applications. The present book is mainly focused on mass-dependent stable isotope fractionation of Ca within these subjects.

1.1 Alkaline Earth Elements

The alkaline earth metals (Table 1), group 2 in the periodic table, comprise the elements beryllium (Be), magnesium (Mg), calcium (Ca), strontium (Sr), barium (Ba) and radium (Ra). In nature, they do not occur in their elemental state as soft metals. Instead, they easily lose or share their two s-valence electrons forming divalent cations in minerals and in solution. Calcium is the most abundant alkaline earth metal in the Earth's crust, while Mg is almost ten times more abundant than Ca in the bulk Earth (Table 2). Due to their high abundance and the relative large mass-difference between their isotopes, many studies focused on Ca and Mg stable isotopes. In contrast, mass dependent Sr and Ba isotope fractionations are less intensively studied,

but resolvable variations were revealed (Fietzke and Eisenhauer 2006; Halicz et al. 2008; Rüggeberg et al. 2008; von Allmen et al. 2010; Krabbenhöft et al. 2010; Shalev et al. 2013; Pearce et al. 2015; Widanagamage et al. 2015). Beryllium has only one stable isotope and all Ra isotopes are radioactive, thus stable isotope fractionation cannot be readily investigated. In Earth sciences, however, cosmic ray induced ^{10}Be provides an important tracer and ^{226}Ra is of relevance in uranium-series dating.

1.2 Calcium and Its Isotopes

Calcium has one major isotope, ^{40}Ca, with an abundance of about 97 %. Calcium-40 is a primary product of O and Si burning in massive stars. Due to its efficient production, Ca is an abundant element in the universe and the Solar System. The interest into Ca isotope studies stems in part to its high abundance in the Earth's crust and to the ubiquitous presence of minerals, in which Ca is the major cation or one of the major cations. In this regard, Ca is of particular relevance to the biomineralization of tests, shells, bones and teeth. Calcium is also one of the major dissolved components of seawater (presently $\sim$400 ppm) and the most abundant cation in rivers ($\sim$20 ppm) (Gibbs 1972). The concentrations of Ca and other alkaline-earth elements in important terrestrial reservoirs are summarized in Table 2.

The relatively high abundance of Ca is also related to the stability of its nucleus, which contains 20 protons. The number of neutrons ranges for all stable and radioactive Ca nuclides between 15 and 38 (i.e. ^{35}Ca–^{58}Ca) (Pfennig et al. 1998), thus Ca has a total of 24 isotopes (Amos et al. 2011). The "magic number" of protons also accounts for the large number of six stable (or extremely long-lived) naturally occurring Ca isotopes (Table 3) that span a large mass range of eight atomic mass units u (or Da for Dalton) corresponding to a relative mass difference of 20 % between the heaviest and the lightest stable Ca isotope (^{48}Ca and ^{40}Ca). With the nuclides $^{40}_{20}Ca$ and $^{48}_{20}Ca$, Ca is the only element having 2

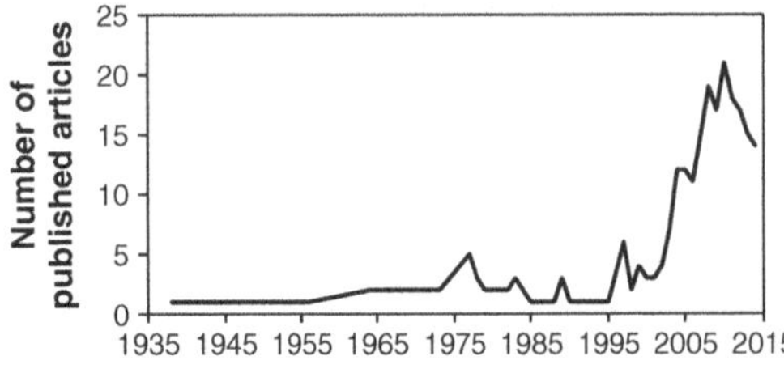

Fig. 1 Number of published articles on Ca isotopes

Table 1 Chemical, physical, and geochemical properties of alkaline earth metals (Lide 1995; Dean 1999)

	Be	Mg	Ca	Sr	Ba	Ra
Atomic number (Z)	4	12	20	38	56	88
Standard atomic weight (g/mol)	9.012182	24.305	40.078	87.62	137.327	226.03
Stable isotopes (nominal mass)	9, (10^a)	24, 25, 26	40^b, (41^a), 42, 43, 44, 46, 48^a	84, 86, 87^b, 88	130, 132, 134, 135, 136, 137, 138	$(226)^a$
Maximum mass difference for stable isotopes in %	(11.1)	8.3	20	4.8	6.2	–
Electron configuration	$[He]2s^2$	$[Ne]3s^2$	$[Ar]4s^2$	$[Kr]5s^2$	$[Xe]6s^2$	$[Rn]7s^2$
Melting point (°C)	1287	650	842	777	727	700
Boiling point (°C)	2471	1090	1484	1382	1897	1737
Density of solid at 25 °C (g/cm^3)	1.848	1.736	1.55	2.63	3.51	5
Oxidation states	2+					
Electronegativity (Pauling units)	1.57	1.31	1	0.95	0.89	0.9
First ionization energy (kJ/mol)	899.5	737.7	589.8	549.5	502.9	509.3
Second ionization energy (kJ/mol)	1757.1	1450.7	1145.4	1064.2	965.2	979
Atomic radius (pm)	111	160	197	215	217	220
Ionic radius (pm):						
[4] coordination	27	57				
[6] coordination	45	72	100	118	136	
[8] coordination		89	112	126	142	148
[12] coordination			135	144	160	170
Atomic volume (cm^3/mol)	4.9	13.97	29.9	33.7	39.24	45.20
Specific heat capacity (J/mol·1/K)	16.443	24.869	25.929	26.4	28.07	27.12
Discovery date	1798	1775	1808	1790	1774	1898
Discovered by	N.L. Vauquelin	J. Black	H. Davy	A. Crawford	C. Scheele	P. & M. Curie
Date of first isolation	1828	1808	1808	1808	1808	1911
Isolated by	F. Wöhler/ A. Bussy	H. Davy	H. Davy	H. Davy	H. Davy	M. Curie

[a] Radioactive isotopes half-lifes (^{226}Ra :1601 years; ^{41}Ca: 103 kyears; ^{48}Ca: 43×10^{18} years; ^{10}Be: 1.387×10^6 years),
[b] radiogenic isotopes (^{40}K $\rightarrow$ ^{40}Ca, see Chapter Analytical Methods; ^{87}Rb $\rightarrow$ ^{87}Sr, 48.8 Gyears)

Table 2 Abundance of Mg, Ca, Sr and Ba in different reservoirs (McDonough and Sun 1995; Lide 1995; McDonough 2001; Rudnick and Gao 2003; Palme and O'Neill 2007)

Reservoir	Mg (wt%)	Ca (wt%)	Sr (ppm)	Ba (ppm)
C I Meteorite	22.17	0.932	7.26	2.41
Bulk silicate Earth	22.8	2.53	20	6.6
Bulk Earth	15.4	1.71	13	4.5
Mantle	22.17	2.6	20	6.7
Continental crust	2.8	4.6	320	456
Average conc. in seawater	0.129	0.041	7.9	0.013
Average conc. in humans	0.027	1.4	4.6	0.3

Table 3 Abundances, masses and nuclear properties of naturally occurring calcium isotopes (Meija et al. 2016; Lide 1995)

Isotope	Atomic mass (u)	Natural abundance (atom %)	Nuclear spin (I)	Magnetic moment (μ/μ_N)	Electric quadrupole moment (b)
^{40}Ca	39.9625909 (2)	96.941 (156)	0	0	0
^{42}Ca	41.958618 (1)	0.647 (23)	0	0	0
^{43}Ca	42.958766 (2)	0.135 (10)	7/2	-1.3173	-0.05
^{44}Ca	43.955482 (2)	2.086 (110)	0	0	0
^{46}Ca	45.95369 (2)	0.004 (3)	0	0	0
^{48}Ca	47.9525228 (8)	0.187 (21)	0	0	0

doubly magic isotopes, i.e. isotopes which have a magic number of protons (20) as well as neutrons (20 and 28). Of all elements, $^{40}_{20}Ca$ is the heaviest stable isotope which has equal numbers of protons and neutrons and ^{40}Ca is also by far the most abundant Ca isotope (Table 3). The heaviest Ca isotope considered as stable (^{48}Ca) is in fact a radioisotope but with a very long half-life (43×10^{18} year). Besides the stable isotopes, Ca has a cosmogenic isotope, ^{41}Ca, which is formed by neutron capture of ^{40}Ca, mainly in the upper few meters of soils. It decays with a half-life of 103 ka to ^{41}K. Radioactive Ca isotopes that are used for medical and research purposes include the aforementioned ^{41}Ca as well as ^{45}Ca and ^{47}Ca, with half-lifes of 163 and 4.54 days, respectively (Pfennig et al. 1998).

In the solid Earth, Ca is a major element in magmatic, metamorphic and sedimentary rocks and a major constituent in minerals from very different mineral groups including carbonates, silicates, sulfates, halogenides and phosphates as well as organo-minerals (e.g. oxalates) (Table 4). Thus, Ca bearing minerals are involved in reactions that take place during weathering, ocean floor alteration, diagenesis, biomineralization, metamorphism, melting and crystallization of rocks. The weathering of Ca bearing rocks, transport of the dissolved Ca into the ocean and precipitation as $CaCO_3$ links tectonic processes to the carbon cycle and hence global climate on geological timescales (Berner et al. 1983).

Calcium plays also an important role in biology, where it is an essential part of biological tissues of animals and plants, and the most abundant cation in products of biomineralization, e.g. as calcite (e.g. foraminifera, coccolithophores, brachiopods), aragonite (e.g. corals), apatite (bones and teeth) and in leaves (Marschner 1995; Weiner and Dove 2004). Due to climate change, modern biomineralization of $CaCO_3$ is affected by ocean acidification.

Calcium is also involved in metabolic processes, like cell signaling and blood coagulation. It is an essential nutrient for plants that plays multiple roles in plant functioning, both at cellular and intracellular scales. For instance, it is important for the formation and stability of the cell walls and the functioning of the cell membranes (Marschner 1995; Taiz and Zeiger 2010).

Some of the chemical reactions involved in biological and geological processes slightly discriminate Ca isotopes depending on their masses. These fractionation patterns can be utilized to study Ca cycling on different temporal and spatial scales or may be used as proxy for different paleo-environmental parameters.

1.3 Notations in Ca Stable Isotope Geochemistry

Before we delve into Ca stable isotope geochemistry, some definitions and notations, for

Table 4 Selected Ca-bearing minerals (Rösler 1988)

Mineral class	Mineral group	Mineral	Chemical composition
Carbonates		Calcite/aragonite/vaterite	$CaCO_3$
		Dolomite	$CaMg(CO_3)_2$
		Ikaite	$CaCO_3 \cdot 6H_2O$
		Monohydrocalcite	$CaCO_3 \cdot H_2O$
		Amorphous calcium carbonate	$CaCO_3 \cdot nH_2O$
Sulphates		Gypsum	$CaSO_4 \cdot 2H_2O$
		Anhydrite	$CaSO_4$
Halogenates		Fluorite	CaF_2
Phosphates		Apatite	$Ca_5(PO_4)_3(OH, F, CO_3)$
Oxides		Perovskite	$CaTiO_3$
Silicates			
Tectosilicates	Feldspars	Anorthite	$CaAl_2Si_2O_8$
	Zeolithes	Heulandite	$(Ca,Na_2,K_2) \, [Al_2 \, Si_7 \, O_{18}] \cdot 6 \, H_2O$
		Chabasite	$Ca \, [Al_2 \, Si_4 \, O_{12}] \cdot 6 \, H_2O$
	Feldspathoids	Lazurite	$(Na,Ca)_8(AlSiO_4)_6(SO_4,S,Cl)_2$
Inosilicates	Pyroxenoid	Wollastonite	$Ca_2Si_2O_6$
	Pyroxenes	Augite	$(Ca,Na)(Mg,Fe,Al,Ti)(Si,Al)_2O_6$
		Diopsite	$MgCaSi_2O_6$
		Hedenbergite	$FeCaSi_2O_6$
		Augite	$(Ca,Na)(Fe^{3+},Mg,Fe^{2+})Si_2O_6$
	Amphibole	Tremolite	$Ca_2Mg_5Si_8O_{22}(OH)$
		Hornblende	$Ca_2(Mg,Fe,Al)_5(Al,Si)_8O_{22}(OH)_2$
		Richterite	$Na_2Ca(Mg,Fe)_5Si_8O_{22}(OH)_2$
		Pargasite	$NaCa_2Mg_3Fe^{2+}Si_6Al_3O_{22}(OH)_2$
Nesosilicates	Garnets	Andradite	$Ca_3Fe_2Si_3O_{12}$
		Grossular	$Ca_3Al_2Si_3O_{12}$
		Uvarovite	$Ca_3Cr_2Si_3O_{12}$
Nesosubsilicates		Titanite	$CaTi[SiO_5]$
Sorosilicates	Epidote	Zoisite/clinozoisite	$Ca_2Al_3(SiO_4)(Si_2O_7)O(OH)$
		Epidote	$Ca_2(Fe^{3+},Al)Al_2(SiO_4)(Si_2O_7)O(OH)$
	Vesuvianite	Vesuvianite	$Ca_{10}Mg_2Al_4(SiO_4)_5(Si_2O_7)_2(OH)_4$
Organo-minerals		Ca oxalate	CaC_2O_4
		Whewellite	$Ca(C_2O_4) \cdot H_2O$
		Weddellite	$Ca(C_2O_4)_2 \cdot 2H_2O$
		Earlandite	$Ca_3(C_6H_5O_2)_2 \cdot 4H_2O$
		Ca tartrate	$C_4H_4CaO_6$
		Ca malate	$C_4H_4CaO_5$

mass-dependent isotope fractionation effects are required. A more detailed discussion of notations for Ca stable isotope geochemistry is given in Chapter Analytical Methods.

Mass-dependent variations of Ca stable isotope ratios in samples are commonly expressed *relative* to a reference standard using the well-known δ-notation (Coplen 2011). Frequently, the ^{44}Ca/^{40}Ca isotope ratio is chosen for the calculation of δ-values:

$$\delta^{44/40}Ca \ = \left(\frac{\left(^{44}Ca/^{40}Ca\right)_{sample}}{\left(^{44}Ca/^{40}Ca\right)_{standard}} - 1 \right) \quad (1)$$

As the variations of the Ca isotope compositions are small, δ-values are typically reported as per mill (‰), i.e. the δ-value obtained by Eq. 1 is then multiplied by 1000. Ratios of heavy over light isotopes are always chosen for the calculation of δ-values; positive δ-values thus refer to samples being enriched in the heavy isotopes relative to the reference standard, negative values denote light isotope enrichments in the sample relative to the standard. Since, Ca has six naturally occurring stable isotopes; many different δ-values can be defined. To avoid ambiguity, Hippler et al. (2003) and Eisenhauer et al. (2004) suggested to denote both, numerator and denominator isotope in the δ-notation, e.g. $\delta^{44/40}$Ca or $\delta^{44/42}$Ca rather than e.g. δ^{44}Ca.

The isotope fractionation factor (α) is defined as the ratio of two isotopes in the substance A divided by the same isotope ratio in substance B:

$$\alpha_{A-B} = \frac{R_A}{R_B} \quad (2)$$

Isotope fractionation effects are temperature dependent in most cases, therefore temperatures need to be reported along with fractionation factors. Knowledge of fractionation factors is central to the discussion of stable isotope data and much effort focusses on the accurate determination of fractionation factors. Fractionation factors can be obtained by calculations based on

either spectroscopic data or ab initio modelling, from laboratory experiments and from natural samples. All three approaches have shortcomings, for example modelling may suffer from incomplete or inaccurate spectroscopic data, force field models and assumptions. Laboratory experiments allow for well-constrained run conditions such as temperature and composition, but may not achieve natural or equilibrium conditions. Fractionation factors determined from natural systems may suffer from conditions that may be less well constrained as in experiments.

The fractionation factor α is related to the δ-values of substances A and B:

$$\alpha_{A-B} = \frac{\delta_A + 1000}{\delta_B + 1000} \quad (3)$$

Note that Eq. 3 applies for δ-values given in ‰. The capital delta notation (Δ) is usually also given in ‰. It describes the offset in the isotope composition between two substances A and B as the difference between their δ-values:

$$\Delta_{A-B} = \delta_A - \delta_B \quad (4)$$

The Δ-values in ‰ can also be obtained from the fractionation factor α according to:

$$\Delta_{A-B} \approx 1000 \ln \alpha_{A-B} \approx 1000 \left(\alpha_{A-B} - 1 \right) \quad (5)$$

Note, δ-values (usually in ‰) relate the isotope composition of a sample to a (measured) reference material, while the fractionation factor α and Δ (usually in ‰) define the fractionation between two substances or compartments, for example between a calcite shell and seawater.

1.4 History of Ca Stable Isotope Research

During the 1920s and 1930s all stable Ca isotopes have been discovered. First, Dempster (1922) reported the discovery and abundances of ^{40}Ca and ^{44}Ca. Aston (1934) reported the discovery of

^{42}Ca and ^{43}Ca followed by the discovery of ^{46}Ca and ^{48}Ca by Nier (1938). A more detailed overview on the discovery of Ca isotopes (^{35}Ca–^{58}Ca) is given by Amos et al. (2011).

The first studies of Ca isotopes attributed the measured variations (with up to 40 ‰ between ^{48}Ca and ^{40}Ca) to analytical artefacts and suggested an absence of fractionation or smaller than the analytical resolution (Backus et al. 1964; Hirt and Epstein 1964; Artemov et al. 1966; Miller et al. 1966; Meshcheryakov and Stolbov 1967; Corless 1968; Letolle 1968; Stahl 1968; Stahl and Wendt 1968; Möller and Papendorff 1971; Heumann and Luecke 1973). Russell et al. (1978) where the first who published a reliable measurement protocol for calcium isotopes using a thermal ionization mass spectrometer (TIMS) and a double spike technique. The main result of their study is that the natural mass-dependent variation of Ca isotopes on Earth is minor (<1.3 ‰ per u) and that high analytical precision and accuracy is necessary to resolve natural differences in Ca isotope ratios. Since then, double-spike thermal ionization mass spectrometry remains the principle method for Ca stable isotope analysis but many improvements were made, resulting in an increasing number of publications from about 2000 onwards (Fig. 1). In addition, precise Ca stable isotope data has also been obtained by MC-ICP-MS (multi collector inductively coupled plasma mass spectrometry) instruments (Halicz et al. 1999) and ion probes (Rollion-Bard et al. 2007). A detailed discussion of these methods is provided in Chapter Analytical Methods.

1.5   Applications of Ca Stable Isotope Geochemistry

The applications of Ca isotopes are ubiquitous and steadily increasing (see Fig. 1). In this book we present the current state of knowledge about this isotope system.

This chapter introduces the fundamental concepts needed for the application of stable isotopes and reports reference data for Ca stable isotope research. Current analytical methodologies are reviewed in Chapter Analytical Methods, including a detailed discussion of sample preparation and isotope analysis.

The two introductionary chapters are followed by chapters that identify and discuss six major areas of research and their current state of the art. Open questions and possible future directions are then identified.

Chapter Calcium Isotope Fractionation During Mineral Precipitation from Aqueous Solution focusses on Ca isotope fractionation related to mineral precipitation and diffusion in aqueous solutions in natural and experimental conditions from an inorganic perspective, which sets the stage for the subsequent Chapter Biominerals and Biomaterial, which is devoted to Ca stable isotope studies that comprise a range of products from biomineralization to studies of ecosystems, paleoclimate and archeology. Both chapters include a discussion on stable isotope fractionation of alkaline earth elements other than Ca for comparison.

Topics in Chapters Earth-surface Ca Isotopic Fractionations and Global Ca Cycles: Coupling of Continental and Oceanic Processes comprise Earth surface processes and global cycling. Chapter Earth-surface Ca Isotopic Fractionations focusses on biotic and abiotic processes that lead to Ca isotopic fractionations in the weathering environment. Chapter Global Ca Cycles: Coupling of Continental and Oceanic Processes describes present-day continental and oceanic budgets as well as past oceanic budget and the causes of its variation through the Phanerozoic.

Chapter High Temperature Geochemistry and Cosmochemistry focusses at stable isotope fractionations related to high-temperature processes in the Earth and on cosmochemical applications of stable Ca isotopes including nucleosynthetic effects. Furthermore, radiogenic ^{40}Ca excesses from ^{40}K decay are briefly summarized. This is also relevant for the interpretation of mass dependent Ca isotope effects as potential complications could arise from the use of isotope ratios involving ^{40}Ca.

Finally in Chapter Biomedical Application of Ca Stable Isotopes, this new and emerging field of application for (Ca) stable isotopes, are presented and discussed.

1.6 Other Applications of Ca Isotopes: Cosmogenic ^{41}Ca and Tracer Studies

Other applications of Ca isotopes exist, they are however not further considered in the present book. These aspects include tracer techniques in medical, earth and life sciences using long and short lived radio-isotopes.

For instance, in medicine, "dual-tracer experiments" have been developed in order to estimate the Ca absorption rate in human gastro-intestinal tract (e.g. Beck et al. 2003; Barger-Lux et al. 1989; López-Huertas et al. 2006). A ^{48}Ca was used to study biomineralization in foraminifers (Lea et al. 1995). In forest studies a ^{42}Ca tracer has been injected under pressure in trees or spread over the ground in order to follow its uptake, its storage in the tree, its return to the soil (litter, leaves) and its fate and behavior after decomposition (Zeller et al. 1998, 2000; Caner et al. 2004; Augusto et al. 2011; van der Heijden et al. 2013).

Short-lived radionuclides of Ca (e.g. ^{45}Ca, ^{47}Ca) were used for instance to study biomineralization of corals and foraminifers (Anderson and Faber 1984), while long-lived radionuclides (^{41}Ca) were used as tracer for medical applications (e.g. Denk et al. 2007; Walczyk et al. 2010; Sharma et al. 2011) and as a proxy for nuclear bomb fallouts (Zerle et al. 1997).

2 Principles of Mass-Dependent Stable Isotope Fractionation

Mass-dependent stable isotope fractionation typically results from isotope exchange reactions under thermodynamic equilibrium or from kinetic processes. Mass-independent isotope fractionation effects as observed for O, S and Hg isotopes were so far not observed for Ca isotopes and are rather unlikely to be of relevance. Nucleosynthetic Ca isotope anomalies, however, are present in some early solar system materials (Chapter High Temperature Geochemistry and Cosmochemistry).

Equilibrium isotope partitioning arises from the change of vibrational frequencies in molecules and condensed phases upon substitution by isotopes of different mass. Equilibrium isotope fractionation is temperature sensitive and becomes negligible at high temperature, which may be at several hundred or larger than 1000 K depending on the magnitude of fractionation and analytical precision.

Kinetic isotope fractionation refers to chemical or physical processes with no or only partial back reaction, i.e. bond breaking chemical reactions or physical transport processes like diffusion or evaporation. Kinetic isotope fractionation due to bond breaking reactions is temperature sensitive too. It results from the effect of isotopic substitution on activation energies required to reach a transition state. Kinetic isotope fractionations accompanying diffusion and evaporation/condensation are due to the differential translational velocities of isotopes or isotopically substituted molecules (isotopologues) of different mass. This class of kinetic transport phenomena can lead to large isotope fractionations even at high temperatures.

The role of kinetic and equilibrium processes in driving Ca stable isotope fractionation is often debated and therefore demands consideration in this introductory text. However, isotope fractionation theory will be simplified and only briefly touched and the reader is referred to other sources (see reference section).

2.1 Equilibrium Isotope Partitioning

2.1.1 Isotope Exchange Reactions

Equilibrium isotope fractionation refers to the partial separation of isotopes between two or more substances in chemical or more precisely in isotopic equilibrium (Urey 1947; Bigeleisen and

Mayer 1947). It is a quantum-mechanical phenomenon driven mainly by differences in the vibrational energies of molecules and crystals containing isotopes of different masses. These differences lead to higher concentrations of the heavy isotopes in substances, in which the vibrational energy is most sensitive to isotope substitution, i.e. those with higher bond force constants (see discussion further below).

For an isotope exchange reaction between two components A and B, where * denotes the component enriched in the heavy isotope and a and b refer to the coefficients to balance the reaction, the general expression is:

$$aA + bB^* \Leftrightarrow aA^* + bB \qquad (6)$$

Note that both sides of the isotope exchange equation give the same chemical species and differ only in the distribution of isotopes. The following equation provides an example where all coefficients (a and b) are unity:

$$^{40}CaCO_3 + {}^{44}Ca^{2+} \Leftrightarrow {}^{44}CaCO_3 + {}^{40}Ca^{2+} \qquad (7)$$

For the general case, the isotope exchange constant K is equal to:

$$K_{eq} = \frac{(A^*)^a (B)^b}{(A)^a (B^*)^b} = \frac{(A^*/A)^a}{(B^*/B)^b} \qquad (8)$$

Concentrations rather than activities are commonly used in Eq. 8, as the difference between activities and concentrations is commonly negligible for isotopically substituted substances. If the isotopes are randomly distributed among all possible sites of the molecule at equilibrium, the fractionation factor α is connected to the equilibrium constant K_{eq} for the isotope exchange (Bigeleisen 1965):

$$\alpha_{A-B} = K_{eq}^{\frac{1}{ab}} \qquad (9)$$

where the product of the stoichiometry coefficients a and b gives the number of atoms exchanged during the reaction. Thus, if isotope exchange reactions are written such that only one atom is exchanged (ab = 1), the equilibrium constant K_{eq} is identical to the fractionation factor α_{A-B} as is the case for the above example (Eq. 7):

$$K_{eq} = \frac{\left(\frac{^{44}Ca}{^{40}Ca}\right)_{CaCO_3}}{\left(\frac{^{44}Ca}{^{40}Ca}\right)_{Ca^{2+}}} = \alpha_{CaCO_3 - Ca^{2+}} \qquad (10)$$

2.1.2 Equilibrium Isotope Fractionation Theory

In the following, we will briefly examine the basics of isotope fractionation theory. For more in-depth treatments, the readers are referred to Criss (1999), Schauble (2004) and citations therein including the classical papers by Urey (1947) and Bigeleisen and Mayer (1947). For isotope exchange reactions, an equilibrium constant can be determined from the free energy difference between reactants and products according to:

$$\Delta G^0 = -RT \ln(K_{eq}) \qquad (11)$$

where ΔG^0 is the Gibbs free energy of the reaction, R is the molar gas constant (~ 8.314 J mol^{-1} K^{-1}) and T the absolute temperature in K. The Gibbs free energy is several orders of magnitude smaller for isotope exchange reactions than for chemical reactions. Pressure volume work is commonly neglected, since isotopic substitution has a negligible effect on volume and because the number of molecules on both sides of the isotope exchange reactions is the same. Thus the Gibbs free energy ΔG^0 can be replaced by the Helmholtz free energy ΔF. Then rearranging Eq. 11 yields:

$$\ln(K_{eq}) = -\left(\frac{\Delta F_{motion}}{RT}\right) \qquad (12)$$

The bond structure and thus potential energy of each molecule is not changed by the isotopic substitution. Only the dynamic energy associated with atomic motions changed (therefore ΔF_{motion}). The dynamic energy includes translational, rotational and vibrational energies (Schauble 2004). Among these, only vibrational energies are significantly large (at about ≥ 100 K) relative to ambient thermal temperatures to drive chemical reactions (Schauble 2004). It is instructive to first consider the simple case of harmonic oscillations of a diatomic molecule like H_2 or O_2, where the vibrational energy is given by:

$$E_{vib} = (n + 1/2)h\nu \qquad (13)$$

with h being Planck's constant (6.626×10^{-34} Js), ν the vibrational frequency and n the vibrational energy level (with n = 0, 1, 2, etc.). With increasing temperature, higher energy levels are more frequently populated. However, with n = 0, atoms vibrate even at absolute zero with $E_{vib,0} = \frac{1}{2}$ hν. This property is known as the zero point energy. Since the vibrational frequencies are a function of the atomic masses in motion, vibrational frequencies are affected by isotopic substitution. In the example for the harmonic oscillation of a diatomic molecule, vibrational frequencies relate to the atomic masses m_i and m_j according to:

$$\nu = \frac{1}{2\pi}\sqrt{\frac{k_s}{\mu}} = \frac{1}{2\pi}\sqrt{k_s\left(\frac{1}{m_i} + \frac{1}{m_j}\right)} \qquad (14)$$

where k_s is the spring or force constant in N/m, which is a measure of bond stiffness, and

$$\mu = m_i * m_j / (m_i + m_j) \qquad (15)$$

is the reduced mass of the molecule. The harmonic oscillation of a diatomic molecule is often envisioned as two spheres (atoms) that are connected by a spring and move towards and away from each other with the frequency ν. Isotopic substitution changes the mass of the bond partner, but not the force constant k_s. Thus, substitution with a heavier isotope leads to slightly lower frequencies (Eq. 14) and hence lower vibrational energies. Therefore, substances with high force constants k_s and thus stiffer bonds will have larger zero-point energy shifts upon heavy

isotopic substitution and will therefore tend to be enriched in the heavy isotopes. (Vibrational) frequencies are commonly measured and tabulated as wavenumbers (ω in cm^{-1}), where $\nu = c*\omega$ and c refers to the speed of light. Vibrational frequencies range from ca. 100 up to about 4000 cm^{-1}, which corresponds to force constants of ~ 50–2000 N/m (Schauble 2004).

While zero point energy shifts are regarded as the principle driver of equilibrium isotope fractionations, an accurate treatment of molecular motions requires the higher energy states to be included. The probability of a molecule to have a distinct energy state follows the Boltzmann distribution law and this is expressed in the partition function ratio Q, which gives the sum of all possible quantum states i:

$$Q = \sum g_i e^{-E_i/kT} \qquad (16)$$

where e refers to the exponential function, k is Boltzmann's constant (1.381×10^{23} J/K), T is the temperature (in K) and g is the degeneracy factor that is sometimes given to refer to quantum states with the same energy level.

For the case of harmonic vibrations, Eq. 13 is inserted into Eq. 16. Doing so and following the approach of Urey (1947), Schauble (2004) derived an expression where the fractionation factor β_{XR-X} between a diatomic or larger molecule and a monatomic gas can be estimated from data for vibrational frequencies of heavy and light isotopologues only:

$$\beta_{XR-X} = \left[\prod_i \frac{\nu^*}{\nu} \times \left(\frac{exp\left[-\frac{h\nu^*}{2kT}\right]}{1 - exp\left[-\frac{h\nu^*}{kT}\right]}\right) \times \left(\frac{1 - exp\left[-\frac{h\nu}{kT}\right]}{exp\left[-\frac{h\nu}{2kT}\right]}\right)\right]^{\frac{1}{n}} \qquad (17)$$

Molecule XR contains n atoms of the element of interest X, and N total atoms. The frequencies ν and ν^* refer to molecule XR containing the light and heavy isotopes, respectively. Diatomic molecules have one vibration, but for polyatomic linear or nonlinear molecules the product is over 3N-5 or 3N-6 vibrational frequencies, respectively. While isotope fractionation relative to a monatomic gas is not of practical significance, the fractionation factor α between two different molecules A and B is simply $\alpha_{A-B} = \beta_A/\beta_B$. Because the rotational and translational terms cancel, β is often referred to as the reduced partition function ratio. Schauble (2004) points out, that, in order to simplify the calculations, this approach assumes harmonic vibrations, rigid-body rotation, a simplified treatment of rotational energies and averaged intra-molecular isotope fractionations. Nevertheless, he concludes that these assumptions are reasonable for elements other than H, C, N, O, and S and temperatures above about 100 K.

Calcium isotope geochemistry rarely deals with gas-phase molecules, but is commonly concerned with condensed phases such as aqueous solutions or crystals. Crystals however contain far more vibrational frequencies that cannot be explicitly treated and dissolved ions or molecules interact with the surrounding water molecules.

Since vibrational frequencies in crystals appear like a continuous spectrum, Kieffer (1982) calculated reduced partition function ratios using an integral:

$$\beta_{crystal-X} = \left(\frac{m}{m^*}\right)^{\frac{3}{2}} \times \exp\left[\frac{1}{n}\frac{\int_0^{v_{max}} ln\left(\frac{\exp\left[-\frac{\hbar\omega}{2kT}\right]}{1-\exp\left[-\frac{\hbar\omega}{kT}\right]}\right)g^*(\omega)d\omega}{\int_0^{v_{max}} ln\left(\frac{\exp\left[-\frac{\hbar\omega}{2kT}\right]}{1-\exp\left[-\frac{\hbar\omega}{kT}\right]}\right)g(\omega)d\omega}\right]$$

$$(18)$$

The above equation, taken from Schauble (2004), gives the fractionation factor β for element X in a crystal relative to that of the monatomic gas, where g and g* are the continuous vibrational densities of state for crystals containing the light and the heavy isotope of the element of interest. The highest vibrational frequency of the crystal is represented by v_{max}.

Equilibrium stable isotope fractionation factors can thus be calculated from vibrational frequencies. The necessary input data may be obtained from spectroscopic measurements. However, spectroscopic data, where available, are commonly incomplete and only available for phases of natural isotope composition. This requires additional estimates to be made to obtain vibrational frequencies for isotopically substituted phases (except for diatomic molecules with $v^* = v(\mu/\mu^*)^{-0.5}$; cf. Eq. 14). If no, imprecise or incomplete vibrational data are available, missing vibrational frequencies need to be modelled, either by force field modelling or by ab initio quantum mechanical calculations (Schauble 2004, 2009).

2.1.3 Summary of General Characteristics of Equilibrium Isotope Fractionation

Following O'Neil (1986) and Schauble (2004) some general rules regarding equilibrium stable isotope fractionation are summarized:

Temperature-dependence

Equilibrium isotope fractionations decrease with temperature (Urey 1947; Bigeleisen and Mayer 1947), usually as $1/T^2$. Over a small temperature interval, the decrease in the isotope fractionation in per mill per K can be estimated as $-2000(\alpha - 1)/T$ (Schauble 2009). At low temperature or for high frequency vibrations where almost all molecules are in the vibrational ground-state, temperature dependence more closely approaches $1/T$ (Criss 1999; Schauble 2004). At higher temperatures, high energy vibrational levels are more frequently populated. Since the energy difference upon isotopic substitution becomes smaller at higher vibrational energy levels, equilibrium isotope fractionation decreases with increasing temperature.

Mass-dependence

Equilibrium isotope fractionation increases with the mass-difference between the heavy and light isotope, but decreases with the absolute mass according to $(m_{heavy} - m_{light})/m_{heavy}^* m_{light}$. This mass term decreases e.g. from 0.0238 for $^{7/6}$Li, 0.0069 for $^{18/16}$O, 0.0032 for $^{26/24}$Mg, 0.0023 for $^{44/40}$Ca to 0.0003 for $^{88/86}$Sr. This indicates that in terms of *mass only*, the isotope fractionation between ^{44}Ca and ^{40}Ca should be about a third of that observed for oxygen and significantly larger than for heavier elements like Sr.

Dependence on bond-stiffness

As shown in Eq. 14, heavy isotopes will tend to be enriched in the phase with the highest force constants, i.e. the stiffest bonds. Bond stiffness is generally larger for short strong bonds, i.e. at high oxidation states, high covalent character of the bond and low coordination numbers. Schauble (2004) cautions that it is presently not clear which chemical properties are most important in regard to bond stiffness especially in regard to "new" isotope systems.

The extent of isotope fractionations will increase if large differences in bond stiffness exist between the phases involved. Calcium isotopes generally do not display fractionations in excess of one or at best a few per mill. This is in accord with the restricted variation in bond environments compared to e.g. O or S, the ubiquitous presence of Ca in the 2+ oxidation state and with the ionic character of bonds involving Ca. This leaves coordination as one of the main driver of equilibrium Ca stable isotope fractionations (Schauble 2004; Gussone 2005).

2.2   Kinetic Stable Isotope Fractionation

Kinetic stable isotope fractionations arise either from (i) chemical bond-breaking dissociation reactions, including most biochemical transformations or (ii) physical transport phenomena, i.e. evaporation, condensation and diffusion though solid, liquid or gaseous media (e.g. Schauble 2004; Richter et al. 2009; Eiler et al. 2014).

If equilibrium between two (or more) components is attained, forward and back reaction rates are equal. In kinetics, the reaction products become isolated from the reactants. Often, kinetic reactions are defined as unidirectional. However, kinetic effects are involved in

disequilibrium or *non-equilibrium* isotope fractionations where the forward reaction overrides the back reaction:

$$\text{reactant(s)} \; \overset{R_f}{\underset{R_b}{\rightleftharpoons}} \; \text{product(s)} \qquad (19)$$

Here, R_f, R_b and R_{net} denote the forward, back and net reaction rates with $R_{net} = R_f - R_b$.

Effective disequilibrium fractionation factors α_{eff} will be intermediate between those describing equilibrium isotope fractionation (α_{eq}) or unidirectional kinetic forward reaction (α_f). This is illustrated in Eq. 20 (DePaolo 2011) and Fig. 2:

$$\alpha_{eff} = \frac{\alpha_f}{1 + \frac{R_b}{R_f}\left(\frac{\alpha_f}{\alpha_{eq}} - 1\right)} = \frac{\alpha_f}{1 + \frac{R_b}{R_{net} + R_b}\left(\frac{\alpha_f}{\alpha_{eq}} - 1\right)}$$
$$= \frac{\alpha_f}{1 + \left(\frac{\alpha_f}{\alpha_{eq}} - 1\right) \Big/ \left(\frac{R_{net}}{R_b} + 1\right)}$$

$$(20)$$

Equilibrium isotope fractionation dominates were R_b/R_f approaches 1 and R_{net}/R_b is very small. If R_b is only a minor fraction of R_f such that the $R_{net} \gg R_b$, kinetic isotope fractionation dominates α_{eff}. For R_{net}/R_b increasing from ~ 0.01 to ~ 100, α_{eff} changes from values near α_{eq} towards α_f (Fig. 2).

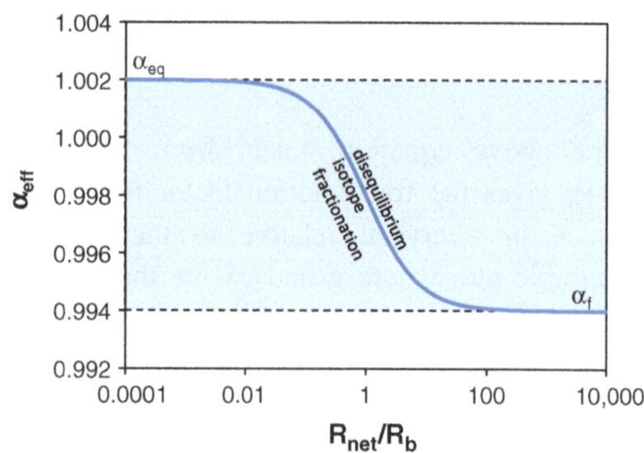

Fig. 2 Dependence of α_{eff} on the ratio of net reaction rate to the back reaction rate as given in Eq. 20. Arbitrary values have been assumed for α_{Eq} (1.002) and α_f (0.994) Redrawn after DePaolo (2011)

DePaolo (2011) derived Eq. 20 to describe Ca isotope fractionation associated with the precipitation of calcite from aqueous solutions at variable precipitation rates. The above description assumes steady state conditions, i.e. no change in the isotope composition of the crystal surface layer or liquid reservoirs with time and furthermore, the absence of transport effects (diffusion) within the crystal or the liquid. Diffusive transport could limit the delivery of ions to the growing crystals surface at high precipitation rates. Such a transport-control could also result in additional kinetic isotope effects. Isotope fractionation associated with transport phenomena and reservoir effects are discussed further below.

Dissociation

The theory of kinetic isotope fractionation associated with the dissociation of molecules has been discussed previously by Bigeleisen (1949), Bigeleisen and Wolfsberg (1958) and Melander (1960). This type of fractionation is due to differences in the reaction rates between isotopes or molecules of different masses (isotopologues). In by far most cases, the lighter isotopes or isotopologues react faster than the heavier ones and thus become concentrated in the product of the reaction. If the reaction pool is limited (e.g. a closed system), the reactant becomes correspondingly enriched in the heavier isotope. There are rare cases of inverse kinetic isotope effects with the reactant being enriched in the heavy isotope that most commonly occur when hydrogen atoms are involved (Bigeleisen and Wolfsberg 1958). Kinetic isotope fractionation can be described with first-order kinetics.

For two competing isotope reactions:

$$A_1 \rightarrow B_1(k_1) \quad \text{and} \quad A_2 \rightarrow B_2(k_2) \tag{21}$$

with k_i ($i = 1, 2$) corresponding to the rate constants for the reaction and 1 and 2 referring to the light and heavy isotope, respectively. The fractionation factor α corresponds to the ratio of the rate constants:

$$\alpha_{A-B} = k_1/k_2 \tag{22}$$

Physical transport processes: evaporation, condensation, diffusion

In gases, the average translational kinetic energy E_{kin} is the same for all molecules or atoms:

$$E_{kin} = \frac{3}{2}kT = \frac{1}{2}mv^2 \tag{23}$$

where k is the Boltzmann constant, m the mass of the molecules or atoms in motion and v is the average velocity. If a gas consists of different isotopes or isotopologues of masses m_1 and m_2 with the same average kinetic energy, the velocity will be slower for heavier isotopes or isotopologues compared to lighter ones, as evident from:

$$E_{kin} = \frac{1}{2}m_1 v_1^2 = \frac{1}{2}m_2 v_2^2 \tag{24}$$

this can be rearranged into:

$$\alpha_f = \frac{v_1}{v_2} = \left(\frac{m_2}{m_1}\right)^{0.5} \tag{25}$$

where α_f is the corresponding kinetic fractionation factor. Equation 25 is often referred to as Graham's law; it shows that kinetic transport effects are not expected to be temperature sensitive. Graham's law is frequently used to predict the fractionation factor for free evaporation into vacuum. However, in experiments, free evaporation commonly yields fractionation factors smaller than those predicted by the square root of the mass dependence aka Grahams law (Richter et al. 2007, 2009). This observation can be accounted for by the introduction of evaporation coefficients γ (e.g. Richter et al. 2007; Zhang et al. 2014):

$$\alpha_{eff} = \frac{\gamma_1}{\gamma_2}\left(\frac{m_2}{m_1}\right)^{0.5} \alpha_{eq} \tag{26}$$

the ratio of evaporation coefficients γ_1/γ_2 then accounts for different probabilities of two isotopes or isotopologues to leave the surface of the condensed phase. Equation 26 also contains the equilibrium isotope fractionation factor α_{eq}. At the high temperatures for evaporation or condensation of Ca $\alpha_{eq} = 1$ can be assumed.

Equation 26 is also expected to be valid for condensation. In this case γ refers to sticking (or condensation) coefficients that give the fraction of atoms or molecules that do not return to the vapor phase after impinging onto the surface of the condensed phase.

Evaporation provides probably the most classical example for kinetic isotope separation. Because Ca is a highly refractory element, Ca isotope fractionation due to evaporation (and condensation) is mostly restricted to calcium-aluminum-rich inclusions (CAIs) in meteorites (Chapter High Temperature Geochemistry and Cosmochemistry). A first thorough experimental investigation of Ca isotope fractionation during evaporation has been carried out recently (Zhang et al. 2014; see also Sect. 2.4 and Chapter High Temperature Geochemistry and Cosmochemistry).

For diffusion in gases or liquids, the isotope fractionation factor α_{diff} equals the ratio of diffusivities, which again is inversely related to mass M_i by a power law, with the exponent $\beta \leq 0.5$ (e.g. Richter et al. 2006; Bourg et al. 2010; see also Eiler et al. (2014) and citations therein):

$$\alpha_{\text{diff}} = \frac{D_1}{D_2} = \left(\frac{M_2}{M_1}\right)^{\beta} \qquad (27)$$

For diffusion of a trace gas through a carrier gas, β may be equal to 0.5 and M_i refers to the reduced mass (Eq. 15 with m_i = mass of trace isotope or isotopologue and m_j = average mass of carrier gas molecules) if both gases interact only by instantaneous binary collisions as suggested for Ca diffusion in H_2, considered relevant in cosmochemistry (Simon and DePaolo 2010).

The above power law (Eq. 27) is also consistent with data for diffusion in liquids, i.e. silicate melts and water (e.g. Bourg et al. 2010). Mass M_i may refer to reduced masses, but more commonly isotopic masses are used. For diffusion in liquids, the exponent β is much lower than 0.5, ranging from 0 to 0.22 (e.g. Richter et al. 2009; Bourg et al. 2010). This reflects chemical interactions with the solvent, but also the use of isotopic masses rather than molecular or reduced masses, as the latter is difficult to assign in the presence of chemical interactions. The mass-dependence of the ratio of diffusivities reflected in the empirical parameter β is thus intermediate between kinetic theory ($\beta = 0.5$) and predictions due to hydrodynamic transport ($\beta = 0$), where the solute is strongly coupled to solvent hydrodynamic modes (Bourg et al. 2010). This is in line with the observation that β-values are inversely related to the residence times of water molecules in the first solvation shells of the diffusing ions (Bourg and Sposito 2007).

Watkins et al. (2011) showed that isotope fractionation related to chemical diffusion in silicate melts, is related to the liquid structure. They observed that isotope separation increases with the ratio of mobility of the cation relative to that of the liquid matrix (e.g. $D_{\text{Ca}}/D_{\text{Si}}$), as the cation is not strongly bound to and hence diffusing through the alumosilicate framework.

Thermal gradients imposed on silicate melts lead to concentration gradients with enrichments of CaO, MgO and FeO on the cold end and SiO_2, K_2O and Na_2O on the hot end (Bowen 1921; Lesher and Walker 1986). This thermal (or Soret) diffusion is balanced by chemical diffusion such that a steady state is approached. Soret diffusion can lead to large stable isotope fractionations (Kyser et al. 1998) and experiments have shown that heavy Ca isotopes get enriched at the cold end (Richter et al. 2009; Huang et al. 2010). One fundamental difference for isotope and chemical fractionation due to Soret and chemical diffusion is that with time chemical diffusion homogenizes the melt, while Soret diffusion approaches a steady-state (e.g. Richter et al. 2009). Dominguez et al. (2011) explained Soret diffusion as a quantum mechanical effect, while Lacks et al. (2012) favor a classical mechanical collision effect. The analysis by Li and Liu (2015) support the classical approach for high temperatures.

See other chapters, in particular Chapters Calcium Isotope Fractionation During Mineral Precipitation from Aqueous Solution and High Temperature Geochemistry and Cosmochemistry for further discussion of kinetic isotope fractionation related to dissociation, evaporation/condensation and diffusion.

2.3 Open System Rayleigh Fractionation and Closed System Equilibrium Fractionation

Rayleigh distillation (or fractionation) occurs when isotopically fractionated reaction products are continuously removed or isolated from the system. The Rayleigh equation (Eq. 28) describes the evolution of the isotope composition of the remaining reservoir R_A as a function of the fraction f of the reactant A that is remaining (Rayleigh 1902; Broecker and Oversby 1971):

$$R_A = R_{A,i} * f^{\alpha_{B-A}-1} \qquad (28)$$

where $R_{A,i}$ refers to the initial isotope ratio and α_{B-A} to the fractionation factor between product B and reactant A. The isotope composition of the instantaneous forming product B_{inst} is simply offset from R_A according to the fractionation factor, thus

$$R_{B,inst} = \alpha_{B-A} * R_{A,i} * f^{\alpha_{B-A}-1} \qquad (29)$$

The isotope composition of the accumulated product $R_{B,accum}$ is dictated by mass balance ($R_{A,i} = f * R_A + (1 - f) * R_{B,accum}$). Rearranged and combined with Eq. 28 this gives:

$$R_{B,accum} = \frac{R_{A,i}(f^{\alpha_{B-A}} - 1)}{f - 1} \qquad (30)$$

The evolution of R_A, $R_{B,inst}$ and $R_{B,accum}$ with progressive transformation of A to B is given in Fig. 3 in the Δ-notation (e.g. $\Delta_{B-A} = 1000 \times (R_B/R_{A,i} - 1)$) as a function of the fraction of the product $(1 - f)$ and arbitrarily assuming $\alpha_{B-A} = 0.997$; i.e. that the product is 3 ‰ lighter than the instantaneous reservoir from which it becomes isolated. Figure 3 reflects the typical characteristics of Rayleigh fractionation: if the reaction proceeds to near completion, the isotope composition of the remaining reservoir

becomes extremely fractionated (here enriched in the heavy isotopes). The same trend is observed for the instantaneously formed product as it is obviously only offset from the contemporaneous isotope composition of the reactant reservoir A (Eq. 29), with the offset defined by the fractionation factor. The accumulated product will eventually attain the initial isotope composition of the reactant as required by mass balance.

Rayleigh fractionation may describe equilibrium, disequilibrium and kinetic fractionations, but requires that the reaction product is continuously isolated and the fractionation factor does not change in the process. For example, Rayleigh evaporation of a water droplet into a dry atmosphere will enrich the remaining droplet progressively in heavy H and O isotopes as kinetic (or disequilibrium) evaporation proceeds. Equilibrium O and H Rayleigh fractionation may occur if water droplets condense in the atmosphere in equilibrium with the surrounding water

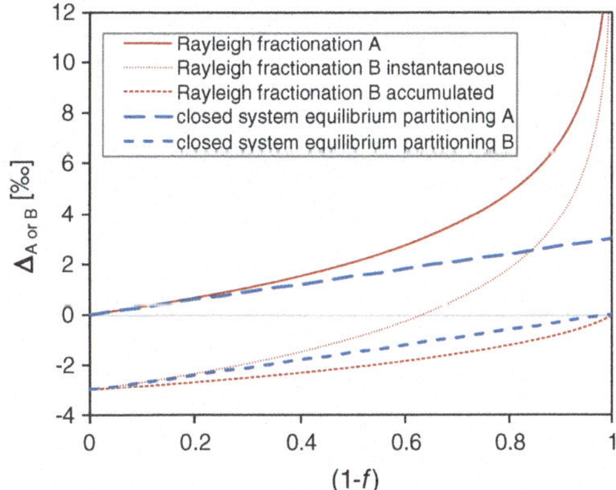

Fig. 3 Isotope ratio evolution for open system Rayleigh fractionation (*red curves*) and closed system equilibrium fractionation (*blue straight lines*) as a function of the elemental fraction in phase B. Given are deviations from $R_{A,i}$ or R_{SYS} as e.g. $\Delta_{B-A} = 1000 * (R_B/R_{A,i} - 1)$ in ‰, assuming $\alpha_{B-A} = 0.997$ in all calculations

vapor. Once big enough, the droplets will rain out, leaving the remaining atmospheric water vapor reservoir progressively enriched in the lighter isotopes. Calcium isotope fractionation during the precipitation of minerals from a restricted liquid reservoir may serve as another example for Rayleigh fractionation.

Figure 3 also shows the isotope evolution during closed-system equilibrium isotope fractionation which relies on mass balance equations. The two pools denoted A and B remain in equilibrium throughout and thus the reactant and product always differ by the same offset $\Delta_{B-A} = 1000 \times (\alpha - 1)$. As above, mass balance requires

$$R_{SYS} = f * R_{A,eq} + (1 - f) * R_{B,eq} \qquad (31)$$

where R_{SYS} refers to the isotope ratio of the bulk system and $R_{A,eq}$ and $R_{B,eq}$ to the isotope composition of both pools, depending on the fraction f of the element in A. Because $\alpha_{B-A} = R_B/R_A$, R_B can be replaced by $\alpha_{B-A} * R_A$ and inserted into Eq. 31:

$$R_{A,eq} = \frac{R_{SYS}}{(f + \alpha(1 - f))} \qquad (32)$$

See Schmitt et al. (2013) for an example for equilibrium Ca isotope fractionation during adsorption on bean roots.

2.4 The Mass-Dependence of Equilibrium and Kinetic Stable Isotope Fractionations

The mass-difference is double for $^{44}Ca/^{40}Ca$ compared to $^{42}Ca/^{40}Ca$. Thus the mass-dependent stable isotope fractionation recorded in the former ratio is about twice as large as that of the latter one. However, the exact relationship between two isotope ratios depends on the fractionation mechanism (e.g. Young et al. 2002). The mass-dependence can be described by mass fractionation laws of the form

$$\alpha_{a/b} = \alpha_{c/d}^{\beta} \qquad (33)$$

where a, b, c, and d refer to the mass numbers of isotopes of element E, $\alpha_{a/b}$ refers to the fractionation factor between two phases for isotope ratio $^aE/^bE$, and $\alpha_{c/d}$ refers to the same fractionation factor for isotope ratio $^cE/^dE$. In most cases b and d are identical as usually the same isotope is used as the denominator in both ratios. The exponent β defines the slope in linearized three isotope plots and is a function of the masses m_i of the isotopes involved (Young et al. 2002). For a heavy element like Ca (compared to oxygen; see Cao and Liu 2011), the slope β for equilibrium isotope exchange should be very close to:

$$\beta_{eq} = \frac{\left(\frac{1}{m_b} - \frac{1}{m_a}\right)}{\left(\frac{1}{m_d} - \frac{1}{m_c}\right)} \qquad (34)$$

for kinetic isotope separation

$$\beta_{kin} = \frac{\ln\left(\frac{M_b}{M_a}\right)}{\ln\left(\frac{M_d}{M_c}\right)} \qquad (35)$$

where mass M_i either refers to the mass of isotopes, isotopologues or reduced masses, depending on the kinetic process. In three isotope plots where the conventional δ-notation is used, fractionation lines according to β are slightly curved. This can be circumvented if the δ'-notation of Hulston and Thode (1965) is applied (Young et al. 2002), e.g.:

$$\delta' = 1000 * \ln\left(\frac{(^{44}Ca/^{40}Ca)_{sample}}{(^{44}Ca/^{40}Ca)_{standard}}\right) \qquad (36)$$

Differences between kinetic and equilibrium fractionation lines are difficult to spot in three isotope plots (see Chapter Analytical Methods, Fig. 8) in particular if the stable isotope fractionation is small. The visual representation can be improved by using a format such as in Fig. 4 (cf. Young and Galy 2004), where the deviation of the data from the equilibrium fractionation line for one isotope ratio is plotted against the δ'-values of another isotope ratio.

The large mass differences between Ca isotopes, especially between ^{40}Ca, ^{44}Ca and ^{48}Ca would make this isotope system ideal to resolve different slopes. Unfortunately, Ca does not commonly display the large isotope fractionations needed to resolve different slopes. Furthermore, three isotopes, ideally with a wide mass differences, must be measured *accurately* at high precision. This is difficult to achieve with MC-ICP-MS due to interferences (especially for ^{40}Ca), but Schiller et al. (2012) report some high precision data for ^{42}Ca/^{44}Ca and ^{48}Ca/^{44}Ca which suggests that different slopes can be resolved even for the small variations displayed in nature. With TIMS, the correction of large progressive mass discrimination requires the use of a double spike, which compromises the ability to analyze three isotopes accurately at high precision (see Chapter Analytical Methods). However, if the fractionation displayed by the samples is sufficiently large, e.g. due to Rayleigh fractionation, different slopes may be resolved.

Figure 4 features a consistent data set with very strongly fractionated Ca isotope compositions obtained by TIMS for residues from Rayleigh evaporation from a perovskite (CaTiO$_3$) melt into vacuum at about 2000 °C (Zhang et al. 2014). Evaporation of Ca into vacuum is expected to be a kinetic process. Calcium forms a monatomic vapor and equilibrium isotope fractionation will be insignificant at the high temperature of evaporation (Zhang et al. 2014). We may therefore expect that the kinetic slope β_{kin} (Eq. 35) calculated from isotopic masses would fit the data. However, the data for the evaporation residues plots onto a trend somewhere between that expected from the equilibrium and kinetic law (with isotopic masses). In fact, the data plots onto a trend that corresponds to the kinetic slope (Eq. 35) calculated with reduced masses (with $m_j = 42$; Eq. 15). This slope, however, agrees with the Rayleigh law, which was therefore considered appropriate for the evaporation process by Zhang et al. (2014). This example shows that the interpretation of fractionation mechanisms, based on the framework of kinetic and equilibrium slopes is not necessarily trivial. Clearly, more suitable data for well-defined processes and advances in theory are desired (cf. Filer et al. 2014).

2.5 Experimental Determination of Equilibrium Isotope Fractionation Factors

The determination of reliable equilibrium fractionation factors in laboratory experiments is difficult, because of the need to demonstrate that thermodynamic equilibrium is achieved (e.g. Young et al. 2015). Time-series experiments represent the classical approach and allow to evaluate the exchange kinetics. To this end, the same experiment needs to be run repeatedly for different durations. For example, in a series of experiments, a mineral powder may be heated in the presence of water for different durations. The fractionation between both phases is measured and displayed as a function of the run duration. Equilibrium is suggested if the same isotope

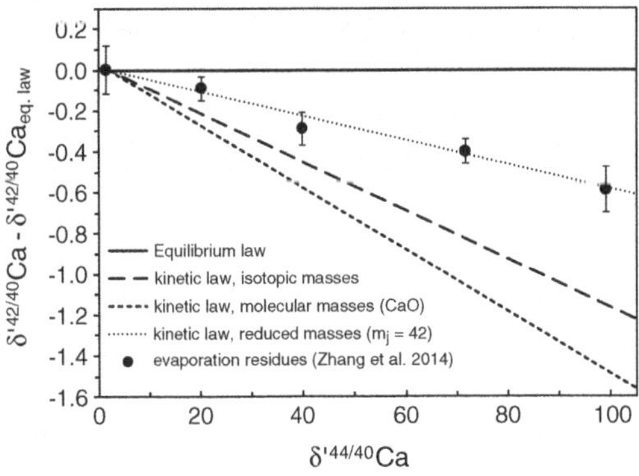

Fig. 4 Representation of three Ca isotope data in the context of the equilibrium and kinetic mass fractionation laws (Young et al. 2002; Young and Galy 2004). The strongly fractionated data refers to evaporation residues reported relative to the starting material (Zhang et al. 2014). For the kinetic law, trends have been calculated with isotopic masses, molecular masses (by adding mass 16 for oxygen to the isotopic masses) and reduced masses with $m_j = 42$ (Eq. 15). The use of $m_j = 42$ fits the data, but this may be because the kinetic law then coincides with the Rayleigh law suggested in Zhang et al. (2014)

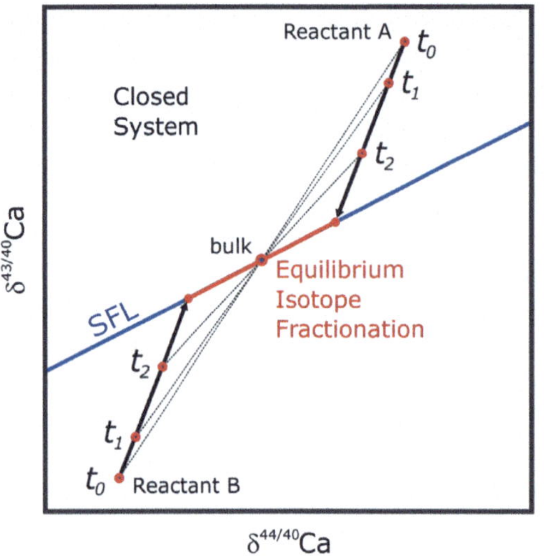

Fig. 5 Principle of three isotope method redrawn from Young et al. (2015). *SFL* denotes the secondary fractionation line

fractionation is observed between the mineral and aqueous phase after several longer experimental runs. The result is substantiated if the experiments were carried out with the starting mineral composition being once heavier and once lighter than the water, respectively. See O'Neil (1986) and Johnson et al. (2004b) for details and citations.

Equilibrium isotope fractionation factors can also be determined from closed system experiments using the three-isotope method of Matsuhisa et al. (1978). By extrapolation, equilibrium isotope fractionation factors can be obtained even if equilibrium is not attained in the experiments. For several non-traditional isotope systems, the three-isotope method has been successfully used to infer equilibrium fractionation factors between minerals and between minerals and water (e.g. Shahar et al. 2008; Li et al. 2011).

Figure 5 illustrates the principle of the three-isotope method. Phase A has been prepared with an enrichment in ^{43}Ca, while phase B has a natural isotope composition. In repeated experiments, both phases are then allowed to react which each other for variable durations (t_1, t_2 …) and the isotope composition is determined for the

run products. With increasing run time, the isotope compositions of both phases will move towards the isotope composition that corresponds to the equilibrium isotope fractionation. The equilibrium isotope compositions have to lie on the secondary fractionation line, the location of which is defined by the bulk composition, which is in turn defined by the tie lines. If equilibrium is not attained, the equilibrium isotope fractionation factor may still be obtained from extrapolation of the time trends of phases A and B to the secondary fractionation line.

References

A selection of texts relevant for stable isotope geochemistry

Allegre CJ (2008) Isotope geology. Cambridge University Press, Cambridge

Baskaran MM (ed) (2011) Handbook of environmental isotope geochemistry. Advances in isotope geochemistry. Springer, Heidelberg

Criss RR (1999) Principles of stable isotope distribution. Oxford University Press, New York

de Groot PA (2004) Handbook of stable isotope analytical techniques, vol 1. Elsevier, Amsterdam

de Groot PA (2009) Handbook of stable isotope analytical techniques, vol 2. Elsevier, Amsterdam

Faure G, Mensing TM (2004) Isotopes: principles and applications. Wiley, New York

Hoefs J (2009) Stable isotope geochemistry. Springer, Heidelberg

Johnson CM, Beard BL, Albarède F (eds) (2004) Geochemistry of non-traditional stable isotopes. Rev Mineral Geochem, 55:454

Melander L, Saunders W (1980) Reaction rates of isotopic molecules. Wiley, New York

Mook WG (2005) Introduction to isotope hydrology. Stable and radioactive isotopes of hydrogen, carbon, and oxygen, vol 25. Taylor&Francis, Abingdon

Sharp ZD (2006) Principles of stable isotope geochemistry. Prentice Hall, New Jersey

Valley JW, Cole DR (eds) (2001) Stable isotope geochemistry. Reviews in mineralogy and geochemistry, vol 43. Mineralogical Society of America, Washington

Zeebe RE, Wolf-Gladrow D (2001) CO2 in seawater: equilibrium, kinetics, isotopes. Elsevier, Amsterdam

Reviews on calcium stable isotope geochemistry

Boulyga SF (2010) Calcium isotope analysis by mass spectrometry. Mass Spectrom Rev 29:685–716

DePaolo DJ (2004) Calcium isotopic variations produced by biological, kinetic, radiogenic and nucleosynthetic processes. In: Johnson CM, Beard BL, Albarède F (eds) Geochemistry of non-traditional stable isotopes. Reviews in Mineralogy and Geochemistry 65. Mineralogical Society of America, Washington, p 255

Eisenhauer A, Kisakürek B, Böhm F (2009) Marine calcification: an alkali earth metal isotope perspective. Elements 5:365–368

Fantle MS (2010) Evaluating the Ca isotope proxy. Am J Sci 310:194–230

Fantle MS, Bullen TD (2009) Essentials of Fe and Ca isotope analysis of natural materials by thermal ionization mass spectrometry. Chem Geol 258:50–64

Fantle MS, Tipper ET (2014) Calcium isotopes in the global biogeochemical Ca cycle: implications for development of a Ca isotope proxy. Earth Sci Rev 129:148–177

Fantle MS, Maher K, DePaolo DJ (2010) Isotopic approaches for quantifying the rates of marine burial diagenesis. In: Reviews in Geophysics 48, RG3002

Nielsen LC, Druhan JL, Yang W et al (2011) Calcium isotopes as tracers of biogeochemical processes. In: Baskaran MM (ed) Handbook of environmental isotope geochemistry. Advances in isotope geochemistry. Springer, Heidelberg, p 105

Schmitt A-D, Vigier N, Lemarchand D et al (2012) Processes controlling the stable isotope compositions of Li, B, Mg and Ca in plants, soils and waters: a review. CR Géosci 344:704–722

Additonal publications cited in this chapter

Amos S, Gross JL, Thoennessen M (2011) Discovery of the calcium, indium, tin, and platinum isotopes. Atomic Data Nucl Data Tab 97:383–402

Anderson OR, Faber WW Jr (1984) An estimation of calcium carbonate deposition rate in a planktonic foraminifer Globigerinoides sacculifer using ^{45}Ca as a tracer: a recommended procedure for improved accuracy. J Foram Res 14:303–308

Artemov YM, Knorre KG, Strizkov VP et al (1966) ^{40}Ca/^{44}Ca and ^{18}O/^{16}O isotopic ratios in some calcareous rocks. Geochemistry (USSR) (English Transl.) 3:1082–1086

Aston FW (1934) Constitution of hafnium and other elements. Nature 133:684

Augusto L, Zeller B, Midwood AJ et al (2011) Two-year dynamics of foliage labelling in 8-year-old Pinus pinaster trees with ^{15}N, ^{26}Mg and ^{42}Ca-simulation of Ca transport in xylem using an upscaling approach. Annals Forest Sci 68:169–178

Backus MM, Pinson WH, Herzog LF et al (1964) Calcium isotope ratios in the homestead and pasamonte meteorites and a devonian limestone. Geochim Cosmochim Acta 28:735–742

Barger-Lux MJ, Heaney RP, Recker RR (1989) Time course of calcium absorption in humans: evidence for a colonic component. Calcif Tissue Int 44:308–311

Beck AB, Bügel S, Stürup S et al (2003) A noval dual radio- and stable-isotope method for measuring calcium absorption in humans: comparison with the whole-body radioisotope retention method. Am J Clin Nutr 77:399–405

Berner RA, Lasaga AC, Garrels RM (1983) The carbonate-silicate geochemical cycle and its effect on atmospheric carbon dioxide over the past 100 million years. Amer J Sci 283:641–683

Bigeleisen J (1949) The relative reaction velocities of isotopic molecules. J Chem Phys 17:675–78

Bigeleisen J (1965) Chemistry of isotopes. Science 147:463–471

Bigeleisen J, Mayer MG (1947) Calculation of equilibrium constants for isotopic exchange reactions. J Chem Phys 15:261–267

Bigeleisen J, Wolfsberg M (1958) Theoretical and experimental aspects of isotope effects in chemical kinetics. Adv Chem Phys 1:15–76

Bourg IC, Sposito G (2007) Molecular dynamics simulations of kinetic isotope fractionation during the diffusion of ionic species in liquid water. Geochim Cosmochim Acta 71:5583–5589

Bourg IC, Richter FM, Christensen JN et al (2010) Isotopic mass dependence of metal cation diffusion coefficients in liquid water. Geochim Cosmochim Acta 74:2249–2256

Bowen NL (1921) Diffusion in silicate melts. J Geol 29:295–317

Broecker WS, Oversby VM (1971) Rayleigh distillation. In: Press F (ed) Chemical equilibria in the Earth, McGraw-Hill, New York. pp 165–167

Caner L, Zeller B, Dambrine E et al (2004) Origin of the nitrogen assimilated by soil fauna living in decomposing beech litter. Soil Biol Biochem 36:1861–1872

Cao X, Liu Y (2011) Equilibrium mass-dependent fractionation relationships for triple oxygen isotopes. Geochim Cosmochim Acta 75:7435–7445

Coplen TB (2011) Guidelines and recommended terms for expression of stable-isotope-ratio and gas-ratio measurement results. Rapid Com Mass Spectrom 25:2538–2560

Corless JT (1968) Observations on the isotopic geochemistry of calcium. Earth Planet Sci Lett 4:475–478

Dean JA (ed) (1999) Lange's Handbook of Chemistry, 15th edn. McGraw-Hill, New York

Dempster A (1922) Positive-ray analysis of potassium, calcium and zinc. Phys Rev 20:631–638

Denk E, Hillegoods D, Hurrell RF et al (2007) Evaluation of 41calcium as a new approach to assess changes in bone metabolism: effect of a bisphosphonate intervention in postmenopausal women with low bone mass. J Bone Miner Res 22:1518–1525

DePaolo D (2011) Surface kinetic model for isotopic and trace element fractionation during precipitation of calcite from aqueous solutions. Geochim Cosmochim Acta 75:1039–1056

Dominguez G, Wilkins G, Thiemens MH (2011) The Soret effect and isotopic fractionation in high-temperature silicate melts. Nature 473:70–73

Eiler JM, Bergquist B, Bourg I et al (2014) Frontiers of stable isotope geoscience. Chem Geol 372:119–143

Eisenhauer A, Nägler T, Stille P et al (2004) Proposal for an international agreement on Ca notations resulting from discussions at workshops on stable isotope measurements held in Davos (Goldschmidt 2002) and Nice (EGS AGU EUG 2003). Geostand Geoanal Res 28:149–151

Fietzke J, Eisenhauer A (2006) Determination of temperature-dependent stable strontium isotope (^{88}Sr/^{86}Sr) fractionation via bracketing standard MC-ICP-MS. Geochem Geophys Geosyst 7:Q08009

Gibbs RJ (1972) Water chemistry of the Amazon river. Geochim Cosmochim Acta 86:1061–1066

Gussone N, Böhm F, Eisenhauer A et al (2005) Calcium isotope fractionation in calcite and aragonite. Geochim Cosmochim Acta 69:4485–4494

Halicz L, Galy A, Belshaw N et al (1999) High-precision measurement of calcium isotopes in carbonates and related materials by multiple collector inductively coupled plasma mass spectrometry (MC-ICPMS). J Anal At Spectrom 14:1835–1838

Halicz L, Segal I, Fruchter N et al (2008) Strontium stable isotopes fractionate in the soil environments? Earth Planet Sci Lett 272:406–411

Heumann KG, Luecke W (1973) Calcium isotope ratios in natural carbonate rocks. Earth Planet Sci Lett 20:341–346

Hippler D, Schmitt A-D, Gussone N et al (2003) Calcium isotopic composition of various reference materials and seawater. Geostand Newslett 27:13–19

Hirt B, Epstein S (1964) A search for isotopic variations in some terrestrial and meteoritic calcium. Trans Am Geophys Union 45:113

Huang FP, Chakraborty CC, Lundstrom C et al (2010) Isotope fractionation in silicate melts by thermal diffusion. Nature 464:396–400

Hulston JR, Thode HG (1965) Variations in the S33, S34, and S36 contents of meteorites and their relation to chemical and nuclear effects. J Geophys Res 70:3475–3484

Johnson CM, Beard B, Albarède F (2004b) Overview and general concepts. In: Johnson CM, Beard B, Albarède F (eds) Geochemistry of non-traditional stable isotopes. Reviews in Mineralogy and GeochemistryMineralogical Society of America, Washington, pp 113–152

Kieffer SW (1982) Thermodynamics and lattice vibrations of minerals: 5. Applications to phase equilibria, isotopic fractionation, and high-pressure thermodynamic properties. Rev Geophys 20:827–849

Krabbenhöft A, Eisenhauer A, Böhm F et al (2010) Constraining the marine strontium budget with natural strontium isotope fractionations (^{87}Sr/^{86}Sr*, $\delta^{88/86}$Sr) of carbonates, hydrothermal solutions and river waters. Geochim Cosmochim Acta 74:4097–4109

Kyser TK, Lesher CE, Walker D (1998) The effects of liquid immiscibility and thermal diffusion on oxygen isotopes in silicate liquids. Contrib Mineral Petrol 133:373–381

Lacks D, Goel G, Bopp CJ et al (2012) Isotope fractionation by thermal diffusion in silicate melts. Phys Rev Lett 108:065901

Lea DW, Martin PA, Chan DA et al (1995) Calcium uptake and calcification rate in the planktonic foraminifer Orbulina universa. J Foram Res 25:14–23

Lesher CE, Walker D (1986) Solution properties of silicate liquids from thermal diffusion experiments. Geochim Cosmochim Acta 50:1397–1411

Letolle R (1968) Sur la composition isotopique du calcium des échantillons naturels. Earth Planet Sci Lett 5:207–208

Li X, Liu Y (2015) A theoretical model of isotopic fractionation by thermal diffusion and its implementation on silicate melts. Geochim Cosmochim Acta 154:18–27

Li W, Beard BL, Johnson CM (2011) Exchange and fractionation of Mg isotopes between epsomite and saturated MgSO$_4$ solution. Geochim Cosmochim Acta 75:1814–1828

Lide DR (ed) (1995) CRC Handbook of Chemistry and Physics, 77th edn. CRC Press, Boca Raton

López-Huertas E, Teucher B, Boza JJ et al (2006) Absorption of calcium from milks enriched with fructo-oligosaccharides, caseinophosphopeptides, tricalcium phosphate, and milk solids. Am J Clinical Nutr 83:310–316

Marschner H (1995) Mineral nutrition of higher plants, 2nd edn. Academic Press, London

Matsuhisa J, Goldsmith JR, Clayton RN (1978) Mechanisms of hydrothermal crystallisation of quartz at 250 °C and 15 kbar. Geochim Cosmochim Acta 42:173–182

McDonough WF (2001) The composition of the Earth. In: Teisseyre R, Majewski E (eds) Earthquake thermodynamics and phase transformations in the Earth's interior. Academic Press, San Diego, pp 3–23

McDonough WF, S-s Sun (1995) The composition of the Earth. Chem Geol 120:223–253

Meija J et al (2016) Atomic weights of the elements 2013. Pure and Applied Chemistry (88 accepted for publication)

Melander L (1960) Isotope effects on reaction rates, Ronald, New York, 181 p

Meshcheryakov RP, Stolbov MY (1967) Measurement of the isotopic composition of calcium in natural materials. Geochemistry (USSR) (English Transl.) 4, 1001–1003

Miller YM, Ustinov VI, Artemov YM et al (1966) Mass spectrometric determination of calcium isotope variations. Geochem Intern 3:929

Möller P, Papendorff H (1971) Fractionation of calcium isotopes in carbonate precipitates. Earth Planet Sci Lett 11:192–194

Nier AO (1938) The isotopic composition of calcium, titanium, sulphur and argon. Phys Rev 53:282–286

O'Neil JR (1986) Theoretical and experimental aspects of isotopic fractionation. In: Valley JW, Taylor HP Jr, O'Neil JR (eds) Stable isotopes in high temperature geological processes. Reviews in Mineralogy, vol 16. Mineralogical Society of America, Washington, p 1

Palme H, O'Neill HStC (2007) Cosmochemical estimates of mantle composition. Treatise on Geochemistry 2 (01): 1–38

Pearce CR, Parkinson IJ, Gaillardet J et al (2015) Reassessing the stable ($\delta^{88/86}$Sr) and radiogenic (^{87}Sr/^{86}Sr) strontium isotopic composition of marine inputs. Geochim Cosmochim Acta 157:125–146

Pfennig G, Klewe-Nebenius H, Seelman-Eggebert W (1998) Karlsruher Nuklidkarte, 6th edn. Marktdienste Haberbeck, Lage/Lippe

Rayleigh JW (1902) On the distillation of binary mixtures. Phil Mag S.6, 4:521–537

Richter FM, Mendybaev RA, Christensen JN et al (2006) Kinetic isotopic fractionation during diffusion of ionic species in water. Geochim Cosmochim Acta 70:277–289

Richter FM, Janney PE, Mendybaev RA et al (2007) Elemental and isotopic fractionation of type B CAI-like liquids by evaporation. Geochim Cosmochim Acta 71:5544–5564

Richter FM, Dauphas N, Teng F-Z (2009) Non-traditional fractionation of non-traditional isotopes: evaporation, chemical diffusion and soret diffusion. Chem Geol 258:92–103

Rollion-Bard C, Vigier N Spezzaferri S (2007) In situ measurements of calcium isotopes by ion microprobe in carbonates and application to foraminifera. Chem Geol 244:679–690

Rösler HJ (1988) Lehrbuch der Mineralogie, VEB Deutscher Verlag für Grundstoffindustrie, Leipzig, 4th ed., 845 p

Rudnick RL, Gao S (2003) Composition of the continental crust. Treatise on Geochemistry 3(01):1–64

Rüggeberg A, Fietzke J, Liebetrau V et al (2008) Stable strontium isotopes ($\delta^{88/86}$Sr) in cold-water corals—a new proxy for reconstruction of intermediate ocean water temperatures. Earth Planet Sci Lett 269:570–575

Russell WA, Papanastassiou DA, Tombrello TA (1978) Ca isotope fractionation on the Earth and other solar system materials. Geochim Cosmochim Acta 42:1075–1090

Schauble EA (2004) Applying stable isotope fractionation theory to new systems. In: Johnson CM, Beard BL, Albarède F (eds) Geochemistry of non-traditional stable isotopes. Reviews in Mineralogy and Geochemistry 65. Mineralogical Society of America, Washington, p 65

Schauble EA, Méheut M, Hill PS (2009) Combining metal stable isotope fractionation theory with experiments. Elements 5:369–374

Schiller M, Paton C, Bizzarro M (2012) Calcium isotope measurement by combined HR-MC-ICPMS and TIMS. J Anal At Spectrom 27:38–49

Schmitt A-D, Cobert F, Bourgeade P (2013) Calcium isotope fractionation during plant growth under a limited nutrient supply. Geochim Cosmochim Acta 110:70–83

Shahar A, Young ED, Manning CE (2008) Equilibrium high-temperature Fe isotope fractionation between fayalite and magnetite: an experimental calibration. Earth Planet Sci Lett 268:330–338

Shalev N, Lazar B, Halicz L et al (2013) Strontium isotope fractionation in soils and pedogenic processes. Proc Earth Planet Sci 7:790–793

Sharma M, Bajzer Z, Hui SK (2011) Development of ^{41}Ca-based pharmacokinetic model for the study of bone remodelling in humans. Clin Pharmacokinet 50:191–199

Simon JI, DePaolo DJ (2010) Stable calcium isotopic composition of meteorites and rocky planets. Earth Planet Sci Lett 289:457–466

Stahl W (1968) Search for natural variations in calcium isotope abundances. Earth Planet Sci Lett 5:171–174

Stahl W, Wendt I (1968) Fractionation of calcium isotopes in carbonate precipitation. Earth Planet Sci Lett 5:184–186

Taiz L, Zeiger E (2010) Plant Physiology, Sinauer Associates Inc., 5th Edn, 782 p

Urey H (1947) The thermodynamic properties of isotopic substances. J Chem Soc (London), pp 562–581

van der Heijden G, Legout A, Midwood A et al (2013) Mg and Ca root uptake and vertical transfer in soils assessed by an in situ ecosystem-scale multi-isotopic (^{26}Mg & ^{44}Ca) tracing experiment in a beech stand (Breuil-Chenue, France). Plant Soil 36:33–45

von Allmen K, Böttcher ME, Samankassou E et al (2010) Barium isotope fractionation in the global barium cycle: first evidence from barium minerals and precipitation experiments. Chem Geol 277:70–77

Walczyk T, Denk E, Hillegonds D et al (2010) Isotopic labeling of the human skeleton using 41Ca for sensitive identification of changes in bone calcium balance in humans. Osteoporosis Int 21:S734–S735

Watkins JM, DePaolo DJ, Ryerson FJ et al (2011) Influence of liquid structure on diffusive isotope separation in molten silicates and aqueous solutions. Geochim Cosmochim Acta 75:3103–3118

Weiner S, Dove PM (2004) An overview of biomineralization processes and the problem of the vital effect. Rev Min Geochem 54:1–29

Widanagamage IH, Griffith EM, Singer DM et al (2015) Controls on stable Sr-isotope fractionation in continental barite. Chem Geol 411:215–227

Young ED, Galy A (2004) The isotope geochemistry and cosmochemistry of magnesium. In: Johnson CM, Beard BL Albarède F (eds) Geochemistry of non-traditional stable isotopes. Reviews in mineralogy and geochemistry, Mineralogical Society of America, Washington, p 197–230

Young ED, Galy A, Nagahara H (2002) Kinetic and equilibrium mass-dependent isotope fractionation laws in nature and their geochemical and cosmochemical significance. Geochim Cosmochim Acta 66:1095–1104

Young ED, Manning CE, Schauble EA et al (2015) High-temperature equilibrium isotope fractionation of non-traditional stable isotopes: experiments, theory, and applications. Chem Geol 395:176–195

Zeller B, Colin-Belgrand M, Dambrine E et al (1998) N-15 partitioning and production of N-15-labelled litter in beech trees following N-15 urea spray. Ann Sci For 55:375–383

Zeller B, Colin-Belgrand M, Dambrine E et al (2000) Decomposition of N-15 labelled beech litter and fate of nitrogen derived from litter in a beech forest. Oecologia 123:550–559

Zerle L, Faestermann T, Knie K et al (1997) The 41Ca bomb pulse and atmospheric transport of radionuclides. J Geophys Res 102(D16):19517–19527

Zhang J, Huang S, Davis AM et al (2014) Calcium and titanium isotopic fractionations during evaporation. Geochim Cosmochim Acta 140:365–380

Analytical Methods

Alexander Heuser, Anne-Désirée Schmitt,
Nikolaus Gussone and Frank Wombacher

Abstract

Despite the large relative mass difference between different Ca isotopes, the isotopic variability of Ca in natural materials is relatively small. Consequently, high-precision Ca isotope analyses are required to accurately resolve Ca isotope fractionation prior to data interpretation. This chapter summarises techniques which are successfully used to digest samples and to purify different types of sample material. The basic principles of different mass spectrometric methods for the accurate and precise determination of Ca isotope compositions are presented with a focus on thermal ionisation mass spectrometry (TIMS) and multi-collector inductively coupled plasma mass spectrometry (MC-ICP-MS). We also present an overview about other, less frequently applied techniques used so far. Additionally, we provide useful information on how to report Ca isotope data (e.g. different notations and reference materials) and explain how to convert literature data based on different reference materials and/or given in different notations.

Keywords

Reference materials · Chemical separation · Digestion · Isotope fractionation · Mass spectrometry · TIMS · MC-ICP-MS · Double spike

A. Heuser
Steinmann-Institut für Geologie, Mineralogie und Paläontologie, Universität Bonn, Bonn, Germany
e-mail: aheuser@uni-bonn.de

A.-D. Schmitt
LHyGeS/EOST, Université de Strasbourg, Strasbourg, France
e-mail: adschmitt@unistra.fr

A.-D. Schmitt
Laboratoire Chrono-Environment, Université Bourgogne Franche-Comté, Besançon, France

N. Gussone (✉)
Institut für Mineralogie, Universität Münster, Münster, Germany
e-mail: nguss_01@uni-muenster.de

F. Wombacher
Institut für Geologie und Mineralogie, Universität zu Köln, Köln, Germany
e-mail: fwombach@uni-koeln.de

© Springer-Verlag Berlin Heidelberg 2016
N. Gussone et al., *Calcium Stable Isotope Geochemistry*,
Advances in Isotope Geochemistry, DOI 10.1007/978-3-540-68953-9_2

1 Introduction

Russell et al. (1978) developed a reliable measurement protocol for calcium isotopes using thermal ionization mass spectrometry (TIMS) and a double spike technique. They observed that the natural fractionation of Ca isotopes is generally low (<1.3 ‰ per amu) and that high analytical precision and accuracy is necessary to resolve natural differences in calcium isotope ratios. Despite further advancements in analytical methods, the precision of stable isotope analysis remains a major limitation for Ca stable isotope research.

Accurate mass spectrometry often requests the purification of the sample before Ca isotope analysis. This purification is commonly achieved by ion exchange using column chemistry. Both, column chemistry and mass spectrometry can induce artefacts that have to be avoided or corrected for in order to accurately resolve the natural or experimentally induced Ca isotope fractionation.

In this chapter we provide the basic principles of the applied techniques along with detailed descriptions of methods applied to analyze Ca isotope ratios in different kinds of geological, biological and extraterrestrial samples. These methods include sample specific preparation (cleaning, digestion, and column chemistry), mass spectrometry protocols and data reduction, including the double spike technique.

2 Notations and Data Presentation

Looking through the Ca isotope literature shows that different authors use slightly different notations to present their data. This can lead to confusion and makes it difficult to compare literature data. We recommend to follow the IUPAC/IUPAP "guidelines recommended terms for expression of stable-isotope-ratio and gas-ratio measurement results" (Coplen 2011).

2.1 δ-Notation

Variations of the Ca isotopic compositions are very small and thus variations of Ca isotopes are normally expressed using the δ-notation. The δ-value refers to the deviation of the isotope composition of a sample from that of the reference material (Coplen 2011):

$$\delta^a X = \left(\frac{\left(^a X /^b X\right)_{sample}}{\left(^a X /^b X\right)_{reference}} - 1 \right) \qquad (1)$$

where $^a X$ and $^b X$ are the isotopes with nominal mass a and b of the element X. By convention, isotope ^{a}X is always heavier than ^{b}X. Therefore, positive δ-values thus refer to samples being enriched in the heavy isotopes relative to the reference standard; negative values denote light isotope enrichment in the sample relative to the standard. As the variations of the isotopic compositions are small, δ-values are typically reported as per mil (‰).

δ values for Ca isotope variations are typically based on either $^{44}Ca/^{40}Ca$ or $^{44}Ca/^{42}Ca$. The latter ratio is commonly used if Ca isotope ratios are determined by MC-ICP-MS, where ^{40}Ar interferes with ^{40}Ca (cf. Sect. 5.3.3).

Using the common δ-notation (Eq. 1) both ratios would be reported as δ^{44}Ca-values. But the stable isotope fractionation in $^{44}Ca/^{42}Ca$ is only about half of the fractionation in $^{44}Ca/^{40}Ca$ (see Eqs. 5 and 7). To circumvent this problem Hippler et al. (2003) and Eisenhauer et al. (2004) defined a $\delta^{44/4x}Ca$ ratio which clearly indicates which Ca isotope ratio is used in the δ-notation:

$$\delta^{44/40}Ca = \left(\frac{\left(^{44}Ca/^{40}Ca\right)_{sample}}{\left(^{44}Ca/^{40}Ca\right)_{reference}} - 1 \right) \qquad (2)$$

$$\delta^{44/42}Ca = \left(\frac{\left(^{44}Ca/^{42}Ca\right)_{sample}}{\left(^{44}Ca/^{42}Ca\right)_{reference}} - 1 \right) \qquad (3)$$

These δ-notations follow the recommendations of the IUPAC (Coplen 2011). Typically,

$\delta^{44/40}Ca$ and $\delta^{44/42}Ca$ values are reported in per mil (‰, parts per thousand) which means that the δ-value obtained by Eqs. 2 and 3 are then multiplied by 1000. δ-values reported as parts per ten thousand (pptt) or even parts per million (ppm) can also be found in the literature.

The existence of two different δ-notations for Ca with different degrees of fractionation hampers a direct comparison of $\delta^{44/40}Ca$ and $\delta^{44/42}Ca$ literature values. Fortunately, it is possible to convert $\delta^{44/40}Ca$ values to $\delta^{44/42}Ca$ values and vice versa. There are three potential problems concerning the conversion of δ-values (e.g. from $\delta^{44/42}Ca$ to $\delta^{44/40}Ca$) which can be ignored in many cases:

(1) The exact relationship between two different δ-values depends on the assumed fractionation mechanism.

   If one assumes kinetic fractionation based on atomic masses of Ca isotopes (see Young et al. 2002 and Chapter "Introduction" for discussion), $\delta^{44/40}Ca$ values can be converted into $\delta^{44/42}Ca$ values and vice versa using the following equation:

$$\delta^{44/40}Ca \approx \delta^{44/42}Ca \times \frac{\ln(m^{44}Ca/m^{40}Ca)}{\ln(m^{44}Ca/m^{42}Ca)} \quad (4)$$

where $m^{x}Ca$ is the exact atomic mass (given in Chapter "Introduction") of the respective isotope. In a good approximation Eq. 4 can be rewritten to:

$$\delta^{44/40}Ca \approx \delta^{44/42}Ca \times 2.05 \quad (5)$$

If one assumes equilibrium isotope fractionation for Ca isotopes:

$$\delta^{44/40}Ca \approx \delta^{44/42}Ca \times \frac{1/m^{44}Ca - 1/m^{40}Ca}{1/m^{44}Ca - 1/m^{42}Ca} \quad (6)$$

with $m^{x}Ca$ being the exact atomic mass of the respective isotope. In a good approximation we obtain:

$$\delta^{44/40}Ca \approx \delta^{44/42}Ca \times 2.10 \quad (7)$$

Given a $\delta^{44/42}Ca$ of 1 ‰, recalculated $\delta^{44/40}Ca$ values amount to 2.05 or 2.1 ‰ respectively depending whether Eq. 5 for kinetic isotope fractionation or Eq. 7 for equilibrium isotope fractionation is used. This difference of 0.05 ‰ appears acceptable given current analytical precisions of typically larger than 0.02 ‰ *per u* and the fact that Ca isotope fractionation in nature is usually <1 ‰ for $\delta^{44/42}Ca$.

(2) The relationship between two δ-values is not linear.

   The above equations assume a linear relationship between $\delta^{44/40}Ca$ and $\delta^{44/42}Ca$, while fractionation lines in conventional three isotope space are curved. However, the error introduced by this linear approximation is negligible. Only extremely large Ca isotope fractionations of more than 6.6 ‰ for $\delta^{44/42}Ca$ would result is systematic errors approaching −0.05 for recalculated $\delta^{44/40}Ca$. The error introduced by this linear approximation is not only negligible, it is also much less than the uncertainty introduced by variations in the fractionation mechanism.

(3) ^{40}Ca can be variable due to ingrowth from radioactive ^{40}K.

   While $\delta^{44/42}Ca$ represents exclusively mass fractionation processes, $\delta^{44/40}Ca$ may also be affected by ^{40}Ca excess (see Sect. 3 in Chapter "High Temperature Geochemistry and Cosmochemistry"). If radiogenic ingrowth is larger than the analytical precision and more significant than differences introduced by the fractionation mechanism, then it is possible to deconvolute both processes by measuring and representing $\delta^{44/40}Ca$ versus $\delta^{44/42}Ca$ (Schmitt et al. 2003a, b; Schmitt and Stille 2005; Ryu et al. 2011).

Because of the ambiguity of the $\delta^{44}Ca$ notation ($\delta^{44/40}Ca$ or $\delta^{44/42}Ca$), Gussone et al. (2005) proposed an alternative notation and introduced a $\delta^{mu}Ca$ ($^{ppm}/_{amu}$) such that δ-values from different Ca isotope ratios can be directly compared without further recalculation. The superscript *mu* refers to the fractionation per one atomic mass unit. $\delta^{mu}Ca$ is defined as:

$$\delta^{mu}Ca\left(^{ppm}/_{amu}\right) = \left(\frac{\left(^{a}Ca/^{b}Ca\right)_{sample}}{\left(^{a}Ca/^{b}Ca\right)_{reference}} - 1\right) \times 10^{6} / (a - b) \tag{8}$$

where a and b refer to the exact masses of the isotopes used. Note that unlike in Eqs. 1–3, the conversion factor is considered in order to report the δ-value in parts per million. As with $\delta^{44/42}Ca$ and $\delta^{44/40}Ca$, $\delta^{mu}Ca$ obtained from different Ca isotope ratios may differ depending on the fractionation mechanism. The use of $\delta^{mu}Ca$ has not become accepted and should be avoided. The same applies to reporting Ca isotope data as $\delta^{44/42}Ca$ in ‰/amu i.e. the calculated $\delta^{44/42}Ca$ is divided by two (Tacail et al. 2014).

2.2 Fractionation Factor (α)

When comparing the isotopic signature of two substances or compartments where the isotopic composition of a compartment remains constant, the usage of the fractionation factor (α) instead of δ-values is preferred. The fractionation factor is defined as the ratio of two isotopes in the compound A divided by the ratio of the same isotopes in the compound B:

$$\alpha_{A-B} = \frac{R_A}{R_B} \tag{9}$$

Normally, the isotope ratios R are not reported in the literature which prevents the direct calculation of fractionation factors from isotope ratios. However, α can be calculated from the reported δ-values (given in ‰):

$$\alpha_{A-B} = \frac{\delta_A + 1000}{\delta_B + 1000} \tag{10}$$

It is obvious that the usage of $\delta^{44/40}Ca$ and $\delta^{44/42}Ca$ in Eq. 10 will result in different fractionation factors. Therefore, fractionation factors calculated from $\delta^{44/42}Ca$ (or $^{44}Ca/^{42}Ca$ in Eq. 9) have to be converted to fractionation factors for $^{44}Ca/^{40}Ca$ or vice versa using the same factors as given in Eqs. 5 and 7.

2.3 Δ-Notation

The capital delta notation (Δ) describes the difference between the δ-values of two phases A and B and relates to the fractionation factor α according to:

$$\Delta_{A-B} = \delta_A - \delta_B \approx 1000 \ln \alpha_{A-B} \tag{11}$$

2.4 ε_{Ca}-Notation for Radiogenic ^{40}Ca Ingrowth

Isotope ^{40}Ca plays a special role within the Ca isotope system as ratios involving ^{40}Ca are not only affected by mass-dependent fractionation, but also by radiogenic ingrowth. Radiogenic ^{40}Ca is produced by a β^- decay of ^{40}K (half-life 1.397×10^9 y, Steiger and Jäger 1977). The K–Ca system can be used to date K-rich rocks, which has been shown by several authors (Heumann et al. 1977, 1979; Marshall and DePaolo 1982, 1989; Marshall et al. 1986; Baadsgaard 1987; Nelson and McCulloch 1989; Shih et al. 1994; Fletcher et al. 1997; Nägler and Villa 2000; Kreissig and Elliot 2005). In analogy to other radiogenic isotope systems (e.g. Hf, Nd, Sr), a ε-notation, representing deviations (excesses) in the abundance of ^{40}Ca from a reference compositions in parts per 10^4 is used. Again, two different ε_{Ca} notations exist, depending on the denominator isotope chosen:

$$\varepsilon_{Ca} = \left(\frac{\left(^{40}Ca/^{42}Ca\right)_{sample}}{\left(^{40}Ca/^{42}Ca\right)_{mantle}} - 1\right) \times 10^4 \tag{12}$$

$$\varepsilon_{Ca} = \left(\frac{\left(^{40}Ca/^{44}Ca\right)_{sample}}{\left(^{40}Ca/^{44}Ca\right)_{mantle}} - 1\right) \times 10^4 \tag{13}$$

The ε_{Ca} values are commonly calculated from isotope ratios that are first corrected for the instrumental mass discrimination by the exponential law, using $^{42}Ca/^{44}Ca = 0.31221$

(Marshall and DePaolo 1982) (see Sect. 5.2.1, and note Sect. 5.2.2 for a routine where both the stable isotope fractionation and ε_{Ca} are determined). Because of the low $^{40}K/^{40}Ca$ ratio of the mantle, radiogenic ingrowth is negligible such that $(^{40}Ca/^{42}Ca)_{mantle}$ can be applied without time-correction. The $(^{40}Ca/^{44}Ca)_{mantle}$ and $(^{40}Ca/^{42}Ca)_{mantle}$ values correspond to about 47.16 and 151.0, respectively. Since the measured absolute Ca isotope ratios vary between different laboratories, the precise reference value for the mantle (=bulk Earth) may be determined by repeated analysis of mantle derived samples in the respective laboratory or even during the particular measurement session (e.g. Kreissig and Elliot 2005; Caro et al. 2010). In contrast to common practice in geosciences, the IUPAC/IUPAP guidelines recommend to use $\delta^{40}Ca^*$ (pptt) as an alternative for ε_{Ca} to denote radiogenic ingrowth of ^{40}Ca (Coplen 2011).

Note that natural Ca stable isotope fractionation can systematically bias ε_{Ca} values if the mass discrimination correction does not accurately describe the mass-scaling of the natural isotope fractionation process. Likewise, ingrowth of ^{40}Ca can affect the determination of $\delta^{44/40}Ca$. These issues are discussed below (Sect. 5.2.2).

For several other non-traditional stable isotope systems, ε is often used to denote mass-dependent stable isotope effects. For example, $\varepsilon^{44/40}Ca$ would refer to the deviation from a reference material in pptt (in analogy to Eq. 2). Furthermore, ε is also applied to denote a separation factor to express stable isotope fractionations in stable isotope systems:

$$\varepsilon = \alpha - 1 \tag{14}$$

2.5 ε- and μ-Notations in Cosmochemistry

In addition to mass-dependent and potentially radiogenic effects, Ca isotopes in extraterrestrial samples also exhibit nucleosynthetic anomalies that attest to the production of Ca isotopes at different stellar environments (Chapter "High Temperature Geochemistry and Cosmochemistry"). Although different formats are in use, they all give deviations from terrestrial Ca isotope compositions (i.e. NIST SRM 915) in pptt or parts per million using ε- or μ-notations:

$$\varepsilon^{43/44}Ca = \left(\frac{\left(^{43}Ca/^{44}Ca\right)_{sample}}{\left(^{43}Ca/^{44}Ca\right)_{reference}} - 1 \right) \times 10^4 \tag{15}$$

$$\mu^{48/44}Ca = \left(\frac{\left(^{48}Ca/^{44}Ca\right)_{sample}}{\left(^{48}Ca/^{44}Ca\right)_{reference}} - 1 \right) \times 10^6 \tag{16}$$

Although in terms of nucleosynthesis all Ca isotope ratios are of interest, ^{44}Ca is commonly chosen as the denominator isotope. Please note, that these ε- and μ-notations do not follow the recommendations of Coplen (2011). Following the recommendations of Coplen $\delta^{43/44}Ca$ and $\delta^{48/44}Ca$ should be used and the results should be reported as either pptt or ppm (cf. Eq. 1).

3 Reference Materials

3.1 Used Reference Materials

The variability of stable isotope ratios is regularly expressed relative to an international standard, to allow the comparison of results obtained from different studies. Unlike many isotopic systems there is no universal agreement on what reference material should be used as international standard for Ca isotopes. Consequently, published Ca isotope data are reported relative to several different reference materials, which add further complications to the different isotope ratios used ($^{44}Ca/^{42}Ca$ and $^{44}Ca/^{40}Ca$). A lookup table (Table 1) summarizing the most frequently used reference materials and equations for the conversion of Ca isotope data from different studies are provided at the end of this section.

In order to achieve better comparability of published Ca isotope data, the IUPAC suggested to use the SRM 915a carbonate standard provided by NIST as a primary standard, and natural

Table 1 Conversion of standards relative to SRM 915a (in ‰)

Standard	$\delta^{44/40}Ca_{SRM915a}$	$\delta^{44/42}Ca_{SRM915a}$
SRM915a	0.00	0.00
Seawater	1.88	0.92
SRM915b	0.72	0.35
CaF_2	1.44	0.70
SRM1486	−1.01	−0.49
BSE	1.03	0.50
$CaCO_3$	1.02	0.50
HPSCa	0.34	0.17
Bone powder	−0.84	0.41

seawater (e.g. IAPSO) as a secondary standard (Coplen et al. 2002). Both reference materials were already isotopically well characterized and had been used in earlier studies. Since then most data are presented relative to one of the two standards, or at least studies provide δ-values of one of these standards to allow renormalization of the data. Nevertheless, the debate concerning the best suited Ca isotope standard is still ongoing. A short overview below provides background information about different standards and discusses their pros and cons:

Zhu and Macdougall (1998) and Schmitt et al. (2001) proposed to use seawater as a common standard, since seawater was the only common and comparable sample reported in several Ca isotopic studies. In particular, it has been shown that, within uncertainties, the Ca isotopic composition of modern seawater is homogenous (Zhu and MacDougall 1998; De La Rocha and DePaolo 2000; Schmitt et al. 2001; Hippler et al. 2003). This is due to the long Ca residence time of 0.5–1 Myr which is long compared to the mixing time of ocean water (10^3 years). Therefore, seawater seems to be a very suitable reservoir to serve as a reference material. Seawater has a relatively high Ca concentration of 400 mg l^{-1} (Taylor and MacLennan 1985) and is furthermore widely available. A minor disadvantage is that a chemical procedure is necessary to separate Ca from the matrix prior to the analysis. Moreover, DePaolo and co-workers (cf. DePaolo 2004) recommend not to use seawater

as a standard because its calcium isotopic composition changes with time (see Chapter "Global Ca Cycles: Coupling of Continental and Oceanic Processes"). Although ^{40}Ca excess from continental weathering is introduced into the ocean, the radiogenic enrichment of ^{40}Ca in ocean water throughout Earth history relative to bulk Earth is smaller than the analytical uncertainties of 0.035 ‰, and thus negligible (Caro et al. 2010, Chapter "High Temperature Geochemistry and Cosmochemistry").

The $CaCO_3$ reference powder of the National Institute of standards SRM 915a, first used by Halicz et al. (1999), is not related to a specific geological reservoir, but has been proposed as future reference standard (Hippler et al. 2003; Eisenhauer et al. 2004; Coplen et al. 2002). However, since 2006 this reference material is out of stock and replaced by NIST SRM 915b. Since this standard is a solid standard, small isotopic differences could occur between different batches. Therefore the use of this standard as an international Ca standard first necessitates verification of its homogeneity (Wombacher et al. 2009). NIST SRM 915b is presently only sparsely employed (Heuser and Eisenhauer 2008; Wombacher et al. 2009; Hindshaw et al. 2011; Silva-Tamayo et al. 2010; Heuser et al. 2011; Reynard et al. 2011; Valdes et al. 2014; Brazier et al. 2015).

Published Ca isotope ratios of other $CaCO_3$ salts used so far as standards can only be considered as internal laboratory values. Russell et al. (1978) furthermore showed that industrially produced $CaCO_3$ salts may have a strongly fractionated composition. This was confirmed by Schmitt et al. (2001) and Hippler et al. (2003) ($\Delta_{Johnson\ Matthey\ Lot\ 4064\text{-}Lot\ 9912} = -11.95$ ‰).

Russell et al. (1978) gave a precisely defined absolute reference value for the $^{42}Ca/^{44}Ca$ of 0.31221 ± 0.00002 (2σ) deduced from two Ca terrestrial standards, two lunar samples and four meteorites. Marshall and DePaolo (1989) proposed for their part to use the $^{40}Ca/^{42}Ca$ value of the mantle, estimated by measuring four terrestrial basaltic lavas and obtained a value equal to 151.016 (Marshall and DePaolo 1982) as a reference. The mantle is indeed considered to have a constant Ca isotopic composition throughout geologic time within reproducibility (see Sect. 3 in Chapter "High Temperature Geochemistry and Cosmochemistry"). This is due to its low K/Ca ratio ($\approx$0.01). More recently, Simon et al. (2009) re-evaluated this value by determining the Ca isotopic composition of oceanic basalts and four differentiated meteorites. They determined a weighted average for $^{40}Ca/^{44}Ca$ equal to 47.1480 $\pm$ 0.0004 (2σ), equivalent to a $^{40}Ca/^{42}Ca$ value of 151.0150 $\pm$ 0.0013 (2σ), consistent with the values of Russell et al. (1978) and Marshall and DePaolo (1989).

Skulan et al. (1997) set an arbitrary but reasonable value for their $^{40}Ca/^{44}Ca$ ultrapure $CaCO_3$ standard ratio of 47.144 and referred all data to this value. They found that their $^{40}Ca/^{44}Ca$ ratios are very close to that of the average value measured in igneous rocks and minerals, and thus suggested that it should be close to that of bulk silicate Earth.

In order to avoid industrially produced $CaCO_3$ salts with fractionated initial isotopic compositions, other studies used a natural CaF_2 as reference material free of technical or biological isotopic fractionation (cf. Russell et al. 1978; Nägler and Villa 2000; Nägler et al. 2000; Heuser et al. 2002).

Added to these more or less recognized standards, additional standards have also been used by some authors, such as NIST SRM 1486 bone meal (Heuser and Eisenhauer 2008), in-house "HPSCa" solution (Blättler et al. 2011), bone powder (Reynard et al. 2011) or an ICP Ca standard solution (ICP1; Channon et al. 2015; Morgan et al. 2012). In the latter studies δ-values were presented using the inhouse standard as reference material. At least the ICP1 based δ-value of SRM915a was presented which allows converting the published data to SRM915a based values. A MORB glass (PH 78-2) has also been used as a standard to express ^{40}Ca excess measurements (Kreissig and Elliott 2005).

A cross calibration of several reference materials between three laboratories (Institute of Geological Sciences of Berne, Geomar from Kiel and Centre de Géochimie de la Surface from Strasbourg) has enabled calibrations independent of laboratory specific biases (Hippler et al. 2003). The main result of this study is that, despite a range of chemical processing protocols, loading techniques and isotopic measurement protocols have been employed, no inter-laboratory bias correction is necessary and that the Ca isotope data are directly comparable with each-other when expressed in $\delta^{44/40}Ca$ notation relative to a common standard.

3.2 Conversion of δ-Values Based on Different Reference Materials

It is possible to convert the δ-values based on exotic or outdated reference materials into δ-values based on commonly used reference materials if the isotopic composition of the commonly used reference material relative to the exotic reference material is known. The δ-value of a sample relative to the "old" reference material is given by

$$\delta_{old} = \left(R_{sample}/R_{ref\_old}-1\right) \cdot 1000 \qquad (17)$$

and the δ-value of a sample based on the "new" reference material is

$$\delta_{new} = \left(R_{sample}/R_{ref\_new}-1\right) \cdot 1000 \qquad (18)$$

The δ-value of the "old" reference material based on the "new" reference material is

$$\delta_{ref} = \left(R_{ref\_old}/R_{ref\_new}-1\right) \cdot 1000 \qquad (19)$$

Solving Eq. 17 for R_{sample} and Eq. 19 for R_{ref_new} and applying to Eq. 18 results in:

$$\delta_{new} = \left(\frac{\frac{\delta_{old}+1000}{1000} \cdot R_{ref\_old}}{\frac{1000}{\delta_{ref}+1000} \cdot R_{ref\_old}} - 1\right) \cdot 1000 \qquad (20)$$

Solving Eq. 20 leads to:

$$\delta_{new} = \frac{\delta_{old} \cdot \delta_{ref}}{1000} + \delta_{ref} + \delta_{old} \qquad (21)$$

If the product $\delta_{old}\cdot\delta_{ref}$ is small (<10) the first term is negligible and δ_{new} is in good approximation:

$$\delta_{new} \approx \delta_{ref} + \delta_{old} \qquad (22)$$

In the following we present convenient ways to convert data expressed relative to different standards used in the literature, relative to NIST SRM 915a, the most commonly used standard and the common reference of all chapters of this book. When no reference is indicated after the conversion equation, it means that the values are compiled from different laboratories-weighted averages.

National Institute of Standards and Technology (NIST) SRM 915a and SRM 915b

NIST SRM 915a is the presently most commonly used standard. As mentioned above this SRM is out of stock and was replaced by SRM 915b.

$$\delta^{44/40}Ca_{sample/NISTSRM915a} = \delta^{44/40}Ca_{sample/NISTSMR915b} + 0.72\,(‰) \tag{23}$$

Seawater standard

Due to the modern seawater $\delta^{44/40}$Ca homogeneity, either an aliquot of seawater available in different laboratories (e.g. seawater from San Diego in Skulan et al. 1997 or from the Atlantic in Schmitt et al. 2001) or IAPSO (International Association for the Physical Sciences of the Ocean) seawater salinity standard available from OSIL (Ocean Scientific International Ltd) have been employed (Hippler et al. 2003):

$$\delta^{44/40}Ca_{sample/NISTSRM915a} = \delta^{44/40}Ca_{sample/sw} + 1.88\,(‰) \tag{24}$$

Bulk Silicate Earth (BSE) standard

This standard is linked to an Earth reservoir that is not subjected to variation through time. The geological significance of this approach is undoubted. Increasing evidence for Ca isotope fractionation occurring at high temperatures and pressures call for a reliable determination of BSEs Ca isotope composition, which is however not a straightforward task and still pending further verification:

$$\delta^{44/40}Ca_{sample/NISTSRM915a} = \delta^{44/40}Ca_{sample/BSE} + 1.03\,(‰) \tag{25}$$

Further used phosphate, CaF_2 and $CaCO_3$ standards

CaF_2 and ultra-pure $CaCO_3$ have only been measured few times and are presently more or less abandoned. Other reference materials reported so far are SRM 1486 (bone ash), bone powder and HFSCa (Heuser and Eisenhauer 2008 (Eq. 28); Reynard et al. 2011 (Eq. 29); Blättler et al. 2011 (Eq. 30); Russel et al. 1978 (Eq. 31)).

$$\delta^{44/40}Ca_{sample/NISTSRM915a} = \delta^{44/40}Ca_{sample/CaF2} + 1.45\,(‰) \tag{26}$$

$$\delta^{44/40}Ca_{sample/NISTSRM915a} = \delta^{44/40}Ca_{sample/CaCO3} + 1.02\,(‰) \tag{27}$$

$$\delta^{44/40}Ca_{sample/NISTSRM915a} = \delta^{44/40}Ca_{sample/NISTSRM1486} - 1.01\,(‰) \tag{28}$$

$$\delta^{44/40}Ca_{sample/NISTSRM915a} = \delta^{44/40}Ca_{sample/bonepowder} - 0.84\,(‰) \tag{29}$$

$$\delta^{44/40}Ca_{sample/NISTSRM915a} = \delta^{44/40}Ca_{sample/HPSCa} + 0.68\,(‰) \tag{30}$$

$$\delta^{44/40}Ca_{sample/NISTSRM915a} = -\delta^{40/44}Ca_{sample/std} + 0.98\,(‰) \tag{31}$$

Factors used for the conversion between different standards are summarized in Table 1.

Another way to rapidly convert Ca isotopic data expressed against one standard to another is to renormalize data from "source" to "target" standard following (Table 2):

$$\delta^{44/40}Ca_{sample/target} = \delta^{44/40}Ca_{sample/source} - (\text{correction value}) \tag{32}$$

For example, renormalisation from seawater to SRM915a: $\delta^{44/40}Ca_{sa/SRM915a} = \delta^{44/40}Ca_{sa/seawater} - (-1.88)‰$ additional standard values can be found in the GeoReM database (http://georem.mpch-mainz.gwdg.de; 2016-01-02).

Table 2 $\delta^{44/40}$Ca conversions between different standards (‰)

To target \ from source	Seawater	SRM 915a	SRM 915b	SRM 1486
Seawater		1.88	1.14	2.88
SRM 915a	−1.88		−0.74	1.01
SRM 915b	−1.14	0.74		1.74
SRM 1486	−2.88	−1.01	−1.74	

4 Sample Preparation

Different preparation protocols have been developed for Ca isotope analysis on various types of samples. The following sections contain detailed descriptions of sample preparation and cleaning protocols for different kinds of materials and matrices.

4.1 Digestion and Cleaning Techniques

4.1.1 Carbonates

Digestion techniques for carbonates depend on the mineralogy and the sample matrix. Pure $CaCO_3$ as found for instance in marine microfossils, or experimentally precipitated carbonates is easily dissolved in diluted acids. Most commonly hydrochloric acid (HCl), nitric acid (HNO_3) or acetic acid (CH_3COOH) are used, depending on the further sample treatment (e.g. taking aliquots for trace-element work, column chemistry and presence of matrix minerals which dissolution or leaching should be avoided). To eliminate any organic impurities the dissolved fraction is treated with a H_2O_2–HNO_3 mixture (Hippler et al. 2003, 2006) or is ultrasonically cleaned with ultrapure water and H_2O_2 (Heuser et al. 2005; Farkaš et al. 2006). Depending on the carbonate minerals and matrix several protocols have been proposed:

_CaCO_3 shells_

Biogenic $CaCO_3$ shells collected from the environment normally need to be cleaned prior to Ca isotope analysis, mainly for two reasons, removing of detritus and organic compounds. Two different methods have been used, depending on the sample material:

H_2O_2–_NaOH_ method: For shell fragments (recommended for planktic and benthonic foraminifers):

The sample is first gently crushed and transferred into acid pre-cleaned 1.5 ml polypropylene (PP) reaction vials and ultrasonicated for 2 min in ultrapure water, the pH of which was elevated to 8–9 by the addition of NH_4OH-solution to prevent partial dissolution of the calcareous samples during the cleaning process. This procedure is repeated twice with ultrapure water, once with methanol and two more times with water. Then the sample is heated to about 80 °C for 30 min in a NaOH–H_2O_2 (0.1 and 0.01 M, respectively) solution in a heated ultrasonic bath. Finally, the tests are washed three times and ultrasonicated with ultrapure water. The cleaned sample is dissolved in 0.5 N HCl (cf. Gussone et al. 2004; Gussone and Filipsson 2010).

NaClO method: suited for powdered samples (e.g. corals, mollusks) and nanofossils (coccolithophores and calcareous dinoflagellates):

Samples are transferred into acid-cleaned polypropylene (PP) reaction vials and bleached for 24 h in a 10 % NaClO solution ($\sim$1 % active chlorine), to remove organic compounds with potentially deviating composition or which might influence the ionization in the mass spectrometer. Samples are ultrasonicated several times during the bleaching. The bleach is subsequently removed and the samples are washed 6 times in distilled water, the pH of which was elevated to 8–9 by the addition of NH_4OH-solution to prevent partial dissolution of the calcareous samples during the cleaning process. The samples are finally dissolved in 0.5 N HCl (cf. Gussone et al. 2006; Böhm et al. 2006).

Dolomite—limestone

Fine grained dolomite powder dissolves readily in 2.5 N HCl or HNO$_3$. Different published protocolls dissolve 1 g of dolomite and limestone in 6 N ultrapure HCl (Fantle and DePaolo 2007; Jacobson and Holmden 2008), 3 M HCl (Halicz et al. 1999), at room temperature or 2.5 N HCl heated at 70 °C (Wang et al. 2012). 2N acetic acid (Brazier et al. 2015) was also proposed. Whatever the protocol, the dissolved samples are centrifuged and the insoluble residues are discarded.

For the selective dissolution of carbonate minerals in the presence of non-carbonate matrix, potential leaching of e.g. clay minerals need to be considered and taken into account for selection of the used type and strength of acid.

4.1.2 Phosphates

Peloidal phosphates

For analyses of sedimentary peloïdal phosphates it can be necessary to remove Sr-rich calcite overgrowths by a treatment with 0.5 N acetic acid and subsequent washing with distilled water prior to dissolution in 6N HCl (Schmitt et al. 2003a, b; Cobert et al. 2011a). An alternative treatment includes a facultative heating to remove any free MgO and CaO, decalcification with tri ammonium citrate (TAC) and washing with deionized water and dissolution in 1.3M HCl (Soudry et al. 2006).

Bone and teeth

Samples of bones (cortex) and teeth (enamel and dentin) are obtained using hand-operated drills. To obtain fresh unaltered material, their surface layer is removed. For digesting the samples, two different techniques are then applied:

(1) The samples are bleached overnight in 2 % NaClO to break down organic molecules, rinsed with water and dissolved in 1.5 mL 2.5 N HCl (Clementz et al. 2003), warm 5 M HNO$_3$ (Chu et al. 2006) or 2 M HCl (Reynard et al. 2010).

(2) The samples are dissolved in 1 mL high-purity concentrated HNO$_3$ and 30 μL of H$_2$O$_2$ over 12 h on a hot plate at 140 °C (Heuser et al. 2011).

4.1.3 Sulfates

Gypsum and anhydrite

Gypsum and anhydrite samples are first washed, depending on their origin and potential contamination, dried and powdered. A few mg are digested in a Teflon beaker in 0.5 ml of 4.5 N HCl. Dissolution takes place within a few days (Hensley 2006).

Barite

Three different methods have been published to digest barite samples:

(1) Na$_2$CO$_3$ digestion method (von Allmen et al. 2010a) for Ba-isotope analysis based on Breit et al. (1985)

Barite powder ($\sim$5 mg) is mixed with 50 mg sodium carbonate (Na$_2$CO$_3$) and 1 ml distilled water in a Teflon beaker, and subsequently heated for 4 h on a hotplate at 95 °C. In order to keep the fluid volume at about 1 ml deionized water is intermittently added. A chemical reaction in the beaker leads to the formation of strongly alkaline Na$_2$SO$_4$ in the liquid phase and solid BaCO$_3$. To ensure complete dissolution of barite, the liquid is decanted and again 50 mg Na$_2$CO$_3$ are added together with 1 ml H$_2$O to the residue and heated for 4 h at 95 °C. After that, the fluid is decanted and the solid residue is several times rinsed with H$_2$O, and finally dissolved in 2.5 N HCl.

(2) HI digestion based on Takano and Watanuki (1972)

About 10–30 mg of powdered barite is mixed with 2 ml of HI in Teflon beakers and placed in pressure bombs in an oven at 180 °C for at least 4 h. The solution is dried down and recovered in 2 ml 2 N HCl. This is repeated (up to 6 times) until all HI is removed. The basic principle of this digestion is the following reaction:

$$BaSO_4 + 8HI + 2H^+ \rightarrow Ba^{2+} + 4H_2O + H_2S + 4I_2 \quad (33)$$

(3) Chelate method (cf. Griffith et al. 2008)

Purified barite (10 mg) is mixed with pure water and 1 ml pre-cleaned cation exchange resin (pH $\sim$5.5) (Mitsubishi Chemical Industries, MCI Gel-CK08P) following

Church (1979) and Paytan et al. (1993) and heated for about 10 days to 90 °C. The cations bind to the resin while the sulfate goes into the fluid phase. To achieve complete dissolution the water is decanted daily. After dissolution, the cations are extracted from the resin by 2 rinses of 2 ml 6 N HCl.

4.1.4 Silicate Minerals, Rocks and Soils

Most digestion protocols use significantly larger amounts of sample material for digestion ($\sim$100 mg on average) compared to what is required and utilised for Ca isotope analysis in the mass spectrometer. On one hand, this procedure minimizes potential blanks during handling, but on the other hand, double spike cannot be added to the sample before digestion. Instead it is added to an aliquot of the sample right after dissolution. This requires a protocol that provides complete dissolution and prevents the formation of insoluble Ca fluorides in the case of silicate samples.

Felsic and mafic rocks

Approximately 50–100 mg whole rock powder are weighed into 15 ml pre-cleaned Teflon screwtop vials and digested with closed caps in different mixtures of HF and HNO_3 (Zhu and MacDougall 1998, 2:1; John et al. 2012, 1:4; Kreissig and Elliott 2005, 3:1; Wiegand et al. 2005, 1:1; Huang et al. 2010, 5:3; Ryu et al. 2011) to break-down any Si-bond, for several days on a hotplate at $\sim$120 °C. Afterwards 1 ml $HClO_4$ is usually added to break down organics and dissolve secondary Ca-fluorites and the solution is evaporated at $\sim$190 °C. The dried sample is re-dissolved in a mixture of 6 N HCl and H_3BO_3 to dissolve secondary Ca-fluorides, heated at 140 °C on a hotplate for 12 h with closed caps. Afterwards, the sample is evaporated at 125 °C. The dried samples is then re-dissolved in 6 ml 6 N HCl and heated with closed caps for 45 min at 100 °C. The cooled solution is then checked for precipitates. While residual refractory compounds (e.g. Al_2O_3, TiO_2...) are not problematic for Ca isotope analyses, because they do not contain significant amount of Ca and can be removed by centrifuging, the presence of any fluorides in the residue is not acceptable for stable isotope analyses.

Besides these table-top protocols, samples can also be dissolved under higher pressure using Parr bombs with a HF-HNO_3 mixture (Simon et al. 2009).

Ultramafic rocks

Ultramafic rock samples can be pre-treated by leaching in 8.8 mol L^{-1} HBr at $\sim$160 °C for 72 h in closed vessels and repeated ultrasonical treatment to promote the release of cations from the crystal cage, before HF–HNO_3 digestion (Amini et al. 2009).

Mineral separates

For Ca isotope work on pure mineral fractions standard procedures for mineral separation are used, e.g. based on the density (heavy liquids), magnetic susceptibility of the mineral grains, and hand-picking (Hindshaw et al. 2011; Ryu et al. 2011). Depending on the studied mineral, one of the above described digestion protocols is applied.

Extra-terrestrial material

Chondrite samples are hand powdered, dissolved in hot HF–$HClO_4$ mixtures, dried down, dissolved again in HNO_3 or HCl and centrifuged (Valdez et al. 2014). Some samples were dissolved under higher pressure in Parr bombs without $HClO_4$ (Simon et al. 2009; Simon and DePaolo 2010). Alternatively, chondrites were dissolved in HF–HNO_3 during 24 h at 120 °C, dried down and fumed in $HClO_4$ to eliminate any CaF_2 precipitate (Caro et al. 2010).

4.1.5 Organic Samples

Plants

The vegetation samples are washed with deionised water. Root samples are especially cleared of soil particles by repeated rinsing and sonication in ultra-pure water, and checked for cleanliness using a binocular microscope (Holmden and Bélanger 2010). Tree-rings can be collected with an increment drill from living trees and dissected based on tree-rings counting (Farkaš et al. 2011). The plant samples are dried in an oven at ~30–60 °C, reduced to powder using either an agate mortar or a tungsten carbide rotary disc mill or a zirconium oxide mixer mill, depending on the sample size. About 50–150 mg powder can be digested within several days in 5 mL concentrated HNO_3 and 1 mL 30 % H_2O_2. In order to avoid a too intense reaction, the dissolution is started at room temperature and later heated to 70 °C. To promote the reaction, the sample acid mixture is regularly ultrasonicated. After evaporation the process is repeated with a concentrated HNO_3–HCl–H_2O_2 mixture, evaporated to dryness, dissolved in dilute HNO_3, and centrifuged (Cenki-Tok et al. 2009; Hindshaw et al. 2011; Cobert et al. 2011b; Bagard et al. 2013; Schmitt et al. 2013). Undissolved siliceous fractions (e.g. phytoliths) are removed by filtration and centrifugation (Chu et al. 2006).

Alternative protocols use hot (~150 °C) concentrated HNO_3 (Bullen et al. 2004; Wiegand et al. 2005; Holmden and Bélanger 2010), or high-pressure microwave HNO_3–HCl digestion in reinforced Teflon (PTFE) vessels for digestion (Hindshaw et al. 2012). The solution is evaporated to dryness, dissolved in ultrapure 5 % HNO_3 (Blum et al. 2008). A further approach is ashing of samples in nickel crucibles in an oven at 500 °C for 8 h, transferring the ashes to Teflon beakers and digesting them in concentrated hot (160 °C) HNO_3 acid.

Animal soft and hard tissues

Soft and hard tissues of organisms are ashed for 12–72 h at 450 °C in acid-washed quartz or platinum crucibles, before being dissolved in 1.5 M HCl. If the dissolution is incomplete, the samples are further treated with $HClO_4$, HNO_3, HCl and HF, before being dried at ~150 °C (Skulan and DePaolo 1999).

Animal/human excretes, blood

Milk and feces are reduced to ash in a muffle furnace at 550 °C for at least 24 h. Then they are digested in various strengths of HNO_3/HCl mixtures (Chu et al. 2006).

Urine samples are digested in a mixture of 2 mL concentrated HNO_3 and 250 μL $HClO_4$ at 150 °C during 12 h. Then the sample-acid mixture is heated to 180 °C. Once evaporated, 1 mL HNO_3 is added to the residue, which is slowly heated up to 180 °C again, until complete dryness. This step should be repeated twice in order to remove relicts of $HClO_4$. Blood samples (plasma or whole blood) are digested similar to urine using an 8:1 mixture of concentrated HNO_3 and $HClO_4$.

4.1.6 Liquid Samples

Liquid samples (e.g. rainwater, seawater, snow, soil solutions, throughfall) are filtered if necessary [with 0.22 μm (Schmitt et al. 2003a; Schmitt and Stille 2005; Bagard et al. 2013; Wiegand and Schwendenmann 2013), 0.40 μm (Holmden and Bélanger 2010) or 0.45 μm (Cenki-Tok et al. 2009; Tipper et al. 2010) cellulose acetate filters] or centrifuged. Water samples are acidified to pH 1 with HCl (Schmitt and Stille 2005) or to pH 2 with HNO_3 (Jacobson and Holmden 2008; Bagard et al. 2013) and stored in precleaned polypropylene (PP), polyethylene (PE) or Teflon bottles. Once evaporated, the solid residues of waters are redissolved in concentrated nitric acid, dried down again and re-dissolved in 3 M HNO_3 (Hindshaw et al. 2013).

4.1.7 Leachates

Carbonates in silicate rock matrix

For carbonate leachates whole rock powders are treated with dilute 1 M CH_3COOH (Ewing et al. 2008; Teichert et al. 2009; Ryu et al. 2011). Centrifugation allows removing any non-carbonate residue, such as quartz, feldspar or clay minerals and organic compounds (Blättler et al. 2011).

Soil sequential extractions

Successive leaching steps are usually performed in order to have access to the nutrient pools that are available to fine roots in soils: generally the soil's exchangeable cations and the acid-leachable fraction. The soil-exchangeable fraction is obtained by leaching 1–5 g soil sample with 1 N NH$_4$OAc (Bullen et al. 2004; Perakis et al. 2006; Page et al. 2008), 0.1 N NH$_4$OAc (Wiegand et al. 2005), 1 N NH$_4$Cl (Hindshaw et al. 2011), or 0.1 N BaCl$_2$ (Holmden and Bélanger 2010; Farkaš et al. 2011).

The soil acid-leachable fraction is obtained by leaching the residue from the soil-exchangeable fraction leachate once rinsed with ultra-pure water with 1 N HNO$_3$ (Bullen et al. 2004; Perakis et al. 2006; Holmden and Bélanger 2010; Farkaš et al. 2011). The same acid is used by Hindshaw et al. (2011) to extract the phyllosilicates. An intermediate step was also introduced by the latter authors: a H$_2$O$_2$/HNO$_3$ mixture allows them to extract organically-bound Ca. After each step the solutions are centrifuged, the supernatant are stored and the solid residues are carried forward to the next step. Less resistant silicate minerals (biotite and hornblende) are leached in a further step using hot (70 °C) 15 N HNO$_3$ acid during several hours (Holmden and Bélanger 2010; Farkaš et al. 2011).

For their part, Hindshaw et al. (2011) published a five-step sequential extraction procedure in order to extract exchangeable, organically-bound, phyllosilicates and two residual soil pools. More recently, Bagard et al. (2013) performed three successive extractions, in order to (1) dissolve carbonates and remove metals adsorbed on the soil particles as outerspheric complexes with 1 N CH$_3$COOH, (2) dissolve remaining adsorbed trace metals and Fe–Mn oxides and hydroxides with 1 N HCl, and (3) digest organic matter with 1 N HNO$_3$.

Leachates must always be interpreted with care because there is always a continuum between leachable and residual phases (Stille and Clauer 1994; Steinmann and Stille 1997, 2006). It can also be noted that BaCl$_2$ is a more powerful ion exchanger than e.g. NH$_4$Cl. As a result, based on ^{87}Sr/^{86}Sr values from Nezat et al. (2010), Farkaš et al. (2011) suggested that BaCl$_2$ can also weather cations from interlayered biotite phases, complicating the interpretation of leachates.

4.2 Chemical Separation

The determination of Ca isotope compositions using mass spectrometry implies a chromatographic clean-up for Ca to remove the sample matrix and elements that form interferences with Ca isotopes (see Sect. 5.3.3). Russell and Papanastassiou (1978) were the first to use such a chemical clean-up. They adapted their elution protocol to separate alkaline and alkaline-earth elements from Tera et al. (1970). The main result of their study is the identification of stable isotope fractionation occurring during the chemical separation; the first eluted Ca fractions are enriched in heavy Ca isotopes, whereas the later eluted Ca fractions are enriched in light Ca isotopes. To avoid this isotope fractionation, digested sample aliquots are mixed with a Ca-double spike before the chemical protocol (see Sect. 5.4). Unspiked aliquots are commonly taken to analyse radiogenic ^{40}Ca or nucleosynthetic anomalies, which implies a near quantitative recovery from the column to avoid any large chromatographically-induced fractionation. Moreover, *pure* Ca carbonates do not need any chemical clean-up and can be directly measured in the spectrometer (Halicz et al. 1999; Nägler et al. 2000; Clementz et al. 2003; Marriott et al. 2004; Fantle and DePaolo 2005, 2007; Hippler et al. 2006; Heuser et al. 2005; Gussone et al. 2007; 2009; Gussone and Filipsson 2010; Reynard et al. 2011).

Calcium purification protocols are based on Ca$^+$–H$^+$ exchange processes between the solution and ion exchange resins. Cation exchange resins consist of sulfonic acid functional groups in the hydrogen form (R–SO$_3$H), attached to a styrene divinylbenzene copolymer of variable crosslinkage.

$$(\text{Eluant})\text{M}^+ + [\text{Resine} - \text{SO}_3]^-\text{H}^+$$
$$\leftrightarrow (\text{Eluant})\text{H}^+ + [\text{Resine} - \text{SO}_3]^-\text{M}^+ \quad (34)$$

The variable affinity of the resin for different cations (Ca, Mg, Fe, Al, Sr…) determines the efficiency of the separation, which is based on the distribution coefficient of the cations between the resin and different acids (HNO$_3$, HCl, HBr). It can be noted that Ca presents high distribution coefficients at high HCl or HBr molarities (Nelson 1964). However such high-molarity acids do not provide a good separation of Ca from Sr, Ba and some REE (e.g. Wombacher et al. 2009). Several different lab-specific Ca purification protocols, which are summarized below, are applied.

Commercially available resins are normally not clean enough for direct usage and need a thorough pre-cleaning, with several H_2O and acid washing steps. The degree of contamination can vary for each batch of resin and the cleaning protocol might be adapted and blanks need to be regularly tested.

Cation exchange resin (AG50W-X8, 200–400 mesh) is often employed for the selective separation of Ca from the other cations, using as eluent diluted HCl (Russell and Papanastassiou 1978; Schmitt et al. 2001, 2003a, b, 2009; Kasemann et al. 2005; Schmitt and Stille 2005; Perakis et al. 2006; Cenki-Tok et al. 2009; Caro et al. 2010; Hindshaw et al. 2011; Heuser et al. 2011; Wiegand and Schwendenmann 2013), HNO_3 (Fig. 1; Fantle and DePaolo 2007; Ewing et al. 2008; Simon et al. 2009) or HBr (Kreissig and Elliott 2005; Heuser and Eisenhauer 2008, 2010). Another particle size (100–200 mesh) of this resin and HCl as an eluent was also employed (Soudry et al. 2006). In addition the higher cross-linked AG50W-X12 (200–400 mesh) with HCl as an eluent was also often used to perform Ca purification (Chang et al. 2004;

Wieser et al. 2004; Chu et al. 2006; Farkaš et al. 2006, 2011; Tipper et al. 2006, 2008a, 2010; Komiya et al. 2008; Teichert et al. 2009; Huang et al. 2010; Reynard et al. 2010; Blättler et al. 2011; Hippler et al. 2009). Macroporous (AGMP50, 100–200 mesh; HCl as eluent) (Jacobson and Holmden 2008; Amini et al. 2009; Holmden and Bélanger 2010; Fantle et al. 2012; Holmden et al. 2012; Lehn et al. 2013) and the Mitsubishi (MCI Gel-CK08P; 75–100 mesh; HCl as eluent) (Zhu and McDougall 1998; Amini et al. 2008, 2009; Griffith et al. 2008; Harouaka et al. 2014) cation exchange resins were also applied.

During these chemical separation protocols Ca and Sr peaks can overlap, which may cause interferences during mass spectrometer measurements. Added to that, Fe and Al need also to be removed. For instance, samples with high Al/Ca ratios cause short-term beam instabilities or lower Ca ionization efficiency (Boulyga 2010). Several techniques have been employed to avoid these drawbacks. Strontium was separated from Ca using a Sr specific resin (Sr spec SPS, 50–100 mesh) (Chang et al. 2004; Chu et al. 2006; Tipper et al. 2006; Simon et al. 2009; Hindshaw et al. 2011; Reynard et al. 2010; Blättler et al. 2011). Another way to avoid Ca–Sr peak overlap is to truncate the elution curve, leading to a recovery yield of about 70 % (Cenki-Tok et al. 2009) or greater than 80 % (Amini et al. 2008). Aluminium, Fe and Ti were removed by passing the eluted aliquot a second time through the same column employing the same cleaning and separating process (Skulan and DePaolo 1999; Kreissig and Elliott 2005; Simon et al. 2009; Simon and DePaolo 2010) or using small cation resin columns and HCl as an eluent (Caro et al. 2010). Hindshaw et al. (2011) proposed for their part to employ AG1-X4 anion resin to retain Fe as $FeCl_4^-$ together with the anion matrix. Aluminium was removed by elution in 0.1 N HF and 1 N HNO_3 through AG50W-X8 cation resin (based on Schiller et al. 2012). Another way to remove Fe and Al is to co-precipitate them with NH_3 at pH 7 (Tipper et al. 2006).

To avoid multiple-step chromatographic clean-up (up to four-step separation chemistry,

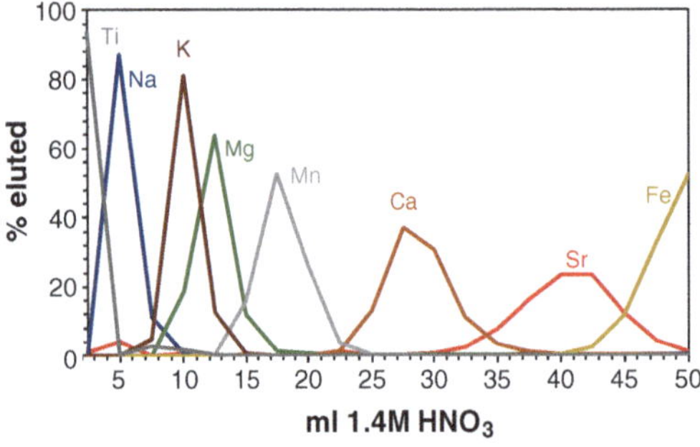

Fig. 1 Example for the chemical separation of Ca from a chondrite sample using 1.4 M HNO_3 and 1 ml of AG50W-X12 cation exchanger. F. Wombacher, unpublished data

Hindshaw et al. 2010), Wombacher et al. (2009) developed a new procedure allowing the combined chemical separation of Mg, Ca and Fe for most studied matrices (water, bone, carbonate and sediment samples, igneous and sedimentary rocks, chondritic meteorites). By using AG50 W-X8 (200–400 mesh) Ca was selectively eluted using either HBr or HCl.

A further approach to isolate Ca from the matrix is the use of a high selectivity automated ionic chromatography separation protocol (Schmitt et al. 2009). It is applicable to multiple natural matrices (waters, mineral and organic samples) and ensures a complete separation of Ca from K, Mg and Sr. This protocol was initially developed to avoid K-tailing into the Ca fraction for vegetation samples, occurring with traditional column procedures. Indeed, K and Sr are the closest elements to Ca in terms of chemical behavior during the various separation steps, so that satisfactorily separation often cannot be achieved for these cations. More recently Romaniello et al. (2015) published a protocol describing a fully automated chromatographic purification of Sr and Ca for isotopic analysis. It relies on a commercially available platform combined to a highly reusable Sr–Ca column. For the moment this protocol has only be tested for rock, bone and seawater standard samples.

In order to oxidize organic functional groups that may have leached from ion-exchange resins during chemical processing, H_2O_2 (Holmden and Bélanger 2010) or HNO_3–H_2O_2 mixtures (Schmitt et al. 2009; Cobert et al. 2011a, b) have been added to the processed samples and heated in closed Teflon (PTFE) vessels. This can help to avoid isobaric interferences or variation in the ionization behavior during mass spectrometric measurements.

Chemical blank values are highly variable from one laboratory to the other, ranging from <1 ng (Caro et al. 2010) to 100 ng (Jacobson and Holmden 2008). Generally these values are low compared to the Ca processed through the columns and can thus be neglected.

5 Mass Spectrometry

5.1 Introduction to Mass Spectrometry for Ca Isotope Analysis

Precise Ca isotope analysis requires mass spectrometry. Mass spectrometers consist of three principle units: an ion source, an analyzer for mass separation and an ion detection unit. Up to now, most Ca isotope analysis have been performed using thermal ionization mass spectrometry (TIMS), followed by multi-collector inductively coupled plasma mass spectrometry (MC-ICP-MS) and occasionally secondary ion mass spectrometry (SIMS). SIMS offers high spatial resolution, as does MC-ICP-MS if coupled to a laser ablation system. Due to matrix effects and interferences, in situ analysis, however, is usually limited to samples that are dominated by Ca, such as calcite or apatite.

Calcium atoms need to be ionized in order to be accelerated and focused inside the mass spectrometer. TIMS, MC-ICP-MS and SIMS differ fundamentally in regard to their ion sources. In TIMS, Ca is ionized at the surface of a metal filament that is heated by an electrical current. The filaments are made from metals characterized by high melting points e.g. W and Re. In SIMS, atoms are sputtered from the sample using a primary ion beam (such as $^{16}O^-$ ions) whereby only a fraction of the analytc atoms liberated is ionized. The Ar plasma source in MC-ICP-MS ionizes Ca atoms very efficiently, but suffers from interferences, in particular from abundant $^{40}Ar^+$ ions (99.6 %) that interfere with $^{40}Ca^+$ ions. In contrast, thermal ionization is rather element specific and thus generates less interferences.

Once formed, Ca^+ ions are accelerated by a negative potential and the ion beam is focused and aligned by a set of electrostatic lenses. When passing through the magnetic field of the mass analyzer, ions are dispersed depending on their mass to charge ratio (m/z) and their kinetic energy (velocity). Unlike thermal ion sources, plasma and secondary ion sources generate ions with a spread in kinetic energy. To compensate for the ion energy spread, MC-ICP-MS and SIMS instruments are commonly equipped with an electrostatic analyzer that is dispersive for ion energy only. In order to compensate the energy discrimination of the magnetic sector field, energy discrimination of the electrostatic analyzer is of the same magnitude, but opposite in sign. Thus, the combination of electrostatic and magnetic analyzer focusses both, ion angles and energy, a property called double focusing.

Precise Ca isotope ratio measurements require that ion beams are collected simultaneously in multiple Faraday detectors (also called Faraday collectors or cups). Multiple collection in Faraday cups takes care of ion beam fluctuations during the course of the analysis, but requires cross calibration of the gain of the attached amplifiers which is possible with accuracies of about 10 ppm. (Multiple) Ion counters are more sensitive, but cross calibration is less precise and prone to drift effects such that the isotope ratios obtained are not sufficiently precise for Ca stable isotope geochemistry. Ion beams generated by thermal ionization can be very stable, which allows ion beams to be collected sequentially in a single Faraday collector with the advantage, that ion beams for different Ca isotopes have identical flight paths. This provides a principle advantage in terms of accurate and reproducible analysis. However, longer analysis times, detector decay and drift issues are the downside of single-collector TIMS (see also Sect. 5.2).

Faraday collectors are commonly attached to high ohmic feedback resistors (typically 10^{11} Ω) where the voltage measured across the resistor is representative of the ion beam intensity. According to Ohms law (I = V/R), 1 V measured across a 10^{11} Ω resistor corresponds to 10^{-11} A or 10 pA. Because 1 A equals 6.241×10^{18} elementary charges per second, 10 pA correspond to 62.41 million ions per second [million counts per second (Mcps)]. Faraday collectors within modern TIMS or MC-ICP-MS instruments equipped with 10^{11} Ω resistors may collect ion beam signals up to about 50 V (500 pA).

Using 10^{10} Ω resistors allows to collect ion beam currents that are ten times larger, which is particular useful for the collection of ^{40}Ca as it allows to detect the less abundant Ca isotopes at higher signal intensities which in tune leads to better isotope ratio precision (provided that sufficiently intense ion beams can be generated). On the other hand, ion currents of less than ~ 0.5 pA may be better detected using higher ohmic (e.g. 10^{12} Ω) resistors as a 10^{12} Ω resistor is characterized by higher signal/noise ratios (e.g. Wieser and Schwieters 2005; Koornneef et al. 2013). According to Ohms law, the voltage measured across 10^{10} (10^{12} Ω) resistor is by a factor of 10 lower (higher) than across a 10^{11} Ω resistor. However, for the ease of users, signals intensities reported as voltages for amplifiers with 10^{10} or 10^{12} Ω resistors are (inaccurately) reported as if 10^{11} Ω resistors were in place.

The inherited noise of the resistors attached to Faraday detectors requires that baselines are taken, usually before measurements (e.g. Schiller et al. 2012). The lower the signal intensities and the better the precision aimed at, the longer should the baseline be taken (Fig. 2).

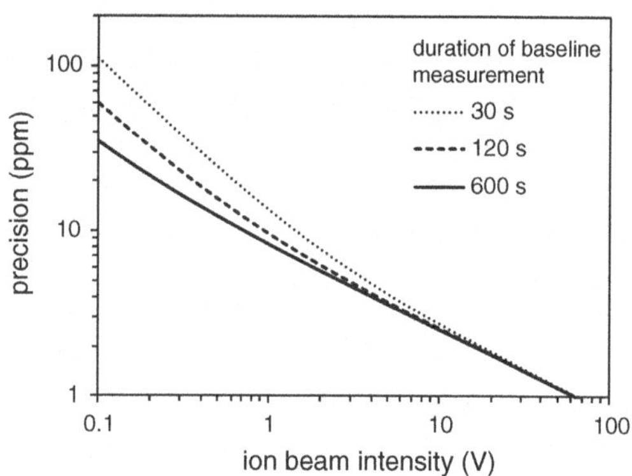

Fig. 2 Calculated effect of signal intensity and duration of baseline measurements on the precision of isotope ratio measurements for an isotope ratio of one, 60 integrations of 8.4 s and Johnson noise for 10^{11} Ω resistors. Drift effects and/or additional uncertainty introduced by the mass discrimination correction procedure are ignored. The figure shows that longer baseline readings are required for small signal intensities where the noise contribution becomes increasingly significant and hints at the signal intensities needed to achieve the precision requirements. See Ludwig (1997) for a thorough discussion of signal versus baseline measurement times

Collected isotope ratios finally need to be corrected online or offline for interferences and mass discrimination and possibly for background contributions and eventually referenced to the Ca isotope standards (e.g. using the δ-notation) that are usually run along with the samples.

5.2 Thermal Ionization Mass Spectrometry (TIMS)

Because the most abundant isotope (^{40}Ca) can be analyzed by TIMS, but not usually by MC-ICP-MS, TIMS remains the method of choice for most geochemists. Several articles have reviewed TIMS measurement techniques (Platzner 2000; Holmden 2005; Fantle and Bullen 2009; Boulyga 2010; Carlson 2014).

5.2.1 Mass Discrimination in TIMS and the Exponential Law Correction

During the course of the measurement, progressive mass dependent fractionation of Ca isotopes occurs. Light Ca isotopes are preferably evaporated from the hot filament, which results in an enrichment of heavy Ca isotopes in the Ca pool remaining on the filament. Thus, the ratio of light to heavy Ca isotopes is decreasing with time during the analysis.

A viable measure for the progressive isotope fractionation in thermal ion sources is the relative (%) fractionation per mass unit:

$$\left[\left(^{X}Ca/^{Y}Ca\right)_{sample} \Big/ \left(^{X}Ca/^{Y}Ca\right)_{reference} - 1\right]$$
$$\times\, 100/(X - Y)$$

$$(35)$$

where ^{X}Ca is the light Ca isotope with mass X and ^{Y}Ca is the heavy Ca isotope with mass Y. For normalization to the reference value, either natural reference ratios or ratios of double spike (Sect. 5.4) compositions are used.

Figure 3 shows a typical evolution path for ^{43}Ca/^{48}Ca during a run. However, reverse fractionation may sometimes be observed if new Ca reservoirs at the filament are tapped. This is a

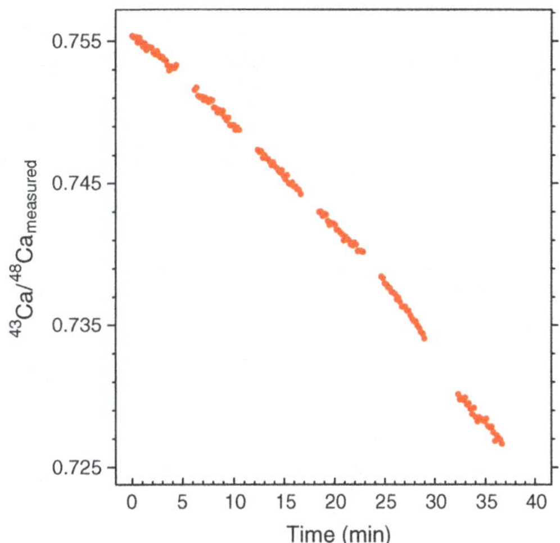

Fig. 3 Typical progressive mass fractionation of ^{43}Ca/^{48}Ca during TIMS analysis

critical issue as the "mixed" ion beam signals from variably fractionated reservoirs are not accurately described by the exponential law (see below) that is commonly used for mass fractionation correction (Hart and Zindler 1989; Fantle and Bullen 2009; Upadhyay et al. 2008; Andreasen and Sharma 2009; Lehn and Jacobson 2015). This observation highlights the importance of reproducible filament loading and heating procedures.

The continuous mass fractionation during TIMS measurements needs to be accurately corrected. As in most cases, natural stable isotope variations are investigated no "true" reference value can be assumed for mass fractionation correction. Therefore, a double spike technique (Sect. 5.4) has to be applied to accurately correct for instrumental mass discrimination.

The so called exponential mass (or isotope) fractionation law (Russell et al. 1978) is widely applied for mass discrimination correction in TIMS as well as MC-ICP-MS. It is commonly applied in double-spike data reduction procedures for stable isotope analysis and for the determination of radiogenic isotope compositions. The exponential law of Russel et al. (1978) has been shown to provide an accurate and suitable description for the progressive Ca isotope discrimination in TIMS (Russell et al. 1978; Hart and Zindler 1989; Schiller et al. 2012), but

Caro et al. (2010), Schiller et al. (2012) and Naumenko-Dèzes et al. (2015) observed small deviations from the exponential law with increasing fractionation. The latter authors applied an empirical second order correction, while Schiller et al. (2012) fitted the generalized power law (Maréchal et al. 1999; Wombacher and Rehkämper 2003) for individual measurement sessions to slightly improve the results.

Below the exponential law is given along with an example that uses the values employed for the correction of mass discrimination for the determination of the radiogenic ^{40}Ca (ε_{Ca}). First the fractionation coefficient f is determined:

$$f = \frac{ln\left(\frac{R_B}{r_B}\right)}{ln\left(\frac{m_3}{m_1}\right)} \quad e.g. f$$

$$= ln\left(\frac{0.31221}{\left(^{42}Ca/^{44}Ca\right)_{measured}}\right) / ln\left(\frac{41.9586183}{43.955481}\right)$$

$$(36)$$

Using the fractionation coefficient f, all other isotope ratios can be corrected:

$$R_A = r_A\left(\frac{m_2}{m_1}\right)^f$$

$$e.g. \quad \left(^{40}Ca/^{44}Ca\right)_{corrected} = \left(^{40}Ca/^{44}Ca\right)_{measured} * \left(\frac{39.9625912}{43.955481}\right)^f$$

$$(37)$$

R_A refers to the mass discrimination corrected isotope ratio and R_B refers to the reference value for the isotope ratio used for normalization; r_A and r_B refer to the measured (fractionated) values of the same isotope ratios as R_A and R_B. The mass of the denominator isotope common to both isotope ratios is denoted m_1, while the numerator isotope of the corrected ratio B is denoted m_2 and the mass of the numerator isotope for the normalizing isotope ratio B is m_3.

Note that some authors used ^{40}Ca/^{42}Ca and ^{44}Ca/^{42}Ca for ratio A and ratio B, respectively, in order to determine the radiogenic ^{40}Ca excess. For samples, the deviation of the corrected ^{40}Ca/^{44}Ca (or ^{40}Ca/^{42}Ca) ratio from a reference value in parts per ten thousand (pptt) is then calculated as ε_{Ca} based on Eqs. 12 or 13.

The exponential law is identical to the kinetic law discussed in Chapter "Introduction", but presented here in a more convenient form with the fractionation coefficient f that is obtained from the properties of the normalizing isotope ratio (cf. Hart and Zindler 1989; Wombacher and Rehkämper 2003). The advantage of this form is that once f is determined from the normalizing isotope ratio, all other isotope ratios can be corrected using Eq. 3. The fractionation coefficient (f) is called β in most publications. This is unfortunate, however, because β also denotes (i) the slope in three isotope plots defined by different isotope fractionation laws, (ii) the exponent used to describe isotope fractionation associated with kinetic transport processes and (iii) reduced partition function ratios (cf. Chapter "Introduction").

5.2.2 Analysis of Radiogenic ^{40}Ca by TIMS

Two data reduction procedures for radiogenic ^{40}Ca excesses have been reported up to now, both based on TIMS analysis.

In the first procedure, unspiked isotope ratios are measured and mass fractionation corrections are performed using Eqs. 36 and 37 above. The accuracy of the correction depends on the precision with which the isotope ratios can be measured (DePaolo 2004). Earlier studies using Finnigan MAT 262 TIMS instruments were able to achieve external reproducibilities of about one ε_{Ca} (Marshall and DePaolo 1982, 1989; Marshall et al. 1986; Nelson and McCulloch 1989). New generation TIMS instruments are able to improve this value below one ε_{Ca} (Kreissig and Elliott 2005; Simon et al. 2009; Caro et al. 2010; Simon and DePaolo 2010). Caro et al. (2010) were able to decrease the external reproducibility down to 0.34 ε_{Ca} by using a multi-dynamic mode and a low-resolution dual source-exit slit assembly. They observed significant deviation from the exponential law (up to 4 ε_{Ca} bias) if the instrumental mass fractionation of the analyses was below -2 ‰/amu. In order to avoid artifacts they filtered out data that was not comprised between -2 and 2 ‰/amu and tightly controlled the amount of Ca loaded onto the filament.

The correction for instrumental mass discrimination also means that natural stable isotope fractionation effects are simultaneously corrected along with the instrumental mass discrimination and therefore natural stable isotope effects should be irrelevant for ε_{Ca}. However, this is true only to the extent that the exponential (=kinetic) law corrects accurately for the commonly small natural mass-dependent isotope fractionation. We can use the difference between the mass-scaling of the equilibrium and kinetic/exponential law to estimate the potential error that can be introduced by inappropriate mass discrimination correction of natural stable isotope fractionations. For any ‰/amu natural Ca isotope fractionation with a mass-scaling according to the equilibrium law (Chapter "Introduction"), the systematic error in ε_{Ca} corresponds to -1 (pptt). It is difficult to estimate the true error in ε_{Ca} introduced by an inappropriate correction of natural stable isotope fractionation as it requires that the mass-dependence for the natural fractionation process is precisely known. However, the above evaluation leads to two important conclusions:

(1) for the determination of precise (better than $\sim \pm 0.5\ \varepsilon Ca$) εCa values, samples that possibly display a large ($\gg 0.1$ ‰/amu) natural stable isotope fractionation should be avoided. Fortunately, most high-temperature samples will not show large Ca stable isotope variations.

(2) If potentially large stable isotope fractionations cannot be avoided, the stable isotope fractionation need to be determined and potential errors on εCa must be evaluated.

The determination of radiogenic ^{40}Ca excesses has been combined with the determination of mass dependent fractionation effects. To this end, $\delta^{44/40}Ca$ and $\delta^{44/42}Ca$ are first determined as described in Sect. 5.2.3, using either one protocol (Schmitt et al. 2003a, b; Schmitt and Stille 2005; Huang et al. 2010, 2011; Hindshaw et al. 2011) or two measurement protocols with two different double spikes in order to improve the

external reproducibility of $\delta^{44/42}Ca$ (Ryu et al. 2011). The ε_{Ca} excess can be visualized by plotting $\delta^{44/40}Ca$ versus $\delta^{44/42}Ca$ in a diagram. Samples that plot below the mass fractionation line defined by Eqs. 38 or 39 display ^{40}Ca excesses. The amount of the excess can be evaluated using following equations, depending either on the equilibrium fractionation law:

$$\varepsilon_{Ca} = \left[\left(\delta^{44/42}Ca \times 2.0995 \right) - \delta^{44/40}Ca \right] * 10 \tag{38}$$

or on the kinetic fractionation law:

$$\varepsilon_{Ca} = \left[\left(\delta^{44/42}Ca \times 2.0483 \right) - \delta^{44/40}Ca \right] * 10 \tag{39}$$

No detectable ^{40}Ca enrichments were recordable in most studies (Schmitt et al. 2003a, b; Schmitt and Stille 2005; Hindshaw et al. 2011) due to low reproducibilities of ε_{Ca} ($\sim 3\ \varepsilon_{Ca}$; Hindshaw et al. 2011) and/or because of young crystallization ages of the studied granite bedrocks (~ 300 Ma; Schmitt et al. 2003b; Hindshaw et al. 2011). Farkaš et al. (2011) and Ryu et al. (2011), who studied old granodiorites (of Precambrian age and of 1.7 Ga, respectively), recorded ^{40}Ca enrichments (see Sect. 3 in Chapter "High Temperature Geochemistry and Cosmochemistry").

Although insignificant in many cases, studies of mass-dependent Ca isotope fractionation effects that rely on $^{44}Ca/^{40}Ca$ isotope ratios need to be aware of potential biases introduced by radiogenic contributions. The effect of ^{40}K decay on the $\delta^{44/40}Ca$ as a function of the $^{40}K/^{44}Ca$ ratio and sample age is illustrated in Fig. 4, demonstrating that radiogenic ingrowth of ^{40}Ca is small in young rocks, even in those with relatively high K/Ca ratios, and it is insignificant in rocks with low K/Ca, irrespective of their formation ages.

5.2.3 Calcium Stable Isotope Analysis by TIMS

Table 3 highlights that there exists no consensus on filament loading techniques among different laboratories and each laboratory has developed

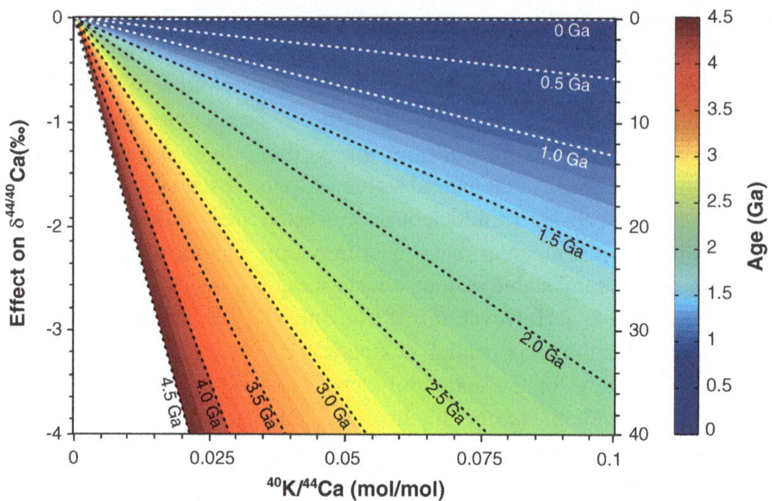

Fig. 4 Effect of ^{40}K decay on the $\delta^{44/40}Ca$ of a rock or mineral, depending on the $^{40}K/^{44}Ca$ ratio and formation age

Table 3 $\delta^{44/40}Ca$ of selected rock standards from the literature

Reference material	Rock type	$\delta^{44/40}Ca$ (‰)	References
BHVO-1	Basalt	0.98	Huang et al. (2010, 2011)
BHVO-2	Basalt	0.87	Magna et al. (2015), Amini et al. (2008), Valdes et al. (2014)
BIR-1	Basalt	0.83	Amini et al. (2008), Wombacher et al. (2009), Valdes et al. (2014)
BIR-2	Basalt	0.99	Valdes et al. (2014)
AGV-2	Andesite	0.77	Valdes et al. (2014)
ATHO-G	Rhyolite	0.87	Amini et al. (2008)
BCR-1	Basalt	0.82	Simon and DePaolo (2010)
BCR-2	Basalt	0.89	Wombacher et al. (2009), Amini et al. (2008), Valdes et al. (2014)
SRM688	Basalt	0.86	Valdes et al. (2014)
JP-1	Peridotite	1.15	Magna et al. (2015)
PCC-1	Peridotite	1.14	Amini et al. (2008)
DTS-1	Dunite	1.54	Amini et al. (2008), Huang et al. (2010)
J-Do1	Dolomite	0.70	Wang et al. (2013)
J-Cp1	Coral	0.63	Wombacher et al. (2009)

its own measurement protocol. Samples are loaded onto a filament in the chloride, nitrate or iodide form, with or without an activator, i.e. a supplementary solution increasing the ionisation efficiency. Calcium isotope analysis is then carried out using a single, double or triple Ta, Re or W filament configuration.

Given the configuration of the TIMS present in the different laboratories, Ca isotope ratios have been measured using either single or multi-collection. The most striking analytical challenge is to obtain the best possible external reproducibility and accuracy for Ca isotope ratios.

In single-collector peak-hopping, individual ion beams are collected sequentially (Russell et al. 1978; Skulan et al. 1997; DeLaRocha et al. Nägler and Villa 2000; Schmitt et al. 2001, 2003a, b; Lemarchand et al. 2004; Schmitt and Stille 2005; Fantle and DePaolo 2005, 2007;

Gopalan et al. 2006). The advantage of this technique is that the measured ion beam intensities remain unaffected by ion optical effects. However it requires a correction of the drift in ion beam intensity between succeeding measurement cycles. Thus time-interpolated Ca isotope ratios may be subject to large uncertainties if the ion beam is not perfectly stable and short ion beam integration times are necessary (1–3 s). Short integration times however render high precision Ca isotope analysis difficult. This is due to counting statistics as Ca is composed of one highly abundant (^{40}Ca) and five minor isotopes (^{42}Ca, ^{43}Ca, ^{44}Ca, ^{46}Ca, ^{48}Ca; see Table 3; Chapter "Indroduction"). It is therefore necessary to generate high intensity ion beams to boost the signal intensity for the low abundance isotopes.

In multi-collection all ion beams are either collected simultaneously (static measurement) or split between alternating measurement cycles (dynamic measurements) (Heuser et al. 2002, 2011; Clementz et al. 2003; Gussone et al. 2003, 2004, 2007, 2010, 2011; Holmden 2005; Kasemann et al. 2005; Böhm et al. 2006; Farkaš et al. 2006, 2007; Amini et al. 2008, 2009; Ewing et al. 2008; Heuser and Eisenhauer 2008, 2010; Cenki-Tok et al. 2009; Schmitt et al. 2009; Simon et al. 2009; Huang et al. 2010; Simon and DePaolo 2010; Cobert et al. 2011a, b; Ryu et al. 2011). So far, the mass difference of 20 % between ^{48}Ca and ^{40}Ca exceeds the mass dispersion of TIMS instruments, thus not all six Ca isotopes can be simultaneously detected. However, Naumenko-Dèzes et al. (2015) report data obtained on a Triton plus with a special Faraday cup array with extended mass dispersion that enables simultaneous collection of all six Ca isotopes without the use of zoom lenses.

In principle, ^{48}Ca only needs to be taken into account if it is chosen as one of the double spike isotopes (see Sect. 5.4), but as the fourth most abundant isotope it is advantageous to be included in any multicollection routine for the comparison of results from different Ca isotope ratios. Ion optical effects due to the large mass difference between Ca isotopes and the resulting wide dispersion of the different ion beams are major obstacles to ameliorate the external precision.

Wide dispersion of Ca ion beams have been accused to cause peak-shape defects with asymmetric peaks at the outermost cup positions (Fletcher et al. 1997; Heuser et al. 2002), which was responsible for the lack of improvement in internal precision for multi-collection over single-collection peak hopping. Holmden (2005) suggested to decrease the imprecision generated by ion optical effects in the mass spectrometer by decreasing the relative mass separation from ~ 10 to 5 %. To attain this objective, he dynamically collected only four ion beams (^{40}Ca, ^{42}Ca, ^{43}Ca, and ^{44}Ca) in three collectors (L1, Axial, H1) on a Triton. Finally, damage to individual Faraday collectors have been reported, linked to ^{40}Ca measurements that poison collector cups (Andreasen and Sharma 2006; Simon et al. 2009; Holmden and Bélanger 2010; Simon and DePaolo 2010). These authors have indeed shown that Faraday cup graphite liners degradation with use can affect efficiencies and thus ^{40}Ca abundances. This can be avoided by replacing the graphite liners for example every 18 months, depending on usage.

5.3 Multiple Collector Inductively Coupled Plasma Mass Spectrometry (MC-ICP-MS)

5.3.1 Basics of MC-ICP-MS

In this section, we first briefly introduce issues that are characteristic for MC-ICP-MS, such as the sample introduction, plasma and interface processes. We then discuss the fundamental problems in MC-ICP-MS: interferences, mass discrimination and matrix effects on mass discrimination. This will be followed by a brief overview centered on key publications where MC-ICP-MS was used for the analysis of Ca isotopes. Note that isotope ratio measurements by (MC-)ICP-MS in general have been reviewed extensively in the past (Halliday et al. 2000; Rehkämper et al. 2001, 2004; Albarède and Beard 2004; Albarède et al. 2004; Vanhaecke et al. 2009; Yang 2009; Epov et al. 2011; Baxter et al. 2012).

The first MC-ICP-MS introduced was the VG Plasma54 (Waldner and Freedman 1992), followed by Nu Instruments NuPlasma, the Micromass IsoProbe, the VG Axiom and the Finnigan Neptune MC-ICP-MS. Note that in most cases, the names of the companies change quicker than authors can adjust. Currently, only the NuPlasma and the Neptune MC-ICP-MS are manufactured, with options and modifications that differ from the first models. Nu Instruments also manufactures a large geometry high-resolution MC-ICP-MS, the Nu Plasma 1700 but only a few instruments have been installed and so far no Ca isotope data are published.

If the MC-ICP-MS is coupled to a laser ablation system, samples can potentially be analyzed for Ca isotope compositions in situ, at the tenth of µm scale without further preparation (e.g. Tacail et al. 2016). However, since the plasma source efficiently ionizes all elements, matrix effects and interferences related to matrix elements (e.g. $^{88}Sr^{++}$ on $^{44}Ca^+$) cause severe problems if present.

The vast majority of Ca isotope studies carried out by MC-ICP-MS were based on sample and standard Ca dissolved in dilute acid (typically HNO_3) with matrix elements commonly separated by ion chromatography beforehand. The solutions are usually aspirated at a rate of about 100 µl min^{-1} and dispersed using a microconcentric nebulizer commonly made of PFA or borosilicate glass. Within the nebulizer, an Ar gas stream of about 1 L min^{-1} passes along the tip of the capillary tubing. This results in a lower pressure at the capillary tip and hence self-aspiration. The aspirated solution is then sprayed into a fine aerosol. If the resulting droplets enter the plasma, the solvent evaporates, the remaining solids are atomized and eventually ionized. Because large droplets of about >10 µm diameter would not dry down completely during the short passage through the plasma, they need to be sorted out by the use of a spray chamber. Therefore, only a few % of the sample and standard solution will make it to the plasma. To improve the delivery of the analyte (i.e. Ca) to

the plasma and thus improve sensitivity, most isotope analysis by MC-ICP-MS employs desolvators such as the Cetac Aridus or the Elemental Scientific Apex. In both systems, the aerosol generated by a microconcentric nebulizer is sprayed into a heated spray chamber to evaporate the solvent. Since too much water vapor would overload the plasma, the vapor must be removed before the Ar gas and analyte enters the plasma. In the Apex, this is facilitated by passing the sample gas through a Peltier cooled condenser. In the Aridus, the solvent vapor diffuses through a heated membrane where it is taken up and removed by a counter flow of Ar gas. The removal of water has the added benefit of reducing oxide and other solvent related interferences. In both desolvators, a small amount of N_2 may be admixed to the dry aerosol to boost the signal further. However, polyatomic interferences containing nitrogen (e.g. $^{14}N_3^+$ and $^{14}N_2^{16}O^+$ on mas 42 and 44) may become more abundant.

The coupling of an efficient Ar plasma ion source, which operates at atmospheric pressure, to a mass spectrometer that needs to be maintained under vacuum in order to prevent high detector noise and collisions between ions and background gas is realized by a water cooled interface. This interface allows extracting the ions generated in the plasma into the mass spectrometer. It consists of a set of two cones made of Ni (or Al or Pt) with orifices of 0.5–1 mm. The cones serve as the inlet and outlet of an expansion chamber which is pumped to achieve a transitional vacuum at a few mbar. The sampler cone "samples" the ions generated in the plasma. The ions then expand into the vacuum of the expansion chamber where the March disk is the region where the ions attain ultrasonic speed. The skimmer cone "skims" the ions from a region slightly before the March disk. At the rear of the skimmer cone, positive ions are accelerated into the mass spectrometer by the negative potential applied to the extraction lens. Ions within the positively charged ion beam (dominated by $^{40}Ar^+$) repel each other. This space charge effect may be responsible for the pronounced mass discrimination in MC-ICP-MS.

5.3.2 Mass Discrimination and Matrix Effects in (MC-)ICP-MS

Mass discrimination or mass bias in MC-ICP-MS is much larger than in TIMS, but far more stable. This is not surprising, since the analyte elements are introduced to the plasma and mass spectrometer under stable and reproducible conditions as long as tuning parameters (gas flow, lens settings, torch position, RF power) are not changed. The fact that the instrumental mass bias drifts only slightly with time allows for stable isotope analysis without the use of double spikes, using the standard-sample-standard bracketing technique. Analyses of samples are simply alternated with the analysis of standards and δ-values are then calculated for samples relative to the average of the preceding and succeeding standard analysis.

The downside of the sample bracketing technique is the possibility that the mass bias changes in response to sample matrices being different than that of the standard. Together with the need to remove interfering ions, this makes chemical separation of Ca a prerequisite for MC-ICP-MS. Fitzke et al. (2004) noted large matrix effects in the per mill range under cool plasma conditions, if samples were measured in different acid matrices. Stürup (2004) observed a matrix related decrease of the ionization rate in urine samples with high Na and K contents. Matrix tests for Mg isotope measurements by MC-ICP-MS showed a memory of the matrix induced mass bias in succeeding matrix element free standard solutions, likely due to depositions on the cone (Wombacher et al. 2009). This hints at the possibility to detect sample matrix effects on mass bias by the matrix related drift induced between the bracketing standards. That the state of the cones is important for stable measurement has been frequently observed. For example, Schiller (2012) analyzed Ca isotopes with high ion beam intensities ($\sim$100 V on ^{44}Ca!). During the first hour of analysis they observed large mass bias drift after cone cleaning that could be avoided by conditioning the cones with Ca during tuning.

Shiel et al. (2009) detected the influence of ion exchange resin derived organics on the mass bias of Zn and Cd isotope measurements by MC-ICP-MS. For Ca isotope analysis organic matter from the ion exchange resin may not be a common problem, as Ca is a major element and is usually separated in sufficient quantity to allow for significant dilution of organic matrix contributions. This can be tested by running a Ca standard through chemistry using an amount of Ca similar or below that of samples.

So far no Ca double spike has been applied for mass bias correction during MC-ICP-MS measurements. This is because the use of a double spike requires that four interference free isotopes can be reliably measured and possibly also because MC-ICP-MS generally consumes more sample Ca and thus double spike than TIMS (Wieser et al. 2004).

Many studies of isotope ratios by ICP-MS use element additions for mass bias drift correction, for example Tl, added to both samples and standards, were used to monitor the mass bias drift in Pb isotopes ratios (Longerich et al. 1987) and Cu isotopes were used for Zn and vice versa (Maréchal et al. 1999). At least potentially, mass bias drift for ratios of ^{42}Ca, ^{43}Ca and ^{44}Ca could likewise be corrected using ^{49}Ti/^{47}Ti.

5.3.3 Interferences in (MC-)ICP-MS

The biggest disadvantage of Ca isotope analysis using ICP-MS is the occurrence of various and often significant isobaric interferences. Table 4 presents a selection of possible isobaric interferences on Ca isotopes.

Several ways exist to deal with these isobaric interferences: (1) chemical purification of the sample, (2) high mass resolution (3) mathematical correction for interfering ions, and (4) suppression of interfering ions by cold (or cool) plasma or collision cell technology.

Purification of the sample via ion chromatography (Sect. 4.2) provides an efficient way to avoid interferences introduced by the sample matrices, e.g. $^{40}K^+$, $^{48}Ti^+$, $^{26}Mg^{16}O^+$ or $^{88}Sr^{2+}$, but cannot remove interferences that are introduced by the Ar gas and its impurities ($^{40}Ar^+$;

Table 4 Potential interferences on Ca isotopes

Ca isotope	Interference	Δm	Required resolution
^{40}Ca	^{80}Kr^{++}	−0.0044	9076
	^{80}Se^{++}	−0.0043	9228
	^{40}Ar^{+}	−0.0002	193,149
	^{40}K^{+}	0.0014	28,378
	^{39}KH^{+}	0.0089	4469
	^{24}Mg^{16}O^{+}	0.0174	2301
	^{20}Ne$_2^{+}$	0.0223	1793
	^{23}Na^{16}OH^{+}	0.0299	1336
^{42}Ca	^{84}Kr^{++}	−0.0029	14,627
	^{84}Sr^{++}	−0.0019	21,993
	^{41}KH^{+}	0.0110	3805
	^{30}Si^{12}C^{+}	0.0151	2770
	^{26}Mg^{16}O^{+}	0.0189	2221
	^{40}ArH$_2^{+}$	0.0194	2161
	^{40}CaH$_2^{+}$	0.0196	2139
	^{28}Si^{14}N^{+}	0.0214	1962
	^{24}Mg^{18}O^{+}	0.0256	1640
	^{25}Mg^{16}OH^{+}	0.0300	1401
	^{14}N$_3^{+}$	0.0506	829
^{43}Ca	^{86}Sr^{++}	−0.0041	10,392
	^{86}Kr^{++}	−0.0035	12,404
	^{42}CaH^{+}	0.0077	5596
	^{27}Al^{16}O^{+}	0.0177	2429
	^{31}P^{12}C^{+}	0.0150	2865
	^{26}Mg^{16}OH^{+}	0.0266	1617
	^{14}N$_3$H^{+}	0.0583	737

Ca isotope	Interference	Δm	Required resolution
^{44}Ca	^{88}Sr^{++}	−0.0027	16,448
	^{43}CaH^{+}	0.0111	3956
	^{28}Si^{16}O^{+}	0.0164	2687
	^{32}S^{12}C^{+}	0.0166	2650
	^{30}Si^{14}N^{+}	0.0214	2058
	^{26}Mg^{18}O^{+}	0.0263	1673
	^{27}Al^{16}OH^{+}	0.0288	1526
	^{12}C^{16}O$_2^{+}$	0.0343	1280
	^{14}N$_2^{16}$O^{+}	0.0456	964
^{46}Ca	^{92}Zr^{++}	−0.0012	39,297
	^{46}Ti^{+}	−0.0011	43,504
	^{92}Mo^{++}	−0.0003	161,524
	^{30}Si^{16}O^{+}	0.0150	3064
	^{32}S^{14}N^{+}	0.0215	2142
	^{29}Si^{16}OH^{+}	0.0255	1799
	^{14}N^{16}O$_2^{+}$	0.0392	1172
^{48}Ca	^{48}Ti^{+}	−0.0046	10,458
	^{96}Mo^{++}	−0.0002	246,860
	^{96}Ru^{++}	0.0013	37,877
	^{96}Zr^{++}	0.0016	29,896
	^{32}S^{16}O^{+}	0.0145	3317
	^{36}Ar^{12}C^{+}	0.0150	3194
	^{24}Mg$_2^{+}$	0.0176	2731
	^{34}S^{14}N^{+}	0.0184	2605
	^{31}P^{16}OH^{+}	0.0240	2000
	^{16}O$_3^{+}$	0.0322	1489

Interferences with $\Delta m < 0$ have masses lighter than the respective Ca isotope, interferences with $\Delta m > 0$ have heavier masses

To fully resolve the Ca and interfering ion beams and achieve a flat plateau region, the resolving power of the instrument R(5, 95 %) should be about a factor of two better than the nominally required resolution given in the table

^{86}Kr^{2+}), the solvent (H in ArH$_2^{+}$) or the N-gas flow of desolvators (N$_2$O^{+}) (Table 4).

Many interferences on Ca isotopes can be avoided by the use of high mass resolution. In the low resolution mode, Ca and interference ion beams overlap significantly, even though their mass is slightly different due to variable mass defects. If ion beams are sufficiently clipped by a narrow source slit in the mass spectrometer, the interference and Ca isotope ion beams can be resolved at the focal plane were the multi collector array is positioned, obviously at the cost of sensitivity which may only be 5 % at high mass resolution. The two currently available MC-ICP-MS instruments, the Nu Instruments Plasma HR and the ThermoScientific Neptune

Plus have the possibility to choose between three different predefined low, medium and high mass resolution modes.

Precise isotope ratio measurements require flat-topped peaks. Flat topped peaks are obtained by the width of the ion beam being significantly smaller than the collector width. In single collector sector field ICP-MS instruments, ion beams of slightly different mass are completely resolved by narrow source and detector slit which commonly results in triangular peaks that are not suitable for precise isotope ratio measurements. In MC-ICP-MS, interferences (most of which are of higher mass than the analyte ions) are screened out at the low mass edge of the Faraday collector. The resolving power R quantifies the ability of the mass spectrometer to resolve ion beams of slightly different mass. Resolving power is commonly determined as the ion beam width at 5 and 95 % peak height: $R_{(5, 95\%)}$ = m/Δm with Δm being the mass at the left hand side peak slope at 5 % peak height minus the mass at 95 % peak height (Weyer and Schwieters 2003; see illustration further below). Because of this definition and the associated ion beam tailing, full resolution between the Ca ion beam and the interfering ion beam requires a resolving power that is about double that given as required resolution in Table 4, in particular, a higher resolving power is needed if interferences display high intensities.

It is possible to eliminate the most significant *polyatomic* interferences from Ca isotope masses using mid or high resolution mode, as these interferences require mass resolutions ranging from $\sim$740 ($^{14}N_3H^+$) to $\sim$3300 ($^{32}S^{16}O^+$) (Table 4).

Atomic interferences like $^{48}Ti^+$ or $^{88}Sr^{2+}$ on $^{48}Ca^+$ and $^{44}Ca^+$ cannot be resolved as this would need resolving powers much larger than 10,000. Furthermore, their mass is lighter than that of the Ca isotopes (Table 4). Therefore, they would have to be resolved on the high mass side peak shoulder where the heavier polyatomic interferences cannot be resolved. Atomic and also some polyatomic isobaric interferences can be handled by mathematical interference corrections, where the contribution of the interfering element is stripped from the measured intensity of the Ca ion beam as shown in the example for the ^{48}Ti interference correction:

$$I_{48Ca} = I_{48Ca + 48Ti} - I_{48Ti} \qquad (40)$$

where I_{48Ca} denotes the signal intensity for ^{48}Ca after the interference correction. The contribution of ^{48}Ti (I_{48Ti}) to the measured intensity $I_{48Ca+48Ti}$ can be determined by monitoring ^{47}Ti during the analysis:

$$I_{48Ti} = I_{47Ti} \times \left(^{48}Ti / ^{47}Ti \right)_{biased} \qquad (41)$$

During analysis, $^{48}Ti/^{47}Ti$ ($\approx$0.7372/0.0744) is affected by the instrumental mass bias; therefore the measured ratio may be by about 5 % higher than in the sample solution. Hence, the interference correction can be much improved if this mass bias is simulated using the exponential law (Eqs. 36 and 37):

$$\left(^{48}Ti / ^{47}Ti \right)_{biased} = \left(^{48}Ti / ^{47}Ti \right)_{nature} \times \left(\frac{47.9479471}{46.9517638} \right)^{-f_{Ca}}$$
$$(42)$$

with f taken from Ca isotope ratios as given in Eq. 36 above. Natural Ca isotope fractionation will affect f_{Ca} and stable isotope fractionation of Ti in nature and during chemical separation will also affect the accuracy of the correction. Furthermore, the natural isotope ratios assumed for $^{48}Ti/^{47}Ti$ and in the determination of f_{Ca} may not be accurate. However all these uncertainties are expected to be about an order of magnitude less significant than the large effect of the instrumental mass discrimination in MC-ICP-MS. Interference corrections obviously work best if the correction is of minor significance. The Ti correction shown in the example above is unfortunate, as the abundance of the monitor isotope ^{47}Ti is about 10 times less than that of ^{48}Ti. As a result, noise recorded by the monitor isotope ^{47}Ti will be amplified to the signal on mass 48 by an order of magnitude. Thus, if the Ti correction is insignificant (which has to be shown) it may better be avoided. For example in TIMS analysis, ionization of Ti is not expected and application of the Ti correction would only induce additional scatter. A further pitfall results from the possible presence of unresolved interferences on the interference monitor mass, e.g. $^{31}P^{16}O^+$ on $^{47}Ti^+$, which would lead to overcorrection.

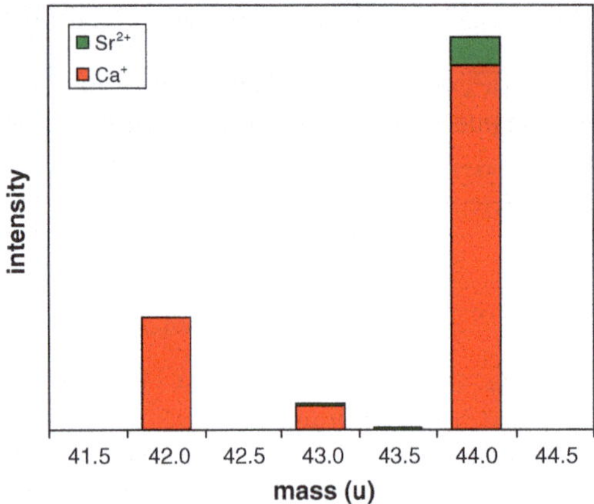

Fig. 5 Schematic illustration of the Sr^{2+}-interference correction

For many Ca isotope measurements interference correction of ^{46}Ti and ^{48}Ti does not play a role as masses ^{46}Ca and ^{48}Ca are commonly not measured.

In analogy to the Ti correction, Sr^{2+}-ions interfering on the masses 42, 43 and 44 can be corrected by measuring $^{87}Sr^{2+}$ on mass 43.5. Using the known isotopic composition of Sr and assuming a mass bias being similar to that of Ca it may be possible to calculate the intensity of $^{84}Sr^{2+}$, $^{86}Sr^{2+}$ and $^{88}Sr^{2+}$ (Fig. 5) and correct the measured intensities on masses 42, 43 and 44. However, the exponential law fractionation coefficient f_{Ca} may not be perfectly suitable for Sr isotopes and the ^{87}Sr abundance in natural samples is somewhat variable (e.g. Hirata et al. 2008). Therefore, Sr isotope corrections should be carefully implemented and large corrections are best avoided by chemical separation.

A similar procedure could be used for the correction of Kr^{2+} ions (monitoring $^{83}Kr^{2+}$ on mass 41.5). As the Ar gas usually only contains traces of Kr, and given the high second ionization potential of Kr (24.36 eV), it may not be possible to reliably monitor $^{83}Kr^{2+}$ at mass 41.5 and the Kr interferences can possibly be neglected or may be subtracted along with the background anyway.

Because large interferences are more difficult to resolve, to correct for, or to be subtracted based on background measurements, interferences should be kept to a minimum. Apart from desolvation and chemical separation, cool plasma techniques and collision cells offer additional interference suppression, such that even ^{40}Ca may be analyzed by MC-ICP-MS. These approaches will be discussed in the next section.

5.3.4 Calcium Isotope Analysis by (MC-)ICP-MS

Several studies applied ICP-MS for Ca isotope analysis and in most cases multiple collector arrays were used (Halicz et al. 1999; Fietzke et al. 2004; Wieser et al. 2004; Steuber and Buhl 2006; Tipper et al. 2006, 2008a, b, 2010; Sime et al. 2007; Hirata et al. 2008; Morgen et al. 2011; Schiller et al. 2012; Tacail et al. 2016; Martin et al. 2015). From the above discussion, it is expected that (i) multi-collection, (ii) chemical separation and (iii) high mass resolution is generally required in order to obtain sufficiently precise and accurate Ca stable isotope data by (MC-)ICP-MS. In the following, published methods for Ca isotope analysis by ICP-MS are discussed, beginning with the pioneering study by Halicz et al. (1999), including methods that aim at the suppression of interferences (*cold plasma and reaction/collision cell technology*) and summarizing first attempts for in situ Ca isotope analysis by *laser ablation*.

Halicz et al. (1999) used a NuPlasma instrument for the first Ca isotope ratio measurements by MC-ICP-MS. Analyzed were ^{42}Ca, ^{43}Ca and ^{44}Ca for commercial Ca reagents and a few natural $CaCO_3$ samples, including speleothems and coral aragonite. The sensitivity was 0.3 V for ^{44}Ca at an uptake rate of ~ 70 µl min^{-1}. Sample and standard solutions containing 20–30 ppm Ca in 0.1 M HNO_3 were introduced into the plasma using a Cetac MCN6000 membrane desolvator (the predecessor of the Cetac Aridus) and analyzed for 10 min. Membrane desolvation was used to reduce polyatomic interferences such as $^{14}N_2^{16}O^+$ and $^{12}C^{16}O_2{}^+$ on $^{44}Ca^+$ and $ArH_2{}^+$ on $^{42}Ca^+$ to low levels. About half of the elevated background levels of about 1.5 mV were attributed to scattered $^{40}Ar^+$ and $^{40}Ca^+$ ions, respectively. This background component was monitored on additional masses (42.5, 43.25, and 44.5) and subtracted from the corresponding Ca ion beam signals. Doubly charged Sr ion interferences on all three Ca isotope masses were corrected based on the background corrected $^{87}Sr^{++}$ signal monitored at mass 43.5. Mass bias was corrected for by bracketing sample solutions with analysis of NIST SRM 915a Ca. Precision assessed from repeated sample analysis ranges from 0.04 for a coral sample (n = 3) to 0.40 for a calcrete sample (n = 5) for $\delta^{44/42}Ca$ (2 sd).

Cool plasma conditions also allow for the reduction of interfering molecular as well as elemental ion formation. Under normal operating conditions the RF (radio frequency) power coupled into the plasma is set between 1200 and 1400 W. Decreased RF power results in a plasma with a lower degree of ionization and temperature and therefore is called "cool plasma" or "cold plasma". Thus elements or molecules with a high first ionization potential are ionized to a much lesser degree. Since the ionization of Ar and the formation of oxides and argides is suppressed, cool plasma conditions provide a reasonable approach for Ca isotope analysis (Fietzke et al. 2004). Several publications report the measurements of Ca isotopes using cool plasma. Most of the published cool plasma Ca measurements (e.g. Patterson et al. 1999, Murphy et al. 2002) are tracer experiments in humans without the need of high precision data and measurement of ^{40}Ca. In addition these studies have been performed using Q-ICP-MS which is not suitable for high precision Ca isotope measurements. Fietzke et al. (2004), however, conducted Ca isotope measurements using cool plasma MC-ICP-MS. Using a VG AXIOM MC-ICP-MS, they set the RF power as low as 400 W, resulting in a ^{40}Ar background decrease from about 2.5×10^7 cps (0.4 V with 10^{11} Ω resistors) to 1.0×10^6 cps (Fig. 6). The rather low $^{40}Ar^+$ background even in hot plasma probably also reflects a narrow source slit that was used to screen out interferences at the high mass edge of the detector at high mass resolution as described in more detail below. Along with $^{40}Ar^+$, the formation of $^{40}ArH^+$, $^{40}ArD^+$ and to some degree $^{14}NO_2^+$ ions decreased (Fig. 6). On

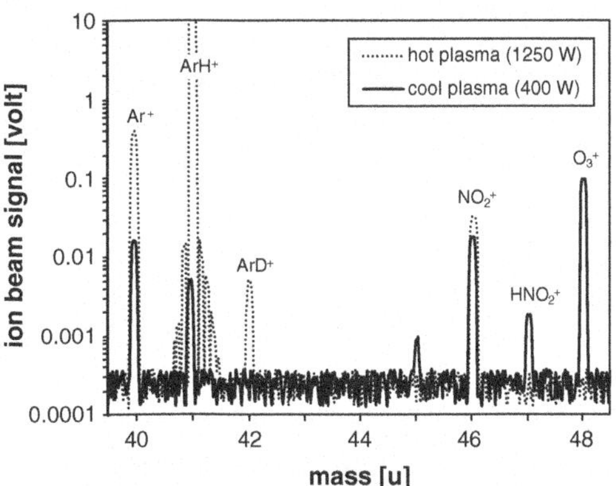

Fig. 6 Comparison of the background spectrum under hot and cool plasma conditions using the VG Axiom MC-ICP-MS. Figure modified from Fietzke et al. (2004)

the other hand the use of cool plasma increased the production of $^{14}N^{16}O_2H^+$ and $^{16}O_3^+$ ions. As the study aimed to measure $^{44}Ca/^{40}Ca$ values, interferences on ^{42}Ca, ^{43}Ca, ^{46}Ca and ^{48}Ca were less relevant. However, masses 42 and 43 were not affected by significant interferences, while mass 46 and 48 were. As the ^{40}Ca signal was about 500 times higher compared to the ^{40}Ar background, drift in the Ar background was considered negligible. Fietzke et al. (2004) reported a RSD of about 0.14 ‰, thus the precision of the $\delta^{44/40}Ca$ data generated by cool plasma MC-ICP-MS is comparable to TIMS measurements. The downside of the cool plasma method were large matrix related shifts in $\delta^{44/40}Ca$ if the acid molarity was not precisely matched between standards and samples. Other reasons why the cool plasma technique has not become widely adopted may be that (1) the production of AXIOM MC-ICP-MS was stopped in 2002 and currently available MC-ICP-MS may not able to set the RF power and thereby reduce the $^{40}Ar^+$ ion beam to such a low level and (2) $\delta^{44/42}Ca$ and other ratios measured using conventional MC-ICP-MS may be considered sufficient for Ca isotope geochemistry. Since cool plasma MC-ICP-MS and the apparent matrix sensitivity is not well investigated, further studies on other instruments would be of interest, perhaps in conjunction with a suitable double-spike which may be able to correct for residual matrix effects.

Fig. 7 Simultaneous peak scan for ^{42}Ca, ^{43}Ca, ^{44}Ca and ^{48}Ca in high resolution. Panel **a** displays the overlapping Ca ion beams for a 10 ppm solution and shows the resolving power $R = m/\Delta m$ with Δm defined as the ion beam width at 5 and 95 % peak height; panel **b** shows the interferences observed in a Ca free background solution. An interference free measurement position (as indicated by the *stippled line*) that is sufficiently distant from the low mass peak edge needs to be identified. Figure modified from Wieser et al. (2004)

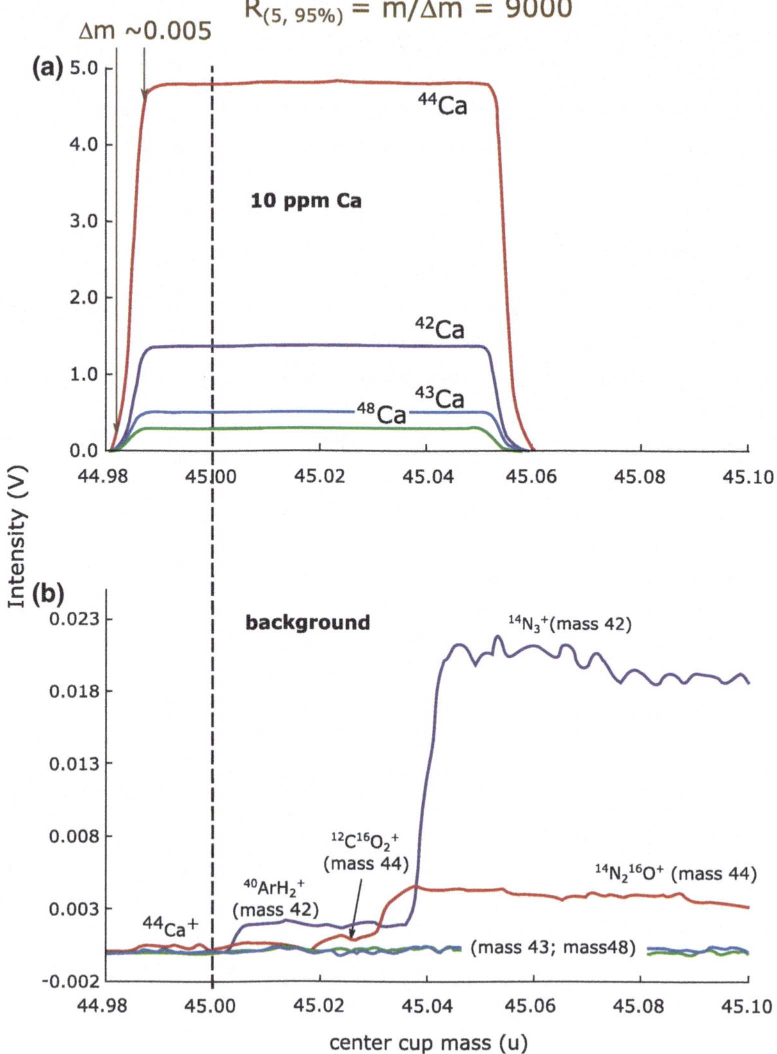

Wieser et al. (2004) determined $\delta^{44/42}$Ca, $\delta^{48/42}$Ca, $\delta^{44/43}$Ca and $\delta^{48/43}$Ca with external reproducibilities of ±0.11, ±0.33, ±0.12 and ±0.28 (2 sd; n = 54) respectively, using a Neptune MC-ICP-MS at hot (1200 W) plasma conditions and high mass resolution (Fig. 7). Data for $CaCO_3$ reference samples were measured relative to seawater Ca. The sample Ca was first separated by cation exchange. Mass bias, which amounts to 5 % per atomic mass unit, was corrected for by bracketing sample solutions with seawater Ca analysis. Background determinations were alternated with sample and standard analysis and background values were subtracted from sample and standard measurements. The ^{40}Ar$^+$ ion beam corresponds to about 8000 V (10^{11} Ω resistor). While scattering of Ar ions was not observed, analysis of ^{40}Ca is obviously impossible under these conditions. The ^{46}Ca signal was too low for meaningful analysis. Sample solutions, containing 10 ppm Ca in 3.5 % HNO_3, were analyzed for two minutes followed by a 120 s rinse. Sample and standard solutions were introduced using an Apex-Q coupled to a Cetac Aridus membrane desolvator. This combination offers a thoroughly dried aerosol which is expressed in a further reduction of the ^{40}ArH$_2^+$ interference down to 2 mV,

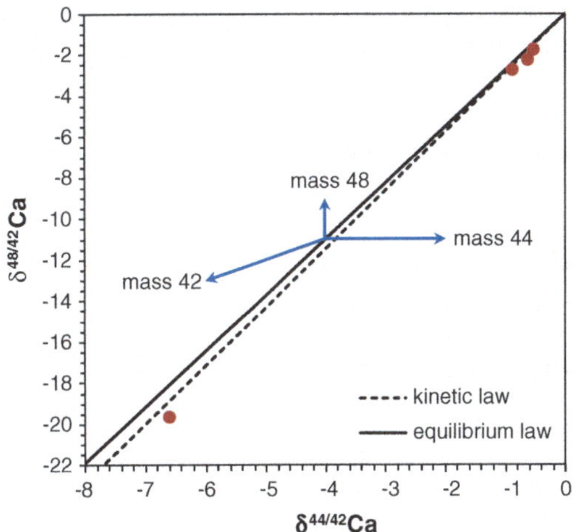

Fig. 8 Three isotope plot featuring data obtained by high-resolution MC-ICP-MS for Ca carbonate samples relative to seawater (Wieser et al. 2004). The size of the symbols roughly corresponds to the precision of the measurements (2 sd). Fractionation curves correspond to the equilibrium and kinetic laws (see Chapter "Introduction"). Arrows indicate 2 ‰ interferences on the masses indicated. The data plots closer to the slope of the kinetic law than that defined by the equilibrium law. At a first glance, this may suggest kinetic isotope fractionation. However, the samples are not related to the zero-delta material (IAPSO seawater) with the possible exception of the least fractionated sample, a modern bivalve shell. Furthermore, the interpretation of kinetic and equilibrium law slopes may not always be straight forward (Chapter "Introduction") and the offset from the reference slopes could also indicate the effects of interferences. For example, the samples could be affected by small interferences on mass 44. Likewise, the seawater reference sample analysis could be compromised by small interferences on mass 42 or 48

compared to 30 mV using the Apex-Q with an Aero Naphion membrane. With this set-up, ^{44}Ca ion beams of 4.8 V were obtained. Due to the large internal volume of the coupled desolvating introduction systems, very stable ion beams are achieved. While the multi-collector array corrects very well for somewhat unstable signals in isotope ratio analysis, unstable ion beam signals impede the precise peak shape tuning needed for high resolution analysis. Three isotope plots were used to constrain the absence of significant interferences (Fig. 8).

Schiller et al. (2012) combined high resolution MC-ICP-MS (Neptune with sensitive jet and x-cones) analysis of ^{42}Ca, ^{43}Ca, ^{44}Ca, ^{46}Ca and ^{48}Ca with TIMS (Triton) analysis for ^{40}Ca, ^{42}Ca, ^{43}Ca and ^{44}Ca in order to measure all stable Ca isotopes in their terrestrial test samples. The method presented was developed with the aim to study nucleosynthetic Ca isotope anomalies in meteorites and their components (Chapter "High Temperature Geochemistry and Cosmochemistry") by high precision analysis. For this reason, no double spike was applied in TIMS analysis, thus no δ-values for mass dependent isotope fractionations were calculated. However, δ-values were obtained from MC-ICP-MS analysis of rock and seawater reference samples from which Ca was chemically separated after acid digestions. Pooled $\delta^{42/44}$Ca, $\delta^{43/44}$Ca, $\delta^{46/44}$Ca and $\delta^{48/44}$Ca values measured relative to NIST SRM 915b yielded typical uncertainties for the mean, corresponding to ±0.03, ±0.01, ±0.04, ±0.05 per mil (2se; n = 10). The data agreed very well with TIMS analysis by Heuser and Eisenhauer (2008), Amini et al. (2009) and Wombacher et al. (2009). For mass bias correction, samples were bracketed by analysis of NIST SRM915b. The instrumental background was carefully monitored and subtracted.

To reveal (mass-independent) nucleosynthetic anomalies, the MC-ICP-MS and TIMS data were internally normalized to ^{42}Ca/^{44}Ca = 0.31221. Results are expressed using a μ-notation, which gives deviations in normalized ^{x}Ca/^{44}Ca from the same value in NIST SRM 915b in parts per million (ppm). The $\mu^{43/44}$Ca, $\mu^{46/44}$Ca, $\mu^{48/44}$Ca values obtained by MC-ICP-MS display uncertainties of ±2, ±45 and ±13 ppm (2se; n = 10). The main reasons for the high precision achieved, besides internal normalization and clean chemical separation, are the high ion beam intensities for the smaller isotopes (~0.15 V for ^{46}Ca; ~100 V for ^{44}Ca) that are possible if ^{40}Ca is omitted, and the lengthy measurement times (10 analysis, each lasting 14 min.). The downside of the high signal intensities are the loss in sensitivity over time due to deposition onto the cones and the degradation of the aperture lens and resolution slits, such that the slits had to be replaced after 150 h of measurement. Furthermore, with increasing Ca ion beam intensities, a non-linear increase of the background between mass 42 and 43 and between mass 47 and 48 up to 300 μV was observed. Effects of this elevated background were minimized by precisely matching the signal intensities between samples and standards.

Sample introduction was facilitated using an Elemental Scientific APEX with an attached ACM membrane for desolvation. Measurements were carried out either in medium or high resolution mode with a resolving power >5000. Before analysis, Ca isotope ratios were measured at slightly different positions at the low mass side of the flat topped peak in order to obtain a measurement position that is interference free, but not too close to the peak edge. This position was typically located about 0.024 amu from the center of the calcium peak. A peak center was conducted prior to every standard analysis. Sensitivity ranged from 6 to 17 V for ^{44}Ca per ppm Ca in solution. Samples and standard solutions contained between 8 and 16 ppm Ca in 2 % HNO_3. Because every sample was analyzed ten times, a total of 250–500 μg Ca was consumed, while 10 μg were consumed during single TIMS analysis.

Schiller et al. (2012) also discuss problems related to the correction of Ti interferences on ^{46}Ca and ^{48}Ca that were corrected based on the monitoring of ^{47}Ti employing a $10^{12}\Omega$ resistor. However, because of difficulties related to the scattered Ca ion beam, Ti intensities were checked for most samples using a secondary electron multiplier equipped with a retarting potential quadrupole filter to avoid scattered ions.

Another way to remove polyatomic interferences and even the ^{40}Ar interference is provided by the use of a collision and reaction cell or a dynamic reaction cell. Reaction cells contain a quadrupole, hexapole or octopol, placed off-axis behind the skimmer cone. Introducing different reaction gases (He, Xe, N_2, O_2, NH_3 or CH_4) into the cell leads to a fragmentation or neutralization of the interfering atomic and polyatomic ions.

Feldmann et al. (1999) used He as a buffer gas and H_2 as the reaction gas. They state that three different processes play a major role for the neutralization of Ar^+-ions:

$$\text{hydrogen atom transfer: } Ar^+ + H_2 \rightarrow ArH^+ + H \tag{43}$$

$$\text{proton transfer: } ArH^+ + H_2 \rightarrow H_3^+ + Ar \tag{44}$$

$$\text{charge transfer: } Ar^+ + H_2 \rightarrow H_2^+ + Ar \tag{45}$$

Boulyga and Becker (2001) used a HEX-ICP-QMS (Micromass "Platform ICP") with He and H as collision gases within the hexapole. They noticed a reduction of ^{40}Ar$^+$ by about three to four orders of magnitude. Stürup et al. (2006) used CH_4 as a reaction gas for their measurements with a dynamic reaction cell (DRC) ICP-MS (Elan DRC-e). They quote three reactions leading to a neutralization of Ar^+-ions:

$$Ar^+ + CH_4 \rightarrow CH_2^+ + H_2 + Ar \tag{46}$$

$$Ar^+ + CH_4 \rightarrow CH_3^+ + H + Ar \tag{47}$$

$$Ar^+ + CH_4 \rightarrow CH_4^+ + Ar \tag{48}$$

Boulyga et al. (2007) used a DRC-ICP-MS (Elan DRC-II) with NH_3 as reaction gas and compared their results with TIMS measurements of the same references materials (in-house standard, SRM 915a and SRM 1486). They argue that the obtained reproducibilities are in the same order of magnitude and the $\delta^{44/40}$Ca values of their lab ICP standard and SRM 1486 relative to SRM 915a are comparable. They concluded that using DRC-ICP-MS seems to be a good alternative to MC-ICP-MS and TIMS measurements of Ca isotopes. However, the reported $\delta^{44/40}$Ca value of SRM 1486 of ca. +3 ‰ (rel. to SRM 915a) is about 4 ‰ heavier than the reported SRM 1486 value of Heuser and Eisenhauer (2008) (Sect. 3.1). As SRM 1486 is a bone meal and bones are normally enriched in the light Ca isotopes (Chapter "Biomedical Application of Ca

Stable Isotopes"), it is unlikely that SRM1486 exhibits a strong enrichment in ^{44}Ca. Although the use of a collision and reaction cell enables the analysis of ^{40}Ca and suppresses the formation of atomic and polyatomic interfering ions, the precision of published single-collector methods is about one order of magnitude worse compared to MC-ICP-MS or TIMS measurements.

The IsoProbe and a few VG Axiom MC-ICP-MS were equipped with a collision cell. Cecil and Ducea (2011) determined radiogenic ^{40}Ca enrichment in K-rich authigenic minerals using a GV IsoProbe and gas-phase reactions in a hexapole collision cell to minimize isobaric Ar interference. A new attempt is currently made by Elliott et al. (2015). They coupled the front end, i.e. the interface and collision cell, from the Thermo Fisher Scientific iCAP Q (a quadrupole single collector ICP-MS) to the double-focusing Neptune mass analyzer.

Laser ablation (LA)-MC-ICP-MS offers the potential for in situ analysis of Ca isotope compositions without chemical separation. The sample (e.g. a thin or thick section, crystal or shell specimen) or standard material is ablated with a laser beam. The resulting aerosol is then transport into the mass spectrometer and analyzed. LA-ICP-MS is more frequently used for concentration determinations and radiogenic isotope systems (Sr, U/Pb). Currently, two published LA-ICP-MS studies on Ca stable isotope fractionation are published.

Santamaria-Fernandez and Wolff (2010) tried to determine the Ca isotopic composition in pharmaceutical packaging for discriminatory purposes. Their study nicely shows that precise and accurate LA-ICP-MS analyses of Ca isotopes are very challenging. The main problems relate to the mass bias correction and interferences. While it is possible to get rid of isobaric atomic interferences in solution analysis by a chemical purification prior to the measurement this is not possible for in situ analyses. Standard-sample bracketing has to be applied for the mass bias correction. This method usually requires that the composition of samples and standard materials are similar for matrix effects to be negligible.

Santamaria-Fernandez and Wolff (2010) used a 213 nm laser ablation system (NewWave UP-213) attached to a Neptune MC-ICP-MS for their Ca isotope measurements with the aim to identify counterfeit pharmaceuticals by comparing the packaging. Cardboard and ink of packaging have high concentrations of Ca mainly due to the use of kaolinite, calcium carbonate, chalk or china clay as coating material. In order to correct the instrumental mass bias relative to standards, Santamaria-Fernandez and Wolff (2010) prepared solid pressed pellets from NIST SRM 915a and 915b calcium carbonates using a KBr press without adding KBr to the pellets. The Ca isotopic composition of both carbonate standards has been characterized previously. Thus standard sample bracketing of both reference materials with SRM 915a being the standard and SRM 915b being the sample allows to detect problems. As both materials are chemically not very different, matrix effects should be negligible.

Santamaria-Fernandez and Wolff report a $\delta^{44/42}$Ca of 1.2 $\pm$ 0.2 ‰ for the SRM 915b relative to SRM 915a. This value is not in agreement with previously reported $\delta^{44/42}$Ca of 0.35 $\pm$ 0.04 to 0.42 $\pm$ 0.02 for SRM 915b relative to SRM 915a obtained by either TIMS or MC-ICP-MS (Heuser and Eisenhauer 2008; Wombacher et al. 2009; Schiller et al. 2012). They also report a $\delta^{44/43}$Ca value, which should be about half the $\delta^{44/42}$Ca value, i.e. 0.6. But their $\delta^{44/43}$Ca value for SRM 915b is 0.2, which better matches the $\delta^{44/43}$Ca of 915b expected from previously published data. Figure 3 of their study also clearly shows serious analytical problems: the ^{42}Ca/^{44}Ca should be highly correlated to the ^{43}Ca/^{44}Ca but both ratios vary independently. This lack of correlation is a typical indication for an isobaric interference affecting at least one of the isotopes used. The extent of this interference varies between the two very similar samples, leading to the observed lack of correlation between ^{43}Ca/^{44}Ca and ^{42}Ca/^{44}Ca. Possibly doubly charged ions interfere on masses 42, 43 and 44; a correction for interferences of Sr^{2+} ions was not carried out.

Tacail et al. (2016) used an Exite 193 nm Photon machines laser ablation system coupled to a Neptune MC-ICP-MS in high resolution mode to measure $\delta^{44/42}$Ca and $\delta^{43/42}$Ca in igneous apatite and modern tooth enamel samples. Matrix matched standard materials were prepared by sintering NIST SRM 1400 bone ash which was used for sample bracketing and two synthetic apatites, in one case with Sr added. The samples were also analyzed by solution mode and good agreement between solution and laser data was obtained for $\delta^{44/42}$Ca for the three sintered standards and tooth enamel samples, but igneous apatites were significantly displaced from the 1:1 correlation line. Moreover, $\delta^{43/42}$Ca from LA analysis was systematically higher than expected from their $\delta^{44/42}$Ca and this is also observed when the $\delta^{43/42}$Ca data obtained by laser ablation is compared with results from solution analysis. This suggests an uncorrected interference on mass 43, likely related to the Sr^{++} interference that is difficult to correct accurately (Tacail et al. 2016). This study was in part successful in as much that $\delta^{44/42}$Ca values that are consistent with solution analysis could be obtained for tooth enamel samples. The precision quoted for repeated analysis of about ±0.7 ‰ for $\delta^{44/42}$Ca (2sd) as judged from repeated analysis of laser ablation data appears insufficient to reveal internal variations in Ca isotope compositions in most samples. Yet, with further improvements, LA-ICP-MS may find its way to the Ca isotope community, especially in regard to Ca dominated materials such as $CaCO_3$, gypsum or apatite. Improvements may result from higher sensitivity e.g. due to modifications on the plasma interface. The use of fs lasers may also reduce the need for near perfect matrix matching (e.g. Horn et al. 2006; Shaheen et al. 2012). Finally, if reintroduced to MC-ICP-MS, collision cells may allow to reduce interferences during laser ablation work (cf. Elliott et al. 2015).

5.4 Double Spike Approach for Stable Isotope Analysis

5.4.1 Basic Principles

As already mentioned it is not possible to correct measurements for the isotope fractionation occurring during measurement without knowing the extent of natural isotope fractionation, i.e. no Ca isotope ratio is fixed. A fractionation correction which is done using a static reference ratio e.g. fractionation of Sr isotopes is corrected using a $^{86}Sr/^{88}Sr$ of 0.1194, also corrects for any natural isotope fractionation. While this is desirable for the determination of radiogenic ingrowth by radioactive decay it is not possible to determine the natural fractionation. This problem can be solved by addition of a double spike to the sample before its measurement or chemical preparation. A double spike is a solution consisting of two isotopes of an element which both are enriched compared to their natural abundances. The ratio of the two enriched isotopes has to be well known and the ratio of double spike/sample has to be high enough so that natural isotope variations of the spiked isotopes become negligible.

Two general requirements for the use of the double spike technique exist: (1) the element must have four or more stable isotopes or long-lived artificial isotopes (two isotopes for the double spike and two isotopes representing the isotopic composition of a sample), and (2) enriched isotopes with a sufficient enrichment are

Table 5 Commercially available enriched Ca isotopes

Isotope	Natural abundance (atom %)	Enrichment (atom %)[a]	Enrichment factor
^{40}Ca	96.97	>99.9	1.03
^{42}Ca	0.64	>93	145.31
^{43}Ca	0.145	>79	544.83
^{44}Ca	2.086	>98.5	47.22
^{46}Ca	0.004	>43	10750
^{48}Ca	0.187	>97	518.72

[a]Data from isotope services, Oak Ridge National Laboratory

available. Table 5 compares natural abundances of Ca isotopes with commercially available enriched isotopes.

Beside these general requirements some more prerequisites for a good working double spike exist: (1) the spike isotopes should be as pure as possible, i.e. high enrichment of the isotope, (2) the enrichment/natural abundance ratio of the used isotopes should be high, (3) by adding a double spike solution the abundance differences between the isotopes in the sample/spike mixture should be small and (4) the mass difference (in u) between double spike isotopes and "sample" isotopes should be similar. As can be seen from Table 5 enriched ^{40}Ca, ^{44}Ca and ^{48}Ca have enrichments >97 atomic-% but ^{40}Ca and ^{44}Ca at the same do not show high enrichment/natural abundance ratios. Therefore it is not advisable to use ^{40}Ca and ^{44}Ca as a double spike. Combinations of two of the remaining four Ca isotopes are suited better for the use as a double spike.

The use of a double spike solution to correct for the instrumentally driven isotope fractionation requires also a correction for the shift in isotopic composition caused by the double spike addition, because commercially available enriched isotopes are not pure, affecting thus to the measured isotope ratio of interest. The double spike deconvolution methods are not Ca isotope specific and generally valid for different isotope systems. Different approaches to correct for the added double spike are published (e.g. Compston and Oversby 1969; Gale 1970; Hamelin et al. 1985; Powell et al. 1998; Johnson and Beard 1999; Galer 1999; Siebert et al. 2001).

5.4.2 Double Spike Calibration

The progressive mass dependent Ca-isotope fractionation during thermal ionization mass spectrometry, equally affects the sample and double spike Ca.

Therefore, the isotope composition of the double spike cannot be directly determined by TIMS. Instead, the double spike can be either calibrated by exact weighing of the two enriched Ca salts, or calibrated against a standard of known isotopic composition. As weighing of small amounts of powder can be difficult due to

electrostatic effects, Ca-double spikes are typically calibrated against a standard. In Fig. 10 we illustrate the basic principle of the Ca double spike calibration following the approach for the Pb-double spike calibration of Galer (1999). This correction algorithm is a three dimensional approach and the isotope fractionation is described in a three dimensional vector space. As an example we show the calibration of a ^{43}Ca/^{48}Ca double spike, using a 3D-vector space, which is defined by three Ca isotope ratios ^{40}Ca/^{48}Ca, ^{44}Ca/^{48}Ca and ^{43}Ca/^{48}Ca. For this calibration, the isotope ratios published for CaF$_2$ by Russell et al. (1978) are used as a fix point.

For the double spike calibration, analyses of the pure spike and a mixture of spike and standard are required. The isotopic fractionation trends (^{40}Ca/^{48}Ca, ^{44}Ca/^{48}Ca and ^{43}Ca/^{48}Ca) of the individual runs are fitted in a 3D vector space with one additional isotope ratio (e.g. the ^{42}Ca/^{48}Ca) as a running parameter. For practical reasons a linear instead of an exponential fractionation trend can be used, if the quality of the fit is sufficiently good and allows for this simplification. The fractionation trends of the spike and the mixture are shown in Fig. 9 as dashed lines. A basic principle of the 3D fractionation correction is that the fractionation trend depends on the isotope composition of the analyzed material, i.e. the fractionation trends of the spike-sample mixture and of the pure spike are not parallel which means that the vector product

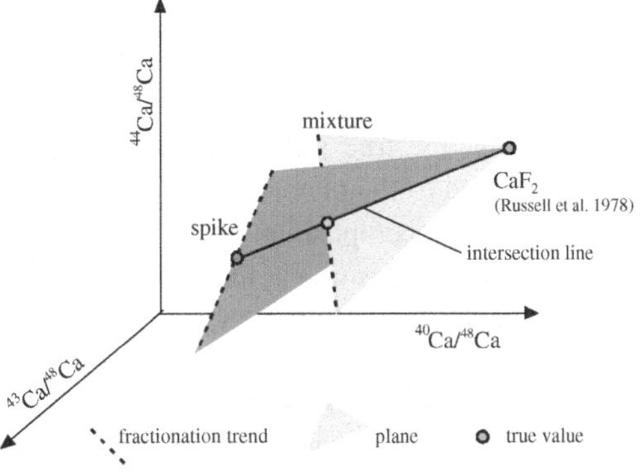

Fig. 9 Schematic illustration of a ^{43}Ca/^{48}Ca-double spike calibration, following the Pb-double spike approach of Galer (1999)

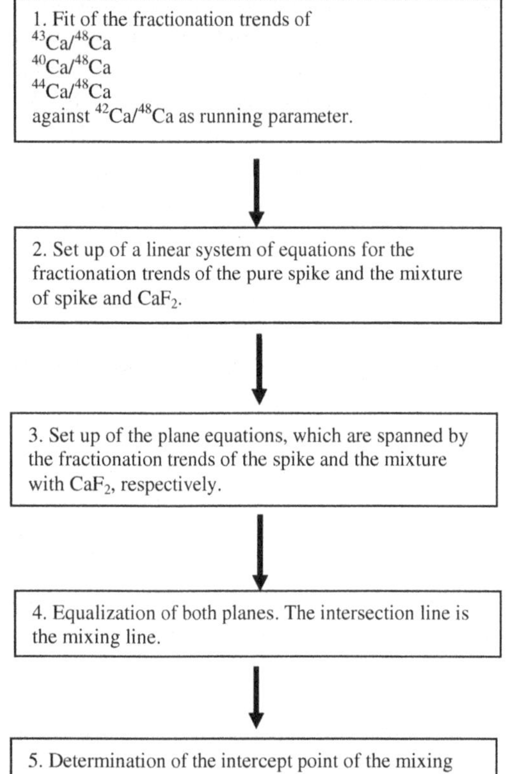

Fig. 10 Flow chart of the Ca-double spike calibration

of the spike-fractionation trend and mix-fractionation trend is unequal to 0 (MxS $\neq$ 0) (Galer 1999). The planes, which are spanned by the spike-fractionation trend and the CaF$_2$ (dark gray plane) and the mixture-fractionation trend and the CaF$_2$ (light grey plane), intersect in an intersection line, which is also the mixing line of all spike—CaF$_2$ mixtures. Consequently, the true isotope compositions of the mixture and of the pure spike lie on this line. The true isotope composition of the spike is in the intercept of the mixing line and the spike fractionation trend. To correct for potential deviations between linear and exponential fractionation law, the obtained isotope ratios can numerically optimized using 'Solver' or similar functions. The general procedure is schematically illustrated in a flow sheet (Fig. 10).

5.4.3 Used Double Spikes

^{42}Ca/^{48}Ca and ^{43}Ca/^{48}Ca double spike

Russell et al. (1978) were the first who used a double spike (^{42}Ca/^{48}Ca) for Ca isotope measurements. The main advantage of using a ^{42}Ca/^{48}Ca or ^{43}Ca/^{48}Ca is that all used isotopes have a low natural abundance combined with a high enrichment factor (cf. Table 5). The mass difference of a $^{(42,43)}$Ca/^{48}Ca double spike well matches the mass difference of the Ca isotope ratio used (^{44}Ca/^{40}Ca). As the fractionation per mass unit is needed for the correction of the double spike, a greater mass difference allows a more precise determination of the fractionation per mass unit. One major disadvantage of such a double spike has to be seen in the fact that it results in a high mass dispersion which is still a problem for modern mass spectrometers although some improvements have been made. The relative mass difference of about 20 % between ^{40}Ca and ^{48}Ca exceeds the mass dispersion of most instruments which are constructed for a maximum mass difference of about 17 % (^{6}Li and ^{7}Li). So far no Ca isotopes data is published using (^{42}Ca or ^{43}Ca)/^{48}Ca double spike and mass spectrometers capable of a 20 % relative mass difference.

^{42}Ca/^{43}Ca double spike

A ^{42}Ca/^{43}Ca double spike also makes use of two minor Ca isotopes with reasonably high enrichment (cf. Table 5). But the mass difference of the used isotope does not well match the difference of the isotope ratio reported for TIMS measurements (^{44}Ca/^{40}Ca). But for TIMS measurements this double spike has the big advantage that works pretty well for a static Ca isotope measurement (^{40}Ca, ^{42}Ca, ^{43}Ca, ^{44}Ca). For ICPMS Ca isotope analyses which normally determine ^{44}Ca/^{42}Ca the use of a ^{42}Ca/^{43}Ca double spike is not possible.

^{43}Ca/^{46}Ca-double spike

During the past years a number of researches tried to make use of a ^{43}Ca/^{46}Ca double spike. ^{46}Ca has the lowest natural abundance of all Ca isotopes but the highest enrichment factor (cf. Table 5). This high enrichment factor makes enriched ^{46}Ca very expensive. On the other hand only small amounts of ^{46}Ca in a ^{43}Ca/^{46}Ca double spike are needed to establish a fixed ratio in the sample double spike mixture.

5.5 Other Instrumentation

Especially in the field of analytical chemistry several other instrumentations for the measurement of stable Ca isotopes have been developed. However, most of these techniques are not suitable for isotope geochemistry as precision and accuracy of these techniques are not sufficient. Some of these instrumentations and methods used either special or unique and even 'home' made instruments. The list below gives a short overview of some of these techniques:

- LD-TOFMS (laser desorption time of flight mass spectrometry, e.g. Koumenis et al. 1995)
- RIMS (resonant ionization mass spectrometry, e.g. Nicolussi et al. 1997)
- FABMS (fast atom bombardment, e.g. Smith 1983)
- MIP-TOFMS (microwave induced plasma time of flight mass spectrometry, Duan et al. 2001)

Some other instrumentations, described below, have either the potential to become a valuable tool in Ca isotope geochemistry (ion microprobe), are an often used tool for several studies of Ca isotopes in the pre-TIMS and pre-ICPMS times (INAA) or are used outside the field of isotope geochemistry (radionuclides).

5.5.1 Ion Microprobe

As with laser ablation ICP-MS, Ca isotope ratios have been determined in situ by secondary ion mass spectrometry (SIMS). Rollion-Bard et al. (2007) and Kasemann et al. (2008) applied SIMS to investigate the spatial variations of Ca isotope ratios in biogenic carbonates. The spatial resolution was 15–20 μm and about 25 μm which is smaller than that reasonably applied in LA-ICP-MS, e.g. Tacail et al. (2016) used a laser beam diameter of 85 μm. Both SIMS studies focused on heterogeneities of Ca isotope ratios in tests of foraminifers. These unicellular calcifiers are of particular interest, since earlier results obtained by conventional TIMS and MC-ICP-MS methods revealed a complex isotope fractionation behavior in these organisms, and indicating that isotope analysis with high spatial resolution might contribute to a better understanding of biomineralization related Ca isotope fractionation effects and their consequences for paleo proxy applications. Both studies indicated a heterogenous distribution of Ca isotope ratios in foraminifer tests. In particular Kasemann et al. (2008) showed systematic differences between different generations of biogenic calcite (ontogenetic and gametogenetic calcite). The difference between both calcite generations is not consistent for the two different species analyzed, as they show variable magnitudes and different signs of isotope fractionation.

Both Ca isotope studies used a Cameca IMS 1270 multicollector secondary ion mass spectrometer with a primary oxygen ion-beam. Calcium isotopes were analyzed with a spatial resolution of $\sim$10 and 25 μm spot-diameter. With count rates on ^{44}Ca of about 8–10 $\times$ 10^{6}cps Rollion-Bard et al. (2007) reported internal errors 0.2 ‰ 2 S.E. and external reproducibility of 0.25–0.50 ‰ (2 S.D.) and found intra-test variation in foraminifer shells of about 1.5 ‰ ($\delta^{44/40}$Ca).

Kasemann et al. (2008) operated with slightly smaller count rates of about 4–9 $\times$ 10^{6}cps and report internal errors of about 0.4 ‰ (2 S.E.) and an external reproducibility of standard and samples of $\sim$1 and 0.4 ‰, respectively. The intra-test variability of the foraminifer tests account to 1.6 and −3.7 ‰ between gametogenetic and ontogenetic calcite in different species. The observed overall variability of Ca isotope ratios within single foraminifer tests is compatible in both studies indicating the potential of SIMS for Ca isotope analysis. Nevertheless both studies also point to challenges that need to be resolved. These are mainly related to matrix effects, interferences and the availability of appropriate standard materials. Presently, there are no certified homogenous Ca isotope standards available that are suited for SIMS. These complications include matrix-matching between samples and standards or structural or chemical/isotopical heterogeneities of standards leading to different fractionation effects or background/signal-ratio of samples and standards. Possible interference include (e.g. $^{12}C^{16}O_2$, $^{24}Mg^{16}O$, $^{88}Sr^{2+}$, $^{86}Sr^{2+}$).

Calcium isotope analysis by SIMS has been frequently applied to minerals (hibonites, old-hamite (CaS) in chondritic meteorite samples (Zinner et al. 1986; Ireland et al. 1990; Lundberg et al. 1994 and Weber et al. 1995). The main focus was on nucleosynthetic ^{48}Ca isotope anomalies and mass-dependent Ca isotope fractionation in CAIs. Beside these geoscientific oriented studies other studies on Ca isotopes exist (Roy et al. 1995; Bushinsky et al. 1990). Both studies used ^{44}Ca for labeling of bone (Bushinsky et al. 1990) and for determination of Ca deposition at the cell walls of apple fruit. As both studies are tracer studies working with high amounts of ^{44}Ca enrichment these studies did not need a precision needed to investigate natural Ca isotope variations.

SIMS has also been used to determine the radiogenic ^{40}Ca ingrowth in K-rich minerals. In order to suppress the K intensities relative to Ca, Harrison et al. (2010) analysed doubly charged K^{++} and Ca^{++} ions. With this technique the authors were able to reconcile K–Ar and K–Ca systematics in alkali feldspars from the Klokken syenite (1166 Ma) in southern Greenland and to unravel complex genetic relations.

5.5.2 Neutron Activation Analysis INAA

Mass dependent Ca isotope fractionation effects were also approached by methods other than mass spectrometry. One of the applied analytical methods was neutron activation analysis (INAA) (Corless 1966, 1968). The basic concept of neutron activation is that certain nuclides are selectively activated by neutron-radiation. To analyze Ca isotope fractionation effects with this methods, the ratio of ^{48}Ca and the total Ca concentration ($^{48}Ca/Ca_{total}$) is determined. The ^{48}Ca is measured indirectly on the basis of the nuclear reaction:

$$^{48}Ca(n,\gamma)^{49}Ca \xrightarrow{\beta^-\,(8.7\,min)} {}^{49}Sc \xrightarrow{\beta-(57.5\,min)} {}^{49}Ti$$

The total Ca concentration is precisely determined for instance by acid EDTA titration (ethylenediaminetetraacetic) in wet chemistry. The results obtained by neutron activation analysis showed large variations of up to 10 ‰ (Corless 1966, 1968), which were, however, not reproduced by subsequent studies.

5.5.3 Determination of Mass Dependent Ca Isotope Fractionation by Radionuclide Tracers

A further non-mass spectrometric approach to investigate mass dependent calcium isotope fractionation effects was the application of radiotracers. For instance, Möller and Papendorff (1971) used a ^{45}Ca tracer ($t_{1/2}$ = 163 d) during calcium carbonate precipitation experiments. Their experiments revealed that Ca isotope fractionation between solution and precipitated $CaCO_3$ is smaller than 0.3 ‰/amu. Based on their results, Möller and Papendorff (1971) proposed that the hydration of the Ca^{2+}-ion may hinder a larger degree of Ca isotope fractionation and that fractionation in Ca^{2+}-diffusion processes may exceed 0.3 ‰/amu. Although this method was not sensitive enough to quantify the degree of Ca isotope fractionation during $CaCO_3$ precipitation, their results are still consistent with recent studies.

5.6 Error Representation

An overview of existing external reproducibilities of Ca isotopic measurements shows that the long-term 2 standard deviation (2SD) is comprised between 0 and 0.50 ‰ for $\delta^{44/40}Ca$ (Table 6). Usually reproducibility is calculated from full procedural replicates on reference material. However, samples present different matrixes than standards and thus react differently during ionization. As a result, some authors proposed to take

Table 6 Summary of different TIMS measurement protocols for Ca isotopes and associated external reproducibilities

Continent	Country	Laboratory	Instrument	Collection	Filament	Loaded with	Activator	Amount of Ca loaded	^{40}Ca peak intensity	Sequence of measurement	Number of scans	±2 SD[a]	References
Australia	Australia	Curtin	Modified VG 354	Multi	Triple zone-refined Re	HI				1/^{40}Ca, ^{41}K, ^{42}Ca; 2/^{42}Ca, ^{43}Ca, ^{44}Ca; 3/^{44}Ca, ^{46}Ca		0.40	Fletcher et al. (1997)
North America	Canada	Saskatchewan	Thermo Triton	Multi	Double Re	HNO$_3$		3–5 µg	8 V	1/^{40}Ca, ^{42}Ca, ^{43}Ca, ^{44}Ca; 2/^{40}Ca, ^{41}K, ^{42}Ca	180	0.10	Holmden (2005)
		Saskatchewan	Thermo Triton	Multi	Single Ta (using parafilm dams)	HNO$_3$	H$_3$PO$_4$	3–8 µg	8–20 V	1/^{40}Ca,^{41}K, ^{42}Ca; 2/^{42}Ca; ^{43}Ca; 3/^{43}Ca, ^{44}Ca	130–180	0.07–0.10	Jacobson and Holmden (2008), Amini et al. (2009), Holmden (2009), Holmden and Bélanger (2010), Ryu et al. (2011)
	USA	Caltech	Lunatic I	Single	Single V-shaped zone refined Ta	HCl, HNO$_3$	Oxidized filament	5–10 µg	4 V	^{40}Ca, ^{42}Ca, ^{44}Ca, ^{48}Ca	200	0.12 –0.50	Russel et al. (1978), Lemarchand et al. (2004)
	USA	Scripps	VG54-A	Single	Single W		Ta$_2$O$_5$ (sandwich technique)	300–600 ng	5.4–6.6 V	^{40}Ca, ^{42}Ca, ^{43}Ca, ^{44}Ca, ^{48}Ca	2 Independent measurements (spiked and not spiked)	0.10–0.20	Zhu and MacDougall (1998)
	USA	Santa Cruz univ.	VG 54/WARP	Multi	Single Re	HNO$_3$	Ta$_2$O$_5$, H$_3$PO$_4$	1 µg		^{40}Ca, ^{42}Ca, ^{44}Ca, ^{48}Ca	100–200	0.10–0.30	Clementz et al. (2003)
		IGGRC, Carleton univ.	Thermo Triton	Multi	Zone-refined double Re	HNO$_3$	H$_3$PO$_4$, taken to dull red	3 µg	10 V	1/^{40}Ca, ^{41}K, ^{42}Ca, ^{43}Ca, ^{44}Ca; 2/^{44}Ca, ^{48}Ca	100	0.15	Farkaš et al. (2006, 2007)
	USA	Berkely	modified VG 354 with one large Faraday bucket collector	Single	Single Ta		H$_3$PO$_4$	3–8 µg	6–9 V			0.10–0.20	Skulan et al. (1997), DeLa Rocha and DePaolo (2000), Fantle and DePaolo (2005, 2007)
		Berkely	VG Sector 54		Single Ta		H$_3$PO$_4$	8 µg		^{40}Ca, ^{42}Ca, ^{44}Ca, ^{48}Ca	100–200		Skulan and DePaolo (1999)
		Berkely	Thermo triton	Multi	Double Re	HNO$_3$	H$_3$PO$_4$	3 µg	20 V	^{39}K, ^{40}Ca, ^{42}Ca, ^{43}Ca,	200	0.14	Ewing et al. (2008), Simon et al. (2009),

(continued)

Table 6 (continued)

Continent	Country	Laboratory	Instrument	Collection	Filament	Loaded with	Activator	Amount of Ca loaded	^{40}Ca peak intensity	Sequence of measurement	Number of scans	±2 SD[a]	References
										^{44}Ca, ^{48}Ca, ^{49}Ti			Simon and DePaolo (2010)
	USA	Stanford	Finnigan MAT 262		Single Ta		H_3PO_4			^{40}Ca, ^{42}Ca, ^{44}Ca, ^{48}Ca	80	0.15	Wiegand et al. (2005)
	USA	Harvard	Isoprobe-T	Multi	Triple Re			5 µg		$1/^{40}$Ca, ^{42}Ca, ^{43}Ca, ^{44}Ca; $2/^{44}$Ca, ^{48}Ca		0.13	Huang et al. (2010)
	USA	North Western University	Thermo Triton	Multi	Single Ta	HNO_3	H_3PO_4	10–16 µg	20 V	$1/^{40}$Ca,^{41}K, ^{42}Ca; $2/^{42}$Ca, ^{43}Ca; $3/^{43}$Ca, ^{44}Ca	90	0.04	Lehn et al. (2013)
Europe	France	Strasbourg	VG Sector	Single	Single Ta	1.5 N HCl	H_3PO_4	10 µg	6 V	^{40}Ca,^{41}K, ^{42}Ca, ^{43}Ca, ^{44}Ca, ^{48}Ca	2 Independent measurements (spiked and not spiked)	0.20	Schmitt et al. (2001, 2003a, b), Schmitt and Stille (2005)
		Strasbourg	Thermo triton	Multi	Single Re	1.5 N HCl	Ta_2O_5, H_3PO_4	5 µg		$1/^{40}$Ca,^{41}K, ^{42}Ca, ^{43}Ca, ^{44}Ca; $2/^{42}$Ca, ^{48}Ca		0.30	Cenki-Tok et al. (2009)
		Strasbourg	Thermo triton	Multi	Single Ta	1.5 N HCl	Filament oxidized in partial vacuum	5 µg	5 V	$1/^{40}$Ca,^{41}K, ^{42}Ca; $2/^{43}$Ca, ^{44}Ca; $3/^{42}$Ca, ^{43}Ca	130–200	0.12	Schmitt et al. (2009). Cobert et al. (2011a, b)
	Germany	Geomar, Kiel	Finnigan MAT 262 RPQ$^+$	Multi	Single zone-refined Re	2.5 N HCl	Ta_2O_5 sandwich technique; sample heated to red	300 ng	4–5 V	$1/^{40}$Ca, ^{41}K, ^{42}Ca, ^{43}Ca; $2/^{44}$Ca, ^{48}Ca	154	0.20–0.50	Heuser et al. (2002), Gussone et al. (2003, 2004, 2005), Heuser and Eisenhauer (2008)
		Geomar, Kiel	Thermo triton	Multi	Single zone-refined Re	2.5 N HCl	Ta_2O_5 sandwich technique; sample heated to red	300 ng		Static ^{40}Ca, ^{41}K, ^{42}Ca, ^{43}Ca, ^{44}Ca or $1/^{40}$Ca, ^{42}Ca, ^{43}Ca, ^{44}Ca; $2/^{43}$Ca, ^{44}Ca, ^{48}Ca		0.13–0.16	Gussone et al. (2006, 2007), Böhm et al. (2006), Amini et al. (2008), Griffith et al. (2008), Heuser and Eisenhauer (2008), Amini et al. (2009), Gussone et al. (2009;

(continued)

Table 6 (continued)

Continent	Country	Laboratory	Instrument	Collection	Filament	Loaded with	Activator	Amount of Ca loaded	^{40}Ca peak intensity	Sequence of measurement	Number of scans	±2 SD[a]	References
													2010), Heuser and Eisenhauer (2010)
	Germany	Münster	Thermo Triton	Multi	Single zone-refined Re	6N HCl	Ta_2O_5	300–400 ng	12–16 V	$1/^{40}Ca,41K,$ $^{42}Ca,$ $^{43}Ca,$ $^{44}Ca;$ $(2/^{43}Ca,$ $^{44}Ca,$ $^{48}Ca)$	180	0.02–0.21	Arning et al. (2008), Gussone et al. (2010, 2011), Heuser et al. (2011)
	United Kingdom	Bristol	Thermo Triton	Multi	Double zone-refined Re	1 N HCl		2 µg		$^{40}Ca,$ $^{41}K,$ $^{42}Ca,$ $^{43}Ca,$ $^{44}Ca,$ ^{48}Ca	90	0.10–0.12	Kasemann et al. (2005)
	Switzerland	Bern	modified AVCO equipped with a Thermolinear ion source	Single	Single Re	2.5 N HCl	Ta_2O_5	0.3–5 µg		$^{48}Ca,$ $^{44}Ca,$ $^{43.5}Sr,$ $^{43}Ca,$ $^{41}K,$ ^{40}Ca		0.15–0.30	Nägler and Villa (2000), Nägler et al. (2000), Hippler et al. (2006, 2009), Silva-Tamayo et al. (2010), von Allmen et al. (2010b)
	Switzerland	ETH, Zürich	Thermo triton	Multi	Single Re		Ta_2O_5	0.7–1 µg		Static $^{40}Ca,$ $^{41}K,$ $^{42}Ca,$ $^{43}Ca,$ $^{44}Ca,$ ^{46}Ca	280	0.07[b]	Hindshaw et al. (2010)

[a]$\delta^{44/40}Ca$
[b]$\delta^{44/42}Ca$

into account repeated measurements of various sample materials (Teichert et al. 2005; Griffith et al. 2008; Gussone et al. 2009; Schmitt et al. 2009, 2013; Cobert et al. 2011a, b). A way to avoid this artifact is to process seawater standard, which is a matrix-rich standard through each measurement session and to process it similarly to samples (Lehn et al. 2013; du Vivier et al. 2015). In doing so, du Vivier et al. (2015) observed no difference between replicate sample and standard measurements. The Saskatchewan group (Holmden and Bélanger 2010) introduced drift corrections for each measurement session. They indeed observed an instrumental drift from about 0.3 ‰ over 2-years. In order to avoid this, they usually performed around 15–30 measurements in 1–2 weeks measurement sessions, including samples and standards. Then they adjusted the isotopic composition of their double spike using an exponential law in each measurement session, to achieve and average constant isotopic difference between two standards (seawater and CaF_2). Similarly, Hindshaw et al. (2011) measured in each turret standards, which values were averaged and corrected to zero. This correction was then applied to all samples run on the same turret. As a result these authors (Hindshaw et al. 2011; Holmden and Bélanger 2010; Ryu et al. 2011) decreased their long-term external 2SD down to $\pm$0.07 %, compared to $\pm$0.10–0.15 ‰ obtained by most other laboratories. More recently Lehn et al. (2013) decreased their long-term external reproducibility down to $\pm$0.04 ‰ by improving the measurement protocol and optimizing the $^{43}Ca/^{42}Ca$ spike mixture. However, in doing so they also increased the amount of Ca loaded on the filament (10–16 µg), which may be limiting for samples very poor in Ca.

To improve the statistical significance of a single $\delta^{44/40}Ca$ measurement, most of presently published Ca isotope data result from the combination of at least two individual measurements. These values and their corresponding errors are presented in different ways, trying to minimize the error, in order to be able to resolve the smallest $\delta^{44/40}Ca$ variations. Thus, direct comparison between different laboratories is difficult.

However, the significance of the numbers is mostly clearly stated in the different articles dealing with Ca isotopes, so that the reader can judge their significance for themselves and recalculate the errors if necessary.

Five main ways of reporting replicate $\delta^{44/40}Ca$ (or $\delta^{44/42}Ca$) values and associated errors can be listed from the literature and are reported hereafter:

(1) Mean $\delta^{44/40}Ca$ (or $\delta^{44/42}Ca$) values and 2SD (95 % confidence level) corresponding to N measurements of one given sample (Skulan et al. 1997; Soudry et al. 2004; Wieser et al. 2004; Fantle and DePaolo 2005, 2007; Immenhauser et al. 2005; Schmitt and Stille 2005; Steuber and Buhl 2006; Kasemann et al. 2008; Komiya et al. 2008; Gussone et al. 2009; Tipper et al. 2010; Reynard et al. 2011).

(2) Mean $\delta^{44/40}Ca$ (or $\delta^{44/42}Ca$) values and 1SD (68 % confidence level) corresponding to N measurements of one given sample (e.g. Halicz et al. 1999; Heuser et al. 2002, 2005; Clementz et al. 2003; Fietzke et al. 2004; Wiegand et al. 2005; Chu et al. 2006; Rollion-Bard et al. 2007; Sime et al. 2007; Ewing et al. 2008; Page et al. 2008; Reynard et al. 2010).

(3) Mean $\delta^{44/40}Ca$ (or $\delta^{44/42}Ca$) values and 2SE corresponding to N measurements of one given sample (e.g. Zhu and Macdougall 1998; Gussone et al. 2003, 2004; Schmitt et al. 2003a, b; Kasemann et al. 2005; Sime et al. 2005; Teichert et al. 2005; Böhm et al. 2006; Tipper et al. 2006; Amini et al. 2008; Griffith et al. 2008; Heuser and Eisenhauer 2008; Heuser et al. 2011).

(4) Mean $\delta^{44/40}Ca$ (or $\delta^{44/42}Ca$) values corresponding to N measurements of one given sample and long-term 2SE calculated from replicates of different samples or standards (e.g. Marriott et al. 2004; Farkaš et al. 2007; Ewing et al. 2008; Jacobson and Holmden 2008; Cenki-Tok et al. 2009; Schmitt et al. 2009, 2013; Cobert et al. 2011a; b; Hindshaw et al. 2010, Farkaš et al. 2011; Bagard

Table 7 Comparison of the way of reporting multiple isotopic measurements of one single sample

		$\delta^{44/40}Ca$	$2SE_{Int}$	N	Mean	Weighted average	2SD	1SD	$2SD_{weighted}$	$2SD_{long\ term}$	2SE	$2SE_{weighted}$	$2SE_{long\ term}$
Replicates of different sample measurements													
Sample 1	Rainwater	0.62	0.12	2	0.59	0.58	0.10	0.05	0.11	0.11	0.07	0.08	0.08
		0.55	0.11										
Sample 2	Rainwater	0.57	0.12	2	0.59	0.59	0.04	0.02	0.10	0.11	0.03	0.07	0.08
		0.60	0.09										
Sample 3	Swamp	0.08	0.09	2	0.02	0.01	0.17	0.08	0.08	0.11	0.12	0.06	0.08
		−0.04	0.08										
Sample 4	Nutritive solution	0.98	0.13	3	0.93	0.93	0.07	0.04	0.12	0.11	0.04	0.07	0.06
		0.91	0.13										
		0.91	0.11										
Sample 5	Carbonate	−0.63	0.12	2	−0.63	−0.63	0.00	0.00	0.11	0.11	0.00	0.08	0.08
		−0.63	0.10										
Sample 6	Carbonate	−1.16	0.13	2	−1.20	−1.21	0.10	0.05	0.11	0.11	0.07	0.07	0.08
		−1.23	0.08										
Sample 7	Granite	−1.36	0.13	2	−1.34	−1.33	0.06	0.03	0.10	0.11	0.04	0.07	0.08
		−1.32	0.09										
Sample 8	Apatite	0.77	0.08	2	0.84	0.84	0.18	0.09	0.08	0.11	0.13	0.06	0.08
		0.90	0.08										
Sample 9	Soil	−1.12	0.09	2	−1.14	−1.14	0.04	0.02	0.08	0.11	0.03	0.06	0.08
		−1.15	0.07										
Sample 10	Soil	−0.93	0.11	2	−1.01	−1.01	0.21	0.11	0.11	0.11	0.15	0.08	0.08
		−1.08	0.11										
Sample 11	Leave	0.15	0.06	2	0.25	0.20	0.27	0.13	0.07	0.11	0.19	0.05	0.08
		0.34	0.10										

(continued)

Table 7 (continued)

		$\delta^{44/40}$Ca	2SE$_{Int}$	N	Mean	Weighted average	2SD	1SD	2SD$_{weighted}$	2SD$_{long\ term}$	2SE	2SE$_{weighted}$	2SE$_{long\ term}$
Sample 12	Cotyledon	−0.36	0.09	2	−0.37	−0.37	0.01	0.01	0.08	0.11	0.01	0.06	0.08
		−0.37	0.07										
Sample 13	Tegument	−0.90	0.11	2	−0.94	−0.96	0.13	0.07	0.08	0.11	0.09	0.06	0.08
		−0.99	0.07										
Sample 14	Root	0.01	0.09	2	0.04	0.02	0.08	0.04	0.11	0.11	0.06	0.08	0.08
		0.07	0.18										
Sample 15	Stem	0.02	0.16	2	−0.07	−0.10	0.25	0.13	0.14	0.11	0.18	0.10	0.08
		−0.16	0.12										
Sample 16	Sap	−0.07	0.09	2	−0.12	−0.14	0.13	0.06	0.06	0.11	0.09	0.04	0.08
		−0.16	0.05										
Sample 17	Trunk	−1.92	0.18	2	−1.88	−1.86	0.13	0.06	0.16	0.11	0.09	0.11	0.08
		−1.83	0.13										
Replicates of standards													
E 5109	Seawater	1.83	0.16	6	1.86	1.79	0.12	0.06	0.12	0.11	0.05	0.05	0.04
		1.90	0.18										
		1.91	0.09										
		1.79	0.15										
		1.69	0.07										
		1.67	0.03										
NIST SRM 915a	Carbonate	−0.03	0.16	4	0.00	0	0.06	0.03	0.12	0.11	0.03	0.06	0.06
		0.00	0.15										
		−0.01	0.09										
		0.04	0.12										

(continued)

Table 7 (continued)

		$\delta^{44/40}$Ca	$2SE_{Int}$	N	Mean	Weighted average	2SD	1SD	$2SD_{weighted}$	$2SD_{long\ term}$	2SE	$2SE_{weighted}$	$2SE_{long\ term}$
CaF$_2$	Fluorure	1.36	0.10	6	1.38	1.38	0.04	0.02	0.12	0.11	0.02	0.05	0.04
		1.37	0.14										
		1.40	0.10										
		1.36	0.10										
		1.39	0.10										
		1.48	0.09										

et al. 2013; Gangloff et al. 2014; Blättler et al. 2015; du Vivier et al. 2015)

(5) Weighted $\delta^{44/40}$Ca and 2SE corresponding to N measurements of one given sample (e.g. Nägler et al. 2000; Hippler et al. 2003, 2006).

As it is widely known, statistically SD and SE do not have the same significance. The standard deviation (SD or σ) says how widely scattered some measurements are. The standard error (SE or σ_{mean}) indicates the uncertainty around the estimate of the mean measurement. Moreover, SE and SD are related by the following equation:

$$SE = \frac{SD}{\sqrt{n}}, \text{ with n: number of replicates} \quad (49)$$

If the data follow a normal (or Gaussian) distribution, then about 68 % of the data values are within one standard deviation of the mean ($\pm$1SD). About 95 % are within two standard deviations ($\pm$2SD). It has also to be noted that, for small number of replicates, the mean values and corresponding errors are usually expressed using weighted values.

In order to check the differences induced by the multiple ways of reporting replicate measurements of $\delta^{44/40}$Ca values and corresponding errors, we have applied all the existing calculations to replicate analysis of samples (Table 7). It can especially be observed that for the presented dataset up to 50 % relative difference can be observed between mean and weighted average $\delta^{44/40}$Ca values, with an average difference of 3 $\pm$ 41 % (2SD; N = 17). For a limited number of replicates, weighted average values are statistically closer to the "true" value than mean ones. However, when only few replicates are available this method accords more weight to a sample measured with a good internal error, with no glance at the accuracy.

Given the small number of replicates usually performed in Ca isotopic studies, 2SD, 2SE and 1SD are statistically not robust. Indeed, when we consider for instance samples 5 or 12, which are measured twice and give each time similar $\delta^{44/40}$Ca values, internal 2SE values, external SD or SE would be close to zero, which makes no

sense. We can deduce from Table 7 that the variation between weighted 2SE (rep. weighted 2SD) and long-term 2SE (resp. long-term 2SD) differ but are in the same order of magnitude. Weighted errors refer to replicates of the same sample and reflect the way the measurement is completed. Long-term errors average short-term instabilities and show how the measurement reproduces over a long period of time. Each method presents its highlights and drawbacks; no optimal method exists for small replicates (2–3). As a result caution should be exercised in drawing conclusions over small differences in samples when comparing data from one laboratory to another without recalculating the $\delta^{44/40}$Ca values and corresponding errors.

References

Albarède F, Beard B (2004) Analytical methods for non-traditional isotopes. In: Johnson CM, Beard BL, Albarède F (eds) Geochemistry of non-traditional stable isotopes., Reviews in mineralogy and geochemistryMineralogical Society of America, Washington, pp 113–152

Albarède F, Telouk P, Blichert-Toft J et al (2004) Precise and accurate isotopic measurements using multiple-collector ICPMS. Geochim Cosmochim Acta 68:2725–2744

Amini M, Eisenhauer A, Böhm F et al (2008) Calcium Isotope ($\delta^{44/40}$Ca) fractionation along hydrothermal pathways, logatchev field (mid-atlantic ridge, 14° 45'N). Geochim Cosmochim Acta 72:4107–4122

Amini M, Eisenhauer A, Böhm F et al (2009) Calcium isotopes ($\delta^{44/40}$Ca) in MPI-DING reference glasses, USGS rock powders and various rocks: evidence for Ca isotope fractionation in terrestrial silicates. Geostand Geoanal Res 33:231–247

Andreasen R, Sharma M (2006) Solar nebula heterogeneity in P-process samarium and neodymium isotopes. Science 314:806–809

Andreasen R, Sharma M (2009) Fractionation and mixing in a thermal ionization mass spectrometer source: implications and limitations for high-precision Nd isotope analyses. J Anal At Spec 285:49–57

Baadsgaard H (1987) Rb–Sr and K–Ca isotope systematics in minerals from potassium horizons in the prairie evaporite formation, Saskatchewan, Canada. Chem Geol Isotope Geosci Sect 66:1–15

Bagard ML, Schmitt AD, Chabaux F et al (2013) Biogeochemistry of stable Ca and radiogenic Sr isotopes in a larch-covered permafrost-dominated watershed of Central Siberia. Geochim Cosmochim Acta 114:169–187

Baxter DC, Rodushkin I, Engström E (2012) Isotope abundance ratio measurements by inductively coupled plasma-sector field mass spectrometry. J Anal At Spec 27:1355–1381

Blättler CL, Jenkyns HC, Reynard LM et al (2011) Significant increases in global weathering during oceanic anoxic events 1a and 2 indicated by calcium isotopes. Earth Planet Sci Lett 309. doi:10.1016/j.epsl.2011.06.029

Blättler CL, Miller NR, Higgins JA (2015) Mg and Ca isotope signatures of authigenic dolomite in siliceous deep-sea sediments. Earth Planet Sci Lett 419:32–42

Blum JD et al (2008) Use of foliar Ca/Sr discrimination and $^{87}Sr/^{86}Sr$ ratios to determine soil Ca sources to sugar maple foliage in a northern hardwood forest. Biogeochem 87:287–296

Böhm F, Gussone N, Eisenhauer A et al (2006) Calcium isotope fractionation in modern scleractinian corals. Geochim Cosmochim Acta 70:4452–4462

Boulyga SF (2010) Calcium isotope analysis by mass spectrometry. Mass Spectrom Rev 29:685–716

Boulyga SF, Becker JS (2001) ICP-MS with hexapole collision cell for isotope ratio measurements of Ca, Fe, and Se. Fresen J Anal Chem 370:618–623

Boulyga SF, Klötzli U, Stingeder G et al (2007) Optimization and application of ICPMS with dynamic reaction cell for precise determination of $^{44}Ca/^{40}Ca$ isotope ratios. Anal Chem 79:7753–7760

Brazier JM et al (2015) Calcium isotope evidence for dramatic increase of continental weathering during the Toarcian oceanic anoxic event (Early Jurassic). Earth Planet Sci Lett 411:164–176

Breit GN, Simmons EC, Goldhaber MB (1985) Dissolution of barite for the analysis of strontium isotopes and other chemical and isotopic variations using aqueous sodium carbonate. Chem Geol Isot Geosci Sect 52:333–336

Bullen TD et al (2004) Calcium stable isotope evidence for three soil calcium pools at a granitoid chrono-sequence. In: Wanty RB, Seal RR (eds) Water-rock interaction, proceedings of the 11th international symposium on water-rock interaction, vol 1. Saratoga Springs/Taylor & Francis, New York/London, July 2004, pp 813–817

Bushinsky DA, Chabala JM, Levi-Setti R (1990) Comparison of in vitro and in vivo ^{44}Ca labeling of bone by scanning ion microprobe. Am J Physiol 259:586–592

Carlson RW (2014) 15.18—thermal ionization mass spectrometry. In: Holland HD, Turekian KK (eds) Treatise on geochemistry (2nd Edn), 337–54. Elsevier, Oxford. http://www.sciencedirect.com/science/article/pii/B9780080959757014273

Caro G, Papanastassiou DA, Wasserburg GJ (2010) $^{40}K–^{40}Ca$ isotopic constraints on the oceanic calcium cycle. Earth Planet Sci Lett 296:124–132

Cecil MR, Ducea MN (2011) K–Ca ages of authigenic sediments: examples from Paleozoic glauconite and applications to low-temperature thermochronometry. Int J Earth Sci 100:1783–1790

Cenki-Tok B, Chabaux F, Lemarchand D et al (2009) The impact of water-rock interaction and vegetation on calcium isotope fractionation in soil- and stream waters of a small, forested catchment (the Strengbach case). Geochim Cosmochim Acta 73: 2215–2228

Chang VTC, Williams RJP, Makishima A et al (2004) Mg and Ca isotope fractionation during $CaCO_3$ biomineralisation. Biochem Bioph Res Co 323:79–85

Channon MB, Gordon GW, Morgan JLL et al (2015) Using natural, stable calcium isotopes of human blood to detect and monitor changes in bone mineral balance. Bone 77:69–74. doi:10.1016/j.bone.2015.04.023

Chu N-C, Henderson GM, Belshaw NS et al (2006) Establishing the potential of Ca isotopes as proxy for consumption of dairy products. Appl Geochem 21:1656–1667

Church TM (1979) Marine barite. In: Burns RG (ed) Marine minerals. Mineralogical Society of America, Washington D.C., pp 175–209

Clementz MT, Holden P, Koch PL (2003) Are calcium isotopes a reliable monitor of trophic level in marine settings? Int J Osteoarchaeol 13:29–36

Cobert F et al (2011a) Biotic and abiotic experimental identification of bacterial influence on Ca isotopic signatures. Rapid Com Mass Spectrom 25:2760–2768

Cobert F, Schmitt AD, Bourgeade P et al (2011b) Experimental identification of Ca isotopic fractionations in higher plants. Geochim Cosmochim Acta 75:5467–5482

Compston W, Oversby VM (1969) Lead isotopic analysis using a double spike. J Geophys Res 74(17):4338–4348

Coplen TB (2011) Guidelines and recommended terms for expression of stable-isotope-ratio and gas-ratio measurement results. Rapid Comm Mass Sp 25:2538–2560

Coplen TB, Hopple JA, Böhlke JK et al (2002) Compilation of minimum and maximum isotope ratios of selected elements in naturally occurring terrestrial materials and reagents. USGS, Reston, Virginia. 01–4222, 98 p

Corless JT (1966) Determination of calcium-48 in natural calcium by neutron activation analysis. Anal Chem 38:810–813

Corless JT (1968) Observations on the isotopic geochemistry of calcium. Earth Planet Sci Lett 4:475–478

De La Rocha CL, DePaolo DJ (2000) Isotopic evidence for variations in the marine calcium cycle over the cenozoic. Science 289:1176–1178

DePaolo DJ (2004) Calcium isotopic variations produced by biological, kinetic, radiogenic and nucleosynthetic processes. Rev Mineral Geochem 55:255–288

Du Vivier ADC, Jacobson AD, Lehn GO et al (2015) Ca isotope stratigraphy across the Cenomanian-Turonian OAE 2: links between volcanism, seawater geochemistry, and the carbonate fractionation factor. Earth Planet Sci Lett 416:121–131

Duan Y, Su Y, Jin Z et al (2001) Measurements of calcium isotopes and isotope ratios: a new method based on helium plasma source "off-cone" sampling time-of-flight mass spectrometry. J Anal At Spec 16:756–761

Eisenhauer A, Nägler TF, Stille P et al (2004) Proposal for an international agreement on Ca notation as result of the discussions from the workshops on stable isotope measurements in Davos (Goldschmidt 2002) and nice (EGS-AGU-EUG 2003). Geostand Geoanal Res 28:149–151

Elliot T, Wehrs H, Coath C et al (2015) Collision cell MC-ICPMS. Goldschmidt Abstr 2015:824

Epov VN, Malinovskiy D, Vanhaecke F et al (2011) Modern mass spectrometry for studying mass-independent fractionation of heavy stable isotopes in environmental and biological sciences. J Anal At Spec 26:1142–1156

Ewing SA, Yang W, DePaolo DJ et al (2008) Non-biological fractionation of stable Ca isotopes in soils of the Atacama Desert. Chile Geochim Cosmochim Acta 72:1096–1110

Fantle MS, DePaolo DJ (2005) Variations in the marine Ca cycle over the past 20 million years. Earth Planet Sci Lett 237:102–117

Fantle MS, DePaolo DJ (2007) Ca isotopes in carbonate sediment and pore fluid from ODP Site 807A: the Ca^{2+}(aq)–calcite equilibrium fractionation factor and calcite recrystallization rates in Pleistocene sediments. Geochim Cosmochim Acta 71:2524–2546

Fantle MS, Bullen TD (2009) Essentials of iron, chromium, and calcium isotope analysis of natural materials by thermal ionization mass spectrometry. Chem Geol 258:50–64

Farkaš J, Buhl D, Blenkinsop J et al (2006) Evolution of the oceanic calcium cycle during the late Mesozoic: evidence from $\delta^{44/40}Ca$ of marine skeletal carbonates. Earth Planet Sci Lett 253:96–111

Farkaš J, Böhm F, Wallmann K et al (2007) Calcium isotope record of Phanerozoic oceans: implications for chemical evolution of seawater and its causative mechanisms. Geochim Cosmochim Acta 71:5117–5134

Farkaš J et al (2011) Calcium isotope constraints on the uptake and sources of Ca^{2+} in a base-poor forest: a new concept of combining stable ($\delta^{44/42}Ca$) and radiogenic (εCa) signals. Geochim Cosmochim Acta 75:7031–7046

Feldmann I, Jakubowski N, Stuewer D (1999) Application of a hexapole collision and reaction cell in ICP-MS Part I: instrumental aspects and operational optimization. Fresen J Anal Chem 365:415–421

Fietzke J, Eisenhauer A, Gussone N et al (2004) Direct measurement of $^{44}Ca/^{40}Ca$ ratios by MC-ICP-MS using the cool plasma technique. Chem Geol 206:11–20

Fletcher IR, Maggi AL, Rosman KJ et al (1997) Isotopic abundance measurements of K and Ca using a wide-dispersion multi-collector mass spectrometer and low-fractionation ionization techniques. Int J Mass Spec Ion Proc 163:1–17

Gale NH (1970) A solution in closed form for lead isotopic analysis using a double spike. Chem Geol 6:305–310

Galer SJG (1999) Optimal double and triple spiking for high precision lead isotopic measurement. Chem Geol 157:255–274

Gangloff S et al (2014) Impact of bacterial activity on Sr and Ca isotopic compositions ($^{87}S/^{86}Sr$ and $\delta^{44/40}Ca$) in soil solutions (the Strengbach CZO). Procedia Earth Planet Sci 10:109–113

Gopalan K, Macdougall D, Macisaac C (2006) Evaluation of a $^{42}Ca–^{43}Ca$ double-spike for high precision Ca isotope analysis. Int J Mass Spectrom 248:9–16

Griffith EM, Schauble EA, Bullen TD et al (2008) Characterization of calcium isotopes in natural and synthetic barite. Geochim Cosmochim Acta 72:5641–5658

Gussone N, Filipsson HL (2010) Calcium isotope ratios in calcitic test of benthic foraminifers. Earth Planet Sci Lett 290:108–117

Gussone N, Eisenhauer A, Heuser A et al (2003) Model for kinetic effects on calcium isotope fractionation ($\delta^{44}Ca$) in inorganic aragonite and cultured foraminifer (orbulina universa and globigerinoides sacculifer). Geochim Cosmochim Acta 67:1375–1382

Gussone N, Eisenhauer A, Tiedemann R et al (2004) Reconstruction of caribbean sea surface temperature and salinity fluctuations in response to the pliocene closure of the central american gateway and radiative forcing, using $\delta^{44/40}Ca$, $\delta^{18}O$ and Mg/Ca ratios. Earth Planet Sci Lett 227:201–214

Gussone N, Böhm F, Eisenhauer A et al (2005) Calcium isotope fractionation in calcite and aragonite. Geochim Cosmochim Acta 69:4485–4494

Gussone N, Langer G, Thoms S et al (2006) Cellular calcium pathways and isotope fractionation in emiliania huxleyi. Geology 34:625–628

Gussone N, Langer G, Riebesell U et al (2007) Calcium isotope fractionation in coccoliths of cultured calcidiscus leptoporus, helicosphaera carteri, syracosphaera pulchra and umbilicosphaera foliosa. Earth Planet Sci Lett 260:505–515

Gussone N, Höhnisch B, Heuser A et al (2009) A critical evaluation of calcium isotope ratios in tests of planktonic foraminifers. Geochim Cosmochim Acta 73:7241–7255

Gussone N, Zonneveld K, Kuhnert H (2010) Minor element and Ca isotope composition of calcareous dinoflagellate cysts of cultured thoracosphaera heimii. Earth Planet Sci Lett 289:180–188

Gussone N, Nehrke G, Teichert BMA (2011) Calcium isotope fractionation in ikaite and vaterite. Chem Geol 285:194–202

Halicz L, Galy A, Belshaw NS et al (1999) High-precision measurement of calcium isotopes in carbonates and related materials by multiple collector inductively coupled plasma mass spectrometry. J Anal At Spec 14:1835–1838

Halliday AN, Christensen JN, Lee DC et al (2000) Multiple-collector inductively coupled plasma mass spectrometry. In: Barshick CM, Duckworth DC, Smith DH (eds) Inorganic mass spectrometry—fundamentals and applications. M. Dekker, Inc, New York, Basel, pp 291–327

Hamelin B, Manhes G, Albarede F et al (1985) Precise lead isotopic measurements by the double spike technique: a reconsideration. Geochim Cosmochim Acta 49:173–182

Harouaka K, Eisenhauer A, Fantle MS (2014) Experimental investigation of Ca isotopic fractionation during abiotic gypsum precipitation. Geochim Cosmochim Acta 129:157–176

Harrison TM, Heizler MT, McKeegan KD et al (2010) In situ ^{40}K–^{40}Ca "double-plus" SIMS dating resolves Klokken feldspar 40 K–40Ar paradox. Earth Planet Sci Lett 299:426–433

Hart SR, Zindler A (1989) Isotope fractionation laws: a test using calcium. Int J Mass Spectrom Ion Process 89:287–301

Hensley TM (2006) Calcium isotopic variation in marine evaporites and carbonates: applications to late miocene mediterranean brine chemistry and late cenozoic calcium cycling in the oceans. PhD thesis, University of California

Heumann KG, Gindner F, Klöppel H (1977) Abhängigkeit des calcium-isotopieeffekts von der elektrolytkonzentration bei der Ionenaustausch-chromatographie. Angew Chem Ger Ed 89:753–754

Heumann KG, Kubassek E, Schwabenbauer W et al (1979) Analytical method for the K/Ca age determination of geological samples. Fresen Z Anal Chem 297:35–43

Heuser A, Eisenhauer A (2008) The calcium isotope composition ($\delta^{44/40}$Ca) of NIST SRM 915b and NIST SRM 1486. Geostand Geoanal Res 32:311–315

Heuser A, Eisenhauer A (2010) A pilot study on the use of natural calcium isotope (^{44}Ca/^{40}Ca) fractionation in urine as a proxy for the human body calcium balance. Bone 46:889–896

Heuser A, Eisenhauer A, Gussone N et al (2002) Measurement of calcium isotopes (δ^{44}Ca) using a multicollector TIMS technique. Int J Mass Spectrom 220:387–399

Heuser A, Eisenhauer A, Böhm F et al (2005) Calcium isotope ($\delta^{44/40}$Ca) variations of neogene planktonic foraminifera. Paleoceanography 20, doi:10.1029/2004PA001048

Heuser A, Tütken T, Gussone N et al (2011) Calcium isotopes in fossil bones and teeth—diagenetic versus biogenic origin. Geochim Cosmochim Acta 75:3419–3433

Hindshaw RS, Reynolds BC, Wiederhold JG et al (2011) Calcium isotopes in a proglacial weathering environment: damma glacier, Switzerland. Geochim Cosmochim Acta 75:106–118

Hindshaw RS, Reynolds BC, Wiederhold JG et al (2012) Calcium isotope fractionation in alpine plants. Biogeochem 112:373–388. doi:10.1007/s10533-012-9732-1

Hindshaw RS, Bourdon B, Pogge von Strandmann PA et al (2013) The stable calcium isotopic composition of rivers draining basaltic catchments in Iceland. Earth Planet Sci Lett 374:173–184

Hippler D, Schmitt A-D, Gussone N et al (2003) Calcium isotopic composition of various reference materials and seawater. Geostand Geoanal Res 27:13–19

Hippler D, Eisenhauer A, Nägler TF (2006) Tropical atlantic SST history inferred from Ca isotope thermometry over the last 140 ka. Geochim Cosmochim Acta 70:90–100

Hippler D, Kozdon R, Darling KF et al (2009) Calcium isotopic composition of high-latitude proxy carrier Neogloboquadrina pachyderma (sin.) Biogeosc 6:1–14

Hirata T, Tanoshima M, Suga A et al (2008) Isotopic analysis of calcium in blood plasma and bone from mouse samples by multiple collector-ICP-mass spectrometry. Anal Sci 24:1501–1507

Holmden C (2005) Measurement of δ^{44}Ca using a ^{43}Ca–^{42}Ca double-spike TIMS technique; in summary of investigations 2005, Volume 1, Saskatchewan geological survey, Sask. Industry Resources, Misc Rep 2005-1, CD-ROM, Paper A-4, 7 p

Holmden C (2009) Ca isotope study of Ordovician dolomite, limestone, and anhydrite in the Williston Basin: implications for subsurface dolomitization and local Ca cycling. Chem Geol 268:180–188

Holmden C, Bélanger N (2010) Ca isotope cycling in a forested ecosystem. Geochim Cosmochim Acta 74:995–1015

Holmden C et al (2012) Tightly coupled records of Ca and C isotope changes during the Hinuantian glaciations event in an epeiric sea setting. Geochim Cosmochim Acta 98:94–106

Horn I, von Blanckenburg F, Schoenberg R et al (2006). In situ iron isotope ratio determination using UV-Femtosecond laser ablation with application to hydrothermal ore formation processes. Geochimica Et Cosmochimica Acta 70:3677–3688. doi:10.1016/j.gca.2006.05.002

Huang S, Farkaš J, Jacobsen SB (2010) Calcium isotopic fractionation between clinopyroxene and orthopyroxene from; mantle peridotites. Earth Planet Sci Lett 292:337–344

Huang S, Farkas J, Jacobsen SB (2011) Stable calcium isotopic compositions of hawaiian shield lavas: evidence for recycling of ancient marine carbonates into the mantle. Geochim Cosmochim Acta 75(17):4987–4997

Immenhauser A, Nägler TF, Steuber T, Hippler D (2005) A critical assessment of mollusk 18O/16O, Mg/Ca, and 44Ca/40Ca ratios as proxies for Cretaceous seawater temperature seasonality. Palaeogeogr Palaeoclimatol Palaeoecol 215:221–237

Ireland TR (1990) Presolar isotopic and chemical signatures in hibonite-bearing refractory inclusions from the Murchison carbonaceous chondrite. Geochim Cosmochim Acta 54:3219–3237

Jacobson AD, Holmden C (2008) δ^{44}Ca evolution in a carbonate aquifer and its bearing on the equilibrium isotope fractionation factor for calcite. Earth Planet Sci Lett 270:349–353

John T, Gussone N, Podladchikov YY et al (2012) Volcanic arcs fed by rapid pulsed fluid flow through subducting slabs. Nat Geosci 5:489–492

Johnson CM, Beard BL (1999) Correction of instrumentally produced mass fractionation during isotopic analysis of Fe by thermal ionization mass spectrometry. Int J Mass Spectrom 193:87–99

Kasemann SA, Hawkresworth CJ, Prave AR et al (2005) Boron and calcium isotope composition in Neoproterozoic carbonate rocks from Namibia: evidence from extreme environmental change. Earth Planet Sci Lett 231:73–86

Kasemann SA, Schmidt DN, Pearson PN et al (2008) Biological and ecological insights into Ca isotopes in planktic foraminifers as a paleotemperature proxy. Earth Planet Sci Lett 271:292–302

Komiya T, Suga A, Ohno T et al (2008) Ca isotopic compositions of dolomite, phosphorite and the oldest animal embryo fossils from the Neoproterozoic in Weng'an, South China. Gondwana Res 14:209–218

Koornneef JM, Bouman C, Schwieters JB et al (2013) Use of 1012 Ohm current amplifiers in Sr and Nd isotope analyses by TIMS for application to sub-nanogram samples. J Anal At Spec 28:749–54. doi:10.1039/C3JA30326H

Koumenis IL, Vestal ML, Yergey AL et al (1995) Quantitation of metal isotope ratios by laser desorption time-of-flight mass spectrometry. Anal Chem 67:4557–4564

Kreissig K, Elliott T (2005) Ca isotope fingerprints of early crust-mantle evolution. Geochim Cosmochim Acta 69:165–176

Lehn GO, Jacobson AD (2015) Optimization of a ^{48}Ca–^{43}Ca double-spike MC-TIMS method for measuring Ca isotope ratios ($\delta^{44/40}$Ca and $\delta^{44/42}$Ca): limitations from filament reservoir mixing. J Anal At Spectrom 30:1571–1581

Lehn GO, Jacobson AD, Holmden C (2013) Precise analysis of Ca isotope ratios ($\delta^{44/40}$Ca) using an optimized ^{43}Ca–^{42}Ca double-spike MC-TIMS method. Int J Mass Spectrometry 351:69–75

Lemarchand D, Wasserburg GJ, Papanastassiou DA (2004) Rate-controlled calcium isotope fractionation in synthetic calcite. Geochim Cosmochim Acta 68 (22):4665–5678

Longerich HP, Fryer BJ, Strong DF (1987) Determination of lead isotope ratios by inductively coupled plasma-mass spectrometry (ICP-MS). Spectrochim Acta, Part B 42:39–48. doi:10.1016/0584-8547(87)80048-4

Ludwig KR (1997) Optimization of multicollector isotope-ratio measurement of strontium and neodymium. Chem Geol 135:325–334. doi:10.1016/S0009-2541(96)00120-9

Lundberg LL, Zinner E, Crozaz G (1994) Search for isotopic anomalies in oldhamite (CaS) from unequilibrated (E3) enstatite chondrites. Meteoritics 29:384–393. doi:10.1111/j.1945-5100.1994.tb00602.x

Magna T, Gussone N, Mezger K (2015) The calcium isotope systematics of mars. Earth Planet Sci Lett 430:86–94

Maréchal CN, Télouk P, Albarède F (1999) Precise analysis of copper and zinc isotopic compositions by plasma-source mass spectrometry. Chem Geol 156:251–273

Marriott CS, Henderson GM, Belshaw NS et al (2004) Temperature dependence of δ^7Li, δ^{44}Ca and Li/Ca during growth of calcium carbonate. Earth Planet Sci Lett 222:615–624

Marshall BD, DePaolo DJ (1982) Precise age determinations and petrogenetic studies using the K–Ca method. Geochim Cosmochim Acta 46:2537–2545

Marshall BD, DePaolo DJ (1989) Calcium isotopes in igneous rocks and the origin of granite. Geochim Cosmochim Acta 53:917–922

Marshall BD, Woodard HH, Krueger HW et al (1986) K–Ca–Ar systematics of authigenic sanidine from Wakau, Wisconsin, and the diffusivity of argon. Geology 14:936–938

Martin JE, Tacail T, Adnet S et al (2015) Calcium isotopes reveal the trophic position of extant and fossil elasmobranchs. Chem Geol 415:118–125. doi:10.1016/j.chemgeo.2015.09.011

Möller P, Papendorff H (1971) Fractionation of calcium isotopes in carbonate precipitates. Earth Planet Sci Lett 11:192–194

Morgan JLL, Gordon GW, Arrua RC et al (2011) High-precision measurement of variations in calcium isotope ratios in urine by multiple collector inductively coupled plasma mass spectrometry. Anal Chem 83:6956–6962. doi:10.1021/ac200361t

Morgan JLL, Skulan JL, Gordon GW et al (2012) Rapidly assessing changes in bone mineral balance using natural stable calcium isotopes. Proc Nat Acad Sci 109:9989–9994

Murphy KE, Long SE, Rearick MS et al (2002) The accurate determination of patassium and calcium using isotope dilution inductively coupled "cold" plasma mass spectrometry. J Anal At Spec 17:469–477

Nägler T, Villa I (2000) In pursuit of the ^{40}K branching ratios: K–Ca and ^{39}Ar–^{40}Ar dating of gem silicates. Chem Geol 169:5–16

Nägler TF, Eisenhauer A, Müller A et al (2000) The δ^{44}Ca-temperature calibration on fossil and cultured globigerinoides sacculifer. Geochem Geophys Geosyst 1: doi:10.1029/2000GC000091

Naumenko-Dèzes MO, Bouman C, Nägler ThF et al (2015) TIMS measurements of full range of natural Ca isotopes with internally consistent fractionation correction. Int J Mass Spectrom 387:60–68. doi:10.1016/j.ijms.2015.07.012

Nelson DR, McCulloch MT (1989) Petrogenic applications of the ^{40}K–^{40}Ca radiogenic decay scheme—a reconnaissance study. Chem Geol 79:275–293

Nelson F et al (1964) Ion exchange procedures. I. Cation exchange in concentration HCl and $HClO_4$ solutions. J Chromatogr 13:503–535

Nezat CA, Blum JD, Driscoll ChT (2010) Patterns of Ca/Sr and ^{87}Sr/^{86}Sr variation before and after a whole watershed $CaSiO_3$ addition at the hubbard brook experimental forest, USA. Geochim Cosmochim Acta 74:3129–3142

Nicolussi GK, Pellin MJ, Calaway WF et al (1997) Isotopic analysis of Ca from extraterrestrial micrometer-sized SiC by laser desorption and resonant ionization mass spectrometry. Anal Chem 69:1140–1146

Page BD, Bullen TD, Mitchell MJ (2008) Influences of calcium availability and tree species on Ca isotope fractionation in soil and vegetation. Biogeochemistry 88:1–13

Patterson KY, Veillon C, Hill AD et al (1999) Measurement of calcium stable isotope tracers using cool plasma ICP-MS. J Anal At Spec 14:1673–1677

Paytan A, Kastner M, Martin EE et al (1993) Marine barite as a monitor of seawater strontium isotope composition. Nature 366:445–449

Perakis SS, Maguire DA, Bullen TD et al (2006) Coupled nitrogen and calcium cycles in forests of the oregon coast range. Ecosystems 9:63–74

Platzner IT (2000) Modern isotope ratio mass spectrometry. Wiley, Chicester

Powell R, Woodhead J, Hergt J (1998) Uncertainties on lead isotope analyses; deconvolution in the double-spike method. Chem Geol 148(1–2):95–104

Rehkämper M, Schönbächler M, Stirling CH (2001) Multiple collector ICP-MS: introduction to instrumentation, measurement techniques and analytical capabilities. Geostand Newsl 25:23–40

Rehkämper M, Wombacher F, Aggarwal JK (2004) Chapter 31—stable isotope analysis by multiple collector ICP-MS. In: de Groot PA (eds) Handbook of stable isotope analytical techniques. Elsevier, Amsterdam, pp 692–725. http://www.sciencedirect.com/science/article/pii/B9780444511140500338

Reynard LM, Henderson GM, Hedges REM (2010) Calcium isotope ratios in animal and human bone. Geochim Cosmochim Acta 74:3735–3750

Reynard LM, Henderson GM, Hedges REM (2011) Calcium isotopes in archaeological bones and their relationships to dairy consumption. J Archaeol Sci 38:657–664

Rollion-Bard C, Vigier N, Spezzaferri S (2007) In situ measurements of calcium isotopes by ion microprobe in carbonates and application to foraminifera. Chem Geol 244:679–690

Romaniello SJ, Field M, Smith HB et al (2015) Fully automated chromatographic purification of Sr and Ca for isotopic analysis. J Anal At Spectrom 30:1906–1912. doi:10.1039/C5JA00205B

Roy S, Gillen G, Conway WS et al (1995) Use of secondary ion mass spectrometry to image 44 calcium uptake in the cell walls of apple fruit. Protoplasma 189:163–172

Russell WA, Papanastassiou DA (1978) Calcium isotope fractionation in ion-exchange chromatography. Anal Chem 50:1151–1154

Russell WA, Papanastassiou DA, Tombrello TA (1978) Ca isotope fractionation on the earth and other solar system materials. Geochim Cosmochim Acta 42:1075–1090

Ryu JS, Jacobson AD, Holmden C et al (2011) The major ion, $\delta^{44/40}$Ca, $\delta^{44/42}$Ca, and $\delta^{26/24}$Mg geochemistry of granite weathering at pH = 1 and T = 25 °C: power-law processes and the relative reactivity of minerals. Geochim Cosmochim Acta 75:6004–6026

Santamaria-Fernandez R, Wolff J-C (2010) Application of laser ablation multicollector inductively couples plasma mass spectrometry for the measurement of calcium and lead isotope ratios in packaging for discriminatory purposes. Rapid Comm Mass Sp 24:1993–1999

Schiller M et al (2012) Calcium isotope measurement by combined HR-MC-ICPMS and TIMS. J Anal Atom Spectrom 27:38–49

Schmitt AD, Stille P (2005) The source of calcium in wet atmospheric deposits: Ca–Sr isotope evidence. Geochim Cosmochim Acta 69:3463–3468

Schmitt AD, Bracke G, Stille P et al (2001) The calcium isotope composition of modern seawater determined by thermal ionisation mass spectrometry. Geostand Newslett 25:267–275

Schmitt AD, Chabaux F, Stille P (2003a) The calcium riverine and hydrothermal isotopic fluxes and the oceanic calcium mass balance. Earth Planet Sci Lett 213:503–518

Schmitt AD, Stille P, Vennemann T (2003b) Variations of the ^{44}Ca/^{40}Ca ratio in seawater during the past 24 million years: evidence from δ^{44}Ca and δ^{18}O values of miocene phosphates. Geochim Cosmochim Acta 67:2607–2614

Schmitt AD, Gangloff S, Cobert F et al (2009) High performance automated ion chromatography separation for Ca isotope measurements in geological and biological samples. J Anal At Spec 24:1089–1097

Schmitt AD, Cobert F, Bourgeade P et al (2013) Calcium isotope fractionation during plant growth under a limiting nutrient supply. Geochim Cosmochim Acta 110:70–83

Shaheen ME, Gagnon JE, Fryer BJ (2012) Femtosecond (fs) lasers coupled with modern ICP-MS instruments provide new and improved potential for in situ elemental and isotopic analyses in the geosciences. Chem Geol 330–331:260–273. doi:10.1016/j.chemgeo.2012.09.016

Shiel AE, Barling J, Orians KJ et al (2009) Matrix effects on the multi-collector inductively coupled plasma mass spectrometric analysis of high-precision cadmium and zinc isotope ratios. Anal Chim Acta 633:29–37. doi:10.1016/j.aca.2008.11.026

Shih C-Y, Nyquist LE, Bogard DD et al (1994) K–Ca and Rb–Sr dating of two lunar granites: relative chronometer resetting. Geochim Cosmochim Acta 58:3101–3116

Siebert C, Nägler TF, Kramers JD (2001) Determination of molybdenum isotope fractionation by double-spike multicollector inductively coupled plasma mass spectrometry. Geochem Geophys Geosyst 2:2000GC000124

Silva-Tamayo JC, Nägler TF, Villa IM et al (2010) Global Ca isotope variations in c. 0.7 Ga old post-glacial carbonate successions. Terra Nova 22:188–194

Sime NG, De La Rocha CL, Galy A (2005) Negligible temperature dependence of calcium isotope fractionation in 12 species of planktonic foraminifera. Earth Planet Sci Lett 232:51–66

Sime NG, De La Rocha CL, Tipper ET et al (2007) Interpreting the Ca isotope record of marine biogenic carbonates. Geochim Cosmochim Acta 71:3979–3989

Simon JI, DePaolo DJ (2010) Stable calcium isotopic composition of meteorites and rocky planets. Earth Planet Sci Lett 289:457–466

Simon JI, DePaolo DJ, Moynier F (2009) Calcium isotope composition of meteorites, earth and mars. Astrophys J 702:707–715

Skulan JL, DePaolo DJ (1999) Calcium isotope fractionation between soft and mineralized tissues as a monitor of calcium use in vertebrates. Proc Nat Acad Sci 96:13709–13713

Skulan JL, DePaolo DJ, Owens TL (1997) Biological control of calcium isotopic abundances in the global calcium cycle. Geochim Cosmochim Acta 61:2505–2510

Smith DL (1983) Determination of stable isotopes of calcium in biological fluids by fast atom bombardment mass spectrometry. Anal Chem 55:2391–2393

Soudry D, Glenn C, Nathan Y et al (2006) Evolution of tethyan phosphogenesis along the northern edges of the Arabian-African shield during the Cretaceous-Eocene as deduced from temporal variations of Ca and Nd isotopes and rates of P accumulation. Earth Sci Rev 78:27–57

Steiger RH, Jäger E (1977) Subcommission on geochronology: convention on the use of decay constants in geo- and cosmochronology. Earth Planet Sci Lett 36:359–362

Steinmann M, Stille P (1997) Rare earth element behavior and Pb, Sr, Nd isotope systematics in a heavy metal contaminated soil. Appl Geochem 12:607–624

Steinmann M, Stille P (2006) Rare earth element transport and fractionation in small streams of a mixed basaltic-granitic catchment basin (Massif Central, France). J Geoch Expl 88:336–340

Steuber T, Buhl D (2006) Calcium isotope fractionation in selected modern and ancient marine carbonates. Geochim Cosmochim Acta 70:5507–5521

Stille P, Clauer N (1994) The process of glauconization: chemical and isotopic evidence. Contrib Mineral Petrol 117:253–562

Stürup S (2004) The use of ICMPS for stable isotope tracer studies in humans: a review. Anal Bioanal Chem 378:273–282

Stürup S, Bendahl L, Gammelgaard B (2006) Optimisation of dynamic reaction cell (DRC)-ICP-MS for the determination of 42Ca/43Ca and 44Ca/43Ca isotope ratios in human urine. J Anal At Spec 21:297–304

Tacail T, Albalat E, Télouk P et al (2014) A simplified protocol for measurement of Ca isotopes in biological samples. J Anal At Spectrom 29:529–535

Tacail T, Télouk P, Balter V (2016) Precise analysis of calcium stable isotope variations in biological apatites using laser ablation MC-ICPMS. J At Anal Spectrom 31:152–162

Takano B, Watanuki K (1972) Strontium and calcium coprecipitation with lead-bearing barite from hot spring water. Geochem J 6:1

Taylor SR, McLennan SM (1985) The continental crust. Its evolution and composition. Blackwell Science, Oxford

Teichert BMA, Gussone N, Eisenhauer A et al (2005) Clathrites: archives of near-seafloor pore-fluid evolution ($\delta^{44/40}$Ca, δ^{13}C, δ^{18}O) in gas hydrate environments. Geology 33:213–216

Teichert BMA, Gussone N, Torres ME (2009) Controls on calcium isotope fractionation in sedimentary pore-waters. Earth Planet Sci Lett 279:373–382

Tera F et al (1970) Comparative study of Li, Na, K, Rb, Cs, Ca, Sr and Ba abundances in achondrites and in Apollo II lunar samples. Proceedings of the Apollo II Lunar Conferences 2:1637–1657

Tipper ET, Galy A, Bickle MJ (2006) Riverine evidence for a fractionated reservoir of Ca and Mg on the continents: implications for the oceanic Ca cycle. Earth Planet Sci Lett 247:267–279

Tipper ET, Galy A, Bickle MJ (2008a) Calcium and magnesium isotope systematics in rivers draining the himalaya-tibetan-plateau region: lithological or fractionation control. Geochim Cosmochim Acta 72:1057–1075

Tipper ET, Louvat P, Capmas F et al (2008b) Accuracy of stable Mg and Ca isotope data obtained by MC-ICP-MS using the standard addition method. Chem Geol 257:65–75

Tipper ET, Gailiardet J, Galy A et al (2010) Calcium isotope ratios in the world's largest rivers: a constraint on the maximum imbalance of oceanic calcium fluxes. Global Biogeochem. Cy. 24, doi:10.1029/2009GB003574

Upadhyay D, Scherer EE, Mezger K (2008) Fractionation and mixing of Nd isotopes during thermal ionization mass spectrometry: implications for high precision Nd-142/Nd-144 analyses. J Anal At Spec 23:561–568. doi:10.1039/b715585a

Valdes MC, Moreira M, Foriel J et al (2014) The nature of earth's building blocks as revealed by calcium isotopes. Earth Planet Sci Lett 394:135–145

Vanhaecke F, Balcaen L, Malinovsky D (2009) Use of single-collector and multi-collector ICP-mass spectrometry for isotopic analysis. J Anal At Spec 24:863–886

von Allmen K, Böttcher M, Samankassou E et al (2010a) Barium isotope fractionation in the global barium cycle: first evidence from barium minerals and precipitation experiments. Chem Geol 277:70–77

von Allmen K, Nägler TF, Pettke et al (2010b) Stable isotope profiles (Ca, O, C) through modern brachiopod shells of T. septentrionalis and G. vitreus: implications for calcium isotope paleo-ocean chemistry. Chem Geol 269:210–219

Walder AJ, Freedman PA (1992) Communication. Isotopic ratio measurement using a double focusing magnetic sector mass analyser with an inductively coupled plasma as an ion source. J Anal At Spec 7:571–575. doi:10.1039/JA9920700571

Wang S, Yan W, Magalhães HV et al (2012) Calcium isotope fractionation and its controlling factors over authigenix carbonates in the cold seeps of the northern South China Sea. Chinese Sci Bull 57:1325–1332

Wang S, Yan W, Magalhães HV et al (2013) Factors influencing methane-derived authigenic carbonate formation at cold seep from southwestern Dongsha area in the northern South China. Environ Earth Sci. doi:10.1007/s12665-013-2611-9

Weber D, Zinner E, Bischoff A (1995) Trace element abundances and magnesium, calcium, and titanium isotopic compositions of grossite-containing inclusions from the carbonaceous chondrite acfer 182. Geochim Cosmochim Acta 59:803–823

Weyer S, Schwieters JB (2003) High precision Fe isotope measurements with high mass resolution MC-ICPMS. Int J Mass Spectrom 226:355–368

Wiegand BA, Schwendenmann L (2013) Sources in small tropical catchments (La Selva, Costa Rica). A comparison of Sr and Ca isotopes. J Hydrol 488:110–117

Wiegand BA, Chadwick OA, Vitousek PM et al (2005) Ca cycling and isotopic fluxes in forested ecosystems in Hawaii. Geophys Res Lett 32:L11404

Wieser ME, Schwieters JB (2005) The development of multiple collector mass spectrometry for isotope ratio measurements. Int J Mass Spectrom 242:97–115

Wieser ME, Buhl D, Bouman C et al (2004) High precision calcium isotope ratio measurements unsing a magnetic sector multiple collector inductively coupled plasma mass spectrometer. J Anal At Spec 19:844–851

Wombacher F, Rehkämper M (2003) Investigation of the mass discrimination of multiple collector ICP-MS using neodymium isotopes and the generalised power law. J Anal At Spectrom 18:1371–1375

Wombacher F, Eisenhauer A, Heuser A et al (2009) Separation of Mg, Ca and Fe from geological reference materials for stable isotope ratio analyses by MC-ICP-MS and double-spike TIMS. J Anal At Spec 24:627–636

Zinner EK, Fahey AJ, Goswami JN et al (986) Large ^{48}Ca anomalies are associated with ^{50}Ti anomalies in Murchison and Murray hibonites. Ap J 311:L103–L107

Yang L (2009) Accurate and precise determination of isotopic ratios by MC-ICP-MS: a review. Mass Spectrom Rev 28:990–1011. doi:10.1002/mas.20251

Young ED, Galy A, Nagahara H (2002) Kinetic and equilibrium mass-dependent isotope fractionation laws in nature and their geochemical and cosmochemical significance. Geochim Cosmochim Acta 66:1098–1104

Zhu P, Macdougall JD (1998) Calcium isotopes in the marine environment and the oceanic calcium cycle. Geochim Cosmochim Acta 62:1691–1698

Calcium Isotope Fractionation During Mineral Precipitation from Aqueous Solution

Nikolaus Gussone and Martin Dietzel

Abstract

Calcium isotopes are fractionated during inorganic precipitation from aqueous solutions according to their relative mass differences. The magnitude and sign of isotope fractionation depend on the composition and structure of the solid and the physicochemical conditions of the aqueous environment. The first part of this chapter focuses on Ca isotope fractionation during precipitation experiments at well-defined conditions, where results for Mg, Sr and Ba are included for the discussion of fractionation mechanisms. Different conceptual models for Ca isotope fractionation between carbonate minerals and aqueous solutions are discussed in the second part. The following parts of this chapter illustrate Ca isotope variability found in inorganic minerals from natural environments and the Ca isotope fractionation related to ion diffusion, exchange and adsorption in aqueous systems.

Keywords

Inorganic precipitation · Experiments · Isotope fractionation · Diffusion · Ion exchange

1 Inorganic Precipitation Experiments

Knowing the thermodynamic and kinetic induced isotope fractionation effects is the essential basis for the interpretation of isotope ratios in natural rocks and (bio)minerals in order to characterise formation conditions of rocks and minerals, geochemical cycles and paleo-environmental changes. Series of precipitation experiments from aqueous solutions were carried out to elucidate the influences of different parameters,

N. Gussone (✉)
Institut für Mineralogie, Universität Münster, Münster, Germany
e-mail: nguss_01@uni-muenster.de

M. Dietzel
Institut für Angewandte Geowissenschaften, Technische Universität Graz, Graz, Austria
e-mail: martin.dietzel@tugraz.at

© Springer-Verlag Berlin Heidelberg 2016
N. Gussone et al., *Calcium Stable Isotope Geochemistry*,
Advances in Isotope Geochemistry, DOI 10.1007/978-3-540-68953-9_3

including temperature, precipitation rate, super saturation, stoichiometry and coordination environment on the mass-dependent isotope fractionation of alkaline earth metals (except Be and Ra) during mineral growth. Most experiments dealing with stable isotope fractionation of alkaline earth elements during precipitation from aqueous solutions into different minerals reveal a tendency of light isotope enrichment in the solid product phases. However, recent studies on hydrous Ca and Mg minerals show enrichments of light or heavy isotopes in the precipitated solid phase, depending on the cation coordination. In general, the degree of isotope fractionation depends on several factors, e.g. the respective incorporated element, the mineral phase, the chemical composition of the fluid (e.g. saturation state, stoichiometry, cation coordination) and the physicochemical conditions during precipitation. In the following sections the characteristic Ca isotope fractionation patterns in different minerals (mainly carbonates and sulfates) are discussed. Available data on earth alkaline stable isotope (EASI) fractionation, besides Ca, are included for comparison.

1.1 Carbonates

EASI fractionation between carbonate minerals and the fluid they precipitated from depends on parameters as temperature, growth rates, fluid chemistry and mineralogical structure of the precipitated mineral. The effects of the different parameters on mass dependent isotope fractionation during precipitation of carbonate minerals have been studied using various experimental approaches. Experimental results reveal characteristic, but also partly discrepant fractionation patterns, for several Ca carbonate phases (calcite, aragonite, vaterite and ikaite), Mg carbonate (magnesite) and Ba carbonate (witherite).

1.1.1 Calcite

The response of Ca isotope fractionation on varying conditions during calcite growth is approached by several studies using different experimental setups. This section provides an overview of the experimental results, while a discussion of isotope fractionation models and ways to reconcile the partly discrepant observations are given in Sect. 2 of this chapter.

The temperature dependence of Ca isotope fractionation during calcite precipitation is determined in pH-stat experiments at constant growth rates by Marriott et al. (2004). They report a fractionation $\left(\Delta^{44/40}\mathrm{Ca}_{Cc-Ca^{2+}}\right)$ of about -0.6 to -1.0 ‰ and a temperature sensitivity of about $+0.015$ ‰/°C, with decreasing fractionation for increasing temperature, similar to observations on synthetic aragonite (Fig. 1). A temperature dependence of this magnitude in carbonate minerals is interpreted by Lemarchand et al. (2004) as growth rate effect, caused by the temperature dependent speciation of dissolved carbonate and subsequent changes in calcite saturation. This hypothesis is based on experiments, with varying CO_3^{2-} concentrations,

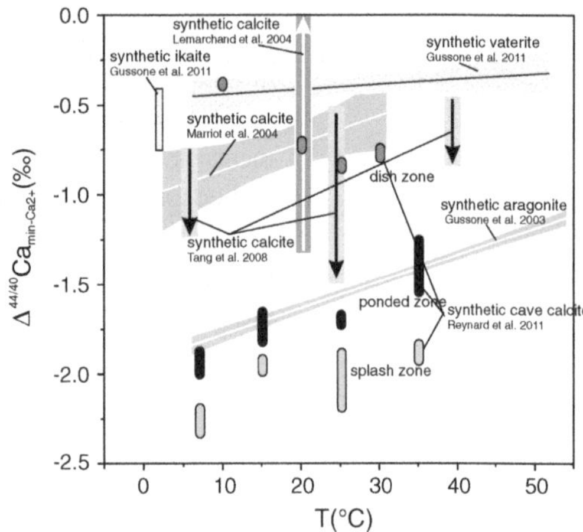

Fig. 1 Temperature dependent Ca isotope fractionation ($\Delta^{44/40}\mathrm{Ca}_{min-Ca^{2+}}$) of different $CaCO_3$ minerals. *Arrows* indicate the effect of increasing precipitation rates on Ca isotope fractionation at a given temperature. Experimentally determined apparent temperature dependencies for aragonite and calcite might be biased by changes in precipitation rate. If the apparent Ca isotope fractionation between the $CaCO_3$ polymorphs is a real mineralogical feature related to Ca coordination or a result of typical precipitation rates, related to solubility and conditions during the precipitation experiments, is not yet finally resolved. Cave-simulating calcite precipitates reveal high degrees of fractionation similar to those observed for aragonite

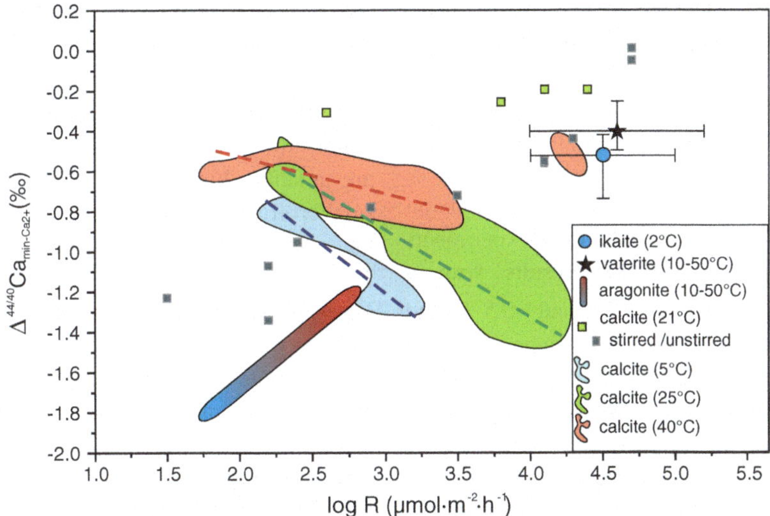

Fig. 2 Calcium isotope fractionation of $CaCO_3$ minerals ($\Delta^{44/40}Ca_{min-Ca^{2+}}$) as a function of precipitation rate. Depending on the experimental setup, different rate dependencies are observed for synthetic calcium carbonate phases. Calcite formed at 5, 25 and 40 °C (Tang et al. 2008a), at 21 °C (Lemarchand et al. 2004), aragonite (Gussone et al. 2003), vaterite and ikaite (Gussone et al. 2011)

leading to variable precipitation rates at a constant temperature of 21 °C. Their results show $\Delta^{44/40}Ca_{Cc-Ca^{2+}}$ between −1.5 and 0 ‰, and decreasing fractionation with increasing precipitation rates in experiments with unstirred solution and values of −0.2 to −0.3 ‰ for stirred solutions (Fig. 2). These data comprise results from experiments with Ca concentrations of 15 and 150 mM, showing no difference between both data sets. An opposite rate dependence is demonstrated by Tang et al. (2008a), who studied the rate dependence at 5, 25 and 40 °C. Their results show $\Delta^{44/40}Ca_{Cc-Ca^{2+}}$ between −0.5 and −1.4 ‰ and increasing fractionation with increasing precipitation rates, with a larger sensitivity at low temperatures and a minor rate dependence at higher temperatures.

In a subsequent study (Tang et al. 2012) it is shown that ionic strength has only a little influence on Ca isotope fractionation during calcite precipitation for ionic strength ranging between 35 and 830 mM (Fig. 3). Both studies demonstrate a strong negative correlation between Sr/Ca partitioning and Ca isotope fractionation. It is suggested that temperature and growth rate are

the main controlling factors and that ionic strength contributes to about 11 % of the apparent isotope fractionation (Tang et al. 2012).

A further set of experiments simulates Ca isotope fractionation during formation of cave deposits at temperatures between 7 and 40 °C

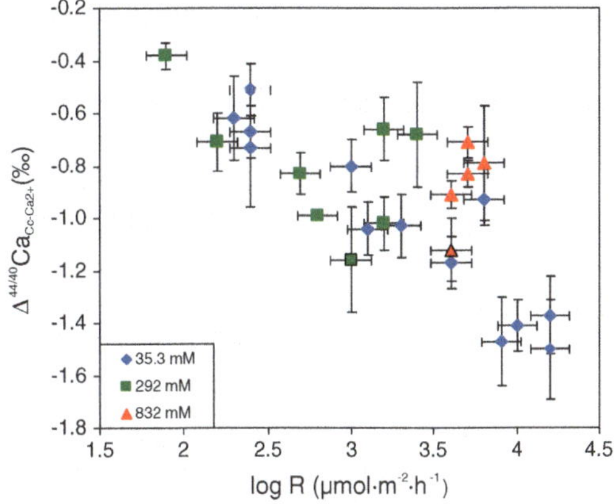

Fig. 3 Calcium isotope fractionation between calcite and dissolved Ca ($\Delta^{44/40}Ca_{Cc-Ca^{2+}}$) as a function of growth rate, related to the ionic strength of the fluid. Differences in ionic strength between 35 and 830 mM have only a minor effect on Ca isotope fractionation of calcite. Modified after Tang et al. (2012)

(Reynard et al. 2011). Overall, in the experiments using seed crystals, Ca isotope fractionation ranges from about -1.2 to -2.0 ‰ in the ponded zone and from -2.0 to -2.3 ‰ in the splash zone. The apparent temperature sensitivities of about 0.010 and 0.016 ‰/°C are similar to those of the batch experiments of calcite of Marriott et al. (2004). The un-seeded dish experiments reveal significantly different results with a $\Delta^{44/40}Ca_{Cc-Ca^{2+}}$ between -0.35 and -0.5 ‰ and an inverse correlation between temperature and isotope fractionation of -0.016 ‰/°C (Fig. 1).

Fractionation of Ca isotopes during calcite growth during experiments on larger scales (10 m, 100 days) is described by Druhan et al. (2013). This study evaluates, if mechanisms described in controlled lab experiments also apply to field scale observations in groundwater systems with active calcite precipitation. In the experiments, $CaCO_3$ precipitation is initiated by injection of acetate into wells. The introduced organic C is transferred by microbial activity to DIC (dissolved inorganic carbon) via sulphate reduction and methanogenesis. During the sampling period of 100 days, Ca concentration decreases from about 6 to >1 mM and $\delta^{44/40}Ca$ values show an increase of about 1.0 to 2.5 ‰ in the groundwater. The results reveal relatively large differences for $\Delta^{44/40}Ca_{Cc-Ca^{2+}}$ between adjacent wells on a few meter scale, indicating a large effect of parameters such as groundwater flow on the precipitation conditions during the experiments. The relation between $[Ca^{2+}]$ decrease and $\delta^{44/40}Ca$ increase does not follow a Rayleigh fractionation trend, which indicates that isotopic equilibrium is not reached during the time of sample collection. Additionally, Druhan et al. (2013) suggest that an isotopic re-equilibration is possible after the chemical equilibrium is reached and that consequently, Ca isotope signals based on kinetic fractionation effects have a little preservation potential in the rock record (see also Mg isotope results on magnesite; Pearce et al. 2012). The evolution of Ca concentration and isotopic composition in the aquifer can be described by a coupled ion by ion and reactive transport model (Nielsen and DePaolo 2013; Druhan et al. 2013). This modelling suggests values for $\Delta^{44/40}Ca_{Cc-Ca^{2+}}$ between -0.6 and -1.0 ‰, with higher degrees of fractionation during earlier stages of precipitation (at higher degrees of super saturation and presumably higher precipitation rates) and smaller fractionation during later sampling (at lower super saturations and slower precipitation). These observations are consistent with the experimental results of Tang et al. (2008b, 2012) and suggest that observations on Ca isotope fractionation obtained in controlled lab experiments can be transferred to natural processes.

Similar to Ca, light Mg isotopes are preferentially incorporated during precipitation of calcite. The isotope fractionation between calcite and dissolved Mg $\left(\Delta^{26/24}Mg_{Cc-Mg^{2+}}\right)$ varies between -1.5 and -3.2 ‰ for calcite grown in controlled precipitation experiments (Fig. 4). Li et al. (2012) present evidence for a temperature dependent Mg isotope fractionation in calcite and high-Mg calcite, precipitated during seeded growth experiments. Between temperatures of 4 and 45 °C, their results show an increase in $\delta^{26/24}Mg$ of 0.011 ‰/°C with increasing temperatures. Experiments of Immenhauser et al. (2010) and Mavromatis et al. (2013) reveal a dependence of $\Delta^{26/24}Mg_{Cc-Mg^{2+}}$ on growth rate (Fig. 4b), with decreasing Mg isotope fractionation at increasing precipitation rate, similar to the observations on Ca isotope fractionation of Lemarchand et al. (2004). Such a clear relation is however not apparent from the data set of Li et al. (2012). The lacking temperature dependence of the experimental data of Saulnier et al. (2012) in the temperature range from 16 to 26 °C (Fig. 4a) might also be related to changes in precipitation rate. Magnesium isotope fractionation seems to be independent from changes in P_{CO_2} in the range between 0.038 and 3 % (Li et al. 2012) and pH between 7.4 and 8.5 (Saulnier et al. 2012). The Mg isotope fractionation is also not related to the absolute Mg concentration of the precipitated calcite (0.8 to 15 mol% Mg), but there is a significant correlation between the partition coefficient of Mg (D_{Mg}) for calcite to

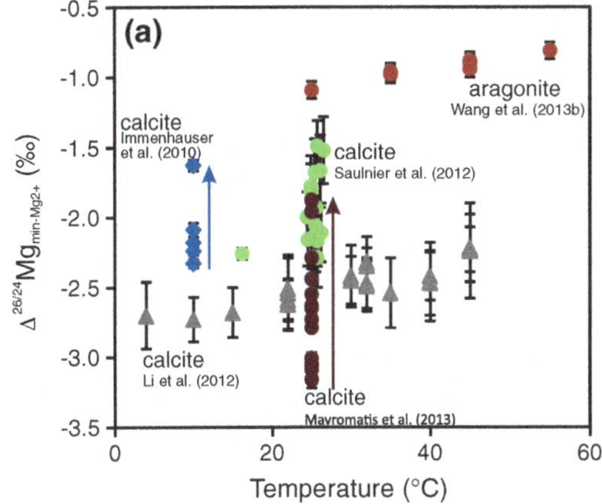

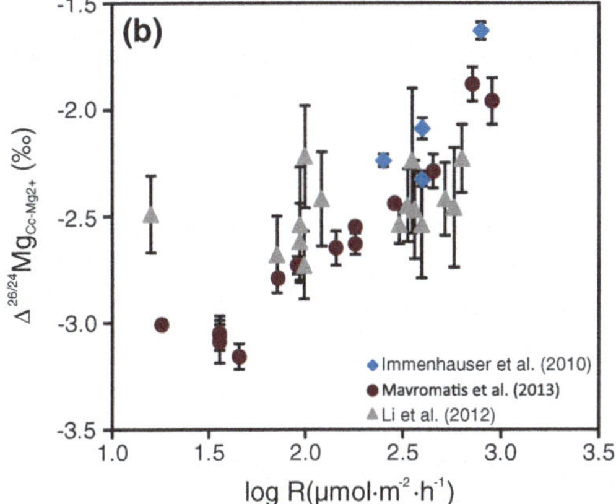

Fig. 4 Magnesium isotope fractionation ($\Delta^{26/24}Mg_{min-Mg^{2+}}$) in synthetic Ca carbonate minerals. **a** Temperature dependent $\Delta^{26/24}Mg_{min-Mg^{2+}}$ in calcite and aragonite. *Arrows* indicate increasing precipitation rates. **b** Effect of precipitation rate on the $\Delta^{26/24}Mg_{Cc-Mg^{2+}}$ value in respect to synthetic calcite

$\Delta^{26/24}Mg$ (Mavromatis et al. 2013), as both directly depend on growth rate.

Magnesium isotope data of Mavromatis et al. (2013) lie in a three isotope plot either on the kinetic mass fractionation line or between the kinetic and equilibrium fractionation line. Calcite grown at high growth rates plot on the kinetic fractionation line, and with decreasing growth rates samples approach the isotope fractionation at equilibrium. By extrapolation of this trend, Mavromatis et al. (2013) suggested that isotope fractionation at equilibrium is -3.5 ± 0.2 ‰ at 25 °C, which is in good agreement to modelling

results of Rustad et al. (2010). This observed Mg isotope fractionation pattern is well explainable by the surface entrapment model (see explanation in Sect. 2), assuming an equilibrium isotope fractionation of -3.5 ‰ and increased incorporation of isotopic un-equilibrated Mg at high precipitation rates. The different isotope fractionation behaviour of Mg relative to Ca and Sr is suggested to be related to the high hydration energy of Mg, compared to Ca and Sr.

An additional isotope fractionation mechanism is suggested by Saulnier et al. (2012), who propose that besides entrapment of Mg in Mg rich surface layers, the incorporation of un-equilibrated Mg from e.g. fluid inclusions can lead to a reduced apparent Mg isotope fractionation in calcite relative to the isotope fractionation at equilibrium.

Strontium isotope fractionation of inorganic calcite $\left(\Delta^{88/86}Sr_{Cc-Sr^{2+}}\right)$ ranges from -0.1 to -0.3 ‰ at 25 °C (Böhm et al. 2012), revealing an increase of Sr isotope fractionation with increasing precipitation rate (Fig. 5a). $\Delta^{88/86}Sr_{Cc-Sr^{2+}}$ shows also good correlation with Sr/Ca partitioning ($D_{Sr} = (Sr/Ca)_{solid}/(Sr/Ca)_{fluid}$) in synthetic calcite (Fig. 5b, c), measured on aliquots of the same samples (Tang et al. 2008). The relationship of Ca and Sr isotope fractionation ($\Delta^{88/86}Sr_{Cc-Sr^{2+}} = 0.19 \cdot \Delta^{44/40}Ca_{Cc-Ca^{2+}}$) is in concert with a kinetic EASI fractionation with a preferential desolvation of the light Ca and Sr isotopes (Böhm et al. 2012).

1.1.2   Aragonite

Calcium isotopes are depleted in ^{44}Ca relative to ^{40}Ca in the precipitated aragonite by about -1.2 to -1.8 ‰ relative to the source fluid (Gussone et al. 2003), at temperatures between 10 and 50 °C (Fig. 1). This degree of fractionation is larger than that for calcite grown at comparable experimental conditions (Marriott et al. 2004; Lemarchand et al. 2004), but similar to calcites in cave calcite-simulating experiments (Reynard et al. 2011) and calcite precipitated at high growth rates (Tang et al. 2008a). The apparent temperature dependence of aragonite is about 0.015 ‰/°C (Gussone et al. 2003), similar to that

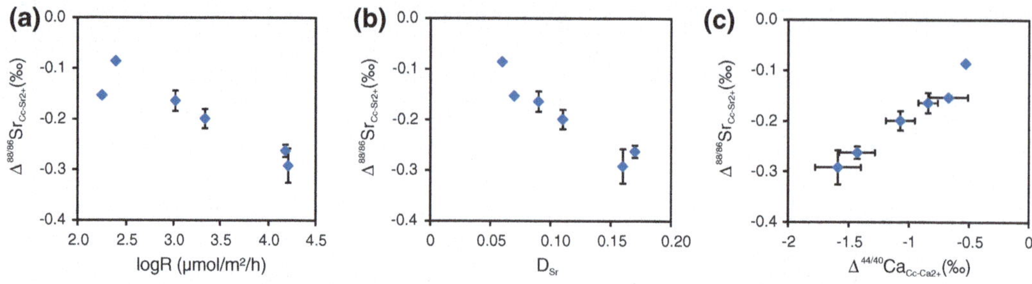

Fig. 5 Mass dependent Sr isotope fractionation in synthetic calcite grown at 25 °C. $\Delta^{88/86}Sr_{Cc-Sr^{2+}}$ is strongly correlated to growth rate (**a**), D_{Sr} (**b**) and Ca isotope fractionation (**c**). Data from Böhm et al. (2012)

of calcite (Marriott et al. 2004). As aragonite precipitation rates changed with temperature in these experiments, it is not fully resolved if the observed dependence is purely temperature controlled or also dependent on the precipitation rate. Trace elements (Sr/Ca and Ba/Ca) analysed on aliquots of the studied aragonite show that Nernst law is fulfilled and that chemical equilibrium is reached (Dietzel et al. 2004). However, Gussone et al. (2005) indicated that there was probably no isotopic equilibration for Ca isotopes in these aragonites. This reasoning is consistent with observations on Ca isotopes in calcite (Druhan et al. 2013) and Mg isotope incorporation in magnesite (Pearce et al. 2012), showing that isotopic re-equilibration continues after reaching chemical equilibrium. Thus, chemical and isotopic equilibration can take place on different time scales.

Mass dependent isotope fractionation during growth of synthetic aragonite is studied in addition to Ca also for the alkaline earth elements Mg and Sr. Aragonite, precipitated under controlled laboratory conditions (Wang et al. 2013b) incorporates preferentially light Mg isotopes, with decreasing fractionation at increasing temperatures. $\Delta^{26/24}Mg_{arag-Mg^{2+}}$ range from -1.1 ‰ (at 25 °C) to -0.8 ‰ (at 55 °C). The temperature dependence is about 0.01 ‰/°C (Fig. 4). In contrast to Ca isotopes, aragonite is less fractionated than calcite in respect to Mg isotopes. Interestingly, the observed $\Delta^{26/24}Mg_{arag-Mg^{2+}}$ values of aragonite are similar to the model calculations of Rustad et al. (2010) for equilibrium

isotope fractionation between the Mg-aqua complex in solution and solid magnesite ($\Delta^{26/24}Mg_{mag-Mg^{2+}}$). Anyhow, the incorporation of Mg in aragonite is still hotly debated as the Mg^{2+} ion does not fit in the cation position of the aragonite structure. The Mg content of aragonite might be rather controlled by crystallographic defects in the aragonite crystal, Mg adsorption and/or fluid inclusions instead of isomorphic substitution of Ca.

The temperature dependent fractionation of the $^{88}Sr/^{86}Sr$ ratio of synthetic aragonite is 0.0054 $\pm$ 0.0005 ‰/°C (Fig. 6). As the degree of fractionation between fluid and solid is not reported, it is not yet resolved if aragonite is more or less fractionated compared to calcite in respect to $\delta^{88/86}Sr$ (Fietzke and Eisenhauer 2006).

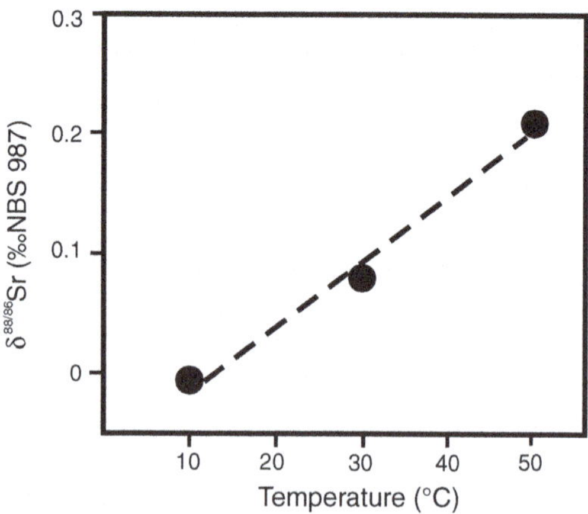

Fig. 6 Temperature dependent incorporation of Sr isotopes in aragonite. The temperature sensitivity of $\delta^{88/86}Sr$ is 0.0054 $\pm$ 0.0005 ‰/°C (Fietzke and Eisenhauer 2006)

1.1.3 Vaterite

Calcium isotope fractionation between synthetic vaterite and fluid ($\Delta^{44/40}Ca_{vat-Ca^{2+}}$) based on different experimental setups ranges from -0.25 to -0.6 ‰. Experiments conducted at high degrees of super saturation resulting in high growth rates, reveal $\Delta^{44/40}Ca_{vat-Ca^{2+}}$ between -0.25 and -0.5 ‰, with no significant temperature dependence between 10 and 50 °C (Fig. 1; Gussone et al. 2011). Precipitation in these experiments takes place at high growth rates, taking advantage of the Ostwald rule of stages to favour precipitation of a more soluble polymorph instead of the more stable calcite or aragonite without addition of inhibitors. These results are consistent with $\Delta^{44/40}Ca_{vat-Ca^{2+}}$ ranging from -0.3 to -0.6 ‰ (Niedermayr et al. 2010), obtained for vaterite grown in CO_2 diffusion experiments, similar to the setup of Tang et al. (2008a). Their results also indicate a dependence of Ca isotope fractionation on the precipitation rate, but significantly less pronounced compared to that observed for calcite (Tang et al. 2008a, 2012). Overall, Ca isotope fractionation of vaterite resembles that of calcite precipitated at high rates in the experiments of Lemarchand et al. (2004). This might suggest that vaterite, grown from highly supersaturated fluids, fractionates according to the model prediction of Lemarchand et al. (2004), i.e. increasing amounts of un-equilibrated Ca from the fluid are incorporated into the crystal by co-precipitation at increasing precipitation rates. As a further explanation, Nielsen et al. (2012) propose that a $\Delta^{44/40}Ca_{min-Ca}$-growth rate dependence as described by Lemarchand et al. (2004) might be related to the precipitation of an amorphous calcium carbonate (ACC) precursor phase, occurring in fluids which are supersaturated with respect to ACC.

1.1.4 Ikaite

Ikaite ($CaCO_3·6H_2O$) is a hydrous Ca carbonate mineral, which is stable at temperatures below about 4 °C or stabilized under special chemical conditions, e.g. in phosphate rich fluids.

$\Delta^{44/40}Ca_{ikaite-Ca^{2+}}$ of synthetic ikaite, precipitated from fluids with high $CaCO_3$ saturation stages and therewith at high precipitation rates, varies between -0.40 and -0.75 ‰ (Gussone et al. 2011).

Experimental setups at high saturation stages are used to favour ikaite precipitation over the less soluble phases calcite and aragonite. The observed Ca isotope fractionation is smaller than that of calcite and aragonite at comparable low temperatures, but similar to that of vaterite and the calcite precipitated by Lemarchand et al. (2004) at similar high precipitation rates (Figs. 1 and 2). This range of $\Delta^{44/40}Ca_{ikaite-Ca^{2+}}$ is in general agreement to Ca isotope fractionation of synthetic ikaite reported by Colla et al. (2013). Their experiments show ikaite enriched in light ^{40}Ca compared to the fluid. During the experiments, a large portion of the dissolved Ca is consumed, which is indicated by the similarity of the $\delta^{44/40}Ca$ of the precipitated ikaite with that of the initial fluid, while the $\delta^{44/40}Ca$ of the fluids at the end of the experiments is strongly depleted in ^{40}Ca. As the Ca consumption during the experiments is not reported, the isotope fractionation factor cannot readily be determined.

1.1.5 Amorphous Calcium Carbonate

The presently available data on amorphous calcium carbonate (ACC) indicate that Ca isotope fractionation during its precipitation ($\Delta^{44/40}Ca_{ACC-Ca^{2+}}$) is significantly smaller compared to aragonite, calcite and vaterite. Niedermayr et al. (2010) report a $\Delta^{44/40}Ca_{ACC-Ca^{2+}}$ of about -0.25 ‰. This value is in general agreement to a qualitative statement of Gagnon et al. (2010) that Ca isotope fractionation during ACC growth is significantly smaller than for calcite over a large range of temperatures and super saturation. They suggest that the rate-limiting step is different for calcite and ACC, and that Ca isotopes might serve as a potential proxy for the identification of crystallisation pathways. Potential applications include biomineralisation processes and occurrence of precursor phases in precipitation experiments.

For instance, Nielsen et al. (2012) suggested that discrepant dependencies on rate dependent Ca isotope fractionation in calcite observed for different experimental setups (Lemarchand et al. 2004; Tang et al. 2008a) might be related to calcite or aragonite formation via precipitation of an ACC precursor.

1.1.6 Magnesite, Dolomite, Dypingite and Nesquehonite

Magnesium isotope fractionation between magnesite ($MgCO_3$) and fluid is studied in respect to precipitation, dissolution and equilibration (Pearce et al. 2012). Similar to Ca isotopes in $CaCO_3$ minerals, light Mg isotopes are enriched in the solid phase. $\Delta^{26/24}Mg_{mag-Mg^{2+}}$ varies between about -1.2 ‰ at 150 °C and -0.8 ‰ at 200 °C, demonstrating decreasing isotope fractionation with increasing temperature and a temperature sensitivity of about 0.008 ‰/°C. This study also provides evidence that Mg stable isotope fractionation takes place not only during precipitation, but also during congruent dissolution, revealing 0.15 to 0.4 ‰ higher $\delta^{26/24}Mg$ in the fluid compared to the dissolving solid. The isotope composition of the fluid also continues to change, after establishing chemical equilibrium, indicating that isotopic equilibration goes on after chemical equilibrium has been reached. These observations show that chemical equilibrium can be reached without isotopic equilibration, and that chemical and isotopic equilibration take place at different time scales. This behaviour documents the dynamic system between mineral and fluid at thermodynamic equilibrium, which means that dissolution and precipitation continue at equal rates.

Magnesium isotope fractionation is also characterised for the hydrous magnesium carbonate phases dypingite ($Mg_5(CO_3)_4(OH)_2 \cdot 5H_2O$) and nesquehonite ($MgCO_3 \cdot 3H_2O$). In precipitation experiments at 25 °C $\Delta^{26/24}Mg_{nesquehonite-Mg^{2+}}$ is determined to be about -0.6 ‰ and $\Delta^{26/24}Mg_{dypingite-Mg^{2+}}$ ranges from -1.2 to -1.5 ‰ (Mavromatis et al. 2012). There is no difference in apparent Mg isotope fractionation between abiogenic experiments and those containing bacteria. In addition, there is no indication for a dependence of Mg isotope fractionation on pH, [Mg^{2+}] and [CO_3^{2-}].

Synthetic dolomite formed by transformation from calcite and aragonite at temperatures between 220 and 130 °C show a range in $\Delta^{26/24}Mg_{dolo-Mg^{2+}}$ from -1.1 to -0.6 ‰ (Li et al. 2015). Sampling of dolomite during the course of experiments (up to 50 days) show that $\Delta^{26/24}Mg_{dolomite-Mg^{2+}}$ changes with time. Magnesium isotope fractionation is larger at the beginning of the precipitation experiments and subsequently converges towards an equilibrium value. As equilibrium values, Li et al. (2015) suggest $\Delta^{26/24}Mg_{dolo-Mg^{2+}}$ of -0.62, -0.83 and -0.93 ‰ at 220, 160 and 130 °C, respectively, showing thus a temperature dependence with decreasing Mg isotope fractionation at increasing temperatures. Extrapolation of this temperature trend towards lower temperatures is consistent with $\Delta^{26/24}Mg_{dolo-Mg^{2+}}$ of natural sedimentary dolomites in the range of -2 to -3 ‰ of Higgins and Schrag (2010) and Fantle and Higgins (2014), while Geske et al. (2015) report smaller fractionation ($\Delta^{26/24}Mg_{dolo-Mg^{2+}}$ 0 to -0.7 ‰).

1.1.7 Witherite and BaMn(CO$_3$)$_2$

Barium isotope fractionation ($\Delta^{137/134}Ba_{min-Ba^{2+}}$) is studied in synthetic witherite ($BaCO_3^-$) by van Allmen et al. (2010) and in synthetic $BaMn(CO_3)_2$, a previously unrecognized double carbonate mineral, by Böttcher et al. (2012). As for Mg, Sr and Ca, light Ba isotopes are preferentially incorporated into the solid carbonate phase during precipitation from aqueous fluids. Barium isotope fractionation between witherite and dissolved Ba ($\Delta^{137/134}Ba_{witherite-Ba^{2+}}$ varies between -0.1 and -0.3 ‰ at temperatures of 21 and 80 °C; Fig. 7). The initial data do not show a clear temperature effect, but suggest that witherite formed at low precipitation rates is more depleted in ^{137}Ba than that at higher precipitation rates (von Allmen et al. 2010). The fractionation of about -0.1 ‰ between $BaMn(CO_3)_2$ and fluid ($\Delta^{137/134}Ba_{BaMn(CO_3)_2-Ba^{2+}}$) precipitated at 21 °C (Böttcher et al. 2012) is similar to that of witherite.

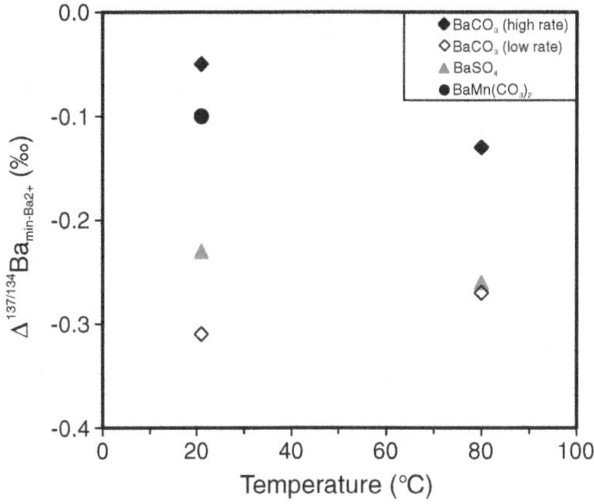

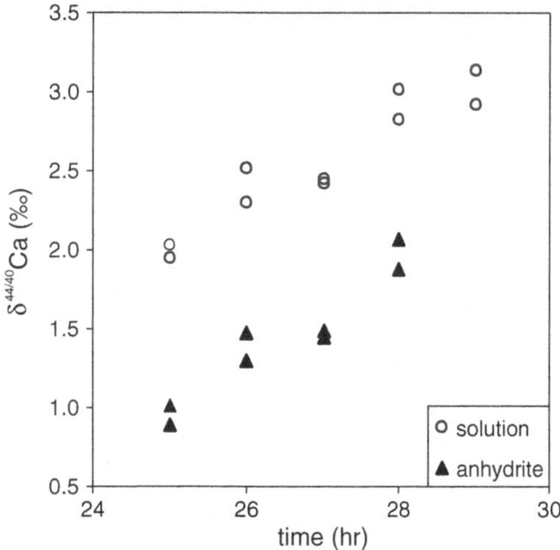

Fig. 7 Stable Ba isotope fractionation ($\Delta^{137/134}Ba_{min-Ba^{2+}}$) in barite, witherite and BaMn$(CO_3)_2$. Barium is preferentially enriched in light isotopes in the solid carbonate and sulfate phase precipitating from an aqueous solution (von Allmen et al. 2010; Böttcher et al. 2012). Initial data suggest a possible Ba isotope fractionation dependent on precipitation rate. Larger fractionation at lower precipitation rates seems to be compatible with the rate dependent Ca isotope fractionation for calcite of Lemarchand et al. (2004)

Fig. 8 Development of $\delta^{44/40}Ca$ in brine and synthetic anhydrite during ongoing precipitation. $\delta^{44/40}Ca$ of brine and anhydrite increase during the evaporation experiments, revealing an isotope fractionation of about -1 ‰ (Hensley 2006)

1.2   Sulfates

1.2.1   Anhydrite

The fractionation of Ca isotopes ($\Delta^{44/40}Ca_{anhyd-Ca^{2+}}$) during precipitation of synthetic anhydrite (CaSO$_4$) is determined to be about -1 ‰ in the temperature range between 30 and 40 °C (Hensley 2006). Precipitation is achieved in this experimental setup by successive evaporation of modified seawater. Samples of fluids and solids taken during ongoing precipitation of anhydrite reveal a Rayleigh type fractionation, with increasing $\delta^{44/40}Ca$ ratios in the fluid and solid during ongoing precipitation (Fig. 8). This development reflects the preferential incorporation of light isotopes in the anhydrite and successive depletion of light isotopes in the fluid.

1.2.2   Gypsum

Calcium isotope fractionation between synthetic gypsum and fluid ($\Delta^{44/40}Ca_{gypsum-Ca^{2+}}$) ranges from -2.25 to -0.80 ‰ in experiments conducted at constant temperature (25 ± 1 °C), pH

(5.6 ± 0.2) and ionic strength (0.5–0.6), but variable gypsum saturation degrees (Ω_{gypsum}: 1.5 to 4.8), aqueous stoichiometry (Ca:SO$_4^{2-}$ ratio from 1:3 to 3:1) and precipitation rates (Harouaka et al. 2014). There are systematic differences in Ca isotope fractionation between gypsum precipitated from stirred and unstirred solutions. While stirred experiments show no clear dependence of Ca isotope fractionation on precipitation rate or Ω_{gypsum} (Fig. 9), the unstirred experiments reveal positive correlations between $\Delta^{44/40}Ca_{gypsum-Ca^{2+}}$ and precipitation rate as well as $\Delta^{44/40}Ca_{gypsum-Ca^{2+}}$ and Ω_{gypsum}. The dependence on precipitation rate for calcium isotope fractionation between synthetic gypsum and fluid resembles that for calcite formation found by Lemarchand et al. (2004), demonstrating decreasing fractionation with increasing precipitation rates. A slight difference in Ca isotope fractionation between crystals with different morphologies suggests that growth on different crystal faces might be associated with shifts in Ca isotope fractionation, possibly related to the incomplete dehydration of the Ca ions being incorporated at the (010) face of the gypsum crystal (Harouaka et al. 2014).

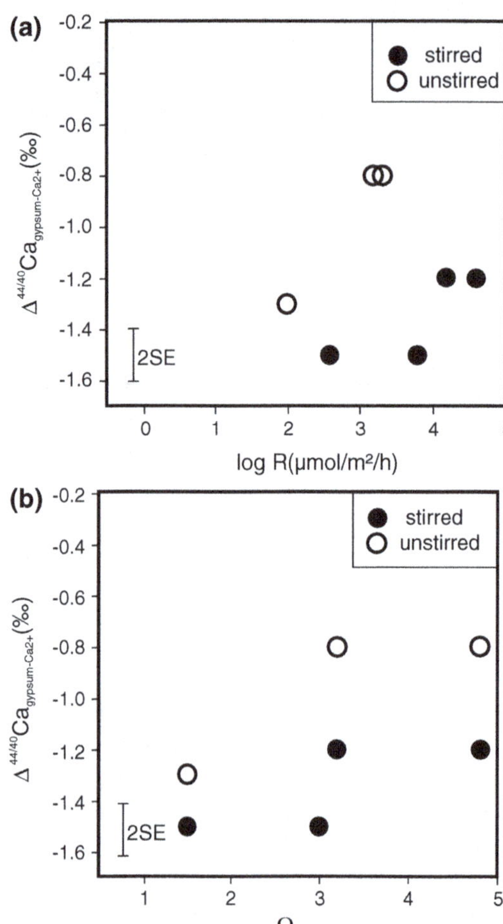

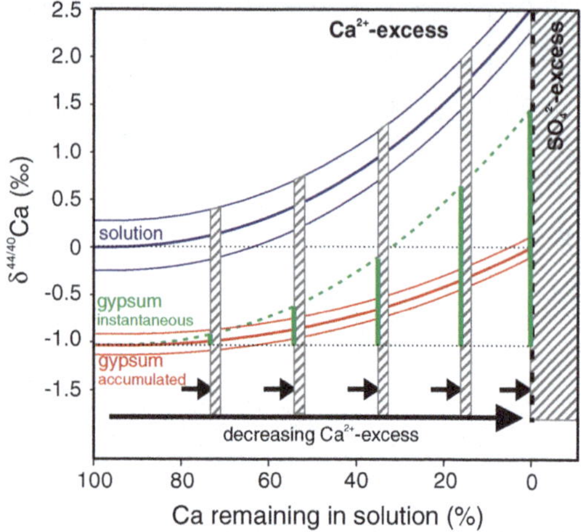

Fig. 10 Rayleigh fractionation during evaporation experiment of Na^+, Cl^-, Ca^{2+} and SO_4^{2-} containing solutions. The degree of Ca consumption prior to termination of gypsum precipitation and initiation of halite precipitation (*grey dashed boxes*) depends on the Ca^{2+}:SO_4^{2-} ratio of the solution. It also controls the development of $\delta^{44/40}Ca$ of the solution (*blue line*) and gypsum (accumulated gypsum in red and instantaneously precipitated gypsum in *green*). The fractionation curve of the instantaneous gypsum defines the maximal range of $\delta^{44/40}Ca$ (*green boxes*) that can be expected in the respective evaporitic sequence. Modified from Blättler and Higgins (2014)

Fig. 9 Fractionation of Ca isotopes between synthetic gypsum and fluid. $\Delta^{44/40}Ca_{gypsum-Ca^{2+}}$ decreases with increasing precipitation rate in unstirred fluids, while the effect is less pronounces in stirred solutions (**a**). Similarly, the influence of gypsum saturation (Ω_{gypsum}) on Ca isotope fractionation is larger in unstirred fluids compared to stirred (**b**). Modified after Harouaka et al. (2014)

While the Ca^{2+}:SO_4^{2-} ratio of the solution seems to have no major effect on $\Delta^{44/40}Ca_{gypsum-Ca^{2+}}$ (Harouaka et al. 2014), it influences the development of brines and minerals in evaporitic sequences, as shown in experiments, where gypsum precipitation is initiated by evaporation of Na^+, Cl^-, Ca^{2+} and SO_4^{2-} containing solutions with different SO_4^{2-}:Ca^{2+} ratios (Blättler and Higgins 2014). If sulfate is in excess over Ca in the solution, large Rayleigh fractionation effects occur in the fluid as nearly all Ca is consumed before gypsum formation terminates and halite formation starts (Fig. 10). The experiments reveal an increase of 2 ‰ in $\delta^{44/40}Ca$ of the fluid during the experiment with sulfate excess over Ca. In contrast, if Ca is in excess over sulfate, the increase is less pronounces. For instance at a SO_4^{2-}:Ca^{2+} of 0.3, the $\delta^{44/40}Ca$ of the fluid increases by only 0.2 ‰. With decreasing Ca excess over sulfate, relatively more Ca is removed from the solution before the sulfate is consumed and gypsum precipitation terminates, and the larger is the increase in $\delta^{44/40}Ca$ of solution and solid at the end of the experiment. The determined $\Delta^{44/40}Ca_{gypsum-Ca^{2+}}$ of these experiments is between -1.0 and -1.3 ‰ and on average about -1.0 ‰, similar to the fractionation observed for anhydrite (Hensley 2006), and within the range of $\Delta^{44/40}Ca_{gypsum-Ca^{2+}}$ revealed by the experiments of Harouaka et al. (2014). The relevance of these experimental results for the interpretation of natural evaporate-sequences is discussed in Sect. 3.3.2.

1.2.3 Barite

Calcium isotope fractionation ($\Delta^{44/40}Ca_{barite-Ca^{2+}}$) of synthetic barite ($BaSO_4$) is -2.4 to -3.4 ‰ relative to the fluid it precipitates from (Griffith et al. 2008). In the range between 5 and 40 °C, $\delta^{44/40}Ca$ depends on temperature with a rate of 0.019 ‰/°C (Fig. 11a). This temperature sensitivity is similar to that shown by synthetic and biogenic carbonate minerals. The experimental results indicate that Ca isotope fractionation during barite precipitation depends on several parameters including temperature, precipitation rate and the chemical composition of the fluid, e.g. barite saturation Ω_{barite}, aqueous stoichiometry ($Ba{:}SO_4$ and Ca: Ba ratio; Fig. 11c). Increasing Ca/Ba activity ratios lead to decreasing $\delta^{44/40}Ca$ of the precipitated barite and thus increased Ca isotope fractionation (Fig. 11b). This dependence likely reflects isotope fractionation due to diffusive Ca^{2+} transport at the crystal fluid interface. At present it is not clear if this effect results from a reduced kinetic fractionation (approaching the assumed equilibrium value) due to the higher Ca activity in the fluid, or alternatively due to an increased impact of the mass difference between ^{44}Ca and ^{40}Ca during diffusive transport at higher Ca/Ba ratios (Griffith et al. 2008).

The authors suggest that a positive relation between $\Delta^{44/40}Ca_{barite-Ca^{2+}}$ and Ω_{barite} of compatible samples (grown at similar temperatures and Ca/Ba ratios) are in general agreement to the fractionation model of Lemarchand et al. (2004) developed for calcite. The latter study demonstrates decreasing isotope fractionation with increasing precipitation rate. Assuming an equilibrium Ca isotope fractionation of ~ 1.6 ‰ [in contrast to the 0 ‰ at equilibrium proposed by Tang et al. (2008a) and Fantle and DePaolo (2007)], this implies increasing deviation from the equilibrium value with increasing precipitation rate.

In contrast to the experimental results, model simulations predict equilibrium Ca isotope

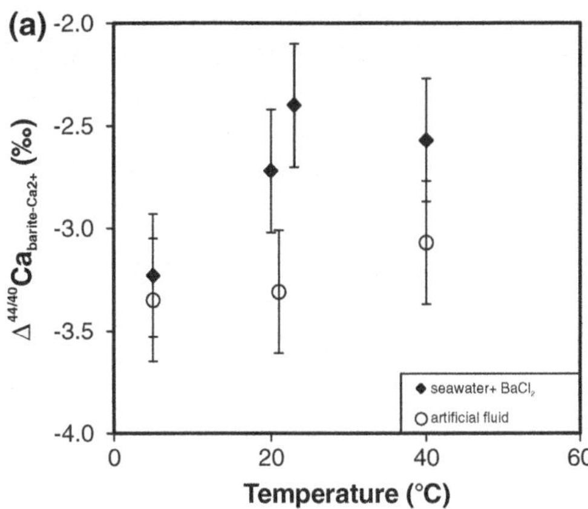

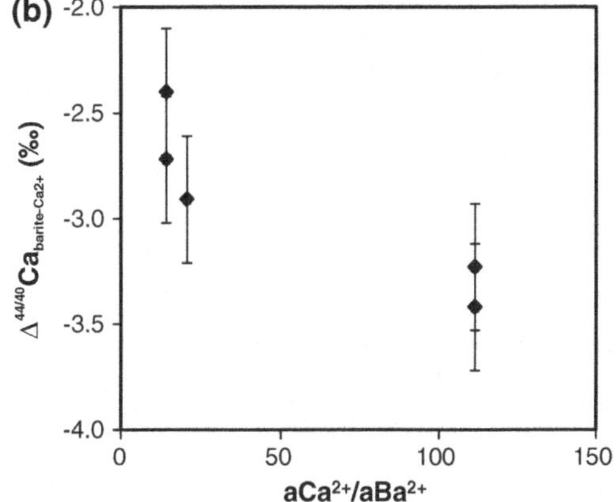

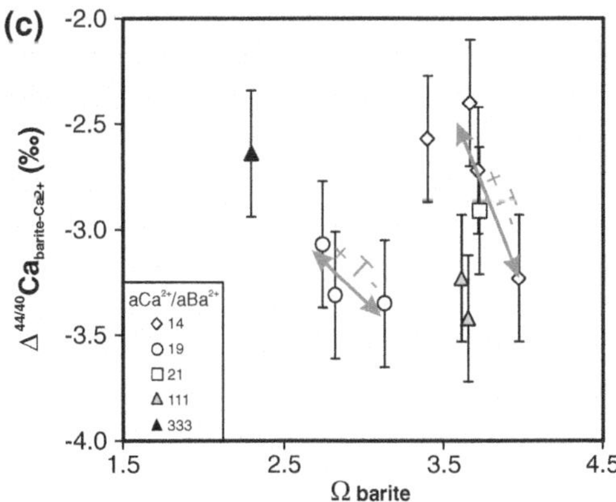

Fig. 11 Calcium isotope fractionation in barite. **a** The temperature dependence of $\Delta^{44/40}Ca_{barite-Ca^{2+}}$ in barite is similar to that observed in carbonate minerals. Barites grown from artificial $BaCl_2$–$CaCl_2$–$NaSO_4$ solution and from seawater are offset, due to differences in fluid chemistry. **b** The Ca/Ba ratio in the fluid is besides temperature a further parameter influencing $\Delta^{44/40}Ca_{barite-Ca^{2+}}$ during precipitation of barite. **c** Calcium isotope fractionation decreases with increasing Ω_{barite} and thus precipitation rate, at otherwise comparable growth conditions (Griffith et al. 2008)

fractionation between fluid and barite of -9 ‰ (at 0 °C) and -8 ‰ (at 25 °C) indicating that the synthetic barites do not reflect equilibrium isotope fractionation, but are largely influenced by kinetic fractionation effects (Griffith et al. 2008).

First data on mass dependent Ba isotope fractionation ($\Delta^{137/134}Ba_{barite-Ba^{2+}}$) during precipitation of barite indicate that, like Ca, the light isotope is preferentially incorporated into the solid phase. $\Delta^{137/134}Ba_{barite-Ba^{2+}}$ lies between -0.24 and -0.27 ‰ (von Allmen et al. 2010). The data show no clear temperature dependence in the range between 21 and 80 °C (Fig. 7). Future work is needed to further elucidate the effect of growth rate, temperature, fluid composition and mineral structure on the Ba isotope fractionation.

1.2.4 Epsomite

Stable magnesium isotope fractionation ($\Delta^{26/24}Mg_{epsomite-Mg^{2+}}$) between epsomite and fluid ($MgSO_4 7H_2O$) is characterised by an enrichment of heavy isotopes of about +0.6 ‰ in the solid relative to the fluid (Li et al. 2011). In the temperature range between 7 and 40 °C, no clear temperature dependence is apparent. The processes leading to the enrichment of heavy Mg isotopes in the solid compared to the fluid are not yet fully understood. However, experiments on Ca isotope fractionation on different Ca bearing hydrous minerals (Colla et al. 2013, see next paragraph) suggest that EASI fractionation during mineral precipitation from aqueous fluids might be related to different cation coordination environments. Following this reasoning, the observed enrichment of heavy Mg isotopes in epsomite (Mg coordination $\sim Mg^{[6]}$) relative to the dissolved Mg^{2+} ($\sim Mg^{[6]}$) in the fluid may be caused by slightly lower Mg coordination numbers or stronger (shorter) Mg–O bonds in the solid epsomite compared to the Mg-aqua complex in the solution.

1.3 Other (Hydrous) Phases

Hydrous Ca-phases ($CaBr_2 \cdot 10H_2O \cdot 2(CH_2)_6N_4$, $Ca(NH_4)PO_4 \cdot 7H_2O$ and $CaKAsO_4 \cdot 8H_2O$) have been studied to explore the influence of the Ca coordination environment on the Ca isotope fractionation (Colla et al. 2013). Hydrous phases are of special interest because of the relatively direct incorporation of aqueous Ca^{2+} ions into the crystal lattice, without complete dehydration of the Ca-aqua complexes. In the selected Ca phases, Ca is differently coordinated: $Ca^{[6]}$: $CaBr_2 \cdot 10H_2O \cdot 2(CH_2)_6N_4$; $Ca^{[7]}$: $Ca(NH_4)PO_4 \cdot 7H_2O$; $Ca^{[8]}$: $CaKAsO_4 \cdot 8H_2O$) and thus allow to study the impact of Ca coordination ($Ca^{[x]}$) on the Ca isotope fractionation. The difference in $\Delta^{44/40}Ca_{min-Ca^{2+}}$ between phases with six fold ($Ca^{[6]}$) and sevenfold ($Ca^{[7]}$) Ca coordination is about 2 ‰, similar to the 1–2 ‰ difference between phases with $Ca^{[7]}$ and $Ca^{[8]}$ (Fig. 12a) and similar to model predictions of 2.6 ‰ difference in $\Delta^{44/40}Ca_{min-Ca^{2+}}$ between $Ca^{[6]}$ and $Ca^{[7]}$ as well as $Ca^{[7]}$ and $Ca^{[8]}$. Depending on the coordination environment of Ca in the respective phase, heavy Ca isotopes are either depleted or enriched in the solid phases compared to the corresponding fluid (Colla et al. 2013). While in most carbonate and sulfate minerals EASI fractionation is characterized by enrichment of light isotopes in the solid phase, heavy Mg isotope are preferentially incorporated into epsomite ($MgSO_4 \cdot 7H_2O$; Li et al. 2011). The enrichment of heavy isotopes is not a general feature of all hydrated minerals, because $\delta^{44/40}Ca$ of gypsum (Harouaka et al. 2014) and ikaite (Colla et al. 2013; Gussone et al. 2011) is lower compared to the fluid. The experiments of Colla et al. (2013) suggest that small Ca coordination numbers, corresponding to small bond length (strong bonds) lead to an enrichment of heavy isotopes, while larger coordination numbers (larger bond length and lower bond strength) are

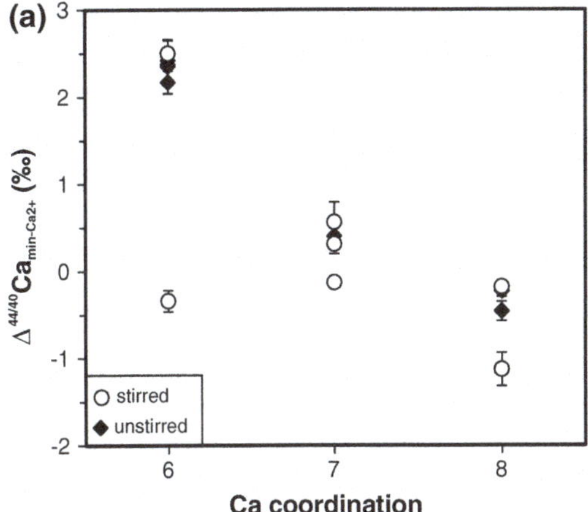

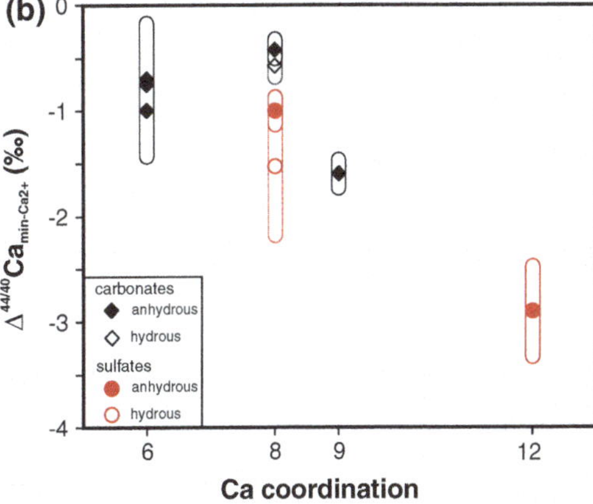

Fig. 12 Calcium coordination and isotope fractionation between fluid and minerals. Calcium isotope fractionation between solid and fluid shows that small Ca coordination numbers are associated with comparatively heavy Ca isotope ratios, while at higher coordination numbers light Ca isotopes are preferentially incorporated. **a** Different hydrous Ca phases from Colla et al. (2013): (Ca[6]: CaBr$_2 \cdot$10H$_2$O$\cdot$2(CH$_2$)$_6$N$_4$; Ca[7]: Ca(NH$_4$)PO$_4 \cdot$7H$_2$O; Ca[8]: CaKAsO$_4 \cdot$8H$_2$O). **b** Anhydrous and hydrous sulphate and carbonate minerals (20–30 °C) show a similar trend of increasing $\Delta^{44/40}$Ca with decreasing Ca coordination. Hydrous and anhydrous carbonates and sulfates are identical within uncertainties. Larger ranges of $\Delta^{44/40}$Ca$_{\text{min}-\text{Ca}^{2+}}$ apparent for certain minerals reflect the influence of factors such as temperature, growth rate and stoichiometry on isotope fractionation. Calcite (Ca[6], Marriott et al. 2004; Lemarchand et al. 2004; Tang et al. 2008, 2012); vaterite (Ca[8], Gussone et al. 2011; Niedermayr et al. 2010), ikaite (Ca[8], at 0 °C, Gussone et al. 2011), aragonite (Ca[9], Gussone et al. 2003) Barite (Ca[12], Griffith et al. 2008), gypsum (Ca[8], Harouaka et al. 2014), anhydrite (Ca[8], Hensley 2006)

associated with enrichment of light vs. heavier isotopes. This observation is in general agreement with increasing Ca isotope fractionation at increasing Ca coordination reported for anhydrous minerals, such as calcite, aragonite and barite (Griffith et al. 2008; Gussone et al. 2005).

However, the change in $\Delta^{44/40}$Ca with changing Ca coordination is larger in the experiments of Colla et al. (2013) (1–2 ‰/amu) compared to carbonates (0.2–1 ‰/amu) and sulfates (0.3–0.4 ‰/amu). Considering the modelled $\Delta^{44/40}$Ca$_{\text{barite}-\text{Ca}^{2+}}$ at equilibrium from −8 to −9 ‰, would reveal ∼2 ‰/amu between barite and anhydrite, which corresponds to the observations of Colla et al. (2013).

The $\Delta^{44/40}$Ca$_{\text{min}-\text{Ca}^{2+}}$ of 8-fold coordinated phases is similar for carbonates, sulfates and CaKAsO$_4 \cdot$8H$_2$O, but while CaBr$_2 \cdot$10H$_2$O$\cdot$2 (CH$_2$)$_6$N$_4$ is enriched in ^{44}Ca, the 6-fold coordinated calcite shows equal or lower values compared to the 8-fold coordinated phases (Fig. 12b). Isotope fractionation during the dehydration of Ca^{2+} ions during crystal growth seems to be not the main process for this difference, as hydrous and anhydrous carbonate and sulfate minerals with identical Ca coordination show the same isotope fractionation within the analytical uncertainty.

Colla et al. (2013) suggest a dependence between Ca isotope fractionation and ionic strength of the fluid, because the H$_2$O activity is decreasing with increasing ionic strength, leading to lower coordination numbers of Ca in the Ca-aqua complex, ranging from [6] to [8]. As Ca isotopes fractionate between different coordination environments, the changing coordination in response to increasing ionic strength is expected to result in a dependency between ionic strength of the fluid and $\Delta^{44/40}$Ca$_{\text{min}-\text{Ca}^{2+}}$. In contrast to this prediction, Tang et al. (2012) found in their experiments only a minor contribution of ionic strength to the Ca isotope fractionation between calcite and fluid. Further work is needed to clarify, if Ca isotope fractionation between solid and fluid can be used as indicator of Ca coordination in fluids.

Magnesium isotope fractionation between brucite $(Mg(OH)_2)$, dissolved Mg and Mg-EDTA, in all of which Mg is octahedral coordinated, demonstrates the effect of bond length (and thus bond strength) on EASI fractionation (Li et al. 2014). Magnesium isotope fractionation between brucite and dissolved Mg is temperature dependent, with $\Delta^{26/24}Mg_{brucite-Mg^{2+}}$ of -0.3, -0.2 and ~ 0 ‰ at 7, 22 and 40 °C, respectively, showing thus decreasing fractionation with increasing temperature. The Mg isotope fractionation between brucite and the Mg-EDTA complex is >2 ‰. These results demonstrate that Mg isotope fractionation takes place between different substances in which Mg is octahedral coordinated. Mg-aqua complexes have the shortest Mg–O bond length, and are enriched in heavy Mg isotopes compared to brucite and Mg-EDTA. Mg-EDTA has the longest Mg–O bonds and is consequently enriched in light Mg isotopes relative to brucite and Mg aqua complexes (Li et al. 2014).

1.4 EASI Fractionation During Mineral Precipitation from Aqueous Fluids

The reported ranges of EASI fractionation of Mg, Ca, Sr and Ba in solids, which are synthetically precipitated from aqueous fluids, demonstrate generally decreasing fractionation per atomic mass unit with increasing denominator isotope mass (Fig. 13a). Likewise, the fractionation of the metal isotopes $(^aMe, ^bMe)$ per amu $(\Delta^{a/b}Me_{min-Me^{2+}}/\Delta m)$ increases with increasing relative mass difference $(\Delta m/m; \; \Delta m = m(^aMe) - m(^bMe))$ (Fig. 13b). $\Delta^{a/b}Me_{min-Me^{2+}}/\Delta m$ is similar for Sr and Ba $(-0.1$ ‰/amu). Both isotope systems have also similar relative mass differences for $^{88}Sr/^{86}Sr$ and $^{137}Ba/^{134}Ba$ of ~ 0.02.

Precipitation experiments for stable isotope fractionation of Ca and Mg demonstrate enrichment and depletion of heavy isotopes in the solid phase relative to the fluid, while for Sr and Ba only depletion of heavy isotopes in the solid has been reported yet.

Experimentally determined Ca isotope fractionation reaches values of up to -0.85 ‰/amu with a relative mass difference of 0.1, while Mg shows a larger degree of fractionation of up to -1.6 ‰/amu at a relative mass difference of 0.08. Modelled values for isotope fractionation of Mg in calcite $(-1.8$ ‰/amu) and Ca in barite $(-2.3$ ‰/amu) are more fractionated. These observations may be biased due to lack of data, but the general observations suggest that EASI fractionation may depend on the relative mass difference $(\Delta m/m)$ of the considered alkaline earth metal isotopes. Future work needs to investigate, if the different behaviour of Mg and

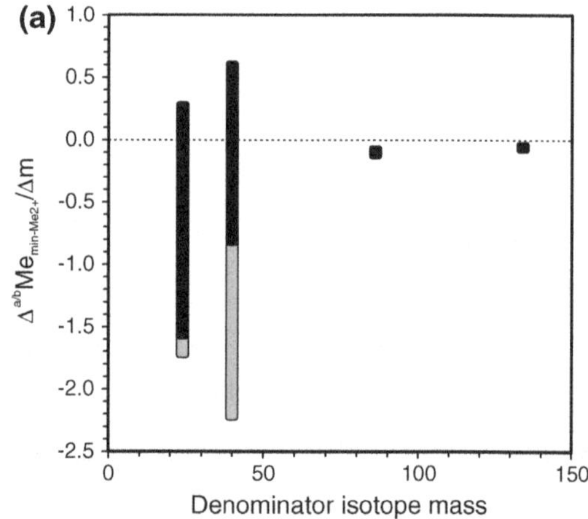

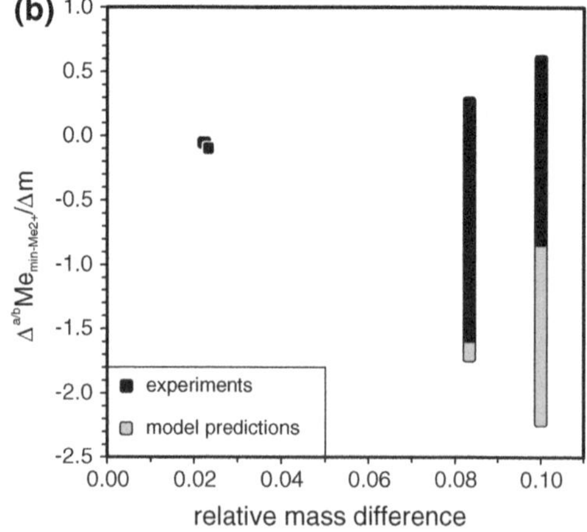

Fig. 13 Comparison of different EASI fractionation (as $\Delta^{a/b}Me_{min-Me^{2+}}/\Delta min$ ‰/amu) in different mineral phases as a function of mass (**a**) and relative mass difference ($\Delta m/m$) (**b**)

Ca might be related to a special fractionation caused by the different precipitation mechanisms (see Reynard et al. 2011) or to element specific differences like hydration shells, individual aqua complexes and cation coordination etc.

2 Calcium Isotope Fractionation Models for Calcium Carbonate Formation

2.1 Principles and Conceptions of Isotope Fractionation Models

In earlier studies various models were developed to explain Ca isotope fractionation during adsorption onto solid surface, ion exchange reactions and precipitation of Ca bearing minerals based on field, experimental, and theoretical studies. Isotopic data reveal that Ca isotope fractionation between solid and fluid phase may be observed at isotopic equilibrium but it may be biased also by kinetic effects (e.g., Eiler et al. 2014). Actually, kinetic effects on stable isotope fractionation may hinder the interpretation of stable isotopic composition of solid phases. Thus, the use of such isotopic signals as environmental proxies, e.g. for estimation the temperature of mineral formation and the grow rate at which a mineral formed, is challenging. On the other side, as isotope fractionation at equilibrium and kinetic effects can be modeled and quantified, novel and advanced techniques and tools to study (paleo)environmental conditions during e.g. calcite formation can be provided (e.g., Tang et al. 2008b; Nielsen and DePaolo 2013).

At first Marriott et al. (2004) presented a model approach for the fractionation of Ca isotopes during inorganic calcite precipitation which is based on isotope fractionation at equilibrium. Their experimental results obtained at a constant calcite growth rate show that the precipitated calcite is isotopically lighter compared to the aqueous Ca of the forming solution. Stronger covalent Ca bonding in aqua complexes than in the carbonate precipitate such as calcite is reflected by lighter Ca isotopes preferentially

bound in the calcite. Their assumption of isotope fractionation at equilibrium during calcite formation is mainly based on the fact that (i) the higher the temperature the less Ca isotope fractionation is observed, and (ii) a similar trend for Ca isotope fractionation during calcite formation is found for calcite from biomineralization by planktonic foraminifera (*Orbulina universa*). They argue that distinct Ca isotope fractionation values by biogenic induced calcite formation could be caused e.g. by active pumping of Ca^{2+} ions to an internal pool, where biomineralization occurs. The Ca isotope fractionation approach of Marriott et al. (2004) is different from that of Gussone et al. (2003), who suggest non-equilibrium Ca isotope fractionation by considering diffusion of a Ca bearing aqua complex. Mass dependent aqueous diffusion at the crystal-solution boundary of heavy $Ca[H_2O]_n^{2+} \cdot xH_2O$ aqua complexes of ~ 600 amu in average instead of "pure" Ca^{2+} ions is considered. Accordingly, Gussone et al. (2003) stress the potential relevance of the aqueous diffusion of Ca-aqua complexes for isotope fractionation, where isotope fractionation mechanisms are based on kinetics. Using this modeling approach various Ca isotope fractionation behaviors of different foraminifera could be explained by diverse species dependent dehydration of Ca-aqua complex before calcification.

Based on their calcite precipitation experiments Lemarchand et al. (2004) argue that Ca isotope fractionation at isotopic equilibrium can be significantly overprinted by kinetic effects. As a result, Ca isotope fractionation varies with precipitation rate at a given temperature (Fig. 2). The different Ca isotope fractionation results from stirred versus unstirred experimental precipitating solutions led them to suggest that diffusive flow of carbonate ions through the crystal-solution interface is a rate controlling mechanism and thus responsible for apparent Ca isotope fractionation during calcite growth: (i) Equilibrium fractionation with $\Delta^{44/40}Ca_{Cc-Ca^{2+}} = -1.5\,\% \cdot (\Delta^{44/40}$ $Ca_{Cc-Ca^{2+}} = \delta^{44/40}Ca_{Cc} - \delta^{44/40}Ca_{Ca^{2+}} \approx (\alpha_{eq} - 1) \times 10^3)$ is valid at the immediate crystal-solution interface and (ii) diffusive carbonate ion flow induces supersaturation with respect to calcite, which results in no Ca isotope fractionation at high degrees of supersaturation. In the latter case, the Ca ion from the bulk solution and not from the zone of isotope equilibrium is preferentially used for the incorporation into the calcite. Their model predicts that the temperature effect on $\Delta^{44/40}Ca_{Cc-Ca^{2+}}$ is due to the change of equilibrium constant values in the CO_2–H_2O system, thus caused by changing the distribution of DIC species.

Fantle and DePaolo (2007) use the Ca isotopic composition of nannofossil ooze and chalk as well as pore fluid samples from ODP Site 807A (Ontong Java Plateau), where calcite is reasonably suggested to be isotopically equilibrated with marine pore fluids within time periods of millions of years, to estimate $\Delta^{44/40}Ca_{Cc-Ca^{2+}}$ at equilibrium to be close to 0 ‰. This finding is highly relevant for further model developments as virtually non-fractionation of Ca isotopes at isotopic equilibrium conditions contradicts former studies (e.g. Lemarchand et al. 2004; Marriott et al. 2004) that proposed a negative $\Delta^{44/40}Ca_{Cc-Ca^{2+}}$ value at equilibrium. Fantle and DePaolo (2007) introduced the term "relative growth rate", where no Ca isotope fractionation occurs at both relative growth rate ≈ 0 (i.e., at equilibrium) and 1 (i.e., at extremely high rates). The developed adsorption-controlled steady state model attributes Ca isotope fractionation to be balanced by attachment and detachment fluxes at the calcite crystal. Their steady state box model considers four Ca reservoirs during the growing of calcite; the bulk solution, interface region, crystal surface, and crystal. The observed Ca isotope fractionation is controlled by the magnitude of the Ca flux to the growing calcite and away from it, thus the precipitation rate of calcite is responsible for Ca isotope fractionation deviation from zero.

The calcite precipitation experiments performed by Tang et al. (2008a) confirm that the Ca isotope fractionation is a function of precipitation rate as formerly suggested e.g. by Lemarchand et al. (2004) and Fantle and DePaolo (2007). However the rate effect is inverse to that observed by Lemarchand's experiments (Fig. 2). Tang et al. (2008a) adapted the surface entrapment model introduced by Watson and Liang (1995) and Watson (2004), which is based on surface entrapment, growth rate and ion diffusion. It is proposed that no Ca isotope fractionation occurs in a regular calcite crystal at isotopic equilibrium, but ^{44}Ca is depleted versus ^{40}Ca in a surface layer during precipitation (Fig. 14). The entrapment of lighter ^{40}Ca in the surface layer might be a result of energetic preference of heavier ^{44}Ca to strong bonds in the solution (see Marriott et al. 2004), preferential adsorption of the lighter ^{40}Ca onto the calcite surface (see Fantle and DePaolo 2007), and/or faster diffusion of the lighter ^{40}Ca at liquid/solid transition (see Gussone et al. 2003). The entrapment factor (F) is suggested to

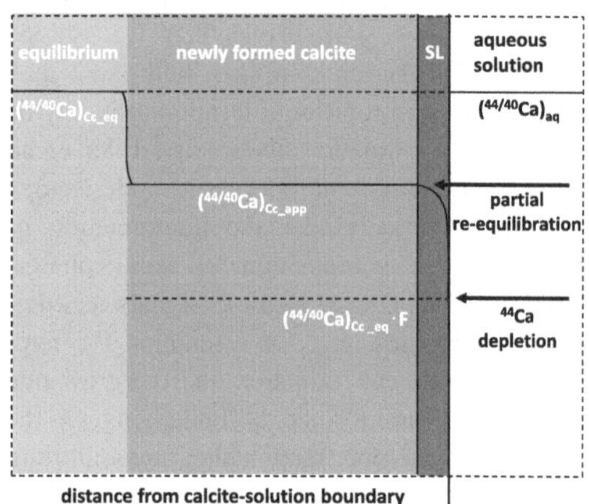

Fig. 14 Schematic view of the Ca isotope fractionation during calcite formation according to the surface entrapment model by Tang et al. (2008a) adapted from Watson (2004). $(^{44/40}Ca)_{Cc\_eq}$, $(^{44/40}Ca)_{Cc\_app}$ and $(^{44/40}Ca)_{aq}$ denote the molar $^{44}Ca{:}^{40}Ca$ ratio of the regular calcite crystal lattice at isotopic equilibrium, of the newly formed apparent crystal lattice, and of the aqueous solution, respectively. For the regular calcite crystal lattice no Ca isotope fractionation at equilibrium is assumed $(\Delta^{44/40}Ca_{Cc-Ca^{2+}} \approx \{(^{44/40}Ca)_{Cc\_eq}/(^{44/40}Ca)_{aq} - 1\} \times 10^3 = 0\%)$. *SL* surface layer, where isotope fractionation occurs. The minimum ^{44}Ca concentration in the SL is referred to the temperature dependent value of the enrichment factor F for ^{40}Ca versus ^{44}Ca. The time provided for isotopic re-equilibration by Ca ion diffusion in the SL during crystal growth controls the apparent Ca isotope fractionation

be related to the physicochemical conditions during calcite growth, e.g. temperature, pH, ionic strength. During crystal growth the ^{44}Ca depleted surface layer (SL) can be entrapped into the crystal lattice at dis-equilibrium conditions, where the degree of Ca isotope fractionation depends on the time for re-equilibration of Ca isotopes within the surface layer via ion diffusion. Thus, limits for isotope fractionation are related to isotopic equilibrium (approached by very low precipitation rates) and isotope entrapment with insufficient time available for re-equilibration via ion diffusion within the SL (very fast precipitation rates). The Ca isotope fractionation limit at equilibrium between a regular calcite crystal lattice and aqueous calcium ions is based on the assumption of Fantle and DePaolo (2007; $\Delta^{44/40}\mathrm{Ca}_{\mathrm{Cc-Ca}^{2+}} \approx 0\,\permil$), and the estimation of the enrichment factors results in $\Delta^{44/40}\mathrm{Ca}_{\mathrm{Cc-Ca}^{2+}} = -1.80$ and $-1.20\,\permil$ for 5 and 40 °C, respectively, for kinetic fractionation limit. By using the surface entrapment model, Tang et al. (2008a) can explain their experimental data for Ca isotope fractionation during calcite formation obtained from CO_2 diffusion method, in particular the influence of both precipitation rate and temperature on apparent $\Delta^{44/40}\mathrm{Ca}_{\mathrm{Cc-Ca}^{2+}}$ values. Interestingly, the surface entrapment model approach can also explain and quantify the incorporation behavior of strontium and stable strontium isotope fractionation during calcite formation with a linear correlation between respective element partitioning and isotope fractionation (see Tang et al. 2008b; Böhm et al. 2012). Thus the approach can be applied to combined multi proxy interpretations, such as the estimation of the $\delta^{44/40}\mathrm{Ca}_{\mathrm{Cc-Ca}^{2+}}$ value of the calcite precipitating solution from the Sr content and $\delta^{44/40}\mathrm{Ca}_{\mathrm{Cc}}$ of the precipitated calcite if the aqueous Sr/Ca ratio is known, but the temperature and precipitation rate are unknown within the given experimental conditions (Tang et al. 2008a, b).

DePaolo (2011) developed a surface reaction kinetic model for isotope and trace element fractionation during calcite formation (Eq. 1). This model is based on the ratio of the net precipitation rate (R_p; usually provided by experiments) and the gross forward precipitation rate (R_f) assuming non rate-limiting ion transport to the growing calcite surface (Fig. 15). R_f values are estimated by using well known (backward) dissolution rates for calcite at various physico-chemical conditions far from equilibrium (R_b) since $R_f = R_p + R_b$. At high precipitation rates ($R_p \gg R_b$) isotope and element fractionation is controlled by kinetics of ion attachment to the calcite surface, e.g. lighter Ca isotopes are preferentially incorporated into the calcite. Thus different precipitation rates approximate intermediate conditions between kinetic and equilibrium control on Ca isotope fractionation during calcite formation. The limiting conditions of "pure" kinetic and equilibrium control are accompanied with individual fractionation factors α_f and α_{eq} from which the apparent Ca isotope fractionation between calcite and aqueous Ca^{2+} can be calculated as shown in Fig. 15. An apparent fractionation factor is obtained from the R_p/R_b ratio according to the equation

$$\alpha_i = \frac{\alpha_f}{1 + \dfrac{R_b}{R_i + R_b}\left(\dfrac{\alpha_f}{\alpha_{eq}} - 1\right)} \tag{1}$$

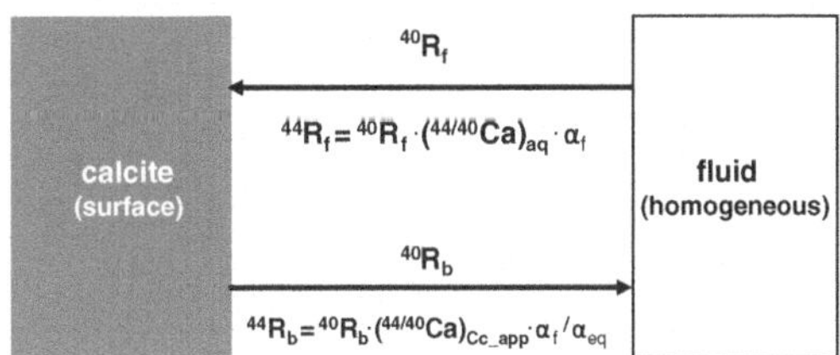

Fig. 15 Illustration of the surface reaction kinetic model for Ca isotope fractionation during calcite formation according to DePaolo (2011). $(^{44/40}\mathrm{Ca})_{\mathrm{Cc\_app}}$ and $(^{44/40}\mathrm{Ca})_{\mathrm{aq}}$ denote the molar $^{44}\mathrm{Ca}{:}^{40}\mathrm{Ca}$ ratio of the apparent calcite and the aqueous solution, respectively. $^{40}\mathrm{R}_f$ and $^{40}\mathrm{R}_b$ represent the Ca fluxes of the forward- and backward-reaction, respectively. α_f and α_{eq} are the individual fractionation factors for the limiting conditions of "pure" kinetic (precipitation rate far from equilibrium) and equilibrium control on Ca isotopic fractionation from which the apparent Ca isotope fractionation ($\Delta^{44/40}\mathrm{Ca}_{\mathrm{Cc-Ca}^{2+}} \approx \{(^{44/40}\mathrm{Ca})_{\mathrm{Cc\_app}}/(^{44/40}\mathrm{Ca})_{\mathrm{aq}} - 1\} \times 10^3$) can be calculated

(DePaolo 2011; $\Delta^{44/40}Ca_{Cc-Ca^{2+}} \approx (\alpha_i - 1) \times 10^3$). This approach is applied to fit experimental data on Ca isotope fractionation and Sr incorporation from Tang et al. (2008a, b), where a good fit of the model with the data is obtained by considering the dissolution rate "constant" for estimation of R_b values to be a function of saturation state conditions in respect to calcite. Accordingly, DePaolo's approach is also successfully applied to model elemental and isotope fractionation for e.g. Sr incorporation and oxygen isotope fractionation during calcite formation (DePaolo 2011) and Si isotope fractionation during precipitation of silicic acid (Oelze et al. 2015; Roerdink et al. 2015). For transport limitation effects in the fluid phase DePaolo (2011) introduced a diffusive boundary layer between the bulk solution (well mixed) and the mineral surface, where growth is taking place. The thickness of the boundary layer is determined by the hydrodynamics of the solution phase.

Nielsen et al. (2012) introduced a self-consistent microscopic theory, where simultaneously both calcite growth rate and Ca isotope fractionation during calcite formation can be followed by ion-by-ion growth model. Input parameters comprise of the solution stoichiometry (molar $Ca^{2+}:CO_3^{2-}$ ratio) and the degree of saturation state in respect to calcite (Ω_{Cc}: saturation degree with respect to calcite) thus both controlling precipitation rate and isotope fractionation. The model includes a large numbers of parameters such as calcite solubility, density, kink height, depth and molecular spacing and kink formation energy. Although the whole parameter set used in their ion-by-ion growth model is not available from calcite precipitation experiments, the modeling results fitted the experimental data of Tang et al. (2008a) well (Fig. 16). An increase of $Ca^{2+}:CO_3^{2-}$ ratio at a given Ω_{Cc} value drives calcite growth to approach isotopic equilibrium. Application of this ion-by-ion crystal growth model to the Ca isotope fractionation during calcite precipitation from natural high-alkaline solutions (Mono Lake, USA) confirms less Ca isotope fractionation at elevated $Ca^{2+}:CO_3^{2-}$ ratios (Nielsen and DePaolo 2013). This approach is also used to explain the competing roles of temperature, pH, and growth rate of the calcite-water oxygen

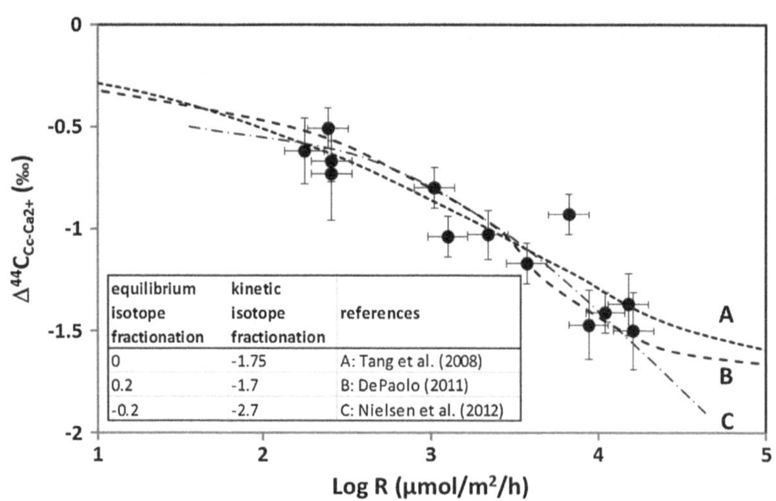

equilibrium isotope fractionation	kinetic isotope fractionation	references
0	-1.75	A: Tang et al. (2008)
0.2	-1.7	B: DePaolo (2011)
-0.2	-2.7	C: Nielsen et al. (2012)

Fig. 16 Modelling results for Ca isotope fractionation during calcite formation by using **a** the surface entrapment model shown in Fig. 14, **b** the surface reaction kinetic model presented in Fig. 15, and **c** the ion-by-ion growth model by Nielsen et al. (2012). In the latter approach the exchange coefficients fitted to Davis (2008) obtuse step speeds were used for modeling assuming 2D nucleation-driven growth. Experimental data at 25 °C from Tang et al. (2008a) are plotted as *solid dots*. Limiting $\Delta^{44/40}Ca_{Cc-Ca^{2+}}$ values (in ‰) are controlled by equilibrium and kinetic isotope fractionation as given in the *inserted table*

isotope fractionation under non-equilibrium conditions (Watkins et al. 2014).

In contrast to calcite, the isotope fractionation mechanisms during the precipitation of other unhydrous and hydrous $CaCO_3$ solids are less explored. Focus on calcite formation is due its dominance in natural surroundings and based on the limits of experimental conditions for the precipitation of e.g. vaterite and amorphous calcium carbonate. The latter $CaCO_3$ formation requires high supersaturation degrees far from thermodynamic equilibrium and transformation to phases like calcite has to be considered. A model concept for the different Ca isotope fractionation during the formation of the $CaCO_3$ polymorphs calcite and aragonite (see Sect. 1) is presented by Gussone et al. (2005). From available data two individual equations for temperature dependence of Ca isotope fractionation for calcite and aragonite are given, where Ca isotopes exhibiting larger fractionation in aragonite than in calcite. The offset of about 0.6 ‰ is suggested to be caused by the distinct crystal structures, e.g. Ca coordination numbers and bond strengths with oxygen in calcite versus aragonite. However, a modeling approach for the combined impact of temperature and precipitation rate on Ca isotope fractionation during aragonite versus calcite is still not well established. For vaterite and ikaite less Ca isotope fractionation versus calcite and aragonite is observed from precipitation experiments at similar temperatures (Gussone et al. 2011; Fig. 1). This offset is suggested to be related to the different precipitation rates and regimes, where Ca isotope fractionation is related to transport limited growth conditions at relatively high growth rates (see also DePaolo 2011). Further studies are required to develop proper models for Ca isotope fractionation for the formation of ikaite and vaterite versus e.g. calcite, where the different mineral structures and precipitation conditions have to be considered. Recently, the Ca isotope fractionation behavior during transformation of amorphous calcium carbonate (ACC) to calcite was experimentally studied by Giuffre et al. (2015). The isotopic composition of the initially formed ACC is modified during transformation to calcite and the isotopic composition of the final calcite records the local solution environment at the time of transformation. The transformation could be best explained by a dissolution-reprecipitation approach and not a solid-state process, where solely dehydration and structural rearrangement into crystalline $CaCO_3$ occur. Their experimental data indicate that the time for transformation of ACC to calcite and the Ca budget of ACC and the surrounding reservoir controls the final isotopic composition of the calcite. Future modeling approaches for calcite formation through ACC pathway should thus account for environmental conditions during the transformation process in terms of kinetics and mass balance, where detailed studies on the evolution of $\Delta^{44/40}Ca_{CaCO_3-Ca^{2+}}$ values during ACC transformation are still missing.

2.2 Comparison of Ca Isotope Fractionation Models and Concluding Remarks

In analogy to the traditional application of stable hydrogen and oxygen isotopic composition of H_2O to validate evaporation and condensation effects (e.g., Kendall and McDonnell 1998; Jouzel and Merlivat 1984), currently used models for Ca isotope fractionation during precipitation of calcite are mostly based on limiting equilibrium and kinetic approaches (e.g. Tang et al. 2008a; DePaolo 2011). Mechanisms which are suggested to cause the kinetically driven Ca isotope fractionation comprise of e.g. ion diffusion, desolvation/dehydration of ions, aqua complex incorporation and surface reactions as discussed above. The kinetically controlled isotope fractionation is typically related to mass-dependent effects, thus isotopically lighter isotopes are preferentially bounded to the precipitate.

Desolvation of Ca^{2+} ions seems to be a plausible explanation of Ca isotope fractionation as desolvation rates of Ca (besides Sr, Ba and others) depend on cation mass by an inverse power law and the predicted isotope fractionations based on molecular dynamics simulation results in $\alpha_{\text{kinetic}} = \alpha_f = 0.9953 \pm 0.0009$ between hydrated and unhydrated Ca ions (corresponding to $\Delta^{44/40}\text{Ca}_{\text{kinetic}} = -4.7 \pm 0.9$ ‰; Hofmann et al. 2012). Thus the kinetically driven isotope fractionation through desolvation of calcium ions to be incorporated into the calcite structure is high enough to result in the documented isotope fractionation during the precipitation of calcite from experiments. Diffusion of aqueous Ca^{2+} ions to a growing calcite may yield in light isotope preferentially bound in the crystal, but this isotope fractionation process is rather limited in value. A small isotope fractionation by diffusion of divalent cations in aqueous fluids is reported for Ca^{2+} and Mg^{2+} ions (e.g., Richter et al. 2006; Bourg et al. 2010). Eiler et al. (2014) claimed that the aqueous diffusion related to e.g. Ca isotope fractionation is too small to explain the experimentally obtained Ca isotope fractionation during calcite growth. Potential isotope fractionation effects by changes in coordination of ligands are documented by experiments and electronic-structure and vibrational frequencies modeling (Colla et al. 2013). The lower the coordination number of Ca in respect to oxygen, the shorter and stronger the Ca–O bondings. Thus, cation coordination and respective environmental changes have to be considered for Ca ions as a process responsible for isotope fractionation during calcite growth. Non-equilibrium conditions might be also induced by crystallographically non-equivalent growth sites at calcite surfaces (e.g., shown for stable carbon and oxygen isotopes; Dickson 1991) and outermost molecular layers of calcite which are different from the bulk considering bond lengths and orientation (Fenter et al. 2000). Although local isotopic equilibrium exists, a structural site that differs e.g. in coordination from the final site in the crystal interior may result in significant isotopic disequilibrium conditions if re-equilibration is not approached. This kind of kinetically forced fractionation effect at the crystal-solution boundary is close to that suggested for the surface entrapment model, but detailed reaction mechanisms and kinetics are still not-well explored.

The surface entrapment model provides a reaction based mechanistic approach, where a temperature dependent precipitation rate effect is considered for Ca isotope fractionation between calcite and aqueous Ca ions. However, it has to be noted that the re-equilibration by Ca^{2+} ion diffusion within a solid surface layer (SL in Fig. 14) is a conceptual approach and values for individual diffusion constants of $^{44}\text{Ca}^{2+}$ and $^{40}\text{Ca}^{2+}$ are not being confirmed. In contrast, DePaolo's model is based on overall measureable and estimated thermo-kinetic parameters. For this non-mechanistic model net precipitation rates and gross forward precipitation rates are used to approach intermediate conditions between close to and far from equilibrium conditions during precipitation of calcite. Nevertheless, in both cases model approximations of experimental data are similar (Fig. 16). For the advanced ion-by-ion growth approach of Nielsen et al. (2012) simultaneously calcite growth rates and Ca isotope fractionation during growth can be modeled and obtained results fit the experimental data well considering 2D nucleation-driven growth. However, isotope fractionation at elevated precipitation rates behaves different compared to the models of Tang et al. (2008a) and DePaolo (2011) as this high degree of super saturation range is not compatible with ion-by-ion growth theory.

Experimental results from simulating cave drip water conditions by Reynard et al. (2011) confirm the same trend of precipitation rate dependence on Ca isotope fractionation during calcite growth as reported by Tang et al. (2008a), but absolute isotope fractionation is about twice as much (Fig. 1). Thus the general Ca isotope fractionation behavior can be explained e.g. by Tang's and DePaolo's model by simply using larger entrapment factors and α_f values, respectively. It is suggested that the rapid growing calcite surface in a stalagmite precipitation environment, in particular in the splash zone, causes an extremely large kinetic effect by entrapment of the lighter Ca isotope on the surface due to short residence time of the drip solution at the growing surface. Thus insufficient time for Ca isotope re-equilibration via depletion of Ca from the interface near the calcite surface is provided.

It is still a matter of debate why the experimental results of Lemarchand et al. (2004) do not fit with those of Tang et al. (2008a). Proposed reasons comprise of different experimental design and conditions during calcite growth. For instance intermediate ACC or vaterite, which are known to be transformed to calcite, might have been occurred during the experiments. Accordantly, the observed inverse $\Delta^{44/40}Ca_{Cc-Ca^{2+}}$ trend by Lemarchand et al. (2004) compared to Tang et al. (2008a) could be caused by different Ca isotope fractionation behavior via distinct precipitation pathways (e.g. less isotope fractionation for vaterite vs. calcite: Gussone et al. 2011). Other effects might be related to different solution compositions stimulating distinct aqua complex formation and/or non-homogenization of the precipitating solution. In the latter case even stirring of the very high Ca concentrated solution in Lemarchand's experiments may have created transport limitations on the precipitation rate, as the critical boundary layer thickness, where isotope fractionation occurs, decreased (DePaolo 2011).

In conclusion, the above mentioned kinetic effects on stable Ca isotope fractionation limits the simple use of the isotopic values of solid $CaCO_3$, e.g. for direct temperature reconstruction as a positive and negative temperature function of $\Delta^{44/40}Ca_{Cc-Ca^{2+}}$ can be obtained caused by its additional dependence on precipitation rates. But this mutual behavior makes Ca isotopic values highly attractive as environmental proxies to study complex physicochemical conditions during growth and re-crystallization in sedimentary and diagenetic as well as in man-made surroundings. As isotope fractionation at equilibrium and kinetic effects on apparent $\Delta^{44/40}Ca_{Cc-Ca^{2+}}$ values can be reasonably modeled and quantified, e.g. using above discussed models, novel and advanced tools are provided to decipher environmental conditions during (trans)formation of carbonates, in particular by combining Ca isotopic values with elemental, (micro)structural and/or further isotopic signals within a multi-proxy approach.

---

3 Inorganic Mineral Precipitation in Natural Environments

3.1 Carbonates

Calcium isotope fractionation in sedimentary carbonates significantly differs between authigenic carbonates formed under net-precipitation conditions and those formed by recrystallization. Authigenic carbonates show an enrichment of light Ca isotopes relative to the fluid, similar to the fractionation revealed by precipitation experiments (Sect. 1.1). In contrast, calcium carbonate recrystallization under near equilibrium conditions demonstrates a ~ 0 ‰ fractionation in marine sediments and in continental aquifers. Calcium isotope systematics of authigenic carbonates in the marine realm, lakes and soils are discussed in Sect. 3.1.1 and carbonate recrystallization in Sect. 3.1.2.

3.1.1 Primary and Authigenic Carbonates

Marine Realm

Authigenic aragonite at cold seeps

Authigenic aragonite formed at colds seeps in close contact with gas hydrates at Hydrate Ridge (Cascadia margin off Oregon) at temperatures of about 5 °C show a range in $\delta^{44/40}Ca$ from about 0.4 to 1.2 ‰ (Teichert et al. 2005). This corresponds to an apparent Ca isotope fractionation relative to modern seawater from -0.7 to -1.4 ‰. They are thus enriched in ^{44}Ca compared to synthetic aragonite formed at comparable temperatures (Gussone et al. 2003). Spatially resolved $\delta^{44/40}Ca$, $\delta^{13}C$ and $\delta^{18}O$ analyses of these aragonite crusts reveal a typical growth pattern, with an increase in $\delta^{44/40}Ca$ and a decrease in $\delta^{18}O$ from the proximal (close to the gas hydrates) to the distal layers of the carbonate crust. Together with low $\delta^{13}C$ values, indicative for methane derived CO_3^{2-}, the isotope data reveal precipitation from a brine which is influenced by gashydrate formation in a restricted pore space. The elevated $\delta^{44/40}Ca$ is consistent with Rayleigh fractionation, as ^{44}Ca is enriched in the pore water due to successive depletion of ^{40}Ca during ongoing aragonite precipitation. Calculated $\delta^{44/40}Ca_{pore\ water}$ range from 2.3 to 2.7 ‰ (0.4–0.8 ‰ higher than present seawater), using the $\Delta^{44/40}Ca_{arag-Ca^{2+}}$ of synthetic aragonite (Teichert et al. 2005; Gussone et al. 2003). These considerations are in agreement to $\delta^{44/40}Ca$ of solid carbonate and pore water samples from the Niger Delta Seep Province. $\delta^{44/40}Ca_{carbonate}$ range from about 0.5 to 1.5 ‰, and $\delta^{44/40}Ca_{pore\ water}$ from about 1.5 to 2.9 ‰, showing a similar enrichment in ^{44}Ca of the pore fluid, related to the formation of authigenic carbonates (Henderson et al. 2006).

Aragonite vein cements

Aragonite veins from the Logatchev hydrothermal field (Mid Atlantic ridge), formed at temperatures between about 3 and 13 °C, demonstrate $\delta^{44/40}Ca$ between about −0.2 and −0.6 ‰, corresponding to a $\Delta^{44/40}Ca_{arag-sw}$ of −1.7 and −1.3 ‰ relative to seawater (Amini et al. 2008). These values are slightly higher (0.2–0.3 ‰) than those of synthetic aragonites at comparable temperatures (Gussone et al. 2003). This offset in $\delta^{44/40}Ca$ might be explained by uncertainties related to temperature estimation, due to $\delta^{18}O_{fluid}$ deviating from seawater (precipitation at higher temperatures than assumed) or by a $\delta^{44/40}Ca_{fluid}$ higher than seawater, e.g. depleted in ^{40}Ca due to Rayleigh type fractionation during precipitation of Ca bearing phases.

Calcite cements

In contrast to aragonite cements, which $\delta^{44/40}Ca$ is significantly lower than seawater, calcite cements in ocean crust basalts have $\delta^{44/40}Ca$ similar to seawater. Böhm et al. (2012) report $\Delta^{44/40}Ca_{Cc-sw}$ of −0.1 to −0.2 ‰ for calcite cements formed at 1–2 °C. Similarly, stable Sr isotopes analysed in this calcite show values slightly lower than seawater, with a fractionation relative to seawater ($\Delta^{88/86}Sr_{Cc-sw}$) of −0.05 ‰.

Both Ca and Sr isotope fractionation is interpreted as the result of slow precipitation rate and near equilibrium conditions (Böhm et al. 2012).

Little Ca isotope fractionation between inorganic calcite and seawater is also suggested by Steuber and Buhl (2006), which report $\delta^{44/40}Ca$ of different kinds of marine calcite cements and biogenic carbonates from the Cretaceous and the Carboniferous. The $\delta^{44/40}Ca$ of Cretaceous (Cenomanian) diagenetic calcite, diagenetic cements and calcitic ooids lie between 1.1 and 1.2 ‰ (SRM 915a). These values are about 0.6 ‰ higher compared to the pristine biogenic carbonates of the same age ($\delta^{44/40}Ca$ 0.2–0.7 ‰). Recrystallised biominerals originally composed of aragonite, reveal intermediate values of about 1.0–1.2 ‰, indicating a shift in $\delta^{44/40}Ca$ of the biominerals during recrystallization towards the inorganic calcite cements. A homogenization of $\delta^{44/40}Ca$ during diagenesis of different sedimentary $CaCO_3$ compounds is also implied by sediment compounds from the Carboniferous (Moscovian), which biominerals are only −0.1 to −0.2 ‰ relative to inorganic cements of the same age. The carbonate cements from the Moscovian sediments show a rather narrow range in $\delta^{44/40}Ca$ and no significant difference between different kinds of cements, e.g. fibrous calcite (1.1–1.2 ‰), saddle dolomite (~ 1.2 ‰), diagenetic calcite (1.2–1.3 ‰), bottroidal cements (~ 1.2 ‰) and micrite, including microbial micrite (~ 1.1 ‰) (Steuber and Buhl 2006).

Depending on the applied $\delta^{44/40}Ca$ seawater reconstruction, the Cenomanean cements are either about −0.5 ‰ (Farkaš et al. 2007) or ~ 0 ‰ (Soudry et al. 2006) relative to contemporaneous paleo seawater. The $\delta^{44/40}Ca$ of calcite cements from the Carboniferous (Moscovian) of Steuber and Buhl are about −0.3 to −0.4 ‰ compared to the seawater curve of Farkas et al. (2007). The deviations from the proposed 0 ‰ fractionation may be caused either by uncertainties in the seawater reconstructions or by precipitation from a fluid, which $\delta^{44/40}Ca$ differs from seawater. A detailed overview about complications related to $\delta^{44/40}Ca_{seawater}$

reconstructions based on different archives, are discussed in Sect. 4.1 of Chapter "Global Ca Cycles: Coupling of Continental and Oceanic Processes".

Aragonitic and calcitic ooids

Most aragonitic ooids formed from present seawater show $\delta^{44/40}$Ca between 0.3 and 0.5 ‰, corresponding to a $\Delta^{44/40}$Ca$_{arag-sw}$ of -1.6 to -1.4 ‰ (Steuber and Buhl 2006; Blättler et al. 2012). This fractionation is within the range of values reported for synthetic aragonite. In contrast to the significant fractionation revealed by recent aragonitic ooids, Cretaceous ooids, composed of primary calcite, rather imply little or no Ca isotope fractionation relative to contemporaneous seawater (Steuber and Buhl 2006). This apparent discrepancy may be caused by several processes. Different amounts of Ca contributed by bioclast nuclei to the ooid may shift its $\delta^{44/40}$Ca towards lower values. Alternatively, the difference between recent aragonitic and Cretaceous calcitic ooids may be caused by recrystallisation or due to different equilibrium isotope fractionation of calcite and aragonite. The latter hypothesis is consistent with observed Ca isotope fractionation of aragonite and calcite cements formed in mid ocean ridge basalts, with $\Delta^{44/40}$Ca$_{Cc-fluid}$ of ± 0 ‰ and $\Delta^{44/40}$Ca$_{arag-fluid}$ of -1.3 to -1.7 ‰ (Böhm et al. 2012; Amini et al. 2008).

Ikaite

Ikaite ($CaCO_3 \cdot 6H_2O$) formed in shallow sediment depth at different locations shows $\delta^{44/40}$Ca values between 1.1 and 1.5 ‰, which corresponds to a $\Delta^{44/40}$Ca$_{ikaite-sw}$ of -0.4 to -0.8 ‰ relative to present seawater (Gussone et al. 2011). The degree of fractionation agrees to values obtained for synthetic ikaite. Samples originating from the continental margin off Uruguay and the Congo-fan show identical values ($\Delta^{44/40}$Ca$_{ikaite-sw}$ mainly between -0.6 and -0.8 ‰). In general samples from the Atlantic are slightly lighter compared to samples from a cold seep setting off Sumatra, which $\Delta^{44/40}$Ca values range between -0.4 and -0.6 ‰. The latter samples experienced already

transformation to calcite. The reduced fractionation compared to the Atlantic samples was suggested to result from either higher $\delta^{44/40}$Ca pore water values at the Sumatra site (smaller sediment depth) or higher growth rates due to the increased alkalinity at the cold seep site. Ikaite grown in Antarctic sea ice brines is about -0.6 ‰ relative to seawater, which is in the range of experimental and sedimentary ikaite (Gussone et al. 2011).

Dolomite

Initial results on dolomite $\delta^{44/40}$Ca values indicate distinct Ca isotope fractionation in primary and recrystallized dolomite, similar to the observations on primary and recrystallized calcite. While a $\Delta^{44/40}$Ca$_{dolomite-fluid}$ of about 0 ‰ is suggested for dolomite recrystallisation (cf. Jacobson and Holmden 2008; Holmden 2009), a significant enrichment of ^{40}Ca relative to the fluid is reported for primary dolomite formation in siliciclastic sediments (Wang et al. 2012, 2013; Teichert et al., submitted).

Calcium isotope ratios of dolomite and pore water of siliciclastic sediments at the Peru continental margin reveal a $\Delta^{44/40}$Ca$_{dolomite-fluid}$ of -0.5 to -0.7 ‰ for primary dolomite formation (Teichert et al., submitted). Similarly, $\Delta^{44/40}$Ca$_{dolomite-seawater}$ of -0.4 to -0.7 ‰ are observed for primary dolomites from two cold seep sites in the South China Sea, which formed in shallow sediment depth near the sea floor (Wang et al. 2012, 2013a). Numerical modelling based on dolomite sequences from on-shore outcrops of the Monterey formation and hypothetic paleo-pore water profiles (Blättler et al. 2015) reveals $\Delta^{44/40}$Ca$_{dolomite-fluid}$ of -0.4 ‰, which is slightly smaller than the measured values. Based on the modelling results, Blättler et al. (2015) suggest that the $\delta^{44/40}$Ca of dolomite depends on its formation depth in the sediment, assuming a constant $\Delta^{44/40}$Ca$_{dolomite-fluid}$ and a static pore water profile with decreasing $\delta^{44/40}$Ca$_{pore\ water}$ towards depth, mainly caused by dissolution of calcite. Consequently, shallow depth of precipitation (pore water close to seawater $\delta^{44/40}$Ca) results in relatively high $\delta^{44/40}$Ca$_{dolomite}$ values, while dolomite formation at greater depth, from a ^{40}Ca enriched pore fluid, results in lower $\delta^{44/40}Ca_{dolomite}$. According to this model, the Ca and Mg isotope variability observed within a dolomite bank is consistent with Rayleigh fractionation during precipitation from a restricted reservoir. However, Ca and Mg isotope ratios of pore waters and dolomites from a sediment core off Peru shows that Rayleigh fractionation alone cannot fully explain the pore water profiles (Teichert et al., submitted) and that additional processes take place in the sediment.

A stronger degree of Ca isotope fractionation ($\Delta^{44/40}Ca_{dolomite-Ca^{2+}}$ from −1.0 to −1.1 ‰) is reported for dolomites formed in lab experiments mediated by microbial activity (Krause et al. 2012). The difference in $\Delta^{44/40}Ca_{dolomite-Ca^{2+}}$ may result from different processes governing Ca isotope fractionation in dolomites from the lab experiments and those of the sedimentary environment, e.g. the contribution of EPS-derived Ca to the crystals or different precipitation rates (see Sect. 1 in Chapter "Biominerals and Biomaterial").

Through Earth history, dolomites of the sedimentary record exhibit a large variability in $\delta^{44/40}Ca$, reflecting different recrystallization histories, $\delta^{44/40}Ca$ of source rocks, $\delta^{44/40}Ca_{fluid}$, and fluid rock ratios during recrystallization (cf. Tipper et al. 2008; Jacobson and Holmden 2008; Fantle and Higgins 2014; Holmden 2009; Komiya et al. 2008). In different settings, the offset between the calcitic source rocks and the dolomite is variable with respect to the absolute value and sign. For instance, Böhm et al. (2011) report dolomite enriched in ^{44}Ca relative to the source rock, while Holmden (2009) reports a depletion in ^{44}Ca relative to surrounding limestones. These observations are not consistent with Ca isotope modelling of Artemov et al. (1967), predicting ^{44}Ca enrichment in dolomite relative to the source calcite, due to diffusive loss of ^{40}Ca, but can be explained by a 0 ‰ fractionation during dolomitization and different $\delta^{44/40}Ca$ of the involved reacting fluids (Holmden 2009).

Lakes

Case studies from two lakes indicate that the $\delta^{44/40}Ca$ of Ca carbonate minerals from lake sediments reveal information about the Ca budget of lakes. Both lakes, Mono Lake (California) and Laguna Potrok Aike (Patagonia) are located in arid areas and have currently no surface outflow. Their main Ca sink is the precipitation of authigenic Ca carbonate, which is reflected by higher $\delta^{44/40}Ca$ of the lake water compared to the lakes' main Ca sources. This successive enrichment in heavy isotopes in the lake water, caused by the preferential removal of light Ca isotopes by the precipitation of isotopically light $CaCO_3$, acts like a small scale analogue for the enrichment of ^{44}Ca in ocean water. Besides these similarities, there are significant differences between both lakes.

In Mono Lake, the apparent isotope fractionation between $CaCO_3$ minerals and lake water is highly variable, ranging from −0.2 to −1.8 ‰, which is explained by the large variability of the $Ca^{2+}:CO_3^{2-}$ of more than 4 orders of magnitude between spring water ($Ca^{2+}:CO_3^{2-} \sim 3$) and the highly alkaline lake water ($Ca^{2+}:CO_3^{2-} \sim 10^{-4}$) (Fig. 17; Nielsen and DePaolo 2013). Lake carbonates are strongly fractionated (up to −1.8 ‰), while shoreline carbonates, influenced by spring water, are less fractionated. The water of Mono Lake is highly super saturated with respect to $CaCO_3$ solid phases, due to high phosphate concentration, inhibiting $CaCO_3$ precipitation. The variability of lake water sampled over a 15 years period, ranges from 1.2 to 3.7 ‰ (SRM915a), while the main sources lie between 0.5 and 1.0 ‰. Excursions in the $\delta^{44/40}Ca_{lake\ water}$ towards higher values correspond to reduced [Ca^{2+}] and are related to large $CaCO_3$ precipitation events, caused by the breakdown of the lake's chemocline. The temporal changes and spatial variability observed in $\delta^{44/40}Ca_{lake\ water}$ and the Ca residence time of 1.5–10 years indicate that the Ca budget of Mono Lake is not in steady state and that Ca isotopes may not record in and output fluxes of the lake at short time scales (Nielsen and DePaolo 2013).

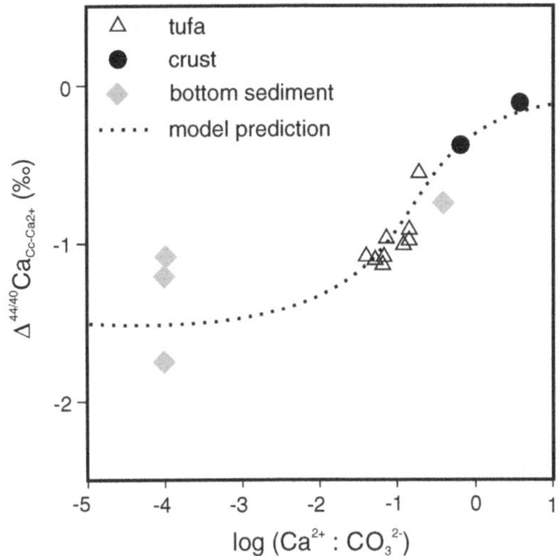

Fig. 17 Ca isotope fractionation of different precipitates from Mono Lake as a function of the molar $Ca^{2+}:CO_3$ ratio of the solution. Crusts and tufa precipitated from spring dominated water with higher $Ca^{2+}:CO_3$ are less fractionated than the bottom sediment carbonates formed within the lakes at low $Ca^{2+}:CO_3$ (redrawn from Nielsen and DePaolo 2013)

The main Ca sources of Laguna Potrok Aike have similar $\delta^{44/40}Ca$ (0.7–1.0 ‰) compared to Mono Lake, and the present lake water (1.30 ‰) is also enriched in ^{44}Ca relative to the source. In contrast to Mono Lake, the $\Delta^{44/40}Ca_{carbonate-Ca^{2+}}$ of Laguna Potrok Aike's sink is relatively uniform. Presently, ikaite is the dominant mineralogy of the Ca carbonates precipitating from the lake water, being −0.8 ‰ relative to the lake water. This fractionation agrees with $\Delta^{44/40}Ca_{carbonate-Ca^{2+}}$ obtained from surface sediments from different parts of the lake ranging from −0.7 to −0.9 ‰ (Oehlerich et al. 2015).

A downcore record from 15 ka towards today reveals $\delta^{44/40}Ca_{carbonate}$ between 0.5 and 0.9 ‰, corresponding to paleo lake water from 1.3 to 1.7 ‰. The $\delta^{44/40}Ca_{lake\ water}$ increases from ∼11 ka to the maximum value at ∼5 ka, followed by a decrease towards the present value. A change in $\delta^{44/40}Ca$ of the source Ca is not compatible with the Ca and Sr isotope signatures of potential sources, and can thus be largely excluded as main reason for the observed $\delta^{44/40}Ca$ variability. A shift in $\Delta^{44/40}Ca$ is also unlikely, because during the investigated time period the calcium carbonate record of Laguna Potrok Aike turned out to be most likely ikaite-derived. Therefore, the variability of $\delta^{44/40}Ca_{sediment}$ reflects predominantly changes in the Ca budget of the lake, with higher degrees of Ca utilization during the period with lowest lake level (Oehlerich et al. 2015).

The increased Ca utilization from ∼11 to 5 ka (with Ca output > input) and reduced utilization since 5 ka is in general agreement to other proxies, although the $\delta^{44/40}Ca$ signal is delayed compared to the maxima of TIC (total inorganic carbon) concentration and carbonate accumulation rates. This delay may be caused by the non-linear dependence of the $\delta^{44/40}Ca$ on the removal of dissolved lake water Ca by $CaCO_3$ precipitation (Oehlerich et al. 2015).

Soils

Calcium carbonate minerals formed in a hyper arid soil of the Atacama Desert show a large variability from −0.3 to +1.2 ‰ over a ca. 1 m depth profile (Ewing et al. 2008). Below a decrease within the upper few centimetres, $\delta^{44/40}Ca$ of the soil carbonates show a trend to higher values at depth. Similar to the coexisting sulfates (see Sect. 3.3.2), this pattern is suggested to be caused by repeated dissolution and precipitation events and preferred downward transport of heavy Ca isotopes. In the upper part of the depth profile, carbonate and sulfate $\delta^{44/40}Ca$ show the same pattern but are offset, with carbonates being +0.1 to +0.6 ‰ relative to sulfate. Towards the bottom of the profile $\delta^{44/40}Ca$ of the carbonates decrease to values below the coexisting sulfate. The offset between sulfate and carbonate might be related either to different fractionation factors (if both precipitate from the same fluid), or the result of transformation, i.e. dissolution-reprecipitation processes. The observation of carbonates being present on crack surfaces rather suggests the latter possibility or precipitation from a later, more evolved fluid (Ewing et al. 2008; see also Sect. 3.3.2.1).

In forested ecosystems, different types of pedogenic carbonates reveal different Ca isotope systematics. $\delta^{44/40}Ca$ of needle fibre calcite differs from inorganically formed calcite cements, formed in the same environment at identical temperatures (Milliere et al. 2014). In addition, three morphological groups of needle fibre calcites vary with respect to their $\delta^{44/40}$Ca. While inorganic cements are significantly fractionated ($\Delta^{44/40}$Ca$_{Cc-Ca^{2+}}$ about −1.0 ‰), the primary needle fibres are only slightly lower (−0.1 ‰) than the soil water. The more evolved morpho-types of NFC have intermediate $\delta^{44/40}$Ca values ranging between the primary needles and the cements. These results support the hypothesis that formation of needle fibre calcite is closely related to fungal hyphae. The Ca isotope fractionation observed for the inorganic calcite cements are consistent with the fractionation between groundwater and travertine of about −0.9 ‰ reported by Tipper et al. (2006).

3.1.2 Carbonates Formed at Elevated Temperatures

Relatively few data exist for carbonate minerals precipitated from fluids at higher temperatures (Fig. 18). Significant ^{44}Ca depletion in carbonates precipitated at higher temperature is inferred from fluids of the Long Valley hydrothermal

system (California). Along a 16 km transect, hydrothermal fluids mix with ground waters and CaCO$_3$ precipitation is initiated presumably due to CO$_2$ degassing (Brown et al. 2013). This is apparent from a decrease in Ca concentration from 20 to 5 ppm and increase in $\delta^{44/40}$Ca$_{fluid}$ from 0.4 to 0.9 ‰ in the first part of the investigated transect. Increasing [Ca] and decreasing $\delta^{44/40}$Ca at the end of this transect is attributed to calcite dissolution. $\Delta^{44/40}$Ca$_{Cc-Ca^{2+}}$ range from −0.1 to −0.5 ‰ at temperatures between 150 and 180 °C. The evolution of [Ca^{2+}] and $\delta^{44/40}$Ca$_{fluid}$ along the transect is well explainable by a Rayleigh distillation model assuming a $\Delta^{44/40}$Ca$_{Cc-Ca^{2+}}$ of −0.35 ‰ (Brown et al. 2013), while reaction with wall rocks can be largely excluded by the relatively invariable ^{87}Sr/^{86}Sr, suggesting rapid channelized fluid flow. The observation of significant Ca isotope fractionation is in agreement with $\Delta^{44/40}$Ca$_{carb-Ca^{2+}}$ of about −0.6 ‰ reported for carbonates from blueschist facies veins from the Tianshan mountains (Fig. 18), formed at temperatures of $\sim$500 °C (John et al. 2012). As the fluids involved in deep-subduction zone processes are not preserved, the calculated $\Delta^{44/40}$Ca$_{carbonate-Ca^{2+}}$ values are based on the assumption, that the Ca isotopic composition of the fluid largely resembles that of the silicate bulk rock. In contrast, calcites formed at the Logatchev hydrothermal fields, formed at temperatures between 130 and 160 °C show $\Delta^{44/40}$Ca$_{cc-sw}$ values of about 0 to +0.3 ‰ relative to seawater (Fig. 18; Amini et al. 2008).

Brown et al. (2013) suggest that the observed $\delta^{44/40}$Ca values are caused by kinetic fractionation, mainly controlled by forward and backward precipitation rates, while temperature plays only a minor role. This hypothesis can explain the similar $\Delta^{44/40}$Ca found over a wide temperature range from ambient temperatures up to 500 °C. The enrichment of ^{44}Ca in the calcites from the Logatchev hydrothermal field are likely related to the precipitation from an evolved fluid, likely seawater, which is depleted in ^{40}Ca by Rayleigh type fractionation due to carbonate precipitation, thus leading to the reduced apparent isotope

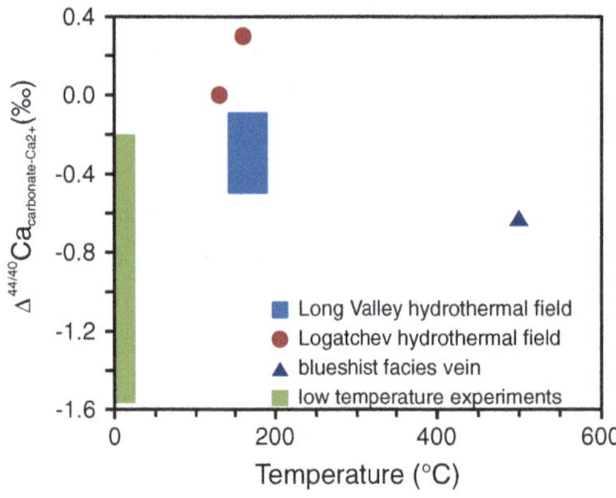

Fig. 18 Apparent Ca isotope fractionation of carbonates precipitated at higher temperatures. $\Delta^{44/40}$Ca range between −0.6 and +0.3 ‰ for temperature between 130 and 500 °C. Hydrothermal calcite from the Logatchev hydrothermal field show high values (0–0.3 ‰) likely as a result of precipitation from an evolved fluid (Amini et al. 2008). Calcite from the Long valley hydrothermal field ranges from −0.5 to −0.1 ‰ (Brown et al. 2013) and that of *blueshist* facies vein from 0.6 to 0.64 ‰ (John et al. 2012)

fractionation (Amini et al. 2008). Alternatively, the elevated $\delta^{44/40}$Ca values might be related to recrystallization, i.e. higher backward reaction rate, compared to the carbonates from the Long Valley hydrothermal system and the Tianshan blueshist.

3.1.3 Carbonate Recrystallization

Calcium isotope fractionation between Ca carbonate minerals and pore fluid is about 0 ‰ during recrystallization under near equilibrium conditions in marine carbonate sediments (Fantle and DePaolo 2007) and long lived terrestrial aquifers (Jacobson and Holmden 2008). In carbonate rich sediments this is evident from converging $\delta^{44/40}$Ca of fluid and solid phase (Fig. 19). The depth of recrystallization and rate of Ca isotope equilibration, apparent from the convergence of $\delta^{44/40}$Ca$_{\text{pore water}}$ and $\delta^{44/40}$Ca$_{\text{solid}}$ depends on the respective sediment characteristics and is used to determine carbonate recrystallization rates in marine sediments (Fantle and DePaolo 2007). In carbonate rich sediments equilibration is achieved in shallow depth of about 80 mbsf (Fantle and DePaolo 2007), while it takes place in greater depth (>500 m) in siliciclastic organic rich sediments, due to reduced carbonate dissolution rates in such sediments (Turchyn and DePaolo 2011). The lack of isotope fractionation during recrystallization is also in general agreement to results obtained from dissolution-reprecipitation experiments showing that no Ca isotope fractionation takes place during the replacement of aragonite by apatite (Kasioptas et al. 2011).

In certain sedimentary settings, the calculation of recrystallization rates from Ca isotopes is complicated as $\delta^{44/40}$Ca of sedimentary pore water is not only influenced by $CaCO_3$ dissolution and recrystallization processes.

The $\delta^{44/40}$Ca$_{\text{pore water}}$ can be affected by the advection and diffusion of deep sources fluids of different origin and Ca isotope signatures. The isotopic composition of hydrothermal fluids was shown to resemble the isotope composition of mid ocean ridge basalt (~ 0.8 to 1 ‰)

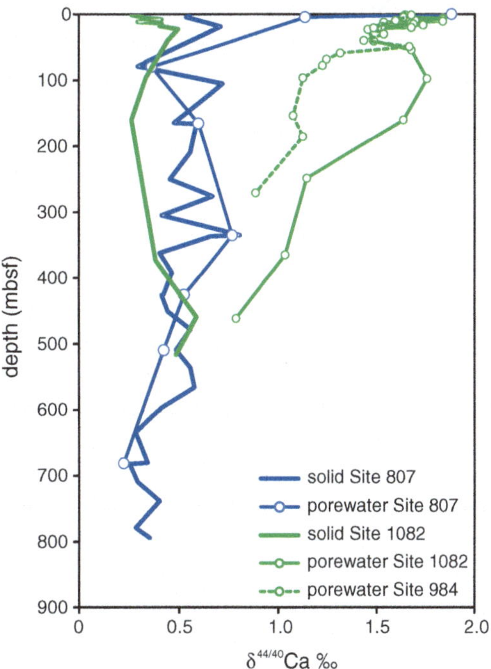

Fig. 19 Calcium isotope ratios of $CaCO_3$ and porewater of marine sediments. $\delta^{44/40}$Ca of sediment and pore waters converge towards depth due to carbonate dissolution and recrystallization. Depth profiles of carbonate oozes (ODP site 807) are best explained by assuming $\alpha \sim 1$ (Fantle and DePaolo 2007). While $\delta^{44/40}$Ca of solid and pore water equilibrate within the upper few tens of meters, in siliciclastic, organic rich sediments (IODP Sites 984 and 1082), solid and pore water equilibrate at greater depth (Turchyn and DePaolo 2011), due to reduced carbonate dissolution. Additional processes taking place in the sediments can affect $\delta^{44/40}$Ca$_{\text{pore water}}$ profiles, e.g. authigenic carbonate formation leading to elevated $\delta^{44/40}$Ca or the release of adsorbed Ca from clay mineral surfaces can shift the $\delta^{44/40}$Ca$_{\text{pore water}}$ towards lower values (Teichert et al. 2009; Ockert et al. 2013)

(Amini et al. 2008; Schmitt et al. 2003a; Teichert et al. 2009). In contrast, brines originating from ancient evaporitic basins are characterized by high $\delta^{44/40}$Ca values and high Ca concentrations (Teichert et al., submitted). Further diagenetic processes that take place in sediments and affect $\delta^{44/40}$Ca$_{\text{pore water}}$ include authigenic mineral precipitation, ion exchange on clay minerals (Sect. 4) and alteration of volcanic ashes (Teichert et al. 2005, 2009; Ockert et al. 2013, 2014).

3.2 Phosphates

Marine phosphate deposits can form in sediments with relative high organic carbon contents, e.g. in areas with high primary production such as upwelling zones at continental margins. The available data suggest that Ca isotope fractionation of Ca phosphates might vary depending on the type of phosphate-deposit. While the reported $\Delta^{44/40}Ca_{phosphate-sw}$ is about -1.0 ‰ for sedimentary peloidal Ca phosphate (Schmitt et al. 2003b), it ranges between -0.4 and -0.5 ‰ for phosphorite crusts (Arning et al. 2009). Paired $\delta^{44/40}Ca$ and $\delta^{18}O$ analyses of Miocene peloidal phosphates indicate a temperature dependent Ca isotope fractionation of about 0.05 ‰/°C, which is in the same order of magnitude as inorganic carbonates (Schmitt et al. 2003b). The reduced apparent Ca isotope fractionation in crusts relative to the peloidal phosphates might be related to different growth conditions, e.g. related to differences in growth rates, $Ca^{2+}:PO_4^{3-}$ stoichiometry etc. The different fractionation of both types of phosphate might also be related to the different environments they form in. As phosphate crusts grow within the sediments, their reduced apparent Ca isotope fractionation is consistent with precipitation from a semi-enclosed pore water reservoir, in which the light Ca isotopes are increasingly depleted during precipitation.

These observations are important to consider, because ancient marine phosphates are used for the reconstruction of $\delta^{44/40}Ca$ seawater records (Schmitt et al. 2003b; Soudry et al. 2004, 2006). Establishing reconstructions of Ca isotope seawater evolution based on different phosphate types, which show specific $\Delta^{44/40}Ca_{apatite-sw}$ can introduce significant artefacts. See also Sect. 4.1 of Chapter "Global Ca Cycles: Coupling of Continental and Oceanic Processes" for discussion on $\delta^{44/40}Ca$ paleo seawater reconstructions.

3.3 Sulfates

3.3.1 Barite

Barite formed in different settings of the marine realm, in the water column (pelagic marine barite),

at cold seep sites and hydrothermal vents, demonstrate typical distinct Ca isotope signatures. Natural marine pelagic barite shows a relatively small variability in $\delta^{44/40}Ca$ of $\sim 0.1 \pm 0.15$ ‰ (relative to SRM915a), which corresponds to an apparent isotope fractionation relative to seawater ($\Delta^{44/40}Ca_{barite-sw}$) of about -2.0 ± 0.15 ‰ (Griffith et al. 2008). The formation of pelagic barite occurs mainly in the upper 700 m of the water column, and is associated to the remineralisation of planktic organic matter. The exact processes taking place in the barite-forming microenvironments are not fully resolved, e.g., if Ca is derived directly from seawater or from the remineralised biomaterial, if reservoir effects might reduce the apparent isotope fractionation and at which growth rates barite is formed. The observed fractionation factor deviates from precipitation experiments ($\Delta^{44/40}Ca_{barite-Ca^{2+}}: -3.4 \text{ to} -2.4$ ‰) and modelling results ($\Delta^{44/40}Ca: -8$ to -9 ‰) (Griffith et al. 2008; Sect. 1.2.3). Based on the available data it is suggested that Ca isotope fractionation of pelagic barite is not controlled by environmental parameters. For instance, no correlation between Ca isotope fractionation and the water temperature of the upper 700 m of the water column is found. Also, the water depth of the site (depth of the sampled coretop, ranging between 3.1 and 4.4 km) does not show a significant control on Ca isotope fractionation. $\Delta^{44/40}Ca_{barite-sw}$ of pelagic barite is also identical for the Atlantic and the Pacific Oceans, showing no basin-specific fractionation. Additionally, it is shown that the sediment composition, the faunal composition and the barite accumulation rates do not control the Ca isotopic composition of the pelagic barites. It also appeared that the barite saturation of the bottom water and several water properties of the surface water (0–700 m) do not correlate with the Ca isotope fractionation.

$\delta^{44/40}Ca$ of barites formed at cold seep sites range from -0.8 to -1.4 ‰ ($\Delta^{44/40}Ca_{barite-sw}: -2.7$ to -3.4 ‰), and barites from hydrothermal vents from -0.8 to -2.1 ‰ ($\Delta^{44/40}Ca_{barite-sw}: -2.7$ to -4.0 ‰). The apparent Ca isotope fractionation relative to seawater is similar to that of synthetic barite, but more fractionated than the pelagic barites and less fractionated than the model-predictions (Griffith et al. 2008). The difference between pelagic, cold seep and hot vent barites might be related to either different fractionation factors (different precipitation rate, reservoir effects, etc.) or different fluid Ca isotope composition. The determination of $\Delta^{44/40}Ca_{barite-Ca^{2+}}$ of the natural barites might involve some uncertainties, because the involved fluids are not available and thus not well determined with respect to $\delta^{44/40}Ca_{fluid}$. In sediment settings with high organic carbon content additional complication might arise from dissolution and reprecipitation of barite due to sulfate reduction.

3.3.2 Anhydrite and Gypsum

Marine and Hydrothermal Ca Sulfates

Anhydrite precipitated in the marine hydrothermal Logatchev field is enriched in light Ca isotopes relative to the fluid. $\Delta^{44/40}Ca_{anhydr-sw}$ is -0.33 and -0.62 ‰ for precipitation temperatures of 234 and 113 °C, respectively (Amini et al. 2008). Combined with experimental results showing a fractionation of about -1.0 ‰ at 30–40 °C (Hensley 2006), the data indicate a temperature dependence with decreasing fractionation with increasing temperature (~ 0.003 ‰/° C), or alternatively differences in growth conditions, such as stoichiometry, precipitation rate or Rayleigh fractionation effects (Amini et al. 2008).

In natural evaporitic sequences, Ca sulfate deposits show a relatively large range in $\delta^{44/40}Ca$ (Blättler and Higgins 2014). This variability is explained by Rayleigh fractionation during ongoing Ca sulfate precipitation, as the $\delta^{44/40}Ca$ of the remaining brine successively increases, due to preferential incorporation of ^{40}Ca into the solid (see Sect. 1.2; Fig. 10). The $\delta^{44/40}Ca$ development of the solids and brines within these sequences is largely controlled by the $Ca^{2+}:SO_4^{2-}$ stoichiometry of the seawater, as it determines the fraction of Ca removed from the brine until Ca sulfate precipitation ceases. In evaporitic sequences formed from solutions with high $SO_4^{2-}:Ca^{2+}$ ratios, generally a large range in $\delta^{44/40}Ca$ is found within a cycle, and high $\delta^{44/40}Ca$ is

apparent at its end. In contrast, a smaller variability within the evaporitic sequence and a less pronounces increase in $\delta^{44/40}Ca$ is typical for successions formed from seawater with low $SO_4^{2-}:Ca^{2+}$, because the Rayleigh fractionation effect is smaller, the larger the Ca excess over sulfate is. Examples from Earth history for evaporitic sequences formed from brines with SO_4^{2-} excess are cycles from the Messinian, with a range in $\delta^{44/40}Ca$ of 1.3 ‰ and maximum $\delta^{44/40}Ca_{solid}$ 0.24 ‰ higher than present seawater and cycles from the Permian showing a variability of up to 1.7 ‰ (Blättler and Higgins 2014). In contrast, evaporitic cycles from the Cretaceous and the Silurian show a smaller range of ~ 0.4 ‰, which is indicative for lower $SO_4^{2-}:Ca$ ratios. The first Ca sulfate formed in a sequence is about -1.0 ‰ relative to the seawater and can help to characterise the isotopic composition of the evaporate basin relative to the open ocean. $\delta^{44/40}Ca$ of anhydrite from the Ordovician indicate on average higher $\delta^{44/40}Ca_{water}$ of the evaporitic basin than contemporaneous ocean water, which is consistent with the preferential removal of ^{40}Ca by the Ca sulfates (Holmden 2009).

Calcium Sulfates in Soils

Calcium sulfates (variably hydrated) from a hyper-arid soil in the Atacama Desert show a range in $\delta^{44/40}Ca$ from -0.8 to 1.7 ‰ SRM915a (Ewing et al. 2008). The correlated fractionation of S, O and Ca isotopes is consistent with isotope fractionation during downward transport by small, infrequent, but regular rainfall events. Internal fractionation processes within the soil are most likely, as dust, the main Ca input to the soil, is relative homogeneous with respect to $\delta^{44/40}Ca$. The depth profile of the $\delta^{44/40}Ca_{sulfate}$ in this soil shows a considerable decrease within the upper few centimetres, which is followed by an increase towards depth (Ewing et al. 2008). This trend is related to repeated precipitation and dissolution of Ca bearing minerals and preferred downward transport of heavy dissolved Ca, due to the incorporation of light Ca in the solid mineral phases. Assuming that the dust-derived Ca is transported downward, the $\delta^{44/40}Ca$ at a given depth reflects the fraction of Ca precipitated at that depth relative to the Ca inventory below it. Following this approach, the observed trend can be reproduced by a model using a $\Delta^{44/40}Ca_{sulfate-Ca^{2+}}$ of about -0.4 ‰ (Ewing et al. 2008), which is smaller compared to the -1.0 ‰ of synthetic anhydrite (Hensley 2006) and -1.5 to -0.8 ‰ for synthetic gypsum (Harouaka et al. 2014; Blättler and Higgins 2014). The model outcome suggests that the upper level of the profile is controlled by smaller rain events and is now in steady-state, which is achieved after 10^5 years. The observed $\delta^{44/40}Ca$ variation apparent in the lower part of the profile is suggested to be determined by episodically occurring larger rain events (Ewing et al. 2008).

The coexisting Ca carbonate and Ca sulfate phases of this soil show distinct trends in their $\delta^{44/40}Ca$ profile. In the upper part, $\delta^{44/40}Ca$ profiles of sulfates and carbonates are almost parallel, with carbonates being overall heavier than sulfates. In the lower part of the profile carbonate and sulfate minerals show opposite trends, leading in the lower third of the profile to higher $\delta^{44/40}Ca$ of sulfates compared to the coexisting carbonates. This observation indicates that coexisting carbonates and sulfates are not cogenetic at the base of the profile. The genetic relation of Ca carbonates and Ca sulfates is not clear, but it is suggested that carbonates crystallize on cracks of the dominant Ca sulfates or that the Ca sulfate replaces carbonates.

4 Diffusion, Exchange and Adsorption of Cations in Aqueous Systems

The dependence of solute diffusion in aqueous solutions on the isotopic composition is well documented for many aspects like marine pore water and terrestrial groundwater evolution, solvent exchange and mineral formation (e.g. DePaolo 2004; La Bolle et al. 2008; Lavastre et al. 2005; Richter et al. 2006; Donahue et al. 2008; Bourg et al. 2010; Beekmann et al. 2011; Wanner and Hunkeler 2015). In contrast to knowledge about isotope fractionation by diffusion in melts, significant gaps exist for the mechanisms and quantification of potential stable isotope fractionation of dissolved ions during diffusion (e.g. Bourg et al. 2010; Chopra et al. 2012 and references therein).

Stable isotope fractionation of dissolved ions and molecules during diffusion can be described by an inverse power law relation with the diffusion coefficient D (in $m^2\ s^{-1}$) being proportional to $m^{-\beta}$, where m denotes the isotopic mass of the solute with exponent β ranging between 0 (predicted by hydrodynamic theory) and 0.5 (kinetic theory) (Bourg and Sposito 2007; Bourg et al. 2010). Accordingly, the fractionation coefficient by aqueous diffusion is obtained by the equation

$$\alpha_{diffusion} = D_i/D_j = \left(m_j/m_i\right)^{\beta} \qquad (2)$$

where i and j denote the respective isotopes and as required the reduced mass in respect to solute hydration spheres has to be used instead of m (e.g. Gussone et al. 2003). Between 25 and 75 °C no temperature dependence was observed for β of different elements (Bourg and Sposito 2007). Generally, β is largest for uncharged noble gases, followed by monovalent ions (e.g. alkali elements) and small values are observed for divalent ions (e.g. alkaline earth elements). In addition, β decreases with the solute radius. While alkali elements show a considerable isotope fractionation during diffusion in aqueous fluids, very small isotope fractionation is reported for Ca and Mg (e.g., Richter et al. 2006; Bourg et al. 2010). Diffusion of aqueous Ca^{2+} ions to a growing calcite may yield in light isotope preferentially bound in the crystal, but strongly limited.

The absence of measureable EASI fractionation during diffusion in aqueous fluids was attributed to the special nature of ionic diffusion in water compared to other liquids (like melts), e.g. solvation and dielectric effects. Two effects determine isotope fractionation in aqueous fluids: (1) collision dominated friction, a process with a strong kinetic mass dependent isotope fractionation. (2) hydrodynamic friction, a process that reduces isotope fractionation, which is important at large solute radius or strong solute solvent interaction. These mechanisms explain the larger fractionation for monovalent compared to divalent cations. Longer residence times of ions in the hydration sphere lead to more collisions with ions in water sphere, rendering the mass dependent isotope fractionation during diffusion and leading to smaller β-values (Bourg et al. 2010). Accordingly, water surrounding dissolved cations within hydration shells is assumed to be an important limiting factor for isotopic fractionation associated with diffusion.

While EASI fractionation is mostly negligible during diffusion in aqueous fluids, significant isotope fractionation is observed for elements migrating through media like gels, resins or soils. Responsible for this isotope fractionation effect is the interaction with charged surfaces, adsorption and desorption processes, e.g. also referred to the formation of complexes at the surface of a solid (see Chapter "Analytical Methods"). The large isotope fractionation effect on chromatographic columns is shown for several elements including alkaline earth metals like Ca and Mg (cf. Russel and Papanastassiou 1978). Such fractionation effects can introduce artificial mass dependent isotope fractionation during Ca purification on chromatographic columns, thus hindering investigations on $\delta^{44/40}$Ca of impure materials (see Chapter "Analytical Methods"), or stable isotope fractionation between precipitates and pore solutions in gel. On the other hand mass dependent Ca isotope fractionation on chromatographic columns is used for artificial isotope enrichments. Several studies are published on the systematics of Ca and Sr isotope fractionation and chemical purification, investigating the behavior of calcium on ion exchange columns using different acids, acidities and complexing ions (cf. Heumann 1972; Heumann et al. 1977, 1982; Heumann and Lieser 1972; Heumann and Klöppel 1979, 1981; Heumann and Schiefer 1980, 1981; Fujii et al. 1985, 2010; Ban et al. 2001; Jepson 1992). The observed Ca isotope fractionation taking place on the column depends on the kind and strength of the elution phase. By changing the properties of the elution phase it is possible to change the magnitude and sign of isotope fractionation.

Ion exchange reactions have the potential to fractionate alkaline earth metal isotopes not only in artificial chromatographic columns, but also in natural environments. Such fractionation effects are of great interest, as alkaline earth metals participate in basic biogeochemical reactions in soils and marine sediment e.g. during diagenesis. Therefore, changing concentrations of dissolved alkaline earth metals in marine pore waters provide important information about early diagenetic reactions. Additional insights into these processes can be gained from variations of the stable isotope composition of alkaline earth metals. Besides radiogenic isotopes, which are well established tools for the characterisation of fluid sources and processes like ash alteration, the stable isotope composition of Ca and Mg can help to identify processes such as authigenic carbonate precipitation, carbonate dissolution, recrystallization and desorption from sediment-particle surfaces as well as fluid sources, e.g. brines and hydrothermal fluids. In Ca carbonate rich sediments, the Ca isotope composition is governed by $CaCO_3$ dissolution and reprecipitation (see Sect. 3.1.3). In siliciclastic sediments, the upper ~ 100 m are often characterised by authigenic carbonate precipitation, leading to a decrease in dissolved Ca by up to 80 %. This decrease in concentration is often not accompanied by the expected depletion in light isotopes in the pore water (Teichert et al. 2005, 2009). A mechanism that is suggested to overwhelm the precipitation signal is the release of Ca which is attached at surfaces of sediment particles, mainly clay minerals (Hindshaw et al. 2011; Teichert et al. 2009). Laboratory experiments on Ca exchange on clay minerals reveal that light isotopes are preferentially adsorbed to the clays and can thus provide a source for light Ca isotopes, but the degree of isotope fractionation depends on the respective mineral (Ockert et al. 2013). Kaolinite shows the largest isotope fractionation during calcium ion adsorption with $\Delta^{44/40}$Ca$_{ads\text{-}fluid}$ of about -1.2 to -3.0 ‰. Calcium isotope fractionation for adsorption onto

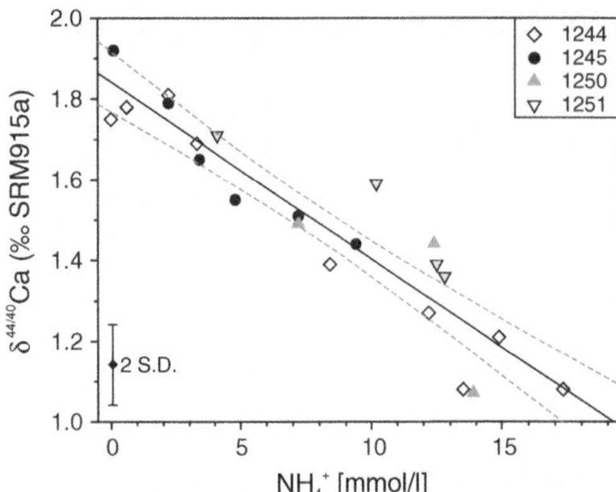

Fig. 20 $\delta^{44/40}$Ca of sedimentary pore water as a function of NH$_4^+$. Sedimentary pore water from the Cascadia margin demonstrates a strong correlation between NH$_4^+$ and $\delta^{44/40}$Ca, suggesting the release of isotopically light Ca from clay mineral surfaces due to increasing competition with NH$_4^+$ for adsorption sites (modified from Teichert et al. 2009)

illite range from -0.5 to -1.2 ‰, and $\Delta^{44/40}$Ca$_{ads\text{-}fluid}$ of montmorillonite is about -0.5 ‰.

In marine sedimentary pore water, Ca is released from clay minerals if either the Ca concentration in the pore water decreases (e.g. due to carbonate precipitation) or if the NH$_4^+$ concentration increases, e.g. as a result of organic matter degradation. As NH$_4^+$ and Ca^{2+} compete for adsorption sites at clay mineral surfaces, increased supply of NH$_4^+$ promotes the release of adsorbed Ca. This process leads to a dependence of $\delta^{44/40}$Ca$_{pore\ water}$ and the ammonium concentration (Fig. 20; Teichert et al. 2009). Because the Ca concentration is determined by the carbonate chemistry, the latter process is not observable in the Ca concentration, but only in the Ca isotopes (Teichert et al. 2009). These observations are compatible to Mg isotope fractionation suggested for Mg adsorption on clay surfaces, indicated by the offset of weathered source rocks and sampled river water (Tipper et al. 2006, 2008).

In accordance with Ca isotope fractionation behaviour, the ligther Cu isotopes are preferential adsorbed onto kaolinite surfaces (Δ^{65}Cu$_{solid-Cu^{2+}} = -0.3$‰; Li et al. 2015) and on the bacterium *P. aureofaciens* (Δ^{65}Cu$_{solid-Cu^{2+}} = -1.2$‰;

$1.8 \leq$ pH ≤ 3.5; Pokrovsky et al. 2008). In addition, the adsorption of lighter Cd isotopes is favored at Mn oxyhydroxide surfaces (birnessite: $\Delta^{114/112}$Cd$_{solid-Cu^{2+}} \approx -0.2$‰; Wasylenki et al. 2014). Contrarily, adsorption of Cu and Zn isotopes onto Al- and Fe-(hydr)oxides yields in preferentially heavier isotopes bound to the solid surface (Pokrovsky et al. 2008; Balistrieri et al. 2008). This discrepant isotope fractionation behaviour during cation adsorption onto solid surfaces can be caused by the distinct adsorption of isotopically different cation species in aqueous solutions (e.g. by considering various cation coordination environments) or/and by kinetic (lighter isotopes preferentially absorbed) versus equilibrium control (heavier isotopes preferentially adsorbed) of the adsorption process. The latter mechanism can be quantified by using the surface reaction kinetic approach by DePaolo (2011) developed for fractionation behavior during calcite formation.

References

Amini M, Eisenhauer A, Böhm F et al (2008) Calcium isotope ($\delta^{44/40}$Ca) fractionation along hydrothermal pathways, Logatchev field (Mid-Atlantic Ridge, 14° 45′N). Geochim Cosmochim Acta 72:4107–4122

Arning ET, Lückge A, Breuer C et al (2009) Genesis of phosphorite crusts off Peru. Mar Geol 262:68–81

Artemov YM, Strizhov VP, Ustinov VI et al (1967) Possible isotope fractionation during dolomitization. Geochemistry (USSR) (English trans.) 4:443–451

Balistrieri L, Borrok DM, Wanty RB et al (2008) Fractionation of Cu and Zn isotopes during adsorption onto amorphous Fe(III) oxyhydroxide: experimental mixing of acid rock drainage and ambient river water. Geochim Cosmochim Acta 72:311–328

Ban Y, Nomura M, Fujii Y (2001) Isotope effects of strontium in crown ether chromatography. Sep Sci Technol 36:2165–2180

Beekman HE, Eggenkamp HGM, Appelo CAJ (2011) An integrated modelling approach to reconstruct complex solute transport mechanisms—Cl and δ^{37}Cl in pore water of sediments from a former brackish lagoon in The Netherlands. Appl Geochem 26:257–268

Blättler CL, Henderson GM, Jenkyns HC (2012) Explaining the Phanerozoic Ca isotope history of seawater. Geology 40:843–846

Blättler CL, Higgins JA (2014) Calcium isotopes in evaporates record variations in Phanerozoic seawater SO$_4$ and Ca. Geology 42:711–714

Blättler CL, Miller NR, Higgins J (2015) Mg and Ca isotope signatures of authigenic dolomite in siliceous deep-sea sediments. Earth Planet Sci Lett 419:32–42

Böhm F, Eisenhauer A, Fietzke J et al (2011) Calcium isotope fractionation during dolomite formation. In: Goldschmidt Conference Abstracts, Mineralogical Magazine, p. 544

Böhm F, Eisenhauer A, Tang J et al (2012) Strontium isotope fractionation of planktic foraminifera and inorganic calcite. Geochim Cosmochim Acta 93:300–314

Böttcher ME, Geprägs P, Neubert N et al (2012) Barium isotope fractionation during experimental formation of the double carbonate $BaMn[CO_3]_2$ at ambient temperature. Isot Environ Health Stud 48:457–463

Bourg IC, Sposito G (2007) Molecular simulations of kinetic isotope fractionation during the diffusion of ionic species in liquid water. Geochim Cosmochim Acta 71:5583–5589

Bourg IC, Richter FM, Christensen JN et al (2010) Isotopic mass dependence of metal cation diffusion coefficients in liquid water. Geochim Cosmochim Acta 74:2249–2256

Brown ST, Kennedy BM, DePaolo DJ et al (2013) Ca, Sr, O and D isotope approach to defining the chemical evolution of hydrothermal fluids: example from Long Valley, CA, USA. Geochim Cosmochim Acta 122:209–225

Chopra R, Richter FM, Watson EB et al (2012) Magnesium isotope fractionation by chemical diffusion in natural settings and in laboratory analogues. Geochim Cosmochim Acta 88:1–18

Colla CA, Wimpenny J, Yin Q-Z et al (2013) Calcium-isotope fractionation between solution and solids with six, seven or eight oxygens bound to Ca (II). Geochim Cosmochim Acta 121:363–373

Davis K (2008) Resolving the nanoscale mechanisms of calcite growth from nonstoichiometric and microbial solutions. Ph.D. thesis, Rice University

DePaolo DJ (2004) Calcium isotopic variations produced by biological, kinetic, radiogenic and nucleosynthetic processes. In: Johnson CM, Beard BL, Albarède F (eds) Geochemistry of non-traditional stable isotopes. Miner Soc Am, Washington, DC, pp 255–288

DePaolo DJ (2011) Surface kinetic model for isotopic and trace element fractionation during precipitation of calcite from aqueous solutions. Geochim Cosmochim Acta 75:1039–1056

Dickson JAD (1991) Disequilibrium carbon and oxygen isotope variations in natural calcite. Nature 353:842–844

Dietzel M, Gussone N, Eisenhauer A (2004) Precipitation of aragonite by membrane diffusion of gaseous CO_2 and the coprecipitation of Sr^{2+} and Ba^{2+} (10° to 50° C). Chem Geol 203:139–151

Donahue MA, Werne JP, Meile et al (2008) Modeling sulfur isotope fractionation and differential diffusion during sulfate reduction in sediments of the Cariaco Basin. Geochim Cosmochim Acta 72:2287–2297

Druhan JL, Steefel CI, Williams KH et al (2013) Calcium isotope fractionation in groundwater: Molecular scale processes influencing field scale behavior. Geochim Cosmochim Acta 119:93–116

Eiler JM, Bergquist B, Bourg I et al (2014) Frontiers of stable isotope geoscience. Chem Geol 372:119–143

Ewing S, Yang W, DePaolo DJ et al (2008) Non-biological fractionation of stable Ca isotopes in soils of the Atacama Desert, Chile. Geochim Cosmochim Acta 72:1096–1110

Fantle MS, DePaolo DJ (2007) Ca isotopes in carbonate sediment and pore fluid from ODP Site 807A: The Ca^{2+}(aq)-calcite equilibrium fractionation factor and calcite recrystallization rates in Pleistocene sediments. Geochim Cosmochim Acta 71:2524–2546

Fantle MS, Higgins J (2014) The effects of diagenesis and dolomitization on Ca and Mg isotopes in marine platform carbonates: implications for the geochemical cycles of Ca and Mg. Geochim Cosmochim Acta 142:458–481

Farkaš J, Buhl D, Blenkinsop J et al (2007) Evolution of the oceanic calcium cycle during late Mesozoic: evidence from $\delta^{44/40}Ca$ of marine skeletak carbonates. Earth Planet Sci Lett 253:96–111

Fenter P, Geissbühler P, DiMasi E et al (2000) Surface speciation of calcite observed in situ by high-resolution X-ray reflectivity. Geochim Cosmochim Acta 64:1221–1228

Fietzke J, Eisenhauer A (2006) Determination of temperature-dependent stable strontium isotope (88Sr/86Sr) fractionation via bracketing standard MC-ICP-MS. Geochem Geophys Geosyst 8. doi:10.1029/2006GC001243

Fujii Y, Hoshi J, Iwamoto H et al (1985) Calcium isotope effects in ion exchange electromigration and calcium isotope analysis by thermo-ionization mass spectrometry. Z Naturforsch 40a:843–848

Fujii Y, Nomura M, Kaneshiki T et al (2010) Mass dependence of calcium isotope fractionations in crown-ether resin chromatography. Isot Environ Health Stud 46:233–241

Gagnon AC, Depaolo DJ, DeYoreo JJ (2010), Calcium Isotope Signature of Amorphous Calcium Carbonate: A Probe of Crystallization Pathway? Abstract PP22A-08 presented at 2010 Fall Meeting, AGU, San Francisco, Calif., 13–17 Dec

Geske A, Goldstein RH, Mavromatis V et al (2015) The magnesium isotope ($\delta^{26}Mg$) signature of dolomites. Geochim Cosmochim Acta 149:131–151

Griffith EM, Schauble EA, Bullen TD et al (2008) Characterization of calcium isotopes in natural and synthetic barite. Geochim Cosmochim Acta 72:5641–5658

Giuffre AJ, Gagnon AC, De Yoreo JJ et al (2015) Isotopic tracer evidence for the amorphous calcium carbonate to calcite transformation by dissolution-reprecipitation. Geochim Cosmochim Acta 165:407–417

Gussone N, Böhm F, Eisenhauer A et al (2005) Calcium isotope fractionation in calcite and aragonite. Geochim Cosmochim Acta 69:4485–4494

Gussone N, Eisenhauer A, Heuser A et al (2003) Model for kinetic effects on calcium isotope fractionation (δ^{44}Ca) in inorganic aragonite and cultured planktonic foraminifera. Geochim Cosmochim Acta 67:1375–1382

Gussone N, Nehrke G, Teichert BMA (2011) Calcium isotope fractionation in ikaite and vaterite. Chem Geol 285:194–202

Harouaka K, Eisenhauer A, Fantle MS (2014) Experimental investigation of Ca isotopic fractionation during abiotic gypsum precipitation. Geochim Cosmochim Acta 129:157–176

Henderson GM, Chu NC Bayon G et al (2006) $\delta^{44/42}$Ca in gas hydrates, porewaters and authigenic carbonates from Niger Delta sediments. Geochim Cosmochim Acta S70:A244

Hensley TM (2006) Calcium isotope variation in marine evaporates and carbonates, applications to late Miocene Mediterranean brine chemistry and late Cenozoïc calcium cycling in the oceans, PhD Scripps institution of oceanography, UC San Diego, http://escholarship.org/uc/item/3fk9j79s

Heumann KG (1972) Calciumisotopieeffekte beim Ionenaustausch an Dowex A1. Z Naturforsch 27a:492–497

Heumann KG, Klöppel H (1979) Abhängigkeit der Calcium-Isotopenseparation bei der Ionenaustauschchromatographie von der HNO3-Elutionsmittel konzentration. Z Naturforsch 34(b):1044–1045

Heumann KG, Klöppel H (1981) Calciumisotopenseparation und Reaktionsenthalpie. Z Anorg Allg Chem 472:83–88

Heumann KG, Lieser KH (1972) Untersuchung von Calziumisotopieeffekten bei heterogenen Austauschgleichgewichten. Z Naturforschung 27b:126–133

Heumann KG, Schiefer HP (1980) Calcium Isotope Separation on an Exchange Resin Having Cryptand Anchor Groups. Angew Chem Int Ed Engl 19:406–407

Heumann KH, Schiefer HP (1981) Calcium Isotopenseparation mit Kryptanden als Komplexbildner. Z Naturforsch 36b:566–570

Heumann KG, Gindner F, Klöppel H (1977) Abhängigkeit des Calcium-Isotopieeffekts von der Elektrolytkonzentration bei der Ionenaustausch-Chromatographie. Angew Chem 89:753–754

Heumann KH, Klöppel H, Sigl G (1982) Inversion der Calcium-Isotopenseparation an einem Ionentauscher durch Veränderung der LiCl-Ionenkonzentration. Z Naturforsch 37(b):786–787

Higgins JA, Schrag DP (2010) Constraining magnesium cycling in marine sediments using magnesium isotopes. Geochim Cosmochim Acta 74:5039–5053

Hindshaw RS, Reynolds BC, Wiederhol JG et al (2011) Calcium isotopes in a proglacial weathering environment: Damma Glacier, Switzerland. Geochim Cosmochim Acta 75:106–118

Holmden C (2009) Ca isotope study of Ordovician dolomite, limestone, and anhydrite in the Williston Basin: implications for subsurface dolomitization and local Ca cycling. Chem Geol 268:180–188

Hofmann AE, Bourg IC, DePaolo DJ (2012) Ion desolvation as a mechanism for kinetic isotope fractionation in aqueous systems. Proc Nat Acad Sci USA 109:18689–18694. doi:10.1073/pnas.1208184109

Immenhauser A, Buhl D, Richter D et al (2010) Magnesium-isotope fractionation during low-Mg calcite precipitation in a limestone cave—field study and experiments. Geochim Cosmochim Acta 74:4346–4364

Jacobson AD, Holmden C (2008) δ^{44}Ca evolution in a carbonate aquifer and its bearing on the equilibrium isotope fractionation factor for calcite. Earth Planet Sci Lett 270:349–353

Jepson BE (1992) Calcium isotope separation by chemical exchange with polymer-bound crown compound. Proceedings of the international symposium on isotope separation and chemical exchange uranium enrichment. Bull Res Lab Nucl Reactors (special issue) 1:307–312

John T, Gussone N, Podladchikov YY et al (2012) Volcanic arcs fed by rapid pulsed fluid flow through subducting slabs. Nat Geosci 5:489–492

Jouzel J, Merlivat L (1984) Deuterium and oxygen 18 in precipitation: modeling of the isotopic effects during snow formation. J Geophys Res 89:11,749–711

Kasioptas A, Geisler T, Perdikouri C et al (2011) Polycrystalline apatite synthesized by hydrothermal replacement of calcium carbonates. Geochim Cosmochim Acta 75:3486–3500

Kendall C, McDonnell JJ (1998) Isotope tracers in catchment hydrology. Elsevier, Amsterdam 839

Komiya T, Suga A, Ohno T et al (2008) Ca isotopic compositions of dolomite, phosphorite and the oldest animal embryo fossils from the Neoproterozoic in Weng'an, South China. Gondwana Res 14:209–218

Krause S, Liebetrau V, Gorb S et al (2012) Microbial nucleation of Mg-rich dolomite in exopolymeric substances under anoxic modern seawater salinity: new insight into an old enigma. Geology 40:587–590

La Bolle E, Fogg GE, Eweis JB et al (2008) Isotopic fractionation by diffusion in groundwater. Water Resour Res 44. doi:10.1029/2006WR005264

Lavastre V, Jendrzejewski N, Agrinier P et al (2005) Chlorine transfer out of a very low permeability clay sequence (Paris Basin, France): ^{35}Cl and ^{37}Cl evidence. Geochim Cosmochim Acta 69:4949–4961

Lemarchand D, Wasserburg GJ, Papanastassiou DA (2004) Rate-controlled calcium isotope fractionation in synthetic calcite. Geochim Cosmochim Acta 68:4665–4678

Li W, Beard BL, Johnson CM (2011) Exchange and fractionation of Mg isotopes between epsomite and saturated MgSO4 solution. Geochim Cosmochim Acta 75:1814–1828

Li W, Chakraborty S, Beard BL et al (2012) Magnesium isotope fractionation during precipitation of inorganic calcite under laboratory conditions. Earth Planet Sci Lett 333:304–316

Li W, Beard BL, Li C et al (2014) Magnesium isotope fractionation between brucite [$Mg(OH)_2$] and Mg aqueous species: implications for silicate weathering and biogeochemical processes. Earth Planet Sci Lett 394:82–93

Li W, Beard BL, Li C et al (2015) Experimental calibration of Mg isotope fractionation between dolomite and aqueous solution and its geological implications. Geochim Cosmochim Acta 157:164–181

Marriott CS, Henderson GM, Belshaw NS et al (2004) Temperature dependence of δ^7Li, δ^{44}Ca and Li/Ca during growth of calcium carbonate. Earth Planet Sci Lett 222:615–624

Mavromatis V, Schmidt M, Botz R et al (2012) Experimental quantification of the effect of Mg on calcite-aqueous fluid oxygen isotope fractionation. Chem Geol 310–311:97–105

Mavromatis V, Gautier Q, Bosc O et al (2013) Kinetics of Mg partition and Mg stable isotope fractionation during its incorporation in calcite. Geochim Cosmochim Acta 114:188–203

Milliere L, Verrecchia E, Gussone N (2014) Stable calcium isotope composition of a pedogenic carbonate in forested ecosystem: the case of the needle fibre calcite (NFC). EGU general assembly conference abstracts, vol 16, p 5812

Niedermayr A, Dietzel M, Kisakürek B et al (2010) Calcium isotopic fractionation during precipitation of calcium carbonate polymorphs and ACC at low temperatures. Geophysical research abstracts, vol 12, p 12448-2

Nielsen LC, DePaolo DJ (2013) Ca isotope fractionation in a high-alkalinity lake system: Mono Lake, California. Geochim Cosmochim Acta 118:276–294

Nielsen LC, DePaolo DJ, De Yoreo JJ (2012) Self-consistent ion-by-ion growth model for kinetic isotopic fractionation during calcite precipitation. Geochim Cosmochim Acta 86:166–181

Ockert C, Gussone N, Kaufhold S et al (2013) Isotope fractionation during Ca exchange on clay minerals in a marine environment. Geochim Cosmochim Acta 112:374–388

Ockert C, Wehrmann LM, Kaufhold S et al (2014) Calcium-ammonium exchange on standard clay minerals and natural marine sediments in seawater. Isotopes in Environmental and Health Studies 50:1–17

Oehlerich M, Mayr C, Gussone N et al (2015) Late Glacial and Holocene climatic changes in south-eastern Patagonia inferred from carbonate isotope records of Laguna Potrok Aike (Argentina). Quatern Sci Rev 114:189–202

Oelze M, von Blanckenburg F, Bouchez J et al (2015) The effect of Al on Si isotope fractionation investigated by silica precipitation experiments. Chem Geol 397:94–105

Pearce CR, Saldi GD, Schott J et al (2012) Isotopic fractionation during congruent dissolution, precipitation and at equilibrium: evidence from Mg isotopes. Geochim Cosmochim Acta 92:170–183

Pokrovsky OS, Viers J, Emnova EE et al (2008) Copper isotope fractionation during its interaction with soil and aquatic microorganisms and metal oxy(hydr) oxides: Possible structural control. Geochim Cosmochim Acta 72:1742–1757

Reynard LM, Day CC, Henderson GM (2011) Large fractionation of calcium isotopes during cave-analogue calcium carbonate growth. Geochim Cosmochim Acta 75:3726–3740

Richter FM, Mendybaev RA, Christensen JN et al (2006) Kinetic isotopic fractionation during diffusion of ionic species in water. Geochim Cosmochim Acta 70:277–289

Roerdink DL, van den Boorn SHJM, Geilert S et al (2015) Experimental constraints on kinetic and equilibrium silicon isotope fractionation during the formation of non-biogenic chert deposits. Chem Geol 402:40–51

Russell WA, Papanastassiou DA (1978) Calcium isotope fractionation in ion-exchange chromatography. Anal Chem 50:1151–1153

Rustad JR, Casey WH, Yin Q-Z et al (2010) Isotopic fractionation of Mg^{2+}(aq), Ca^{2+}(aq), and Fe^{2+}(aq) with carbonate minerals. Geochim Cosmochim Acta 74:6301–6323

Saulnier S, Rollion-Bard C, Vigier N et al (2012) Mg isotope fractionation during calcite precipitation: an experimental study. Geochim Cosmochim Acta 91:75–91

Schmitt AD, Chabaux F, Stille P (2003a) The calcium riverine and hydrothermal isotopic fluxes and the oceanic calcium mass balance. Earth Planet Sci Lett 213:503–518

Schmitt AD, Stille P, Vennemann T (2003b) Variations of the ^{44}Ca/^{40}Ca ratio in seawater during the past 24 million years: Evidence from δ^{44}Ca and delta δ^{18}O values of Miocene phosphates. Geochim Cosmochim Acta 67:2607–2614

Soudry D, Segal I, Nathan Y et al (2004) ^{44}Ca/^{42}Ca and ^{143}Nd/^{144}Nd isotope variations in Cretaceous-Eocene Tethyan francolites and their bearings on phosphogenesis in the southern Tethys. Geology 32:389–392

Soudry D, Glenn CR, Nathan Y et al (2006) Evolution of Tethyan phosphogenesis along the northern edges of the Arabian-African shield during the Cretaceous-Eocene as deduced from temporal variations of Ca and Nd isotopes and rates of P accumulation. Earth Sci Rev 78:27–57

Steuber T, Buhl D (2006) Calcium-isotope fractionation in selected modern and ancient marine carbonates. Geochim Cosmochim Acta 70:5507–5521

Tang J, Dietzel M, Böhm F et al (2008a) Sr^{2+}/Ca^{2+} and ^{44}Ca/^{40}Ca fractionation during inorganic calcite formation: II. Ca isotopes. Geochim Cosmochim Acta 72:3733–3745

Tang J, Köhler SJ, Dietzel M (2008b) Sr^{2+}/Ca^{2+} and ^{44}Ca/^{40}Ca fractionation during inorganic calcite formation: I. Sr incorporation. Geochim Cosmochim Acta 72:3718–3732

Tang J, Niedermayr A, Köhler SJ et al (2012) Sr^{2+}/Ca^{2+} and ^{44}Ca/^{40}Ca fractionation during inorganic calcite formation: III. Impact of salinity/ionic strength. Geochim Cosmochim Acta 77:432–443

Teichert BMA, Meister P, Ockert C et al (submitted) Primary dolomite formation—insights from Ca-isotopes

Teichert BMA, Gussone N, Eisenhauer A et al (2005) Clathrites—archives of near-seafloor pore fluid evolution ($\delta^{44/40}$Ca, δ^{13}C, δ^{18}O) in seep environments. Geology 33:213–216

Teichert BMA, Gussone N, Torres ME (2009) Controls on calcium isotope fractionation in sedimentary porewaters. Earth and Planetary Science Letters 279:373–382

Tipper ET, Galy A, Bickle MJ (2006) Riverine evidence for a fractionated reservoir of Ca and Mg on the continents: implications for the oceanic Ca cycle. Earth Planet Sci Lett 247:267–279

Tipper ET, Galy A, Bickle MJ (2008) Calcium and magnesium systematics in rivers draining the Himalaya-Tibetan-Plateau region: lithological or fractionation control? Geochim Cosmochim Acta 72:1057–1075

Turchyn AV, DePaolo DJ (2011) Calcium isotope evidence for suppression of carbonate dissolution in carbonate-bearing organic-rich sediments. Geochim Cosmochim Acta 75:7081–7098

von Allmen K, Böttcher M, Samankassou E et al (2010) Barium isotope fractionation in the global barium cycle: first evidence from barium minerals and precipitation experiments. Chem Geol 277:70–77

Wang SH, Yan W, Magalhães HV et al (2012) Calcium isotope fractionation and its controlling factors of authigenic carbonates in the cold seeps from the Northern South China Sea. Chin Sci Bull 57:1325–1332

Wang S, Yan W, Magalhães HV et al (2013a) Factors influencing methane-derived authigenic carbonate formation at cold seep from southwestern Dongsha area in the northern South China. Environ Earth Sci. doi:10.1007/s12665-013-2611-9

Wang Z, Hu P, Gaetani G et al (2013b) Experimental calibration of Mg isotope fractionation. Geochim Cosmochim Acta 102:113–123

Wanner P, Hunkeler D (2015) Carbon and chlorine isotopologue fractionation of chlorinated hydrocarbons during diffusion in water and low permeability sediments. Geochim Cosmochim Acta 157:198–212

Wasylenki L, Swihart JW, Romaniello S (2014) Cadmium isotope fractionation during adsorption to Mn oxyhydroxide at low and high ionic strength. Geochim Cosmochim Acta 140:212–226

Watkins JM, Hunt JD, Ryerson FJ et al (2014) The influence of temperature, pH, and growth rate on the $\delta^{18}O$ composition of inorganically precipitated calcite. Earth Planet Sci Lett 404:332–343

Watson EB (2004) A conceptual model for near-surface kinetic controls on the trace-element and stable isotope composition of abiogenic calcite crystals. Geochim Cosmochim Acta 68:1473–1488

Watson EB, Liang Y (1995) A simple model for sector zoning in slowly grown crystals: implications for growth rate and lattice diffusion, with emphasis on accessory minerals in crustal rocks. Am Mineral 80:1179–1187

Biominerals and Biomaterial

Nikolaus Gussone and Alexander Heuser

Abstract
Biominerals are important archives for various paleo-environmental proxies, and consequently, the understanding of biomineralisation related element and isotope fractionation is vital for reliable climate reconstructions. The Ca isotope composition of biominerals has been investigated to explore its potential for paleo-environmental reconstructions and to better understand biomineralisation processes. The overall range of Ca isotope fractionation reported for biominerals resembles that of inorganic minerals, but shows different responses to changes of environmental parameters such as fluid composition and growth rates. In this chapter, we review Ca isotope fractionation characteristics of biominerals from different taxa with respect to biomineralisation processes and potential proxy applications. Available data of alkaline earth metal isotope systems (Mg, Sr) are included for comparison.

Keywords
Biomineralisation · Trophic level effect · Foraminifers · Coccolithophores · Dinoflagellates · Corals · Sclerosponges · Coralline algae · Brachiopods · Mollusks

Biominerals form during the interplay of inorganic and biological processes. The element Ca is in this context of great interest, as it is a compound of biominerals and also vital for metabolic processes in organisms (e.g. cell signal transduction etc.). For biomineralisation, Ca plays a special role, because it is a main constituent of prevalent biominerals, including apatite, calcite, aragonite and Ca oxalate. The calcium carbonate minerals ($CaCO_3$—calcite, aragonite, vaterite) are the most important group

N. Gussone (✉)
Institut für Mineralogie, Universität Münster, Münster, Germany
e-mail: nguss_01@uni-muenster.de

A. Heuser
Steinmann-Institut für Geologie, Mineralogie und Paläontologie, University of Bonn, Bonn, Germany
e-mail: aheuser@uni-bonn.de

© Springer-Verlag Berlin Heidelberg 2016
N. Gussone et al., *Calcium Stable Isotope Geochemistry*,
Advances in Isotope Geochemistry, DOI 10.1007/978-3-540-68953-9_4

of biominerals based on the number of taxa, by which they are precipitated, and based on the global total mass accumulation rate. The Ca isotope composition of biominerals has been investigated, to explore its potential for paleo environmental reconstructions and to better understand biomineralisation processes. The main focus of this chapter lies on biominerals formed during biologically controlled biomineralisation by eucariota (Sects. 2 and 3) and possible applications of Ca isotopes as indicators for ecological and paleoenvironmental questions (Sect. 4), excluding constraining the global Ca budget and biomedical applications which are subject of Chapters "Global Ca Cycles: Coupling of Continental and Oceanic Processes" and "Biomedical Application of Ca Stable Isotopes", respectively.

Species used for Ca isotope studies have been selected based on their importance in respect to global element-cycling, their applicability as archives for paleoclimatic reconstructions or because of special biomineralisation features or availability. This chapter reviews fractionation characteristics of biominerals, their implications and applications, while technical aspects e.g. cleaning procedures, ion chromatographic-purification and mass spectrometry methods appropriate for the respective kind of sample material are described in Chapter "Analytical Methods" (Sect. 4).

1 Procaryota—Microbial Induced Biomineralisation

In the marine realm precipitation of microbially induced minerals, which are formed by the interaction of metabolic waste products with ions from the environment, is widespread and important for sediment properties, such as stability, porosity and permeability. Nevertheless, due to the similar appearance and geochemical composition of biogenically induced and inorganically precipitated minerals, a clear separation between both is often not straightforward. Therefore, Ca isotope fractionation of natural occurring, likely microbial induced carbonates

are discussed in Chapter "Calcium Isotope Fractionation During Mineral Precipitation from Aqueous Solution" (Sect. 3) together with natural inorganically formed carbonates.

Microbial activity has been suggested to promote the formation of primary dolomite, as its direct inorganic precipitation from seawater-like fluids at ambient temperatures is kinetically suppressed, by factors such as sulphate concentration or the hydration shell of Mg ions.

Microbial dolomite formation under anoxic conditions and associated Ca isotope fractionation has been studied in lab cultures at 21 °C with the sulphate reducing bacteria *Desulfobulbus mediterraneous* (Krause et al. 2012). Their experiments show dolomite crystals associated with exopolymeric substances (EPS) forming first dolomitic nano-spherules which later aggregate to micro-spherules. The EPS biofilm is enriched in Ca relative to the surrounding seawater-like fluid (Mg/Ca ~ 5), as Ca and Mg are present at similar concentrations in the biofilm. Isotopic analyses of Ca from the EPS reveal $\delta^{44/40}$Ca about -0.5 ‰ compared to the fluid. The dolomite crystals are about -1.0 to -1.1 ‰ relative to the fluid. This observation is interpreted as the result of a two-step fractionation, the first step associated with uptake into or adsorption on the EPS and the second step during precipitation of the dolomite from the pre-fractionated Ca of the EPS (Fig. 1).

The experimentally determined $\Delta^{44/40}$Ca of -1.1 ‰ demonstrates a stronger fractionation compared to the values reported for modern primary dolomites sampled from the South China Sea (Wang et al. 2012, 2013a), which $\Delta^{44/40}$Ca range between about -0.7 and -0.4 ‰ and from the Peru Margin showing $\Delta^{44/40}$Ca of about -0.5 ‰ (Teichert et al. submitted). The reason for this difference in apparent isotope fractionation is yet unclear, but might be related to different involved bacteria, fluid composition, growth rates or temperature. The smaller apparent isotope fractionation observed in the marine realm, could also be caused by reservoir effects (high degree of Ca consumption) in diffusion limited pore water spaces. Alternatively, it is possible that during ongoing dolomite

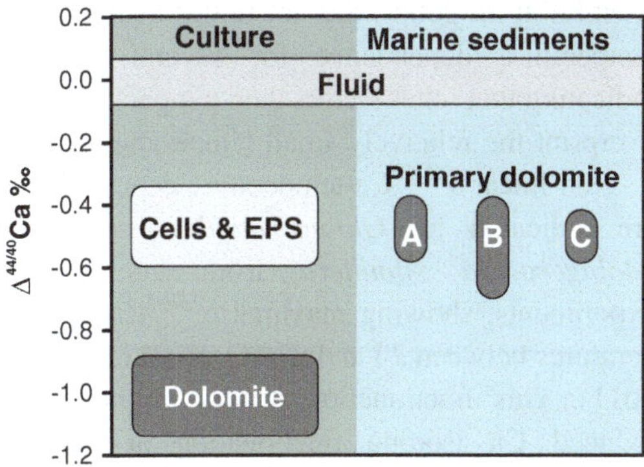

Fig. 1 Fractionation of Ca isotopes during microbial induced dolomite precipitation in lab cultures and in natural primary dolomites from the marine realm. Culture experiments of Krause et al. (2012) suggest a two-step Ca isotope fractionation of dolomite crystals ($\Delta^{44/40}$Ca: about -1.0 ‰) formed in and on bacterial EPS (-0.5 ‰ relative to the culture medium). In contrast, recent natural marine primary dolomites suggest $\Delta^{44/40}$Ca between -0.4 and -0.7 ‰, *A* South China Sea (Wang et al. 2013a), *B* South China Sea (Wang et al. 2012), *C* Peru Margin (Teichert et al. submitted)

precipitation over longer time periods, the dominant Ca source might change from solely EPS derived to bulk fluid dominated. This hypothesis would suggest that dolomite precipitation is initiated by the Ca and Mg supply of the microbial EPS, and at a later stage, un-fractionated Ca from the adjacent fluid becomes more important and eventually the dominant Ca source.

2 Protista

A comparatively large number of studies dealing with Ca isotope fractionation focus on biominerals formed by protista, in particular foraminifera and coccolithophores. These unicellular organisms largely contribute to the world oceans $CaCO_3$ export production, the main sink for the global oceanic Ca cycle. Knowledge of their fractionation behaviour is therefore important for defining the average Ca isotope fractionation of marine biogenic carbonates, an essential parameter for modelling the oceanic Ca budget through Earth history (see Chapter "Global Ca Cycles: Coupling of Continental and Oceanic Processes").

2.1 Foraminifera

Foraminifera are unicellular amoeba, which are known from the sedimentary record since the Early Cambrian (Pawlowski et al. 2003). Most species described originate from the marine realm, but few non-calcifying species are present in freshwater and soils (Meisterfeld et al. 2001). The majority of recent species life in marine benthonic habitats and comparatively few species are planktic. Besides by their habitat, foraminifera can be classified by the properties of their shells, the so called foraminifer tests. Some species surround themselves with organic scales, some with debris, building 'agglutinated' tests (with organic or $CaCO_3$-cement), while other species secret tests of $CaCO_3$. The latter group is further divided according to their test-structure into perforate and imperforate foraminifers, or hyaline and porcelaneous tests (ter Kuile 1991; Erez 2003), with species secreting low magnesium calcite (LMC), high magnesium calcite (HMC) or aragonite.

Foraminifers secreting $CaCO_3$ are widely used archives for paleo-environmental changes. For instance stable isotope ratios, element ratios and changes in faunal assemblages are used for the reconstruction of parameters such as temperature, salinity, global ice volume, pH, productivity and stratigraphic ages. Because of the importance of foraminifers as proxy archives, empirical calibrations of isotope fractionation and element partitioning behaviour are accompanied by the development of models in order to gain a fundamental understanding of biomineralisation and cellular element transport (Erez 2003; de Nooijer et al. 2014).

With respect to their Ca isotope fractionation, foraminifera are comparatively well studied, as the Ca isotope composition of foraminifers was initially proposed as potential recorder for past temperatures (cf. Zhu and Macdougal 1998; Nägler et al. 2000) and alternatively for changes in the oceanic Ca budget (Skulan et al. 1997; De La Rocha and DePaolo 2000). The discrepant observations regarding the apparent temperature dependence of Ca isotope fractionation triggered a series of studies that focussed on the characterisation and understanding of foraminiferal $\delta^{44/40}Ca$ in response to environmental parameters.

2.1.1 Planktic Foraminifers

Reported $\delta^{44/40}Ca$ values of recent planktic foraminifer tests span a range of nearly 4 ‰, from about −2.0 to +1.8 ‰ (relative to SRM 915a), while the majority of samples lie between 0.2 and 1.2 ‰. This corresponds to an apparent isotope fractionation ($\Delta^{44/40}Ca$) of about −0.7 to −1.7 ‰ relative to seawater (Fig. 2). Several environmental parameters, including temperature, salinity and carbonate chemistry of the surrounding fluid have been tested for their influence on Ca isotope fractionation of planktic foraminifers.

Although temperature is one of the most intensively studied factors influencing Ca isotope fractionation in foraminifers, its effect on the $\delta^{44/40}Ca$ of foraminifer tests is still not fully understood and subject of ongoing discussion. To clarify the question, if and how strong Ca isotope fractionation is influenced by temperature, different sample materials have been used, including foraminifers grown in culture experiments, natural samples caught by plankton nets from the ocean water, tests recently deposited in sediment traps or at the seafloor (from core top samples) or specimens obtained from down core records.

Most analysed species show either no resolvable $\delta^{44/40}Ca$-temperature dependence or sensitivities ≤0.03 ‰/°C (Fig. 2; Table 1), similar to that observed in inorganic precipitation experiments (Chapter "Calcium Isotope Fractionation During Mineral Precipitation from Aqueous Solution") or biominerals of other taxa. Significant $\delta^{44/40}Ca$-temperature dependencies are reported for *Orbulina universa* from controlled culture experiments (∼0.019 ‰/°C; Zhu 1999; Gussone et al. 2003, 2009), several recent foraminifer species retrieved from sediment surfaces (Griffith et al. 2008) and cross calibrated foraminifers from the Eocene (Kasemann et al. 2008). On the other hand, several studies that investigated modern foraminifers from core top samples (Sime et al. 2005; Chang et al. 2004; Kasemann et al. 2008) and sediment traps

(Griffith et al. 2008), did not reveal a significant temperature dependence for several species, indicating that other factors then temperature, can overprint the relatively small temperature effect.

Non-linear $\delta^{44/40}Ca$-temperature dependencies are indicated by *Globigerinoides ruber* and *Globigerinella sifonifera* from lab culture experiments, showing maximal $\delta^{44/40}Ca$ at temperatures between 24 and 27 °C (Kisakürek et al. 2011). This fractionation pattern might reflect reduced Ca isotope fractionation at optimal growth conditions. However, this observation is pending further verification, as the total range of $\delta^{44/40}Ca$ values found for each species is only about 0.2 ‰ and most data overlap within uncertainty.

Two species that show special fractionation behaviours are *Globigerinoides sacculifer* and *Neogloboquadrina pachyderma* (sin). *Globigerinoides sacculifer* (including data reported as *G. trilobus and G. quadrilobatus*) exhibits an enigmatic bimodal temperature dependence with two considerably different temperature sensitivities differing by one order of magnitude, ∼0.02 ‰/°C and ∼0.2 ‰/°C (Fig. 2; Table 1). Both trends are reported for specimens obtained from culture experiments (Nägler et al. 2000; Gussone et al. 2009), and natural samples, retrieved from either plankton nets (Hippler et al. 2006), sediment traps (Griffith et al. 2008), coretop sediments (Chang et al. 2004; Sime et al. 2005; Griffith et al. 2008; Kasemann et al. 2008) and downcore records (Nägler et al. 2000; Gussone et al. 2004; Hippler et al. 2006; Heuser et al. 2005; Sime et al. 2007). This suggests that the bimodal temperature response is not related to analytical artifacts or preservation, but likely caused by fractionation processes during biomineralisation, which are yet not fully understood.

Studies on *N. pachyderma* (sin.) reveal a complex fractionation pattern (Fig. 2), with one array at temperatures between 2 and 13 °C, showing a slightly smaller temperature dependence than *G. sacculifer*, and an "anomaly" at temperatures between −1.5 and 4 °C, with elevated $\delta^{44/40}Ca$, which might be related to lower carbonate ion concentrations or calcite saturation

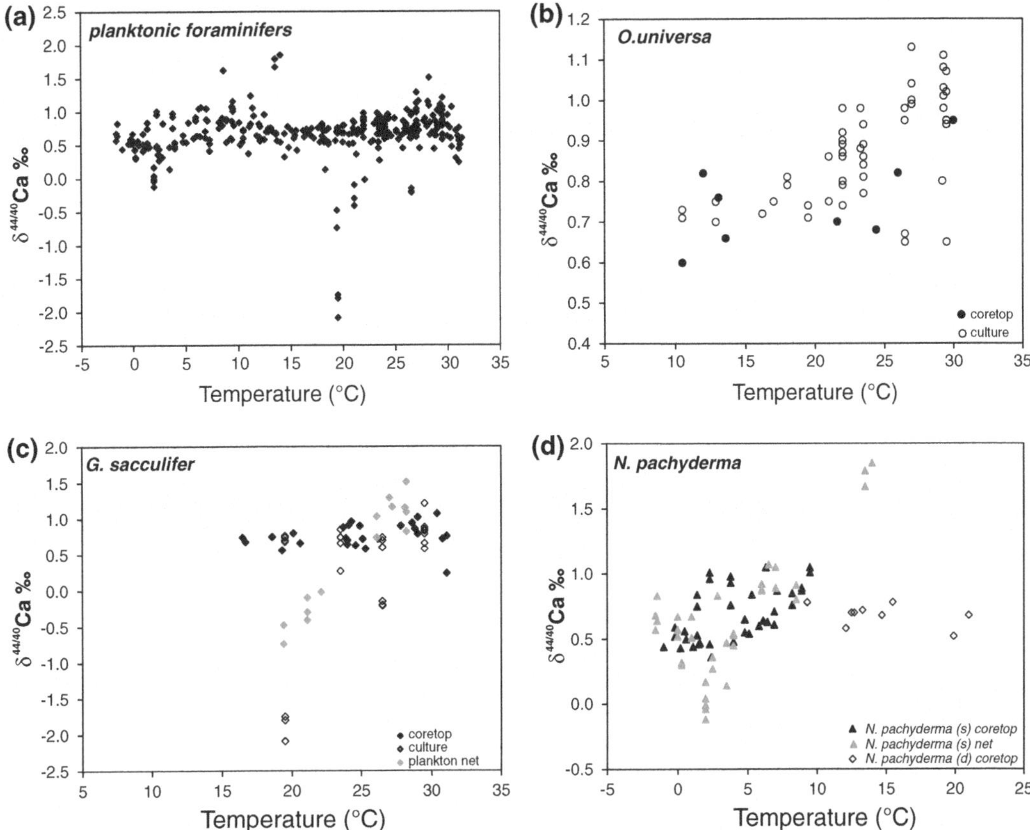

Fig. 2 Temperature dependent Ca isotope fractionation in planktic foraminifers. **a** $\delta^{44/40}$Ca of planktic foraminifers, from culture experiments and natural samples shown as a function of temperature. **b** *Orbulina universa* specimens from culture experiments and coretops reveal a small temperature dependent Ca isotope fractionation of ~0.02 ‰/°C (Gussone et al. 2003, 2009; Kasemann et al. 2008; Sime et al. 2005). **c** Cultured and natural individuals of *Globigerinoides sacculifer* reveal a bimodal temperature sensitivity in their Ca isotope fractionation, differing by an order of magnitude (Nägler et al. 2000; Chang et al. 2004; Hippler et al. 2006; Sime et al. 2005; Gussone et al. 2009). **d** *Neogloboquadrina pachyderma* sinistral (Hippler et al. 2009a; Gussone et al. 2009) exhibits a relatively large temperature dependence at higher temperatures and an anomaly at low temperatures, while *N. pachyderma* dextral (Sime et al. 2005) shows no indication for a temperature dependent Ca isotope fractionation

states in these water masses (Hippler et al. 2009a; Gussone et al. 2009) (see section on benthonic foraminifers).

The impact of the carbon chemistry of the surrounding water on $\delta^{44/40}$Ca in planktonic foraminifer tests was studied in culture experiments of *Globigerinoides ruber*, *Globigerinella sifonifera* (Kisakürek et al. 2011) and *Orbulina universa* (Gussone et al. 2003). All of the three investigated species show little variability in $\delta^{44/40}$Ca isotope ratios and no clear correlation between carbonate ion concentration and Ca isotope fractionation (Fig. 3). The results strongly differ from experiments that investigated inorganically precipitated calcite (Lemarchand et al. 2004), but resemble the slightly negative correlation of $\delta^{44/40}$Ca and carbonate ion concentration found in cultured coccolithophores (Sect. 2.2). As the latter dependence is found to be non-linear, with a stronger gradient at low carbonate concentrations, it is suggested that possibly the interference of temperature and carbonate chemistry (carbonate ion concentration or calcite saturation) could be responsible for the

Table 1 Temperature sensitivities of Ca isotope fractionation in tests of planktic foraminifers

Species	Temp. sensitivity (‰/°C)	Culture	Sediment traps	Plankton nets	Coretop	Downcore
Globigerina bulloides	≤0.03		13		5, 7, 13	9
Globigerinella aequilateralis	≤0.03				7	
Globigerinella siphonifera	≤0.03	16			12	9
Globigerinoides conglobatus	≤0.03				5, 12	
Globigerinoides ruber	≤0.03	16	13		5, 7, 12, 13	9
Globigerinoides sacculifer/trilobus	≤0.03/~0.2	3, 15	13	8	1, 5, 7, 12, 13	1, 2, 3, 6, 8, 9, 11
Globorotalia hirsula	≤0.03				5	
Globorotalia inflata	≤0.03				5, 7, 13	
Globorotalia menardii	≤0.03				5, 12	
Globorotalia scitula	≤0.03				7	
Globorotalia truncatulinoides	≤0.03				5, 7, 13	
Globorotalia tumida	≤0.03				12, 13	
Hastigerina pelagica	≤0.03				5	
Neogloboquadrina dutertrei	≤0.03		13		5, 7, 13	
Neogloboquadrina pachyderma (dextral)	≤0.03				7	
Neogloboquadrina pachyderma (sinistral)	~0.2			14, 15	1, 14, 15	
Orbulina universa	≤0.03	2, 4, 15			5, 7, 12, 15	11
Pulleniatina obliquiloculata	≤0.03				7, 12	
Sphaeroidinella dehiscens	≤0.03				12	
Turborotalia quinqueloba	≤0.03				10	

1: Zhu and Macdougall (1998); 2: Zhu (1999); 3: Nägler et al. (2000); 4: Gussone et al. (2003); 5: Chang et al. (2004); 6: Gussone et al. (2004); 7: Sime et al. (2005); 8: Hippler et al. (2006); 9: Heuser et al. (2005); 10: Gussone et al. (2005); 11: Sime et al. (2007); 12: Kasemann et al. (2008); 13: Griffith et al. (2008); 14: Hippler et al. (2009a); 15: Gussone et al. (2009); 16: Kisakürek et al. (2011)

apparent temperature dependent Ca isotope fractionation of *N. pachyderma* (sin.) in the Arctic Ocean (Hippler et al. 2009a; Gussone et al. 2009). This hypothesis is still pending verification, as the carbonate chemistry of the respective water masses from which the foraminifers were retrieved is unknown.

The response of Ca isotope fractionation in foraminifer tests on salinity revealed from culture experiments is not uniform (Fig. 4). A salinity increase from 32 to 44 psu leads to a 0.15 ‰ decrease of $\delta^{44/40}$Ca in *Globigerinoides ruber* while $\delta^{44/40}$Ca of *Globigerinella sifonifera* increases by about 0.1 ‰ (Kisakürek et al.

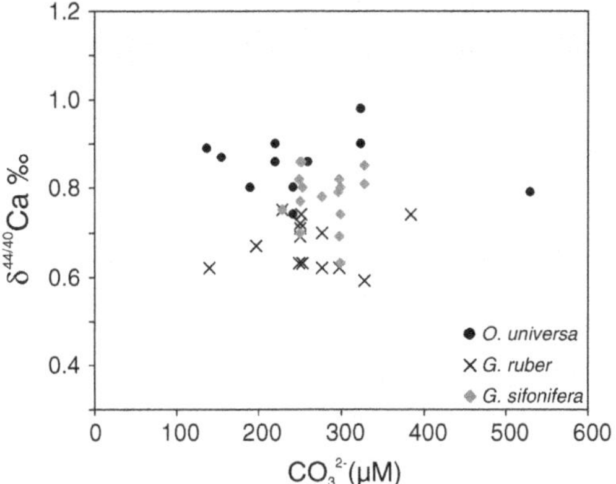

Fig. 3 Calcium isotope fractionation in planktic foraminifers as a function of ambient carbonate ion concentration. In contrast to inorganic calcite, planktic foraminifers show no significant dependence between $\delta^{44/40}$Ca and $[CO_3^{2-}]$ of the culture medium. Calcite data from Lemarchand et al. (2004), *Orbulina universa* from (Gussone et al. 2003), *Globigerinoides ruber* and *Globigerinella sifonifera* from Kisakürek et al. (2011)

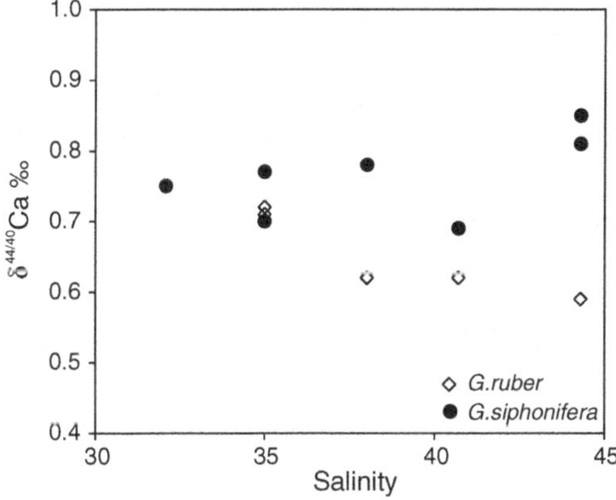

Fig. 4 Calcium isotope fractionation of planktic foraminifers as a function of salinity. $\delta^{44/40}$Ca shows little variability related to salinity in tests of *Globigerinoides ruber* and *Globigerinella sifonifera* (Kisakürek et al. 2011)

2011). Overall, this variability related to salinity is small and close to the analytical uncertainty. Similarly, $\delta^{44/40}$Ca of *Orbulina universa* does not show a clear salinity-dependence; possibly culture experiments at salinities of 33 and 36 indicate a small increase in temperature sensitivity with increasing salinity (Gussone et al. 2003, 2009). The influence of salinity on the Ca isotope fractionation of *G. sacculifer* is ambiguous.

Cultures at 36 psu reveal a large temperature sensitivity of ∼0.2 ‰/°C (Nägler et al. 2000), while experiments at 33 psu indicate for most specimens a small temperature sensitivity of ∼0.02 ‰/°C. Only those specimens at 33 psu that died prior to gametogenesis are consistent with a large temperature gradient (Gussone et al. 2009), as they show Ca isotope ratios close to the temperature dependent fractionation trend of Hippler et al. (2006). A correlation between salinity and temperature dependence is however not directly seen in *G. sacculifer* individuals from the natural environment (cf. Sime et al. 2005; Griffith et al. 2008). This indicates that changes in biomineralisation, rather than salinity itself are responsible for the observed fractionation effects in this species.

The suggestion that the formation of gametogenesis may influence the Ca isotope composition of foraminifers is supported by $\delta^{44/40}$Ca heterogeneities revealed by secondary ion mass spectrometry in tests of *Globorotalia inflata*, *Globorotalia truncatulinoides* and *Globorotalia tumida* (Rollion-Bart et al. 2007; Kasemann et al. 2008). In particular, Kasemann et al. (2008) show significant offsets between ontogenetic and gametogenetic calcite in *G. truncatulinoides* and *G. tumida* of up to ∼4 ‰ which are, however, opposite in sign for both species. Further research in this area will reveal new insight into biomineralisation and associated Ca isotope fractionation effects.

The understanding of EASI fractionation during foraminiferal biomineralisation is brought further by reported $\delta^{26/24}$Mg values. The overall range of $\Delta^{26/24}$Mg in planktic foraminifer tests lies between −5.2 ‰ and −3.2 relative to seawater (Chang et al. 2004; Pogge von Strandmann 2008; Wombacher et al. 2011). $\Delta^{26/24}$Mg obtained from the different studies show some systematic differences, e.g. regarding certain species including *O. universa* and *G. sacculifer*, but all studies consistently demonstrate that $\delta^{26/24}$Mg of planktic foraminifers is stronger fractionated than inorganic calcite (−1.5 to −3.2 ‰, Chapter "Calcium Isotope Fractionation During Mineral Precipitation from Aqueous Solution"). The presently available data do not indicate a resolvable temperature dependent Mg isotope fractionation in planktic foraminifers.

$\delta^{88/86}$Sr of planktonic foraminifers determined on *G. sacculifer* and *G. ruber* retrieved from core top sediments and samples from the last glacial maximum cluster around 0.14 ‰ (relative to NBS 987). This value corresponds to a fractionation of −0.25 ‰ relative to present seawater. Initial results suggest no significant dependence on temperature (Krabbenhöft 2010; Böhm et al. 2012). Instead, it was suggested that Sr isotope fractionation may relate to the precipitation rate, thus reflecting comparatively rapid precipitation in (semi)enclosed spaces with a low degree of Ca (and Sr) consumption.

Calcium isotope ratios of planktonic foraminifers have already been applied as recorder for $\delta^{44/40}$Ca$_{SW}$ variability (Chapter "Global Ca Cycles: Coupling of Continental and Oceanic Processes") as well as paleo sea surface temperature proxy (Sect. 4.3). Calcium isotope derived SST reconstructions as part of multi proxy approaches reveal an overall good agreement to Mg/Ca- and δ^{18}O-based temperatures (Nägler et al. 2000; Gussone et al. 2004; Hippler et al. 2006), while on the other hand planktonic foraminifer based $\delta^{44/40}$Ca records (including *G. sacculifer/trilobus*) seem to reflect predominantly changes in seawater Ca isotope composition (Sime et al. 2007; Heuser et al. 2005). As the origin of this enigmatic bimodal temperature dependence of *G. sacculifer* and its controlling factors are still not understood, it is important to consider this complication when interpreting planktonic foraminifer based Ca isotope records.

2.1.2 Benthonic Foraminifers

The Ca and Mg isotope composition of benthonic foraminifer tests have been studied on species secreting different test structures (hyaline and porcelaneous) and mineralogy (low Mg calcite, high Mg calcite and aragonite) from marine sediments, e.g. *Glabratella ornatissima*, *Alveolina* sp., *fusulina* sp., *Cibicidoides wuellerstorfi*, *Cibicides kullenbergi*, *Uvigerina peregrina*, *Cassidulina laevigata*, *Gyroidinoides* spp., *Elphidium* spp., *Pyrgo* spp.,

Quinqueloculina spp. and *Hoeglundina elegans* (Skulan et al. 1997; Zhu and Macdougall 1998; De La Rocha and DePaolo 2000; Chang et al. 2004; Gussone and Filipsson 2010; Gussone et al. 2016). The calcitic foraminifers reveal an overall $\Delta^{44/40}$Ca of −0.5 to −1.5 ‰ relative to seawater, and show no significant difference between test structures (hyaline and porcelaneous tests) and chemical composition [high Mg calcite (HMC) and low Mg calcite (LMC)]. The apparent temperature dependent Ca isotope fractionation pattern is characterised by a positive correlation of $\delta^{44/40}$Ca to temperature at higher temperature (>5 °C) similar to planktonic foraminifers and by an enrichment of heavy Ca isotopes at low temperatures ($\sim$2–3 °C) (Fig. 5). This enrichment of heavy isotopes at low temperatures is suggested to reflect a decrease in Ca isotope fractionation at low calcite saturation ($\Omega_{calcite}$) of the ambient fluid (Fig. 6), an effect that is demonstrated for cultured coccolithophores (Sect. 2.2).

While Ca isotope fractionation is similar in tests composed of HMC and LMC (Gussone et al. 2016), there is a considerable difference in Mg isotope fractionation between both foraminiferal groups. While LMC tests exhibit $\Delta^{26/24}$Mg between −4.3 and −3.6 ‰, which is in the range of planktic foraminifers, the HMC tests show $\Delta^{26/24}$Mg from −2.9 to −1.7 ‰, similar to inorganic calcite (Wombacher 2011; Pogge von Strandmann 2008; Chang et al. 2004; Yoshimura et al. 2011).

The observed EASI fractionation patterns of calcitic benthonic and planktic foraminifers are in agreement to the basic concept of foraminiferal biomineralisation of Erez (2003) and Elderfield et al. (1996). It suggests that Ca is introduced into the cell by two processes, uptake of seawater-filled vesicles and transmembrane Ca transport (e.g. by Ca channels), but the amount of Ca contributed to the test by each transport mechanism is still under debate.

Although Ca isotope fractionation of planktic and benthonic foraminifers found in different studies is similar, the data are partly ambiguous and led to considerably different hypotheses. For instance, Kisakürek et al. (2011) proposed based

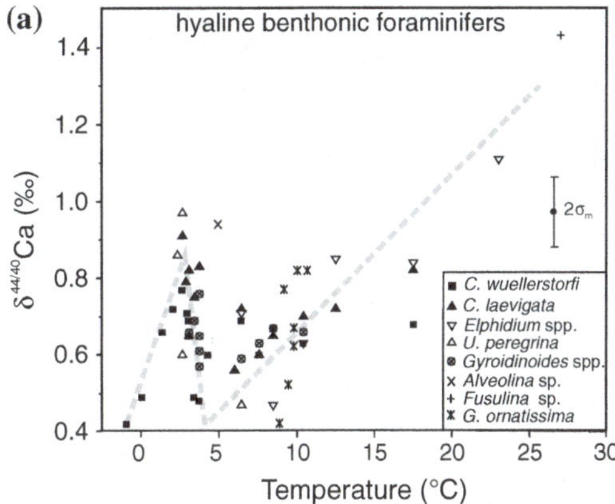

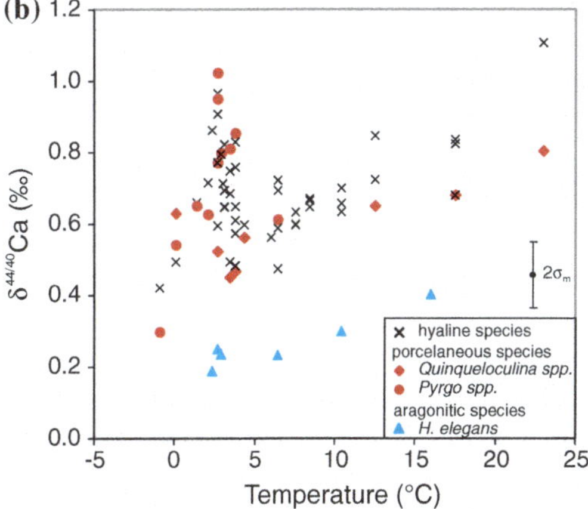

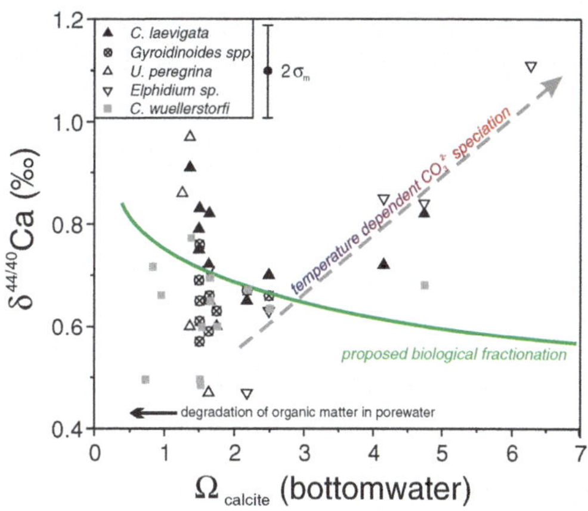

Fig. 6 Calcium isotope composition of benthic foraminifers as a function of bottom water calcite saturation. The relative high $\delta^{44/40}$Ca of benthonic foraminifers at low Ω_{Cc} (<2) is consistent with the proposed biological Ca isotope fractionation trend inferred from *E. huxleyi* (Gussone et al. 2007). At high Ω_{Cc} (>2) $\delta^{44/40}$Ca of benthonic foraminifers resemble the fractionation trend of inorganic calcite (Lemarchand et al. 2004). However, this apparent relation in benthonic foraminifers for Ω_{Cc} > 2 is presumably caused by covarying bottom water temperatures. Modified after Gussone and Filipsson (2010)

Fig. 5 Calcium isotope composition of benthonic for aminifers as a function of bottom water temperatures. **a** $\delta^{44/40}$Ca of hyaline benthonic foraminifers: *Alveolina* sp. and *Fusulina* sp. (Skulan et al. 1997), *G. ornatissima* (De La Rocha and DePaolo 2000), *C. wuellerstorfi, U. peregrina, C. laevigata, Gyroidinoides* spp. (*G. soldanii* and *G. neosoldanii*) and *Elphidium* spp. (Gussone and Filipsson 2010). **b** $\delta^{44/40}$Ca of Porcelaneous and aragonitic species: porcelaneous calcitic foraminifers show similar $\delta^{44/40}$Ca as hyaline species, while $\delta^{44/40}$Ca of the aragonitic *H. elegans* is about 0.4 ‰ lower than those of calcitic species (Gussone and Filipsson 2010; Gussone et al. 2016)

on $\delta^{44/40}$Ca and Sr/Ca systematics of planktic foraminifers that biomineralisation resembles inorganic calcite growth in a relatively open system, with either little Ca consumption (<25 %) from a seawater like Ca source or up to 40 % Ca consumption from a pre-fractionated fluid (0.2 ‰ lower than seawater). Griffith et al. (2008) suggested higher degrees of Ca consumption (85 %) during calcite precipitation

from a significantly pre-fractionated fluid (about −0.8 ‰ relative to seawater). As additional process Gussone et al. (2009) proposed temperature dependent mixing of differently fractionated reservoirs, to explain both temperature dependencies observed in the planktic foraminifer *G. sacculifer*. For benthonic foraminifers, Gussone and Filipsson (2010) and Gussone et al. (2016) propose that due to the high degree of Ca consumption during calcite formation in the calcifying space, foraminifer tests mainly reflect the $\delta^{44/40}$Ca of the pre-fractionated fluid. They also suggest for LMC and HMC secreting species a dominant contribution of transmembrane Ca transport relative to seawater vacuolisation to the formation of foraminifer tests. The latter hypothesis is based on the apparent temperature dependence of benthonic foraminifers, and does not contradict the observation that foraminifers largely follow the $\delta^{44/40}$Ca versus Sr/Ca systematic of inorganic calcite (Tang et al. 2008, 2012), as biological processes can mimic inorganic isotope and element fractionation patterns (Taubner et al. 2012).

The observation that Mg isotopes in planktic foraminifers are stronger fractionated than inorganic calcite has been explained by reduced Mg activity relative to that of Ca in the calcifying fluid due to Rayleigh-removal of Mg from vacuoles (Pogge von Strandmann 2008). However, Wombacher et al. (2011) argued that Rayleigh removal of Mg from vacuoles should result in a pronounced temperature dependence. They suggested a kinetic model, where planktic foraminifera may be able to synthesize biomolecules that increase the energetic barrier for Mg incorporation.

In contrast to Ca isotopes, Mg isotope fractionation is considerably different for benthonic foraminifers secreting either LMC or HMC. While benthonic foraminifer tests composed of HMC demonstrate Mg isotope fractionation, similar to inorganic calcite, benthonic species secreting LMC are stronger fractionated than inorganic calcite, in agreement to $\delta^{26/24}$Mg of planktic foraminifers. These observations are consistent with less reduction of Mg activity in the calcifying fluid of foraminifers precipitating HMC tests.

The aragonitic *Hoeglundina elegans* exhibits $\delta^{44/40}$Ca values, which are about 0.4 ‰ lower compared to the calcitic foraminifer tests (Fig. 5b). This difference is similar to the stronger fractionation in aragonite compared to calcite during inorganic mineral precipitation, but the low Sr concentrations in *H. elegans* tests indicate that biomineralisation of *H. elegans* cannot resemble inorganic aragonite precipitation; instead additional processes may take place during test formation, such as Ca transmembrane ion transport or a precursor phase to aragonite (Gussone et al. 2016).

Due to their habitat on or in the surface sediment, benthonic foraminifers record bottom water conditions and global environmental signals with little short-time climate variability. However, the complex Ca isotope fractionation pattern restricts the applicability of Ca isotopes in benthic foraminifers as proxy. The small temperature dependence of Ca isotope fractionation at temperatures above 5 °C does not allow reconstructions of paleo bottom water temperatures, given the current analytical precision. However, benthonic foraminifers seem to be useful as recorder for $\delta^{44/40}$Ca$_{sw}$ changes, as long as bottom water $\Omega_{calcite}$ does not decrease below ∼1.5 and/or temperatures below 4 °C.

2.2   Coccolithophores

Coccolithophores are unicellular marine phytoplankton, belonging to the taxon Haptophyta and are characterized by an exoskeleton composed of small calcite platelets, the 'coccoliths'. Coccolithophores are important primary producers and contribute about half the total marine $CaCO_3$ export production (Milliman 1993); therefore, they are an important link between global calcium (Ca) and carbon (C) cycling. Since coccoliths are one of the major Ca sinks of the marine realm, their Ca isotopic composition plays an important role in determining the mean isotopic fractionation between seawater and biogenic carbonate sediments (Δ_{sed}).

Coccolithophore species that are studied in respect to EASI fractionation include *Coccolithus braarudii* (Müller et al. 2011), *Coccolithus pelagicus* (Krabbenhöft et al. 2010), *Calcidiscus leptoporus*, *Helicosphaera carteri*, *Syrocosphaera pulchra* and *Umbilicosphaera foliosa* (Gussone et al. 2007), but focussed on *Emiliania huxleyi* (De La Rocha and DePaolo 2000; Gussone et al. 2006; Langer et al. 2007; Ra et al. 2010; Krabbenhöft et al. 2010; Müller et al. 2011). In general, EASI fractionation in *E. huxleyi* is characterised by a depletion of heavy Mg, Ca and Sr isotopes in coccoliths relative to the growth medium.

$\delta^{44/40}$Ca values of cultured coccoliths of *E. huxleyi*, *C. leptoporus*, *H. carteri*, *S. pulchra* and *U. foliosa* mainly range between 0.3 and 0.8 ‰ (SRM 915a), corresponding to a $\Delta^{44/40}$Ca of −1.1 to −1.6 ‰ (Gussone et al. 2006, 2007; Langer et al. 2007; De La Rocha and DePaolo 2000). Two cultures of *S. pulchra* show reduced fractionation as low as −0.6 ‰ ($\Delta^{44/40}$Ca) which origin is yet unclear. Overall, Ca isotope fractionation of coccoliths is similar for the different so far analysed taxa. Temperature dependencies

Table 2 Temperature sensitivity of cultured coccolithophores

Species	Isotope system	Temperature sensitivity (‰/°C)
E. huxleyi	$\delta^{44/40}$Ca	0.027 ± 0.006
H. carteri	$\delta^{44/40}$Ca	0.005 ± 0.005
S. pulchra	$\delta^{44/40}$Ca	0.029 ± 0.013
E. huxleyi	$\delta^{26/24}$Mg	0.06–0.1

$\delta^{26/24}$Mg: Ra et al. (2010), $\delta^{44/40}$Ca: Gussone et al. (2006, 2007)

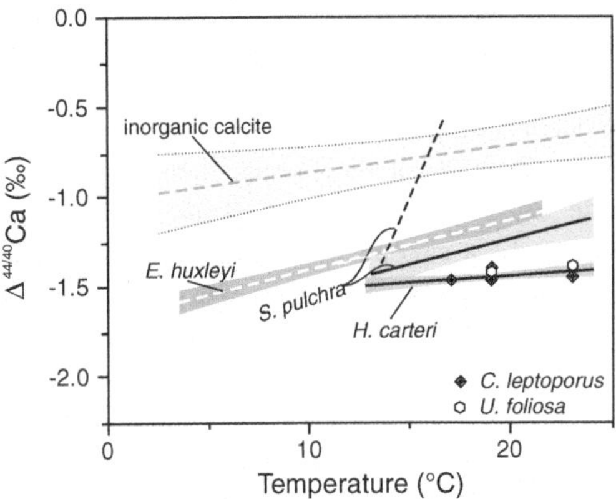

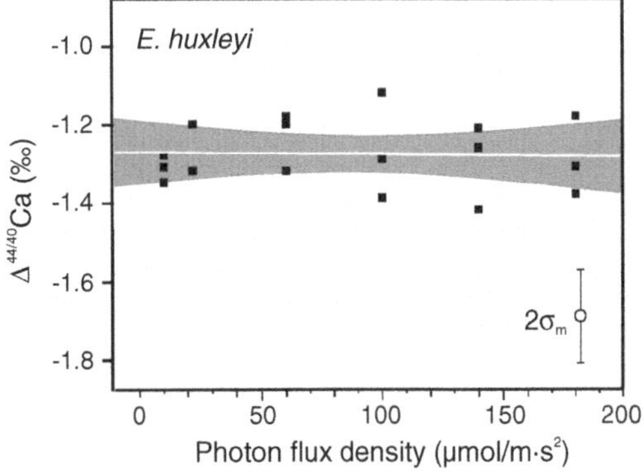

Fig. 7 Calcium isotope ratios of different cultured coccolithophore species as a function of temperature. The species *E. huxleyi*, *C. leptoporus*, *H. carteri*, *U. foliosa* and *S. pulchra* cover a similar range in $\delta^{44/40}$Ca but show some taxon specific difference in their temperature dependence (Table 2, Gussone et al. 2006, 2007)

Fig. 8 Calcium isotope ratios of *E. huxleyi* as a function of photon flux density (*A*) and salinity (*B*). Both environmental parameters have a large impact on coccolithophore growth rates, but do not affect Ca isotope fractionation (Langer et al. 2007)

range from 0.005 ‰/°C in *H. carteri* to 0.027 and 0.029 ‰/°C in *E. huxleyi* and *S. pulchra*, respectively (Table 2; Fig. 7) and resemble those for other marine taxa and inorganic precipitated carbonates (Chapter "Calcium Isotope Fractionation During Mineral Precipitation from Aqueous Solution"). Culture experiments reveal that in contrast to inorganic $CaCO_3$ precipitates (Lemarchand et al. 2004; Tang et al. 2008), $\delta^{44/40}$Ca of coccoliths does not depend on growth rate, either induced by salinity or photon flux density (Langer et al. 2007, Fig. 8). Consequently, the observed $\delta^{44/40}$Ca-temperature dependence in coccolithophores is most likely a real temperature effect and not caused by changes in growth- or precipitation rate.

Variation of Ca^{2+} and carbonate concentration of the growth media reveal a small effect on the Ca isotope fractionation at high [Ca^{2+}] and [CO_3^{2-}], and an increasing effect towards lower concentrations. Combining the data sets indicate a logarithmic decrease of $\delta^{44/40}$Ca with increasing calcite saturation ($\Omega_{calcite}$) (Fig. 9; Gussone et al. 2007). The underlying mechanism is not yet clear, but might be related to ^{40}Ca depletion in the proximity of the cell surface at low $\Omega_{calcite}$.

$\Delta^{26/24}$Mg of cultured *E. huxleyi* and *C. braarudii* coccoliths are −0.4 to −1.7 ‰ relative to the culture medium (Ra et al. 2010; Müller et al. 2011), which is less fractionated than inorganic calcite (cf. Galy et al. 2002). Chlorophyll extracted from *E. huxleyi* shows $\Delta^{26/24}$Mg between +0.5 and −1.4 ‰, about 0.3–0.6 ‰ higher compared to the corresponding coccolith calcite. The apparent temperature dependence of *E. huxleyi* coccoliths is 0.06 and 0.1 ‰/°C for cultures in the late exponential and stationary growth phase, respectively (Ra et al. 2010). At identical temperatures, Mg isotope composition of coccoliths is lighter in the stationary phase compared to the exponential growth phase. Ra et al. (2010) suggest that a

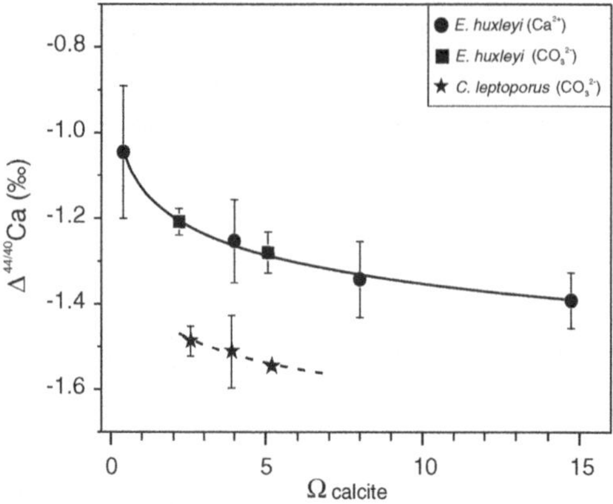

Fig. 9 Calcium isotope ratios of coccolith as a function of calcite saturation of the culture medium. Coccoliths of *E. huxleyi* and *C. leptoporus* show increasing $\delta^{44/40}Ca$ with decreasing Ω_{calcite}, of the culture medium in experiments with changing $[CO_3^{2-}]$ and $[Ca^{2+}]$ concentrations (Langer et al. 2007; Gussone et al. 2007)

growth rate dependent Mg isotope fractionation during the initial Mg acquisition of the coccolithophore is responsible for this effect, because the $\delta^{26/24}Mg$ of both, coccoliths as well as the chlorophyll is correlated to changing growth rate.

The effect of changing Mg/Ca ratios of the surrounding seawater on $\delta^{26/24}Mg$ and $\delta^{44/40}Ca$ is demonstrated by Müller et al. (2011), showing little variability in $\delta^{26/24}Mg$ for coccoliths of *E. huxleyi* and *C. braarudii* at Mg/Ca ratios of 5 mol/mol Mg/Ca (similar to the present day seawater value) and below. In contrast, elevated Mg/Ca of the fluid of up to 10 mol/mol result in significantly lighter $\delta^{26/24}Mg$ values, caused by so far unknown reasons. Calcium isotope fractionation is not significantly influenced by the Mg/Ca ratio of the surrounding fluid for both species (Müller et al. 2011).

Few data exist for the Sr isotope fractionation in coccoliths. Initial results show $\delta^{88/86}Sr$ values of about 0.26 ‰ (relative to NBS 987) for the coccolithophore species *E. huxleyi* and *Coccolithus pelagicus* (Krabbenhöft et al. 2010). This value corresponds to a fractionation of −0.13 ‰ relative to seawater. So far there is no information available on controlling environmental factors affecting Sr isotope fractionation during coccolith formation.

Coccolithophores precipitate $CaCO_3$ in intracellular enclosed so-called coccolith vesicles. Modelling Sr and Ca transport from the surrounding seawater through the cell into the coccolith calcite, suggests a steady flux of Ca and Sr into the vesicle, as the small fluid reservoir in the vesicle compared to the final coccolith suggests more than 20,000 refills for the build-up of one coccolith (Langer et al. 2006). The observed Ca isotope fractionation can therefore not take place inside the vesicle at the crystal surface but needs to occur during cellular Ca transport, likely at or in Ca channels at the plasma membrane (Gussone et al. 2006; Langer et al. 2007). The observed Ca isotope fractionation patterns in *E. huxleyi* presently serve as a first order approximation of transmembrane Ca transport.

The observation that Mg isotope fractionation of the LMC coccolith are enriched in heavy isotopes relative to inorganic calcite, while LMC foraminiferal tests are depleted, indicates that foraminifers and coccolithophores have different strategies of establishing low Mg activities in the calcifying fluid.

The relative uniform Ca isotope fractionation found within single coccolithophore species suggests that in general they may be suited to record $\delta^{44/40}Ca_{sw}$ values and the Mg isotopic composition in coccoliths might provide a tool for reconstruction of paleo SST or productivity (Ra et al. 2010). However, the downcore applicability is restricted by differences between cultured coccolith and coccolith oozes in $\delta^{26/24}Mg$, as coccolith oozes from high sea-surface temperature areas (Wombacher et al. 2011) exhibit an inverse apparent temperature dependence compared to the cultured coccolithophores (Fig. 10). This difference might reflect either an overall non-linear Mg isotope fractionation with maximal $\delta^{26/24}Mg$ at around 25 °C, or diagenetic alteration of the coccolith oozes. Differences between cultured coccoliths and coccolith oozes are also described for $\delta^{44/40}Ca$, which might relate to faunal and floral compositions of the studied oozes (Gussone et al. 2007). Additional complications for the use as proxy archives are related to the small size of the coccoliths and include the difficulty of isolating pure

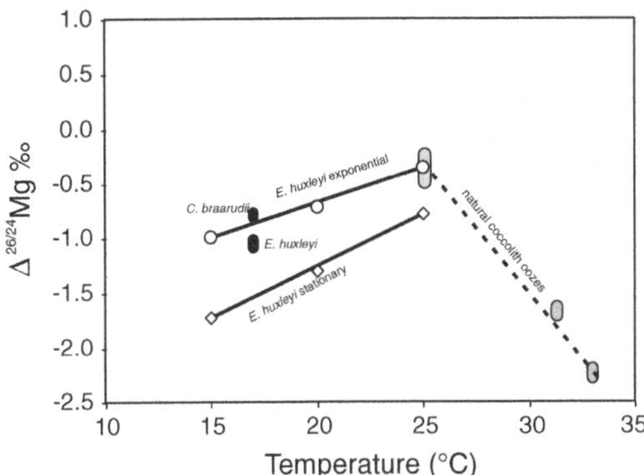

Fig. 10 $\Delta^{26/24}$Mg of cultured coccolith and natural coccolith oozes as a function of temperature. Cultured coccoliths of *E. huxleyi* and *C. braarudii* reveal a positive correlation to temperature between 15 and 25 °C (*open symbols* Ra et al. 2010; *black symbols* Müller et al. 2011). Coccoliths of *E. huxleyi* of the late exponential and stationary growth phase are offset by about 0.3 ‰, but temperature dependencies are similar ($\sim$0.1 ‰/°C) for both growth phases (Ra et al. 2010). Natural coccolith oozes show a temperature dependence opposite in sign above 25 °C (*grey fields*, Wombacher et al. 2011)

mono-specific fractions. Records based on impure coccolith separates however, comprise substantial uncertainties related to species specific fractionation and changing floral assemblages through time.

2.3 Calcareous Dinoflagellates

Dinoflagellates are unicellular algae, which occur abundantly in marine surface waters. As part of their vegetative or sexual life cycle, several dinoflagellate species can form cysts, composed of organic material, silicate or calcium carbonate. The calcium carbonate secreting dinoflagellates are called calcareous dinoflagellates and provide an archive for past environmental changes. In the modern ocean, the dominant calcareous dinoflagellate is *Thoracosphaera heimii*, a primary producer that has a wide geographic distribution. It has a large temperature and salinity tolerance and is already present in the fossil record from the Cretaceous onwards (Hildebrand-Habel and Willems 2000), making it a good candidate for environmental reconstructions throughout the Cenozoic.

$\delta^{44/40}$Ca of cultured *T. heimii* cysts range from 0.5 to 0.9 ‰, corresponding to a $\Delta^{44/40}$Ca from -1.0 to -1.4 ‰ (Fig. 11). This lies in the range of reported marine biogenic calcium carbonates and inorganically precipitated calcite. There is no difference between the two analysed strains from the Mediterranean and the equatorial Atlantic. In the temperature range from 12 to 30 °C, *T. heimii* is not significantly dependent on temperature (0.005 ± 0.005 ‰/°C). In addition, $\delta^{44/40}$Ca seems to be independent from the pH value of the growth medium in the range between 7.9 and 8.4 and from the cyst yield during the experiments (Gussone et al. 2010).

Calcium isotope fractionation (and likewise Mg and Sr incorporation) in *T. heimii* cysts are governed by two phases of calcite biomineralisation (Inouye and Pienaar 1983). Firstly calcite crystals are intracellularly formed and attached to the cell surface. In a second phase the crystals continue growing under the influence of a seawater-like fluid. The contribution of both growth steps is suggested to lead to a mixed signal of element ratios and Ca isotopes in the cysts, likely causing the minor apparent temperature dependence of $\delta^{44/40}$Ca.

Because of the limited variability of $\delta^{44/40}$Ca in calcareous dinoflagellate cysts, they seem to be well suited to record past ocean Ca isotope ratios. Besides that, calcareous dinoflagellates are promising multi-proxy carrier, recording in their

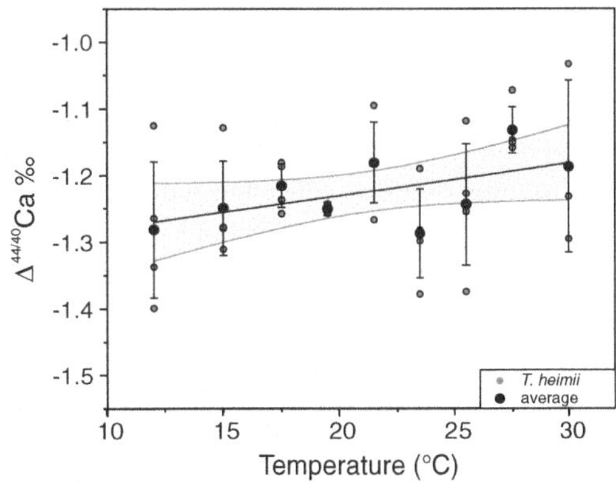

Fig. 11 $\delta^{44/40}$Ca of *T. heimii* cysts. Cultured calcite cysts of *T. heimii* do not show a significant dependence on temperature

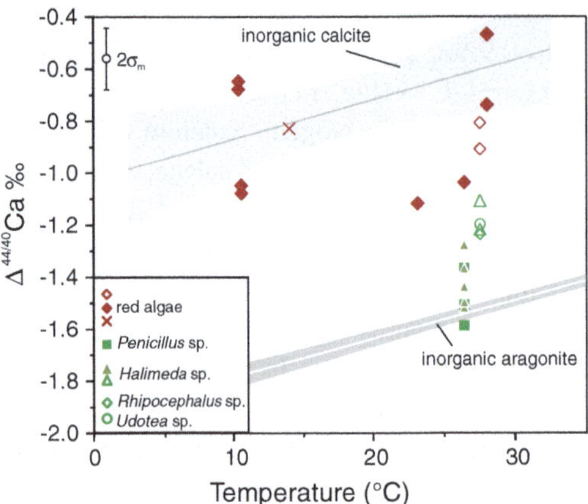

Fig. 12 Calcium isotope fractionation of coralline algae. Calcitic *red algae* and aragonitic *green algae* largely mimic the offset in $\delta^{44/40}$Ca of the inorganic $CaCO_3$ polymorphs calcite (Marriott et al. 2004) and aragonite (Gussone et al. 2003). *Filled symbols* (Blättler et al. 2012), *Open symbols* (Holmden et al. 2012), x (Gussone et al. 2005)

δ^{18}O and Sr/Ca information about temperature and δ^{18}O$_{water}$ (global ice volume and salinity). Its cysts form throughout the year at a stable position within the water column, particularly at the deep chlorophyll maximum depth and they are resistant to dissolution (Zonneveld 2004). The individual cysts are small, with a mean individual weight of about 1 ng, but they can be isolated from marine sediments by a combined sieving and density separation technique (Zonneveld 2004).

2.4 Coralline Algae

Coralline algae are important calcifiers which contribute to the global oceanic Ca sink. Taxa studied with respect to EASI fractionation belong to the groups of green algae and red algae.

Red algae belong to the Rhodophyta and form skeletons of high magnesium calcite. Calcium isotope ratios of different taxa of red algae, originating from Abu Dhabi, Bermuda, Scotland and the Bahamas, covering temperatures between 10 and 28 °C, demonstrate $\Delta^{44/40}$Ca from -1.1 to -0.5 ‰ with no significant correlation between Ca isotope fractionation and temperature (Blättler et al. 2012). Their average value of -0.85 ‰ is in agreement to the $\Delta^{44/40}$Ca values ranging from

-0.8 to -0.9 ‰ (14–27 °C) reported by Holmden et al. (2012) and Gussone et al. (2005). The reason for the relatively large scatter of the red algae data is so far unknown. Potential factors include species specific fractionation factors or local effects including ground water discharge (Holmden et al. 2012).

Green algae belong to the Chlorophyta and form aragonitic skeletons that precipitate mainly in semi-enclosed intercellular spaces. Precipitation occurs at the outer cell surface due to the increase in carbonate ion concentration by photosynthetic activity (CO_2 removal) in the cell. Supersaturation in the intercellular spaces is reached and maintained by diffusion limitation. Specimens of *Rhipocephalus* sp., *Udothea* sp., and *Halimeda* sp. from Ponte Maroma (Mexico) grown at about 27 °C show $\Delta^{44/40}$Ca of -1.1 to -1.2 ‰ with an average value of -1.2 ‰ relative to seawater (Holmden et al. 2012). $\Delta^{44/40}$Ca of green algae from the Bahamas (~ 26 °C) range from -1.4 to -1.6 ‰ for *Penicillus* sp. and -1.3 to -1.5 ‰ for *Halimeda* sp. (Blättler et al. 2012), with an average $\Delta^{44/40}$Ca of green algae of -1.45 ‰ (Fig. 12). Culture experiments of *Halimeda* sp. reveal that most specimens (7 out of 8) precipitate aragonite even in culture media which Mg/Ca ratios would favour inorganic calcite precipitation (Blättler et al. 2014). These specimens show $\Delta^{44/40}$Ca between -1.3 to -1.6 ‰ with an average value of -1.44 ‰, similar to the Ca isotope fractionation found in natural samples. One sample with reduced calcification had formed a calcitic skeleton with a $\Delta^{44/40}$Ca of -0.6 ‰, similar to inorganic calcite.

While the results of Blättler et al. (2012) show a 0.6 ‰ offset between calcitic and aragonitic coralline algae, similar to the inorganic minerals, the offset suggested by Holmden et al. (2012) is only 0.35 ‰. Combining all available coralline algae data reveals an offset between calcitic and aragonitic taxa of ~ 0.5 ‰ which is compatible with the offset between inorganic calcite and aragonite (Chapter "Calcium Isotope Fractionation During Mineral Precipitation from Aqueous Solution"), which is consistent with a relatively basic biomineralisation in the coralline algae. These observations are in agreement to the independence of calcification on Ca channel blockers (Blättler et al. 2014) suggesting that Ca is mainly derived by diffusion of seawater into the intercellular spaces.

Biomineralisation close to inorganic precipitation is supported by Mg isotope fractionation ($\Delta^{26/24}$Mg) of red algae, ranging between -2.4 and -2.2 ‰ (Hippler et al. 2009b; Wombacher et al. 2011), which is in the range of inorganic calcite (Chapter "Calcium Isotope Fractionation During Mineral Precipitation from Aqueous Solution").

3 Metazoa

3.1 Sclerosponges

Sponges belong to the metazoans (multi cellular organisms) and their paleontological records reach far back in time. Consequently, they are interesting as paleo environmental archives. Their biomineralisation is considered as relatively basic, and geochemical signatures are close to inorganic $CaCO_3$, suggesting limited vital effects (cf. Haase-Schramm et al. 2003). $\Delta^{44/40}$Ca of sclerosponges covers a total range from -0.7 to -1.7 ‰ (Fig. 13). The high magnesium calcite precipitating *Acanthochaetetes*

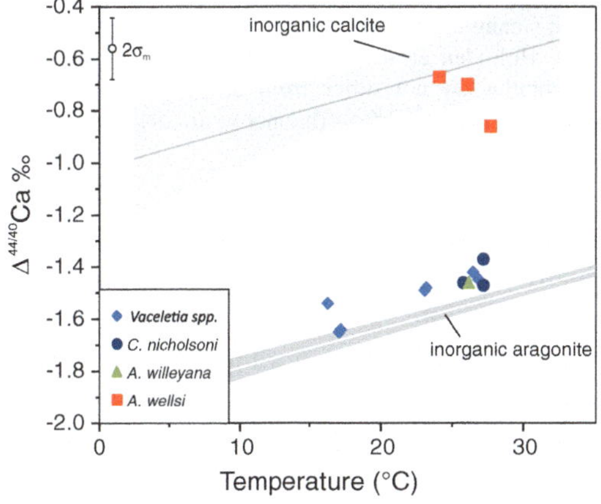

Fig. 13 Calcium isotope fractionation in sclerosponges. Calcium carbonate precipitated by sclerosponges shows the typical offset in $\delta^{44/40}$Ca of about 0.6 ‰ between calcite and aragonite

wellsi shows $\Delta^{44/40}$Ca values from -0.7 to -0.9 ‰ at temperatures between 24 and 28 °C, similar to inorganic calcite (Marriott et al. 2004). The aragonite forming species *Vaceletia* spp., *Ceratoporella nicholsoni* and *Astroclera willeyana* are more fractionated, with $\Delta^{44/40}$Ca between -1.7 and -1.4 ‰, which resembles inorganic aragonite (Gussone et al. 2003). Calcium carbonate precipitated by sponges shows the typical offset in Ca isotopes between calcite and aragonite of ~ 0.6 ‰ (Gussone et al. 2005).

The typical inorganic fractionation pattern is also apparent in the Mg isotopic composition of some of the investigated sclerosponges (Wombacher et al. 2011). The $\Delta^{26/24}$Mg of *A. wellsi* (-2.4 to -2.2 ‰) is close to inorganic calcite (Galy et al. 2002). While the aragonitic *A. willeyana* and *C. nicholsoni* ($\Delta^{26/24}$Mg: -1.0 to -0.7 ‰) are similar to inorganic aragonite (Wang et al. 2013b), *Vaceletia* spp. shows a comparatively large variability from -2.4 to -1.0 ‰, spanning the range between inorganic calcite and aragonite (Wombacher et al. 2011).

3.2 Corals

Understanding factors influencing coral growth is of great interest, because coral have considerable social-economic relevance e.g. in coastal protection, due to the stabilisation of shore areas. In addition, they are extremely valuable for scientific research, since they record paleoclimatic information. Studying coral biomineralisation in response to changing environmental conditions is relevant for estimating the impact of rising temperatures and pCO_2 on future growth of reef corals, as well as for a reliable interpretation of coral based paleo-environmental reconstructions. Important paleoclimatic information recorded in the chemical and isotopic composition of coral skeleton include changes in sea-level height, water temperatures and salinity.

Observations on EASI fractionation in different types of coral skeleton provide insights into biomineralisation of corals and indicate potential proxy applications.

Aragonitic corals

Calcium isotope fractionation of aragonitic coral skeletons in response to environmental factors (temperature, salinity) and taxon specific differences is studied on recent natural samples, downcore records and individuals grown in lab cultures. The overall range in $\delta^{44/40}$Ca is 0.0 to 1.2 ‰, with most specimens lying between 0.4 and 1.0 ‰ (SRM 915a). This corresponds to $\Delta^{44/40}$Ca of about −0.9 to −1.5 ‰ (Fig. 14). The overall $\Delta^{44/40}$Ca of coral aragonite is considerably offset to inorganic aragonite, with corals being enriched in ^{44}Ca by about 0.5 ‰.

Systematic investigation of Ca isotope fractionation in cultured *Acropora* sp. and natural *Pavona clavus* and *Porites* sp. of Böhm et al. (2006) reveal a significant temperature dependence of ∼0.02 ‰/°C between 21 and 29 °C, which is identical to that of cultured *Porites australiensis* (Inoue et al. 2015). These calibrations are in agreement to results on natural corals including *Acropora* sp., *Porites* sp., *Millepora* sp. and other partly un-identified taxa (Chang et al. 2004; Holmden et al. 2012; Blättler et al. 2012; Pretet et al. 2013; Halicz et al. 1999; Zhu and Macdougall 1998). Considering all aragonitic reef coral data reveals a temperature sensitivity of 0.02 ‰, which is identical to that of Böhm et al. (2006). Calculation of individual temperature calibrations reveals for most species similar temperature dependencies. The combined calibration of *Acropora* spp. for natural and cultured specimens has a slope of 0.025 ‰/°C (Chang et al. 2004; Blättler et al. 2012; Pretet et al. 2013; Böhm et al. 2006), natural *Porites* spp. 0.004 ‰/°C (Holmden et al. 2012; Blättler et al. 2012; Pretet et al. 2013; Böhm et al. 2006) and *Pavona clavus* 0.023 ‰/°C (Böhm et al. 2006). In contrast, the available data for *Millepora* spp. suggest a steeper gradient of 0.12 ‰/°C (Holmden et al. 2012; Blättler et al. 2012). Future work needs to evaluate if this is rather an artifact or an indicator for coral species with different temperature dependent Ca isotope fractionation.

Deep sea corals composed of aragonite (Blättler et al. 2012) have similar $\delta^{44/40}$Ca compared to aragonitic reef corals. The available data

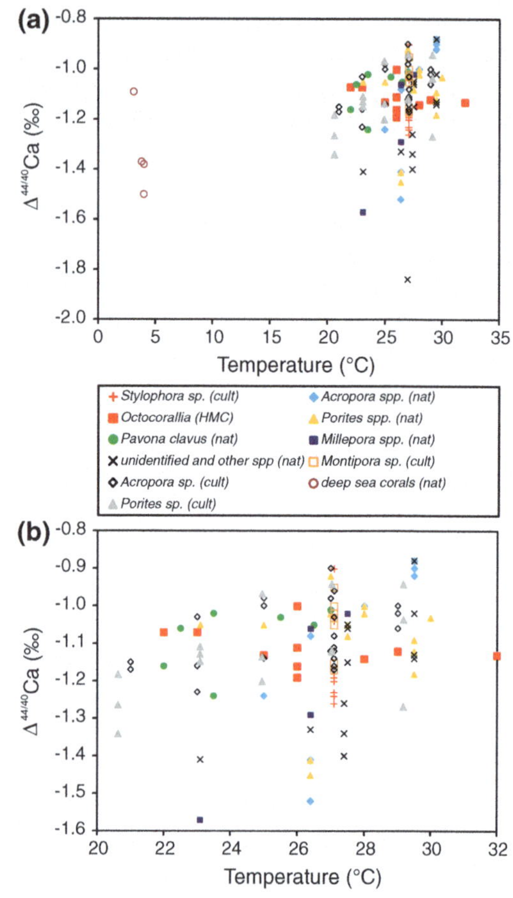

Fig. 14 Calcium isotope ratios of cultured and naturally grown corals. The temperature sensitivity of about 0.02 ‰/°C resembles inorganic precipitated aragonite, but the fractionation array of corals is significantly offset by about +0.5 ‰ (Böhm et al. 2006; Zhu and Macdougall 1998; Chang et al. 2004; Pretet et al. 2013; Blättler et al. 2013; Holmden et al. 2012; Inoue et al. 2015). Calcitic octocorallia do not differ from aragonitic corals with respect to their $\Delta^{44/40}$Ca (Taubner et al. 2012)

suggest at temperatures between 3 and 4 °C an inverse correlation between $\delta^{44/40}$Ca and temperature, with a gradient of −0.4 ‰/°C. Although the data basis for this slope is small, it is interesting to note, as the low-temperature anomaly reported for benthonic foraminifers falls in the same temperature range and shows a similar inverse temperature gradient (e.g. −0.5 ‰/°C in *Cibicides wuellerstorfi*).

Controlled culture experiments suggest slightly different responses of Ca isotope

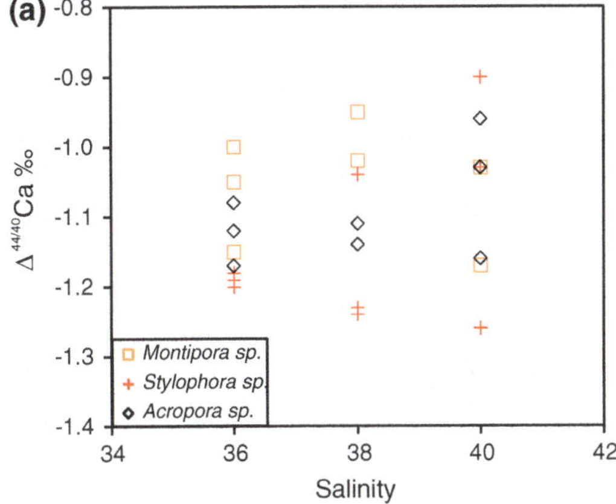
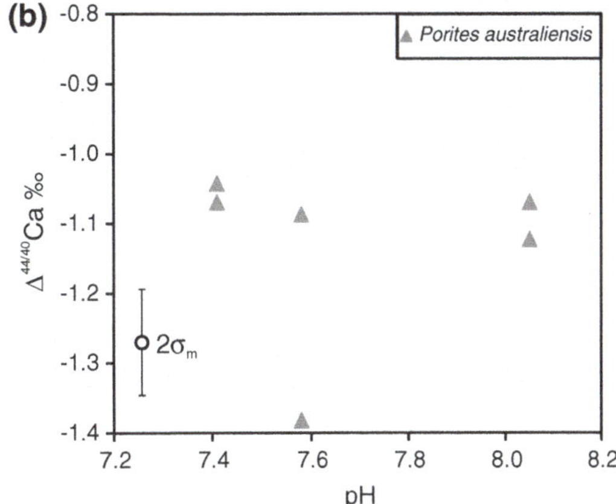

Fig. 15 Effect of salinity (**a**) and pH (**b**) on $\delta^{44/40}$Ca of cultured reef corals. **a** In culture experiments, *Montipora verrucosa* and *Acropora* sp. do not show changes in $\delta^{44/40}$Ca with changing salinity, while *Stylophora pistillata shows* higher $\delta^{44/40}$Ca associated with larger scatter at higher salinities (modified after Pretet et al. 2013). **b** $\delta^{44/40}$Ca in *Porites australiensis* from culture experiments does not depend on the ambient pH value (Inoue et al. 2015)

fractionation on salinity of different coral species. While *Montipora verrucosa* and *Acropora* sp. do not show changes in $\delta^{44/40}$Ca at salinities between 36 and 40, *Stylophora pistillata* exhibits a tendency to higher $\delta^{44/40}$Ca at a higher salinity, associated by a considerably larger scatter at higher salinity (Fig. 15a; Pretet et al. 2013). In addition, changes in ambient pH values (Fig. 15b) and light intensity have no effect on the $\delta^{44/40}$Ca of *Porites australiensis* grown in controlled culture experiments (Inoue et al. 2015). These experiments also reveal that there is no correlation between growth rate and $\delta^{44/40}$Ca.

The independence of Ca isotope fractionation from coral growth rates is supported by the observation that there is no correlation between linear extension rate and $\delta^{44/40}$Ca in down core records of *Porites* spp. from Tahiti (Pretet et al. 2013). The overall variability of ~ 0.35 ‰ observed in this record cannot be explained by the seasonal temperature fluctuation and a $\delta^{44/40}$Ca temperature dependence of about 0.02 ‰/°C. Pretet et al. (2013) suggested instead that so far unidentified processes might influence Ca isotope fractionation in coral skeleton beside temperature changes.

There are no significant differences in the range of $\Delta^{44/40}$Ca of different coral taxa. Natural samples from the Maledives suggest however that a smaller variability in $\Delta^{44/40}$Ca can be found in corals at the reef crest compared to those from the lagoon and the fore reef (Pretet et al. 2013). The reason responsible for this observation is not clear yet, but might be related to differences in light availability, water motion or suspended sediment particles.

The Ca isotope fractionation pattern of reef corals is suggested to be controlled rather by biological processes then by inorganic crystal formation. Similar to coccolithophores (Sect. 2.2) Ca transmembrane transport might cause the observed $\delta^{44/40}$Ca in coral skeleton (Böhm et al. 2006). This hypothesis is based on the small $\delta^{44/40}$Ca-temperature dependence, the 0.5 ‰ offset relative to inorganic aragonite and the similarity to the fractionation found in coccolithophores (cf. Langer et al. 2007). Additional evidence is given by the observation that $\delta^{44/40}$Ca of coral skeleton does not correlate with growth rate induced by changes in pH of the culture medium or light intensity (Inoue et al. 2015). It is further supported by the lacking correlation between linear extension rate and Ca isotope fractionation revealed by *Porites* sp. from Tahiti (Pretet et al. 2013).

Octocorallia

Alcyonarian corals have a marine shallow tropical habitat. They form intracellular spicules composed of high Mg calcite (HMC). *Rhythisma fulvum* cultured in Red Sea seawater at varying temperatures and pH values demonstrates $\delta^{44/40}$Ca between 0.7 and 0.9 ‰, with no significant temperature dependence between 19 and 32 °C (Taubner et al. 2012). In the pH range between 8.15 and 8.44, Ca isotopes show an increasing fractionation with increasing pH (and CO_3^{2-}), which is general agreement to the 'Ω-effect' proposed for coccolithophores (Sect. 2.2). Calcium isotope fractionation of octocorallia spicules seems to be independent of growth rates (Taubner et al. 2012), which is consistent with the lacking growth rate dependence of aragonitic reef corals (Inoue et al. 2015) and coccolithophores (Langer et al. 2007). $\Delta^{44/40}$Ca found in octocorallia spines is between −1.2 and −1.0 ‰, which is in the same range as aragonitic reef corals (Fig. 14). Based on the similarities of aragonitic and calcitic corals, Taubner et al. (2012) suggest that the observed Ca isotope fractionation takes place during Ca transmembrane transport, thus masking the inorganic

mineralogy dependent Ca isotope fractionation. The authors further conclude that the biological fractionation mechanism can mimic inorganic element partitioning patterns, because Mg/Ca and Sr/Ca ratios of octocorallia spicules are in general agreement with a predominant inorganically controlled crystal growths.

In contrast to Ca isotopes, Mg isotope fractionation of coral skeleton shows the typical offset between aragonite and calcite of about 1.5 ‰ (Fig. 16a). $\Delta^{26/24}$Mg of aragonitic corals range from about −0.7 to −1.2 ‰ (Chang et al. 2004; Wombacher et al. 2011; Yoshimura et al. 2011) including reef corals (*Acropora* sp., *Porites* sp.) and deep sea corals (*Lophelia* sp.). Coral skeleton composed of HMC (*Corallium* sp., *Keratoisis* sp.) show $\Delta^{26/24}$Mg in the range of −2.7 to −2.3 ‰ resembling the trend of inorganic calcite (cf. Galy et al. 2002).

A small temperature dependence in Mg isotope fractionation is indicated by aragonitic corals from the natural environment (Chang et al. 2004; Wombacher et al. 2011). Calcitic corals reveal an apparent temperature dependent Mg isotope fractionation of 0.08 ‰/°C, which might alternatively also be attributed to the carbonate

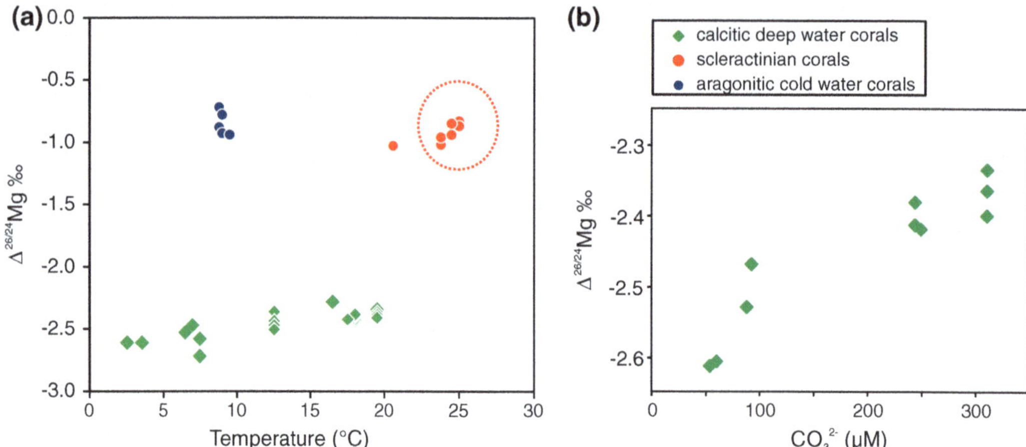

Fig. 16 Magnesium isotope fractionation of corals. **a** $\Delta^{26/24}$Mg of coral skeletons as a function of temperature. In contrast to Ca isotope systematic, aragonitic and calcitic corals show an offset in $\delta^{26/24}$Mg typical for inorganic calcite and aragonite (*symbols* Wombacher et al. 2011; Yoshimura et al. 2011; *dashed ellipsis* range reported from Chang et al. 2004). Scleractinian corals and calcitic cold-water corals indicate a significant apparent temperature dependent Mg isotope fractionation. **b** $\Delta^{26/24}$Mg of calcitic corals demonstrates a positive apparent dependence on of the carbonate ion concentration of the seawater (*redrawn* from Yoshimura et al. 2011)

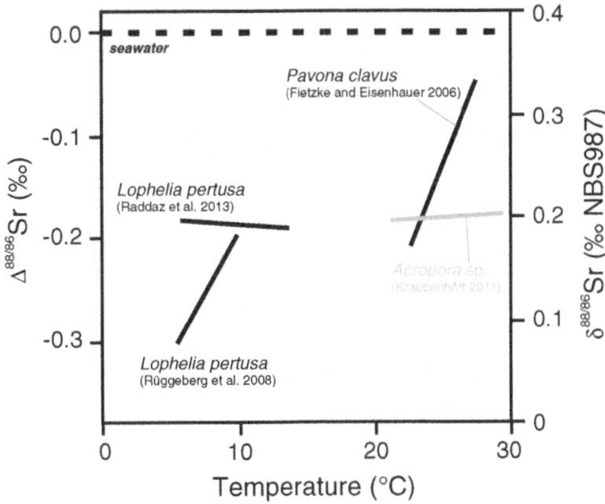

Fig. 17 Mass dependent Sr isotope fractionation in corals as a function of calcification temperature. Different apparent temperature sensitivities of $\delta^{88/86}$Sr are obtained from different studies (Fietzke and Eisenhauer 2006; Rüggeberg et al. 2008; Krabbenhöft 2011; Raddaz et al. 2013)

ion concentration of the ambient water (Figs. 16a, b; Yoshimura et al. 2011).

Similar to Ca and Mg, light Sr isotopes are preferentially incorporated into the coral skeleton. The total published range of apparent Sr isotope fractionation ($\Delta^{88/86}$Sr in aragonitic coral skeleton is −0.05 to −0.30 ‰, showing major discrepancies between different studies, with respect to magnitude and temperature dependence (Fig. 17). The first empirical calibration curve of *Pavona clavus* reveals $\Delta^{88/86}$Sr ranging from −0.05 to −0.2 ‰ between 23 and 27 °C, and a temperature dependence of about 0.03 ‰/° C (Fietzke and Eisenhauer 2006). Subsequent investigations on the cold water coral *Lophelia pertusa* show $\Delta^{88/86}$Sr between −0.15 and −0.30 ‰ in the temperature range from 5 to 9 °C and a similar temperature dependence of 0.026 ‰/°C (Rüggeberg et al. 2008). In contrast, Raddaz et al. (2013) present for the same taxon and temperature range $\Delta^{88/86}$Sr of about −0.2 ‰ and a negligible temperature dependence. Similarly, $\Delta^{88/86}$Sr of about −0.2 ‰ and insignificant temperature dependence is demonstrated for the skeleton of *Acropora* sp. between 17 and 25 °C (Krabbenhöft 2011). The reason for these discrepancies are yet unclear, but might be related to the different applied analytical methods, as the

latter two studies applied a TIMS double spike technique instead of a MC-ICPMS bracketing standard method. Future research is therefore necessary to verify the empirical calibrations by adding further controls on parameters like carbonate chemistry and growth rates. This is important, as the initial temperature dependence revealed by Fietzke and Eisenhauer (2006) is sufficiently large to be exploited as a paleo temperature proxy.

3.3 Molluscs

3.3.1 Bivalves

Shells of bivalve are frequently used as archives for high resolution paleo climate reconstructions. The understanding of their biomineralisation and related element partitioning and isotope fractionation pattern and their controlling factors is consequently of great interest, in particular, because their shells consist of parts with different structures and in some taxa different mineralogy. Studies on Ca and Mg isotope fractionation of bivalve include several different taxa and specimens from lab cultures, recent individuals from natural environments and fossil samples.

Oysters

Crassostrea gigas (Giant Pacific oyster) collected in the North Sea, shows a uniform Ca isotopic composition of 0.68 ± 0.16 ‰ for different shell structures and ontogenetic stages (Ullmann et al. 2013), corresponding to an average apparent fractionation ($\Delta^{44/40}$Ca) between shell and seawater of −1.2 ‰ (Fig. 18). $\delta^{44/40}$Ca of the chalky substance of the shell and foliate layers are 0.68 ± 0.04 and 0.70 ± 0.04 ‰, respectively and thus identical within uncertainty, while the trace element concentrations of both shell parts show significant variability. In the sessile phase of the oyster, calcite is the dominant mineral with only minor aragonite contents, in contrast to the planktic larvae stage, in which aragonite is the dominant mineral (Stenzel 1964). The relative invariant Ca isotope fractionation of *C. gigas* of about −1.2 ± 0.16 ‰ (Ullmann et al. 2013) for water temperatures

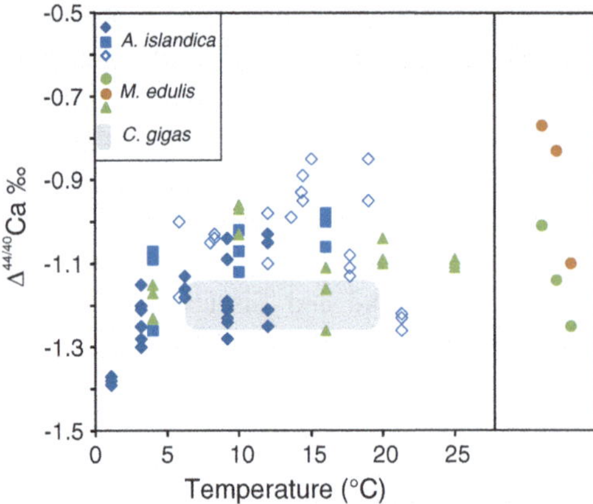

Fig. 18 Calcium isotope fractionation of bivalve shells as a function of temperature. Aragonitic *Arctica islandica* is marked by blue symbols, lab cultures as *filled diamonds* (Hippler et al. 2013) and *filled squares* (Hiebenthal 2009) and field samples as *open diamonds* (Hippler et al. 2013). Calcite layers of *Mytilus edulis* are marked by *green symbols*, lab cultures as *triangles* (Hiebenthal 2009) and field samples grown at unknown temperatures as *green dots* (Heinemann et al. 2008). Aragonite layers of *Mytilus edulis* are shown as *orange dots* (Heinemann et al. 2008). Range of $\delta^{44/40}$Ca of oysters from the natural environment are indicated by the *grey box* (Ullmann et al. 2013)

between 0 and 20 °C and salinities between 27 and 31, is consistent with $\Delta^{44/40}$Ca oyster values of about −1.2 and −1.4 ‰ (Skulan and DePaolo 1999; Steuber and Buhl 2006). As possible reason for the lower observed $\delta^{44/40}$Ca reported by Steuber and Buhl (2006), Ullmann et al. (2013) propose that the $\delta^{44/40}$Ca of the local coastal seawater may be reduced by Ca freshwater input to the marginal sea.

Mytilus edulis

The shell of *Mytilus edulis* consists of different layers composed of calcite and aragonite. For the better understanding of bivalve biomineralisation and exploring potential proxy applications both parts of the shell were studied in respect to Ca isotopes. For temperatures ranging from 4 to 25 °C and salinities from 15 to 25, $\Delta^{44/40}$Ca of the calcitic part of cultured *M. edulis* shells varies from −1.3 to −1.0 ‰, with an average of −1.1 ‰ (Hiebenthal 2009), no clear dependence on temperature (Fig. 18) and a minor dependence on salinity. The dependence of $\Delta^{44/40}$Ca on

salinity is proposed to result from increased contribution of Ca pumping at low salinities into the extrapallial fluid, leading to increased Ca isotope fractionation at lower salinities (Hiebenthal 2009).

$\Delta^{44/40}$Ca of prismatic calcite layers of natural *M. edulis* from the North Sea, Kiel-Fjord and Schwentine Estuary range between −1.1 and −0.8 ‰ (Heinemann et al. 2008). The aragonite layers are more fractionated compared to the calcite layers, with $\Delta^{44/40}$Ca between −1.3 and −1.0 ‰. The difference between aragonite and calcite layers of an individual is between 0.16 and 0.3 ‰, which is smaller than the difference between inorganic calcite and aragonite Chapter ("Calcium Isotope Fractionation During Mineral Precipitation from Aqueous Solution").

Chemical and isotopic analyses of the extrapallial fluid (EPF) show $\delta^{44/40}$Ca, Sr/Ca and Mg/Ca ratios higher compared to seawater, which is suggested to result from the preferential removal of light Ca and exclusion of Sr and Mg from the carbonate phase, thus leading to the observed enrichment of ^{44}Ca and Sr and Mg in the extrapallial fluid (Heinemann et al. 2008). This observation contrasts the conceptual model for *Arctica islandica*, which suggests fractionation of Ca during transmembrane transport into the EPF and is also not consistent with the mechanism explaining the potential dependence of $\Delta^{44/40}$Ca on salinity in *M. edulis* (Hiebenthal 2009). While the $\Delta^{44/40}$Ca of the aragonite and calcite parts of the shell does not exhibit the full offset between the inorganic polymorphs, the trace element composition is also relatively similar in the aragonite and calcite parts of *M. edulis*, showing Sr/Ca and Mg/Ca more typical for calcite than for aragonite. The high Mg/Ca of the aragonite part might be caused by Mg sticking to organic templates, but the low Sr/Ca is not compatible with a primary aragonite precipitation. A possible explanation is the precipitation of a precursor phase, e.g. ACC (Foster et al. 2008).

Information on the Ca isotopic composition of the soft tissue of *M. edulis* is given by Skulan and DePaolo (1999). $\Delta^{44/40}$Ca of soft tissues relative to seawater changes with time, as the soft tissue of a freshly caught specimen is almost identical to seawater, whereas an individual kept for 12 h out of the water has a $\Delta^{44/40}$Ca of about -0.5 ‰ approaching the isotopic composition of the shell ($\Delta^{44/40}$Ca of about -0.7 ‰). Consequently, parts of the shell can be dissolved from the organism to regulate its ionic budget (Skulan and DePaolo 1999). This observation is important also for other taxa in terms of high resolution climate archives, as remodeling of the shell might alter short term proxy signals.

$\Delta^{26/24}$Mg of specimens studied in field culture experiments range from about -4.4 to -2.6 ‰, which is consistent with the stronger Mg isotope fractionation in low magnesium calcite compared to biogenic HMC and inorganic calcite (Hippler et al. 2009b). Similar to $\Delta^{44/40}$Ca, there are indications that $\Delta^{26/24}$Mg of *M. edulis* may depend on the salinity of the ambient water (Hippler et al. 2009b).

Arctica islandica

The calcium isotope fractionation of *A. islandica* shells is characterized by specimens from lab culture experiments and field samples. The aragonite shells of specimens from lab cultures show $\Delta^{44/40}$Ca between -1.4 and -1.0 ‰ (Hiebenthal 2009; Hippler et al. 2013). Data of both studies overlap, but those of Hippler et al. (2013) are on average 0.15 ‰ lower (Fig. 19). $\Delta^{44/40}$Ca of field samples range from -1.3 to 0.8 ‰ (Hippler et al. 2013) and are on average less fractionated than the cultured individuals. The aragonite shell of *A. islandica* is about 0.5 ‰ less fractionated compared to inorganic aragonite, thus resembling the fractionation array of corals (cf. Böhm et al. 2006; Inoue et al. 2015), which is proposed to reflect isotope fractionation during Ca transmembrane transport. The apparent dependence of $\Delta^{44/40}$Ca on temperature of *A. islandica* is almost linear between 1 and 15 °C with a slope of 0.021 ‰/°C (Hippler et al. 2013) and 0.011 ‰/°C (Hiebenthal 2009), similar to that of inorganic aragonite, corals and other biogenic carbonates. At temperatures above 15 °C, $\Delta^{44/40}$Ca decrease with increasing

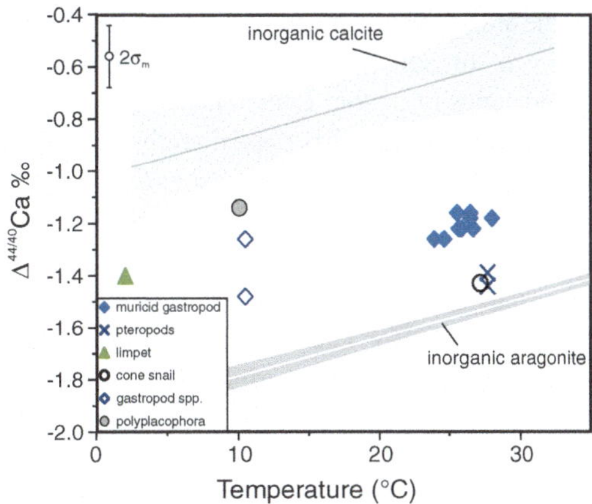

Fig. 19 Calcium isotope fractionation of modern gastropods and polyplacophora as a function of temperature. Pteropods (*blue* X, Gussone et al. 2005) and a cone snail (*open black circle*, Skulan et al. 1997) are close to the inorganic fractionation array of aragonite (Gussone et al. 2003). A muricid gastropod (*blue filled diamonds*, Steuber and Buhl 2006), a limpet (*green triangle*, Skulan et al. 1997) and an undetermined gastropod species (*blue open diamonds*, Blättler et al. 2013) are less fractionated with respect to Ca isotopes, and thus more similar to aragonitic bivalves. Inorganic calcite data from Marriott et al. (2004)

temperatures, resulting in a maximum $\delta^{44/40}$Ca at 15 °C. Salinity has a minor impact on Ca isotope fractionation, with an overall gradient of -0.037 ± 0.004 ‰/psu (Hippler et al. 2013), while data of Hiebenthal (2009) suggest no change in Ca isotope fractionation between 15 and 35 psu.

While Ca isotope fractionation of the field-grown samples alone does not reveal evidence for a dependence on growth rate, the lab culture data and the combined data set suggest a positive correlation between $\delta^{44/40}$Ca and growth rates at low rates (0.005–0.04 mm/day) while at linear extension rates above 0.04 mm/day Ca isotope fractionation is almost constant (Hippler et al. 2013).

The observed Ca isotope fractionation pattern can be linked to conceptual models of bivalve biomineralisation. The shell is precipitated from the extrapallial fluid (EPF), which is not only supersaturated with respect to aragonite, but also serves as regulator for the organisms' pH and ionic budget. The Ca used for biomineralisation is mainly transported by Ca channels into the extrapallial space and the aragonite precipitates from the EPS, on the surface of organic templates. As the growth rate dependencies of neither Lemarchand et al. (2004) nor Tang et al. (2008, 2012) are directly applicable to the fractionation patterns of *A. islandica*, Hippler et al. (2013) suggest that Ca isotope fractionation takes place at the Ca channels during transmembrane transport, similar to the fractionation mechanism proposed for coccolithophores (Sect. 2.2) and corals (Sect. 3.2). Nevertheless, although Ca isotope fractionation of the aragonitic corals and *A. islandica* is similar, Sr/Ca ratios differ significantly for *A. islandica* ($\sim$0.2 mmol/mol; cf. Hiebenthal 2009) and corals ($\sim$9 mmol/mol, cf. Allison et al. 2010), demonstrating that there are considerable differences in ion supply and aragonite precipitation during biomineralisation of bivalves and corals.

Other bivalves

For other bivalve species only less systematic data exist. *Tridacna* sp. exhibits $\Delta^{44/40}$Ca of -1.1 ‰ (Steuber and Buhl 2006) and an unknown bivalve species ranges from -1.2 to -1.0 ‰ (Blättler et al. 2013), which is in the range of the above discussed species.

Rudists, reef-forming bivalves from the Cretaceous are studied with respect to Ca isotopes from Immenhauser et al. (2005) and Steuber and Buhl (2006). The degree of isotope fractionation between sea water and rudists shells cannot be directly determined as rudists are extinct and because the Ca isotopic composition of Cretaceous seawater is still under discussion. While cross calibration with Mg/Ca ratios and δ^{18}O of Immenhauser et al. (2005) indicate a temperature dependence similar to that of the planktic foraminifer *G. sacculifer* ($\sim$0.2 ‰/°C), Steuber and Buhl (2006) suggest based on the correlation with δ^{18}O a slope of 0.012 ‰/°C (Sect. 4.3), similar to inorganic calcite. The latter result is compatible with observations on recent bivalve taxa, which reveal only a minor dependence on temperature. In this context, Ullmann et al. (2013) suggest that oysters might be useful as recorders for paleo seawater Ca isotope reconstructions.

3.3.2 Gastropods and Polyplacophora

A recent muricid gastropod covering a temperature range of about 24–28 °C shows $\Delta^{44/40}$Ca between -1.2 and -1.3 ‰ (Steuber and Buhl 2006). These values define a small temperature sensitivity of about 0.02 ‰/°C, which is similar to inorganic carbonates and other marine calcifiers. The overall range in isotope fractionation is similar to aragonitic bivalves, and offset from inorganic aragonite (Fig. 19).

Pteropods, planktic gastropods, secreting aragonitic shells show at 27 °C $\Delta^{44/40}$Ca of about -1.4 ‰ (Gussone et al. 2005), which is identical to a cone snail (27 °C) with $\Delta^{44/40}$Ca of -1.4 ‰ (Skulan et al. 1997) and close to the inorganic aragonite array, but significantly more fractionated compared to the muricid gastropod (Steuber and Buhl 2006).

Similar degrees of fractionation but at lower temperature are demonstrated by snails that calcified at 10 °C ($\Delta^{44/40}$Ca of -1.5 to -1.3 ‰, Blättler et al. 2013), a limpet (-1.4 ‰, 2 °C, Skulan et al. 1997) and a conch (-1.3 ‰, Russell et al. 1978) at unknown temperatures. One Polyplacophora specimen exhibits $\Delta^{44/40}$Ca of -1.1 ‰ at 10 °C (Steuber and Buhl 2006). This degree of fractionation is slightly smaller compared to snails grown at similar temperature (Blättler et al. 2013).

3.3.3 Cephalopods

While cephalopods are important stratigraphic tools and important geochemical archives, comparatively few studies investigate recent cephalopod Ca isotope fractionation pattern.

Recent *Nautilus* sp. hard tissues composed of aragonite show $\Delta^{44/40}$Ca between -1.2 and -1.0 ‰, with no apparent dependence on temperature ($\sim$7 – 14 °C) or differences between shell parts (shell, septum, septal neck), while *Spirula* sp. shells range from -1.4 to -1.2 ‰ for temperatures between about 6 to 8 °C (Middelberg et al. unpublished). This degree of fractionation is smaller compared to inorganic aragonite, but compatible to aragonitic bivalves.

For belemnites, Farkaš et al. (2007) propose $\Delta^{44/40}$Ca of about -1.4 ‰ relative to contemporary seawater. This value is derived by cross calibration of belemnites with brachiopods originating from the same stratigraphic horizons of the Kimmerigium. In these sediments a difference between both archives is apparent, with belemnites being about 0.5 ‰ lighter compared to the brachiopods. Applying the present day $\Delta^{44/40}$Ca of brachiopods from Farkaš et al. (2007) results in the $\Delta^{44/40}$Ca of -1.4 ‰ assumed for the belemnites. Possible uncertainties of this calibration are related to the variability in the Ca isotope fractionation pattern of brachiopods (Sect. 3.4).

3.4 Brachiopods

Brachiopods are animals living in the marine realm that have two valves, which are composed of chitin and apatite or calcite. In particular the calcitic (low Mg-calcite) brachiopods shells are important archives for the reconstruction of paleo sea-water δ^{18}O, ^{87}Sr/^{86}Sr and Ca isotope ratios (e.g. Veizer et al. 1999; Farkaš et al. 2007). Nevertheless, so far comparatively few studies characterise the fractionation systematics of Ca in shells of brachiopods. Specimens of different brachiopod species from different water temperatures have been analysed, covering a range from $\sim 7 - 27$ °C (Fig. 20). The observed range of $\Delta^{44/40}$Ca of brachiopods is about -1.0 to -0.4 ‰, with systematic differences between species. *Gryphus vitreus* has a shell composed of three layers and a mean $\Delta^{44/40}$Ca of -0.56 ‰ relative to seawater. In a specimen grown at relative constant temperatures (13–14 °C), $\Delta^{44/40}$Ca shows a relatively large range between about -0.4 and -0.8 ‰ (von Allmen et al. 2010). The results suggest that the primary layer has lower $\delta^{44/40}$Ca and higher Sr concentrations compared to the rest of the shell. As this data deviates from the $\Delta^{44/40}$Ca-D$_{Sr}$ array for inorganic calcite (Tang et al. 2008, 2012), a pure growth rate effect is unlikely as explanation for the variability in the Ca isotopic composition. The low $\delta^{44/40}$Ca present in the umbonal region

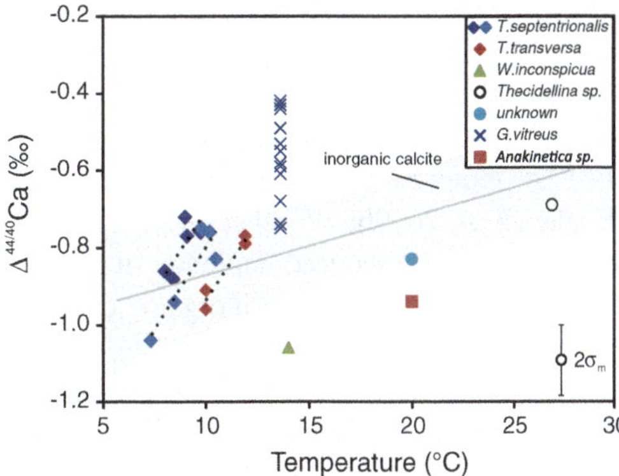

Fig. 20 Calcium isotope fractionation of brachiopods as a function of temperature. *Terebratulina septentrionalis* indicates a potentially large temperature dependence, while *Gryphus vitreus* suggests intra shell variability possibly related to ontogenetic effects and shell structure (von Allmen et al. 2010). *Anakinetica* sp. (Steuber and Buhl 2006), *Thecidellina* sp., *Waltonia inconspicua*, an undetermined species (Gussone et al. 2005) and *Terebratalia transversa* (Farkaš et al. 2007; Gussone et al. 2005) show similar degrees of isotope fractionation, but may also follow independent fractionation trends. Inorganic calcite data from Marriott et al. (2004)

of the shell is consistent with increased fractionation at high growth rates of inorganic calcite (Tang et al. 2008), because growth rate is higher in juvenile than in adult stages of brachiopod ontogeny.

Terebratulina septentrionalis indicates a substantial apparent temperature dependence of about 0.085 and 0.078 ‰/°C in two different specimens (von Allmen et al. 2010), which is identical to the 0.082 ‰/°C of *Terebratalia transversa* (Gussone et al. 2005; Farkaš et al. 2007). The distribution of $\delta^{44/40}$Ca in *Terebratulina septentrionalis* is rather homogeneous, with potentially little influence of ontogeny and growth rates (von Allmen et al. 2010). A combined trend of the species *Anakinetica* sp. (Steuber and Buhl 2006), *Thecidellina* sp., *Waltonia inconspicua* and an unknown species (Gussone et al. 2005) suggests a small temperature dependence of 0.029 ‰/°C in the temperature range from 14 to 27 °C, which is similar to those of other marine calcifiers. The available brachiopod data suggest relatively large differences between species, possibly related to shell structure.

With respect to $\delta^{26/24}$Mg, brachiopod shells have relatively high values, compared to other LMC shells, except coccoliths. $\Delta^{26/24}$Mg ranges from -1.5 to -1.1 ‰ for brachiopods that calcified at temperatures between 9 and 12 °C (Hippler et al. 2009b; Wombacher et al. 2011). The reason for the reduced depletion of ^{24}Mg in brachiopod shells relative to inorganic calcite is still unresolved.

While initial data on brachiopods suggested a small temperature dependence and little overall variability in Ca isotope fractionation (Gussone et al. 2005; Steuber and Buhl 2006; Farkaš et al. 2007), the results of von Allmen combined with the older data suggest a larger variability of the Ca isotope fractionation factor, which needs to be considered when using brachiopods as archives for $\delta^{44/40}$Ca of past seawater (Chapter "Global Ca Cycles: Coupling of Continental and Oceanic Processes").

3.5   Other Taxa

3.5.1   Echinoderms

Calcium isotope fractionation of echinoderm hart tissues, composed of high magnesium calcite, demonstrate an overall range from -1.5 to -0.9 ‰ relative to seawater. While a starfish sample has a $\Delta^{44/40}$Ca of -0.9 ‰ (Skulan et al. 1997), an echinoid spine shows a $\Delta^{44/40}$Ca of -1.1 ‰ (Holmden et al. 2012). $\Delta^{44/40}$Ca of several sea urchins (~ 10 °C) ranges from -1.5 to -1.3 ‰ (Blättler et al. 2013). These values are in the range of biogenic LMC and HMC skeleton and inorganic calcite. With respect to Mg isotope fractionation, Hippler et al. (2009b) and Wombacher et al. (2011) report for sea urchins $\Delta^{26/24}$Mg between -2.0 and -1.6 ‰. *Echinocyamus pusillus* grown at temperature between 9 and 20 °C shows $\Delta^{26/24}$Mg between -2.0 and -1.8 ‰ (Hippler et al. 2009b), with no clear dependence on temperature. $\Delta^{26/24}$Mg of *Diadema setosum* grown at 22 °C ranges from about -1.7 to -1.6 ‰, showing no offset between different parts of the shell, namely ambulacral plates and interambulacral plates (Wombacher et al. 2011).

3.5.2   Vertebrates

The skeleton of vertebrates is composed of mineralized tissues containing water, organics and apatite. The mineralized phase is often and best described as a carbonated hydroxylapatite (HAP) with the general formula $Ca_5(CO_3, PO_4)_3(OH)$ (Boskey 2007; Pasteris et al. 2008). The composition of bones, the CO_3^{2-} content as well as the density varies from species to species and also within an organism depending on the location of the bone. Apatite in bones is the major store of Ca (~ 99 %), P (~ 80 %) and Mg (~ 50 %) (Pasteris et al. 2008) of the body. The Ca isotopic composition of bones typically lies between -2.7 and 0.7 ‰ (Fig. 25). The Ca isotopic composition of bones can be used to investigate biomineralization processes and is closely related to the Ca metabolism of the organism as well as the Ca isotopic composition of the diet being the only external source of Ca for the body.

The transport of Ca to the sites of ossification, promoting bone mineralisation, slightly favours the light Ca isotopes resulting in an enrichment of light isotopes in the mineralized tissue. The magnitude of fractionation between soft tissue and mineralized tissue was reported to be 1.3 ‰ (Skulan and DePaolo 1999), but this value may overestimate the actual fractionation. A closer look into the data of the latter study shows that the fractionation between blood and bone is in average ~ 0.7 ‰ (horse: 0.53 ‰, fur seal: 0.95 ‰, chicken: 0.56 ‰). The fractionation between soft tissue and mineralized tissue of 1.3 ‰ reported by Skulan and DePaolo was calculated by comparing the Ca isotopic composition of the diet and bones under the assumption that no Ca fractionation occurs during absorption of dietary Ca (see Fig. 1 of Skulan and DePaolo 1999). From a physiological point of view blood and soft tissue are different as blood is the main transporter for Ca in the body. Calcium from bones and organs is exchanged with Ca from extracellular fluids (ECF) which comprises blood. Currently, there is no study which systematically investigates the fractionation between blood, soft tissue and mineralized tissue.

Bone resorption is proposed to be a non-fractionating process, i.e. the Ca isotopic composition of the mineralized tissue is transferred one-to-one into the blood. As a consequence, the Ca isotopic composition is lowered during bone resorption as isotopically light Ca is released.

Based on this basic Ca isotope transport model several studies aimed to use the Ca isotopic composition of blood and urine as a marker of the actual bone mineral balance which led to refinements of the transport model (Chapter "Biomedical Application of Ca Stable Isotopes").

4 Applications, Ecosystems and Climate Change

4.1 Monitor of Trophic Levels

The trophic level effect

As a consequence of the fractionation during bone growth, or more generalized, during formation of mineralized tissues in organisms, a so called trophic level effect (TLE) of Ca isotopes has been observed (Skulan et al. 1997; Clementz et al. 2003; DePaolo 2004; Heuser et al. 2011; Martin et al. 2015). This trophic level effect describes that within a food chain the Ca isotopic composition of mineralized tissue becomes lighter with increasing trophic level. Per trophic level, the Ca isotopic composition changes by a about −1 ‰. Similarly, a trophic level effect is also apparent in the stable isotopes of Mg from mammal bones and teeth, which shows however, in contrast to $\delta^{44/40}$Ca, increasing $\delta^{26/24}$Mg with increasing trophic level (Martin et al. 2014).

Consequently, the average Ca isotopic composition of mineralized tissues of a herbivore is lighter than the plants it feeds on, and the mineralized tissue of carnivores is more enriched in light isotopes than the mineralized tissue of the herbivores they eat. This TLE has been described for terrestrial as well as marine ecosystems. However, a few complications which have to be considered when dealing with trophic level effects or reconstructing paleo food chains are discussed below.

Mineralized tissue has to be part of the diet for the next trophic level

Present transport models for Ca isotopes in vertebrates imply a 'zero-permill' fractionation between diet and "soft tissue" (incl. blood, Fig. 21). Consequently, for carnivores that completely feed on soft tissue of herbivores, the Ca isotopic composition of the diet should be similar to that of herbivores, i.e. the plants eaten by the herbivores. The δ^{44}Ca$_{diet}$ is only significantly changed towards lower values, if mineralized tissue is digested by the carnivore. Then, as a consequence, δ^{44}Ca$_{min\text{-}tissue}$ becomes lighter in the carnivore compared to the herbivore (Fig. 22; Clementz et al. 2003; Heuser et al. 2011). Modeling results indicate that about 1 % of the diet has to be mineralized tissue, to develop a TLE (Fig. 23). This calculated value

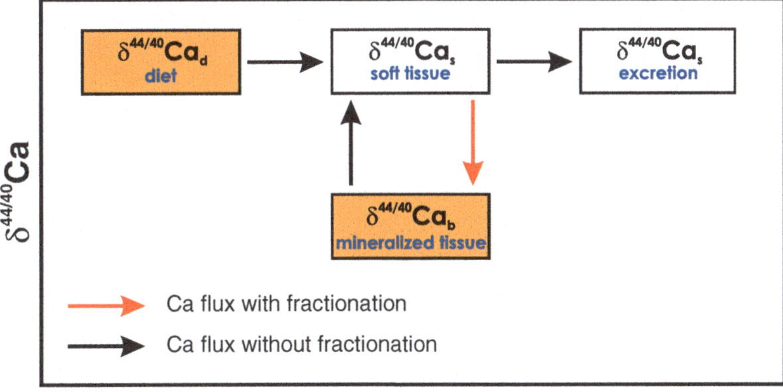

Fig. 21 The basic Ca isotope transport model of Skulan and DePaolo (1999)

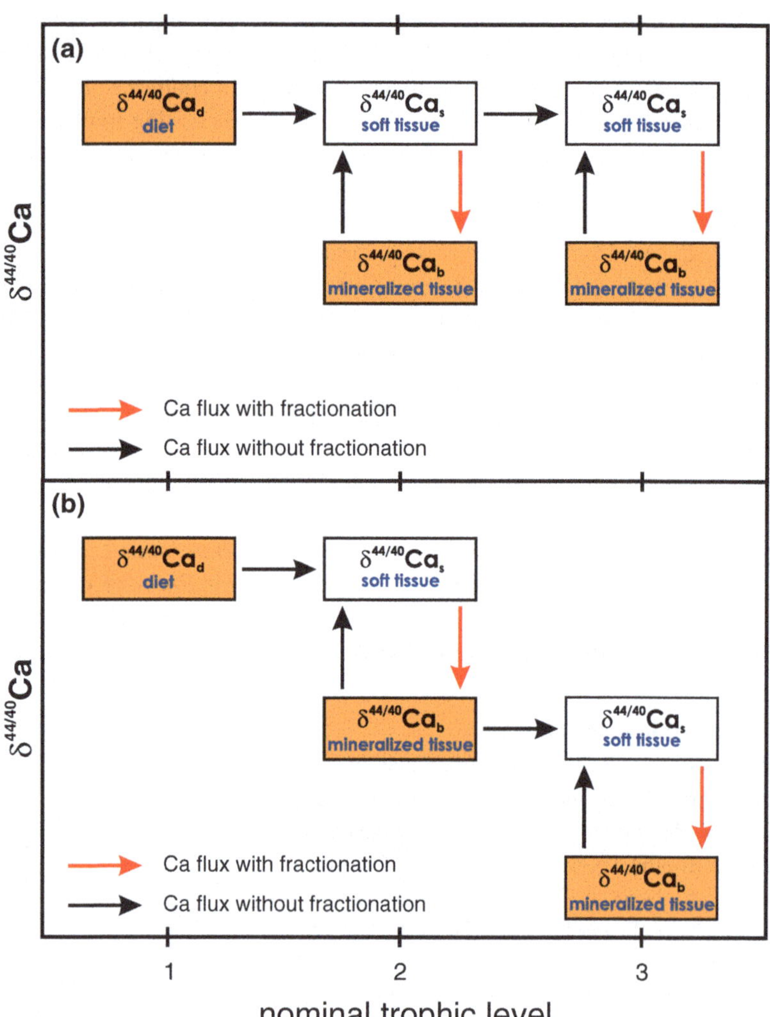

Fig. 22 A TLE between carnivore (nominal trophic level 2) and their prey (nominal trophic level 1) does not develop if only soft tissue is consumed (**a**). Only the consumption and digestion of mineralized tissues lead to a further decrease in the Ca isotopic composition of mineralized tissues of the carnivores (**b**)

refers to the amount of bioavailable mineralized tissue, which might be smaller than the total amount of ingested mineralized tissue. Therefore, even bone consuming carnivores not necessarily show lower $\delta^{44}Ca_{min\text{-}tissue}$ values than herbivores.

TLE only existing within a food chain

The ranges of $\delta^{44}Ca_{min\text{-}tissue}$ of plants (trophic level 1), herbivores (trophic level 2) and carnivores (trophic level 3) are relatively large and partially overlap, when all existing data of $\delta^{44}Ca_{min\text{-}tissue}$ of terrestrial vertebrates is drawn as a function of the nominal trophic level (Fig. 24). The observed ranges are 1.2 ‰ in plants, 3.3 ‰ in herbivores and 2.8 ‰ in carnivores. As a consequence of this large variability, it is not possible to reconstruct the trophic level of an organism exclusively from the Ca isotopic composition of mineralized tissue alone. The lack of clear TLE between herbivores and carnivores is most likely caused by a lack of mineralized tissue in the diet of the carnivores.

Differences between mammals, birds and reptiles

Besides the varying amounts of digested bone Ca, there are additional reasons for the relatively large variability within one trophic level. Calcium isotope ratios of hard tissues of vertebrates indicate systematic differences between different

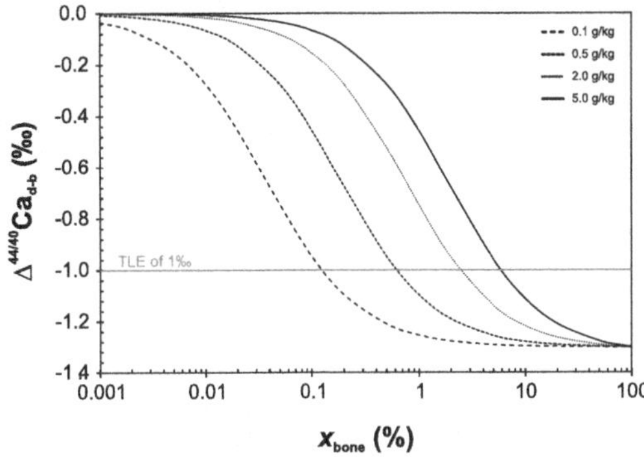

Fig. 23 Modeling the Ca isotopic composition of the diet of a carnivore for different soft-tissue Ca concentration (g/kg) and the true digested portion of mineralized tissue. Assuming Ca concentrations of soft tissue >0.5 g/kg about 1 % of the diet of the carnivore has to be mineralized tissue in order to develop a difference of about 1 ‰ between herbivore and carnivore mineralized tissues

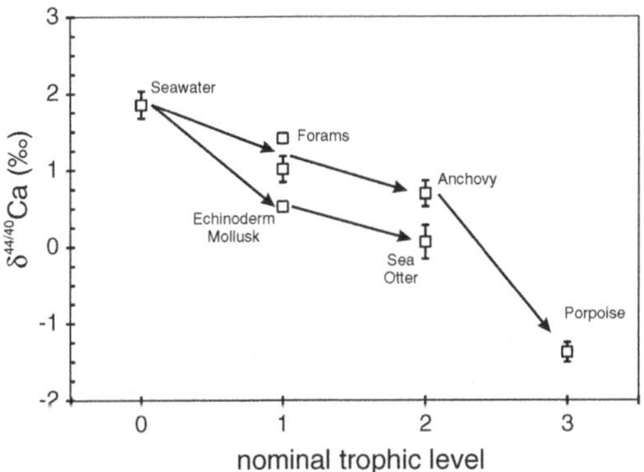

Fig. 24 Trophic level effect. Ca isotope ratios in mineralized tissues decrease with increasing trophic level within a food chain

genera. In general, $\delta^{44}Ca_{bone}$ of mammals tend to be lower than $\delta^{44}Ca_{bone}$ of birds and dinosaurs. $\delta^{44}Ca_{bone}$ of dinosaurs are normally higher than $\delta^{44}Ca_{bone}$ of birds (Skulan et al. 1997; Reynard et al. 2008; Chu et al. 2006; Heuser et al. 2011). This systematic of Ca isotopes in mineralized tissue (see Fig. 25) may be attributed to physiological differences between these different groups of vertebrates. Further investigations are needed for a proper explanation of this pattern.

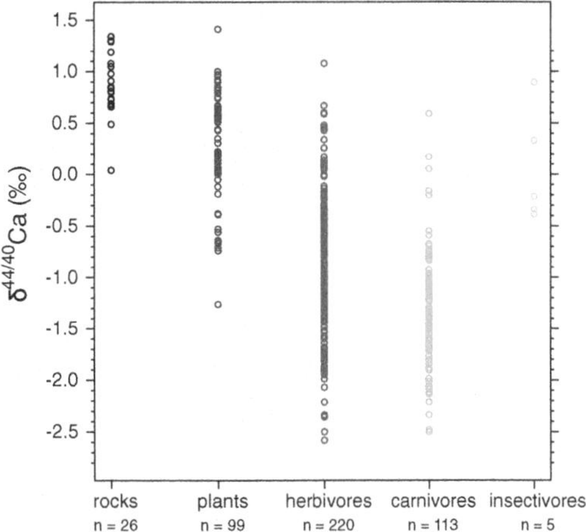

Fig. 25 $\delta^{44/40}Ca$ data compilation of rocks, plants, herbivore and carnivore bones and insectivores. The data of herbivore and carnivore bones are overlapping which makes it difficult to use $\delta^{44/40}Ca_{bone}$ to distinguish between herbivores and carnivores. Data were taken from Amini et al. (2009), Broska (2011), Chu et al. (2006), Cobert et al. (2011), DePaolo (2004), Heuser et al. (2011), Heuser (unpublished), Hirata et al. (2008), Holmden and Bélanger (2010), Jochum et al. (2006), John et al. (2012), Page et al. (2008), Reynard et al. (2010,2011), Skulan and DePaolo (1999), Skulan et al. (1997), Wiegand et al. (2005), Wombacher et al. (2009), Zhu and Macdougall (1998). Data from human bones were included into the carnivore dataset

4.2 Archaeology

Archaeology is one field of application for geochemical techniques especially for stable isotope analysis. Besides the classical isotope system like oxygen ($\delta^{18}O$), carbon ($\delta^{13}C$) or nitrogen ($\delta^{15}N$) also 'non-traditional' isotope systems are applied for archaeological studies. By using stable isotope geochemical techniques it is possible to gain information which otherwise with traditional archaeological techniques are not available. In this context Ca isotopes may be a powerful tool to reconstruct changes in food or recognizing (and dating) the adoption of new foods for humans in the archaeological record.

Table 3 $\delta^{44/40}$Ca of different food samples of human diet

Sample	$\delta^{44/40}$Ca (‰ SRM 915a)	±1SD	Reference
Milk	−1.37	0.14	Chu et al. (2006)
Whey	−1.23	0.12	Chu et al. (2006)
Curd	−1.27	0.06	Chu et al. (2006)
Kefir	−1.21	0.2	Chu et al. (2006)
Yoghurt	−1.44	0.04	Chu et al. (2006)
Commercial milk	−0.98	0.04	Chu et al. (2006)
UHT milk	−1.33	0.08	Heuser (unpublished)
Cheese (gouda)	−1.18	0.05	Heuser (unpublished)
Cheese (Emmentaler)	−1.06	0.05	Heuser (unpublished)
Cheese (Brie)	−1.37	0.09	Heuser (unpublished)
Rice (uncooked)	0.40	0.14	Heuser (unpublished)
Pasta (uncooked)	−1.51	0.06	Heuser (unpublished)
Nachos	−0.66	0.04	Heuser (unpublished)
Egg yolk	0.93	0.13	Skulan and DePaolo (1999)
Egg white	2.81	0.01	Skulan and DePaolo (1999)
Chicken muscle	0.00	0.16	Skulan and DePaolo (1999)

All values have been converted to $\delta^{44/40}$Ca (‰ SRM 915a)

Chu et al. (2006) were the first investigating the human diet and found that dairy products are outstanding with regards to their importance for a normal western diet and their Ca isotopic composition (cf. Table 3). It is possible to roughly estimate the $\delta^{44/40}$Ca of the normal western diet by applying Eq. 1:

$$\delta^{44/40}\text{Ca}_{diet} = \frac{\sum_i \left([\text{Ca}]_i \cdot \delta^{44/40}\text{Ca}_i \cdot x_i\right)}{\sum_i \left([\text{Ca}]_i \cdot x_i\right)} \quad (1)$$

with x_i = relative contribution of the food i to the diet, $[\text{Ca}]_i$ = Ca concentration of the food i and $\delta^{44/40}\text{Ca}_i$ = Ca isotopic composition of food i (Table 4).

From Table 4 it can be seen that the Ca isotopic composition of normal European diet is dominated by dairy products. Assuming a $\Delta^{44/40}\text{Ca}_{\text{bone-diet}}$ of about −1.3 ‰, modern bones should have $\delta^{44/40}$Ca values of about −2.3 ‰. In contrast, a diet free of dairy products has an average $\delta^{44/40}$Ca of about −0.6 ‰ and thus bones should then have $\delta^{44/40}$Ca of about −1.9 ‰ being isotopically heavier than modern bones. This difference in $\delta^{44/40}\text{Ca}_{\text{bone}}$ between modern and ancient bones should make it possible to use Ca isotopes as a proxy for "dairy consumption and weaning practices in past human cultures" (Chu et al. 2006).

Reynard and coworkers did some detailed studies of the Ca isotopic composition of human and animal bones from archeological sites in order to check if dairy consumption can be detected (Reynard et al. 2010, 2011, 2013). All studies showed that the individual Ca metabolism in humans causes variations in the $\delta^{44/40}\text{Ca}_{\text{bone}}$ which are exceeding the shift caused by the change from a non-dairy to a dairy diet. Therefore, the use of Ca isotopes to date the rise of dairy consumption of humans is at least not straightforward.

4.3 Paleoclimate

One of the motivations for the increase in Ca isotope studies since the late 1990ies was the observation of Zhu and Macdougall (1998) that Ca isotope fractionation in calcite tests of certain marine benthonic and planktic foraminifers (*Cibicides kullenbergi*, *Neogloboquadrina pachyderma* and *Globigerinoides sacculifer*),

Table 4 Average Ca concentration, relative contribution of the food to a normal European diet, and the Ca isotopic composition of each group (Chu et al. 2006; Heuser and Eisenhauer 2010)

Group	[Ca]^a (mg/100 g)	Fraction (%)^b	$\delta^{44/40}$Ca (‰ SRM 915a)^c
Diary products	500	14	−1.20
Vegetables	100	21	−0.68
Fruit	20	14	−0.68
Crop products	35	24	−0.56
Meat	10	6	−0.08
Fats	7	2	−1.00
Water	5	20	0.68

These data result in a typical dietary Ca input of about −1 ‰ for a normal European diet
^{a}corresponds to [Ca]$_i$ in Eq. 1; ^{b}corresponds to X$_i$ in Eq. 1; ^{c}corresponds to $\delta^{44/40}$Ca$_i$ in Eq. 1

seems to depend on the water temperature. A series of subsequent calibration studies on various different species, revealed evidences for and against large temperature sensitivities of $\delta^{44/40}$Ca (Sect. 2.2). Temperature dependent Ca isotope fractionation, which is large enough to allow for useful paleotemperature reconstructions, is described so far only in few foraminifer species as well as Cretaceous reef-building molluscs (rudists). However, as discussed earlier (Sects. 2.1 and 3.3.1), ambiguous information regarding the $\delta^{44/40}$Ca-temperature dependence of rudists and foraminifers exist.

Rudists

The temperature dependence of Ca isotope fractionation in rudists, Cretaceous reef-building mollusks, is controversially discussed. While Immenhauser et al. (2005) found covariation of $\delta^{44/40}$Ca with Mg/Ca and δ^{18}O (Fig. 26), indicating a temperature dependent Ca isotope fractionation in well preserved rudists, Steuber and Buhl (2006) did not find such a relation. The reasons for this discrepancy are yet unclear, possibly related to either the influence of additional so far unrecognized environmental parameters or preservation of the used sample material.

Foraminifera

From the numerous investigated foraminifer species, a large temperature effect is apparent only for the two planktic species, *G. sacculifer* and *N. pachyderma* (s), whereas all other studied species reveal a temperature dependencies

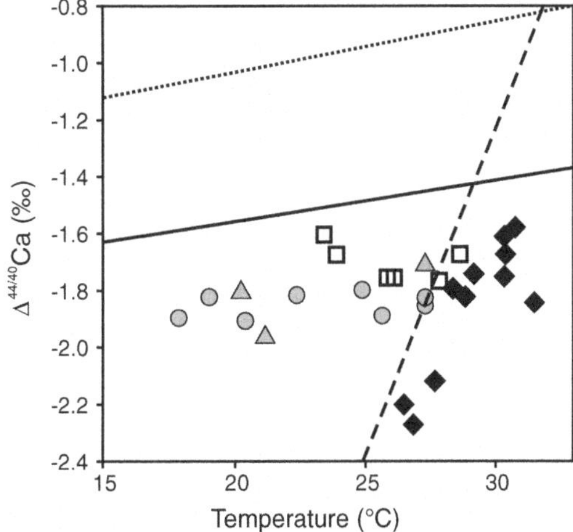

Fig. 26 $\delta^{44/40}$Ca of Cretaceous rudist as a function of temperature. While some rudists show a substantial apparent temperature sensitivity (*black diamonds*, Immenhauser et al. 2005) similar to the planktic foraminifer *G. sacculifer* (*dashed line*), other rudists (*grey triangels* and *dots* and *open squares*, Steuber and Buhl 2006) exibit a temperature response similar to inorganic aragonite (*solid line*) and *O. universa* (*dotted line*, Gussone et al. 2003)). As Cretaceous seawater $\delta^{44/40}$Ca is not known, rudist data are given relative to present seawater, which results in an offset compared to the recent foraminifers and synthetic carbonates

<0.03 ‰/°C. Besides species specific differences, considerable discrepancies are reported within single species.

Globigerinoides sacculifer seems to exhibit a bimodal temperature sensitivity with either ∼0.2 ‰/°C or ∼0.02 ‰/°C (Nägler et al. 2000; Chang et al. 2004; Gussone et al. 2004, 2009; Sime et al. 2005; Hippler et al. 2006; Kasemann et al. 2008; Griffith et al. 2008). This

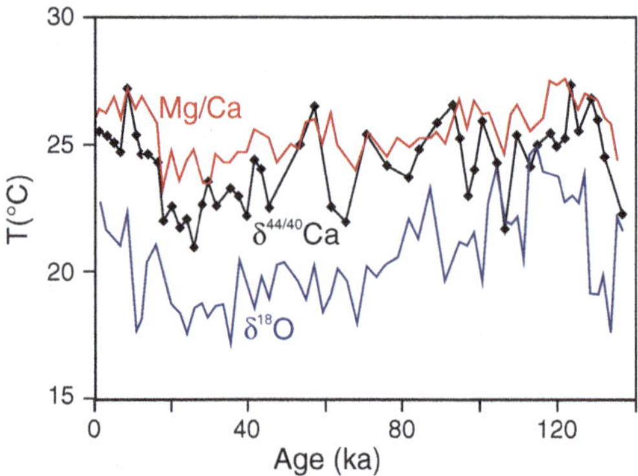

Fig. 27 Calcium isotope ratios as paleo temperature proxy from the geological past. The Quaternary record of *G. sacculifer* from the tropical Atlantic (GeoB 1112, Hippler et al. 2006; Nägler et al. 2000) shows a good agreement of temperature changes reconstructed from Mg/Ca, $\delta^{18}O$ and Ca-isotopes

fractionation pattern of *G. sacculifer* is proposed to relate to changes in the cellular Ca budget and involved Ca transport processes (seawater vacuolization and Ca TMT) in response to environmental changes, and possibly to gametogenesis, but is not yet understood.

Calcium isotope ratios of *Neogloboquadrina pachyderma* (s) demonstrate a linear dependence on temperature of about 0.2 ‰/°C at temperatures above 4 °C and at low temperatures an anomaly towards higher $\delta^{44/40}Ca$ (Gussone et al. 2009; Hippler et al. 2009a).

The possible $\delta^{44/40}Ca$-temperature effect suggested for the benthonic foraminifer *C. kullenbergi* (Zhu and Mcdougall 1998) is in agreement to a relatively large apparent temperature dependence between −1 and 3 °C observed in coretop samples of benthonic foraminifers including *Cibicides wuellerstorfi*, *Pyrgo* spp. and *Oridorsalis umbonatus* (Sect. 2.1.2) (Gussone and Filipsson 2010; Gussone et al. 2016). The anomalies in $\delta^{44/40}Ca$ of foraminifer tests at low temperatures seen in the planktic foraminifer *N. pachyderma* (s) and benthonic species are however no real temperature responses, but suggested to result from the interference of temperature and carbonate chemistry on Ca isotope fractionation.

A few case studies tested the applicability of the $\delta^{44/40}Ca$ thermometer, before the

complications of the temperature dependent Ca isotope fractionation became apparent in its total complexity. In tropical Atlantic and Caribbean sites, *G. sacculifer* shows a relative good agreement between $\delta^{44/40}Ca$ and Mg/Ca based SST (Nägler et al. 2000; Hippler et al. 2006; Gussone et al. 2004). For instance a temperature reconstruction based on $\delta^{44/40}Ca$ of *G. sacculifer* from the tropical Eastern Atlantic (site GeoB 1112) shows a temperature increase from the LGM towards the Holocene similar to reconstructions based on Mg/Ca and $\delta^{18}O$ (Hippler et al. 2006) (Fig. 27). In contrast, other downcore records of *G. trilobus/sacculifer* seem to show no significant temperature dependency (Heuser et al. 2005; Sime et al. 2007).

Although some promising initial case studies indicate a large potential for Ca-isotope paleothermometry, its applicability is strongly limited. The sporadic and not permanent occurrence of the large temperature sensitivity needs to be better understood before Ca isotopes might be useful as a paleo temperature proxy for special environments.

References

Allison N, Finch AA, EIMF (2010) $\delta^{11}B$, Sr, Mg and B in a modern Porites coral: the relationship between calcification site pH and skeletal chemistry. Geochim Cosmochim Acta 74:1790–1800

Amini M, Eisenhauer A, Böhm F et al (2009) Calcium Isotopes ($\delta^{44/40}Ca$) in MPI-DING reference glasses, USGS rock powders and various rocks: evidence for Ca isotope fractionation in terrestrial silicates. Geostand Geoanal Res 33:231–247

Blättler CL, Henderson GM, Jenkyns HC (2012) Explaining the Phanerozoic Ca isotope history of seawater. Geology 40:843–846

Blättler CL, Stanley SM, Henderson GM et al (2014) Identifying vital effects in Halimeda algae with Ca isotopes. Biogeosci Discuss 11:3559–3580

Böhm F, Gussone N, Eisenhauer A et al (2006) Calcium isotope fractionation in modern scleractinian corals. Geochim Cosmochim Acta 70:4452–4462

Böhm F, Eisenhauer A, Tang J et al (2012) Strontium isotope fractionation of planktic foraminifera and inorganic calcite. Geochim Cosmochim Acta 93:300–314

Boskey AL (2007) Biomineralization of bones and teeth. Elements 3:385–391

Broska J (2011) Ernährungsrekonstruktion für terrestrische Wirbeltier. Diploma thesis, Universität Mainz (unpublished)

Chang VT-C, Williams RJP, Makishima A et al (2004) Mg and Ca isotope fractionation during $CaCO_3$ biomineralisation. Biochem Bioph Res Co 323:79–85

Chu N-C, Henderson GM, Belshaw NS et al (2006) Establishing the potential of Ca isotopes as proxy for consumption of dairy products. Appl Geochem 21:1656–1667

Clementz MT, Holden P, Koch PL (2003) Are calcium isotopes a reliable monitor of trophic level in marine settings? Int J Osteoarchaeol 13:29–36

Cobert F, Schmitt A-D, Bourgeade P et al (2011) Experimental identification of Ca isotopic fractionations in higher plants. Geochim Cosmochim Acta 75:5467–5482

De La Rocha CL, DePaolo DJ (2000) Isotopic evidence for variations in the marine calcium cycle over the cenozoic. Science 289:1176–1178

de Nooijer LJ, Spero HJ, Erez J et al (2014) Biomineralization in perforate foraminifera. Earth-Sci Rev 135:48–58

Elderfield H, Bertram CJ, Erez J (1996) A biomineralization model for the incorporation of trace elements into foraminiferal calcium carbonate. Earth Planet Sci Lett 142:409–423

Erez J (2003) The source of ions for biomineralization in foraminifera and their implications for paleoceanographic proxies. Rev Min Geochem 54:115–149

Farkaš J, Buhl D, Blenkinsop J et al (2007) Evolution of the oceanic calcium cycle during the late Mesozoic: evidence from $\delta^{44/40}Ca$ of marine skeletal carbonates. Earth Planet Sci Lett 253:96–111

Fietzke J, Eisenhauer A (2006) Determination of temperature-dependent stable strontium isotope ($^{88}Sr/^{86}Sr$) fractionation via bracketing standard MC-ICP-MS. Geochem Geophys Geosyst 7. doi:10.1029/2006GC001243

Foster L, Finch A, Allison N et al (2008) Mg in aragonitic bivalve shells: seasonal variations and mode of incorporation in *Arctica islandica*. Chem Geol 254:113–119

Galy A, Bar-Matthews M, Halicz L et al (2002) Mg isotopic composition of carbonate: insight from spelothem formation. Earth Planet Sci Lett 201:105–115

Griffith EM, Paytan A, Kozdon R et al (2008) Influences on the fractionation of calcium isotopes in planktonic foraminifera. Earth Planet Sci Lett 268:124–136

Gussone N, Filipsson HL (2010) Calcium isotope ratios in calcitic tests of benthic Foraminifers. Earth Planet Sci Lett 290:108–117

Gussone N, Eisenhauer A, Heuser A et al (2003) Model for kinetic effects on calcium isotope fractionation ($\delta^{44}Ca$) in inorganic aragonite and cultured planktonic foraminifera. Geochim Cosmochim Acta 67:1375–1382

Gussone N, Eisenhauer A, Tiedemann R et al (2004) $\delta^{44}Ca$, $\delta^{18}O$ and Mg/Ca reveal caribbean sea surface temperature and salinity fluctuations during the pliocene closure of the Central-American Gateway. Earth Planet Sci Lett 227:201–214

Gussone N, Böhm F, Eisenhauer A et al (2005) Calcium isotope fractionation in calcite and aragonite. Geochim Cosmochim Acta 69:4485–4494

Gussone N, Langer G, Thoms S et al (2006) Cellular calcium pathways and isotope fractionation in *Emiliania huxleyi*. Geology 34:625–628

Gussone N, Langer G, Geisen M et al (2007) Calcium isotope fractionation in coccoliths of cultured *Calcidiscus leptoporus, Helicosphaera carteri, Syracosphaera pulchra and Umbilicosphaera foliosa*. Earth Planet Sci Lett 260:505–515

Gussone N, Hönisch B, Heuser A et al (2009) A critical evaluation of calcium isotope ratios in tests of planktonic foraminifers. Geochim Cosmochim Acta 73:7241–7255

Gussone N, Zonnefeld K, Kuhnert H (2010) Minor element and Ca isotope composition of calcareous dinoflagellate cysts of *Thoracosphaera heimii*. Earth Planet Sci Lett 289:180–188

Gussone N, Filipsson HL, Kuhnert H (2016) Mg/Ca, Sr/Ca and Ca isotope ratios in benthonic foraminifers related to test structure, mineralogy and environmental controls. Geochim Cosmochim Acta

Haase-Schramm A, Böhm F, Eisenhauer A et al (2003) Sr/Ca ratios and oxygen isotopes from sclerosponges: temperature history of the Caribbean mixed layer and thermocline during the Little Ice Age. Paleoceanography 18. doi:10.1029/2002PA000830

Halicz L, Galy A, Belshaw NS et al (1999) High-precision measurement of calcium isotopes in carbonates and related materials by multiple collector inductively coupled plasma mass spectrometry. J Anal At Spec 14:1835–1838

Heinemann A, Fietzke J, Eisenhauer A et al (2008) Modification of Ca isotope and trace metal composition of the major matrices involved in shell formation of Mytilus edulis. Geochem Geophys Geosyst 9: Q01006. doi:10.1029/2007GC001777

Heuser A, Eisenhauer A, Böhm F et al (2005) Calcium isotope ($\delta^{44/40}Ca$) variations of neogene planktonic foraminifera. Paleoceanography 20:PA2013. doi:10.1029/2004PA001048

Heuser A, Eisenhauer A (2010) A pilot study on the use of natural calcium isotope ($^{44}Ca/^{40}Ca$) fractionation in urine as a proxy for the human body calcium balance. Bone 46:889–896

Heuser A, Tütken T, Gussone N et al (2011) Calcium isotopes in fossil bones and teeth—diagenetic versus biogenic origin. Geochim Cosmochim Acta 75:3419–3433

Hiebenthal C (2009) Sensitivity of *A. islandica* and *M. edulis* towards environmental changes: a threat to the bivalves—an opportunity for palaeo-climatology? PhD thesis, Christian-Albrechts-Universität Kiel, pp 1–138

Hildebrand-Habel T, Willems H (2000) Distribution of calcareous dinoflagellates from the Maastrichtian to middle Eocene of the western South Atlantic Ocean. Int J Earth Sci 88:694–707

Hippler D, Eisenhauer A, Nägler TF (2006) Tropical Atlantic SST history inferred from Ca isotope thermometry over the last 140ka. Geochim Cosmochim Acta 70:90–100

Hippler D, Kozdon R, Darling KF et al (2009a) Calcium isotopic composition of high-latitude proxy carrier *Neogloboquadrina pachyderma* (sin.). Biogeosciences 6:1–14

Hippler D, Buhl D, Witbaard R et al (2009b) Towards a better understanding of magnesium-isotope ratios from marine skeletal carbonates. Geochim Cosmochim Acta 73:6134–6146

Hippler D, Witbaard R, van Aken HM et al (2013) Exploring the calcium isotope signature of *Arctica islandica* as an environmental proxy using laboratory- and field-cultured specimens. Palaeogeogr Palaeoclimatol Palaeoecol 373:75–87

Hirata T, Tanoshima M, Suga A et al (2008) Isotopic analysis of calcium in blood plasma and bone from mouse samples by multiple collector-ICP-mass spectrometry. Anal Sci 24:1501–1507

Holmden C, Bélanger N (2010) Ca isotope cycling in a forested ecosystem. Geochim Cosmochim Acta 74:995–1015

Holmden C, Papanastassiou DA, Blanchon P et al (2012) $\delta^{44/40}$Ca variability in shallow water carbonates and the impact of submarine groundwater discharge on Ca-cycling in marine environments. Geochim Cosmochim Acta 83:179–194

Immenhauser A, Nägler TF, Steuber T et al (2005) A critical assessment of mollusk ^{18}O/^{16}O, Mg/Ca, and ^{44}Ca/^{40}Ca ratios as proxies for Cretaceous seawater temperature seasonality. Palaeogeogr Palaeoclimatol Palaeoecol 215:221–237

Inoue M, Gussone N, Koga Y et al (2015) Controlling factors of Ca isotope fractionation in scleractinian corals evaluated by temperature, pH and light controlled culture experiments. Geochim Cosmochim Acta 167:80–92

Inouye I, Pienaar RN (1983) Observations on the life cycle and microanatomy of *Thoracosphaera heimii* (Dinophyceae) with special reference to its systematic position. S AfrJ Bot 2:63–75

Jochum KP, Stoll B, Herwig K et al (2006) MPI-DING reference glasses for in situ microanalysis: new reference values for element concentrations and isotope ratios. Geochem Geophys Geosyst 7:1–44. doi:10.1029/2005GC001060

John T, Gussone N, Podladchikov YY et al (2012) Volcanic arcs fed by rapid pulsed fluid flow through subducting slabs. Nat Geosci 5:489–492

Kasemann SA, Schmidt DN, Pearson PN et al (2008) Biological and ecological insights into Ca isotopes in planktic foraminifers as a palaeotemperature proxy. Earth Planet Sci Lett 271:292–302

Kisakürek B, Eisenhauer A, Böhm F et al (2011) Controles on calcium isotope fractionation in cultured planktic foraminifera, *Globigerinoides ruber* and *Globigerinella siphonifera*. Geochim Cosmochim Acta 75:427–443

Krabbenhöft A (2011) Stable strontium isotope ($\delta^{88/86}$Sr) fractionation in the marine realm: a pilot study. PhD thesis, Christian-Albrechts-Universität Kiel

Krabbenhöft A, Eisenhauer A, Böhm F et al (2010) Constraining the marine strontium budget with natural strontium isotope fractionations (^{87}Sr/^{86}Sr*, $\delta^{88/86}$Sr) of carbonates, hydrothermal solutions and river waters. Geochim Cosmochim Acta 74:4097–4109

Krause S, Liebetrau V, Gorb S et al (2012) Microbial nucleation of Mg-rich dolomite in exopolymeric substances under anoxic modern seawater salinity: new insight into an old enigma. Geology 40:587–590

Langer G, Gussone N, Nehrke G et al (2006) Coccolith strontium to calcium ratios in *Emiliania huxleyi*: the dependence on seawater strontium and calcium concentrations. Limnol Oceanogr 51:310–320

Langer G, Gussone N, Nehrke G et al (2007) Calcium isotope fractionation during coccolith formation in *Emiliania huxleyi*: independence of growth and calcification rate. Geochem Geophysi Geosys 8:Q05007. doi:10.1029/2006GC001422

Lemarchand D, Wasserburg GJ, Papanastassiou DA (2004) Rate-controlled calcium isotope fractionation in synthetic calcite. Geochim Cosmochim Acta 68:4665–4678

Marriott CS, Henderson GM, Belshaw NS et al (2004) Temperature dependence of δ^{7}Li, δ^{44}Ca and Li/Ca during growth of calcium carbonate. Earth Planet Sci Lett 222:615–624

Martin JE, Vance D, Balter V (2014) Natural variation of magnesium isotopes in mammal bones and teeth from two South African trophic chains. Geochim Cosmochim Acta 130:12–20

Martin JE, Tacail T, Adnet T et al (2015) Calcium isotopes reveal the trophic position of extant and fossil elasmobranchs. Chem Geol 415:118–125

Meisterfeld R, Holzmann M, Pawlowski J (2001) Morphological and molecular characterization of a new terrestrial allogromiid species: *Edaphoallogromia australica* gen. et spec. nov. (Foraminifera) from Northern Queensland (Australia). Protist 152:185–192

Middelberg U, Gussone N, Becker RT et al (unpubl) An evaluation of cephalopod shells as geochemical proxy archives

Milliman JD (1993) Production and accumulation of calcium carbonate in the ocean: budget of a nonsteady state. Global Biogeochem Cy 7:927–957

Müller MN, Kisakürek B, Buhl D et al (2011) Response of the coccolithophores *Emiliania huxleyi* and *Coccolithus braarudii* to changing seawater Mg^{2+} and Ca^{2+} concentrations: Mg/Ca, Sr/Ca ratios and $\delta^{44/40}$Ca, $\delta^{26/24}$Mg of coccolith calcite. Geochim Cosmochim Acta 75:2088–2102

Nägler TF, Eisenhauer A, Müller A et al (2000) δ^{44}Ca-temperature calibration on fossil and cultured *Globigerinoides sacculifer*: new tool for reconstruction of past sea surface temperatures. Geochem Geophys Geosyst 1. doi:10.1029/2000GC000091

Page BD, Bullen TD, Mitchell MJ (2008) Influences of calcium availability and tree species on Ca isotope fractionation in soil and vegetation. Biogeochemistry 88:1–13

Pasteris JD, Wopenka B, Valsami-Jones E (2008) Bone and tooth mineralization: why apatite? Elements 4:97–104

Pawlowski J, Holzmann M, Berney C et al (2003) The evolution of early foraminifera. PNAS 100:11494–11498

Pogge von Strandmann PAE (2008) Precise magnesium isotope measurements in core top planktic and benthic foraminifera. Geochem Geophys Geosyst 9:Q12015. doi:10.1029/2008GC002209

Pretet C, Samankassou E, Felis T et al (2013) Constraining calcium isotope fractionation ($\delta^{44/40}$Ca) in modern and fossil scleractinian coral skeleton. Chem Geol 340:49–58

Ra K, Kitagawa H, Shiraiwa Y (2010) Mg isotopes in chlorophyll-a and coccoliths of cultured coccolithophores (*Emiliania huxleyi*) by MC-ICP-MS. Mar Chem 122:130–137

Raddaz J, Liebetrau V, Rüggeberg A et al (2013) Stable Sr-isotope, Sr/Ca, Mg/Ca, Li/Ca and Mg/Li ratios in the scleractinian cold-water coral *Lophelia pertusa*. Chem Geol 352:143–152

Reynard LM, Hedges REM, Henderson GM (2008) Stable calcium isotope ratios ($\delta^{44/42}$Ca) in bones and teeth for the detection of dairying by ancient humans. Geochim Cosmochim Acta 12:A790

Reynard LM, Henderson GM, Hedges REM (2010) Calcium isotope ratios in animal and human bone. Geochim Cosmochim Acta 74:3735–3750

Reynard LM, Henderson GM, Hedges REM (2011) Calcium isotopes in archaeological bones and their relationships to dairy consumption. J Archaeol Sci 38:657–664

Reynard LM, Pearce JA, Henderson GM et al (2013) Calcium isotopes in juvenile milk-consumers. Archaeometry 55:946–957

Rollion-Bard C, Vigier N, Spezzaferri S (2007) In situ measurements of calcium isotopes by ion microprobe in carbonates and application to foraminifera. Chem Geol 244:679–690

Rüggeberg A, Fietzke J, Liebetrau V et al (2008) Stable strontium isotopes ($\delta^{88/86}$Sr) in cold-water corals—a new proxy for reconstruction of intermediate ocean water temperatures. Earth Planet Sci Lett 269:570–575

Russell WA, Papanastassiou DA, Tombrello TA (1978) Ca isotope fractionation on the Earth and other solar system materials. Geochim Cosmochim Acta 42:1075–1090

Sime NG, De La Rocha CL, Galy A (2005) Negligible temperature dependence of calcium isotope fractionation in 12 species of planktonic foraminifera. Earth Planet Sci Lett 232:51–66

Sime NG, De La Rocha CL, Tipper ET et al (2007) Interpreting the Ca isotope record of marine biogenic carbonates. Geochimica et Cosmochimica Acta 71:3979–3989

Skulan JL, DePaolo DJ (1999) Calcium isotope fractionation between soft and mineralized tissues as a monitor of calcium use in vertebrates. Proc Nat Acad Sci 96:13709–13713

Skulan J, DePaolo DJ, Owens TL (1997) Biological control of calcium isotopic abundances in the global calcium cycle. Geochim Cosmochim Acta 61:2505–2510

Stenzel HB (1964) Oysters: composition of the larval shell. Science 145:155–156

Steuber T, Buhl D (2006) Calcium-isotope fractionation in selected modern and ancient marine carbonates. Geochim Cosmochim Acta 70:5507–5521

Tang J, Dietzel M, Böhm F et al (2008) Sr^{2+}/Ca^{2+} and $^{44}Ca/^{40}Ca$ fractionation during inorganic calcite formation: II. Ca isotopes. Geochim Cosmochim Acta 72:3733–3745

Tang J, Niedermayr A, Köhler SJ et al (2012) Sr^{2+}/Ca^{2+} and $^{44}Ca/^{40}Ca$ fractionation during inorganic calcite formation: III. Impact of salinity/ionic strength. Geochim Cosmochim Acta 77:432–443

Taubner I, Böhm F, Eisenhauer A et al (2012) Uptake of alkaline earth metals in Alcyonarian spicules (Octocorallia). Geochim Cosmochim Acta 84:239–255

Teichert BMA, Meister P, Mavromatis V et al (submitted) Early diagenetic, primary dolomite formation on the Peru Margin—insights from Ca and Mg isotopes. Submitted to Geochim Cosmochim Acta

ter Kuile BT (1991) Mechanisms for calcification and carbon cycling in algal symbiont-bearing foraminifera. In: Lee JL, Anderson OR (eds) Biology of foraminifera. Academic Press, London, pp 74–89, 73–90

Ullmann CV, Böhm F, Rickaby REM et al (2013) The Giant Pacific Oyster (Crassostrea gigas) as a modern analog for fossil ostreoids: isotopic (Ca, O, C) and elemental (Mg/Ca, Sr/Ca, Mn/Ca) proxies. Geochem Geophys Geosyst 14:4109–4120. doi:10.1002/ggge.20257

Veizer J, Ala D, Azmy K et al (1999) $^{87}Sr/^{86}Sr$, $\delta^{13}C$ and $\delta^{18}O$ evolution of Phanerozoic seawater. Chem Geol 161:59–88

von Allmen K, Nägler TF, Pettke T et al (2010) Stable isotope profiles (Ca, O, C) through modern brachiopod shells of *T. septentrionalis* and *G. vitreus*: implications for calcium isotope paleo-ocean chemistry. Chem Geol 269:210–219

Wang SH, Yan W, Magalhães HV et al (2012) Calcium isotope fractionation and its controlling factors of authigenic carbonates in the cold seeps from the Northern South China Sea. Chinese Sci Bull 57:1325–1332

Wang S, Yan W, Magalhães HV et al (2013a) Factors influencing methane-derived authigenic carbonate formation at cold seep from southwestern Dongsha area in the northern South China. Environ Earth Sci. doi:10.1007/s12665-013-2611-9

Wang Z, Hu P, Gaetani G et al (2013b) Experimental calibration of Mg isotope fractionation between aragonite and seawater. Geochim Cosmochim Acta 102:113–123

Wiegand BA, Chadwick OA, Vitousek PM et al (2005) Ca cycling and isotopic fluxes in forested ecosystems in Hawaii. Geophys Res Lett 32:L11404. doi:10.1029/2005GL022746

Wombacher F, Eisenhauer A, Böhm F et al (2011) Magnesium stable isotope fractionation in marine biogenic calcite and aragonite. Geochim Cosmochim Acta 75:5797–5818

Yoshimura T, Tanimizu M, Inoue M et al (2011) Mg isotope fractionation in biogenic carbonates of deep-sea coral, benthic foraminifera, and hermatypic coral. Anal Bioanal Chem. doi:10.1007/s00216-011-5264-0

Zhu P (1999) Calcium isotopes in the marine environment. PhD-thesis, University of California

Zhu P, Macdougall D (1998) Calcium isotopes in the marine environment and the oceanic calcium cycle. Geochim Cosmochim Acta 62:1691–1698

Zonneveld KAF (2004) Potential use of stable oxygen isotope composition of *Thoracosphaera heimii* for upper water column (thermocline) temperature reconstruction. Mar Micropal 50:307–317

Earth-Surface Ca Isotopic Fractionations

Anne-Désirée Schmitt

Abstract

Studying Ca biogeochemical cycle in the critical zone using Ca isotopes has revealed Ca isotope fractionation mechanisms associated to biological (Ca uptake by roots and within-tree translocation, Ca recycling by vegetation) or abiotic (secondary mineral precipitation, adsorption) processes. Ca isotopes have thus shown their potential to trace Ca cycling processes within the tree, Ca uptake mechanisms within the rhizosphere, Ca recycling by the vegetation; Ca isotopes can also be used as a time integrated vegetal turnover marker or as a hydrological tracer.

Keywords

Calcium biogeochemical cycle · Critical zone · Calcium isotopes · Isotopic fractionation · Biotic · Abiotic

1 Introduction

Determining the nutrient element fluxes, e.g. of calcium (Ca), to the oceans is a key question in Earth Sciences dealing with surface processes. Indeed, during the weathering of continents, the dissolution of Ca, via the reaction of atmospheric CO_2 with calcium silicates, extracts carbon from the ocean-atmosphere reservoir by precipitation of calcium or magnesium carbonates in the ocean. This mechanism reduces the greenhouse effect of CO_2. As such, the study of present and past oceanic Ca budget can provide informations on past variations in the carbon cycle and hence on the parameters that control it, such as climatic variations (Berner et al. 1983; Lasaga et al. 1985; Raymo et al. 1988).

The major Ca flux to the ocean is brought by the rivers, which implies the necessity to understand the mechanisms that control the continental weathering and the transport of calcium by world rivers. Therefore, the variability of Ca

A.-D. Schmitt (✉)
LHyGeS/EOST, Université de Strasbourg,
Strasbourg Cedex, France
e-mail: adschmitt@unistra.fr

A.-D. Schmitt
Laboratoire Chrono-Environnement, Université
Bourgogne Franche-Comté, Besançon, France

© Springer-Verlag Berlin Heidelberg 2016
N. Gussone et al., *Calcium Stable Isotope Geochemistry*,
Advances in Isotope Geochemistry, DOI 10.1007/978-3-540-68953-9_5

fluxes in natural waters as well as the physico-chemical parameters that control them have to be identified at global and local catchment scales. Ca dissolved in river waters has two distinct sources: continental bedrocks, including igneous and sedimentary rocks (Aubert et al. 2001; Dijkstra and Smits 2002; Dijkstra 2003; Drouet et al. 2005) and the atmosphere (Miller et al. 1993; Chadwick et al. 1999; Likens et al. 1998; Drouet et al. 2005; Schmitt and Stille 2005). Nutrient recycling from leaf and needle litter contributes also to the Ca availability for trees (Likens et al. 1998).

Until recently the Ca budget was approached by using Sr isotope ratios, routinely measured for a long time and recognized as reliable Sr source tracers (Graustein and Armstrong 1983; Åberg et al. 1990; Miller et al. 1993; Capo et al. 1998; Kennedy et al. 2002; Bullen and Bailey 2005; Bélanger and Holmden 2010). Sr and Ca are indeed alkaline-earth elements with similar ionic radius and charge, which are supposed to behave similarly in natural ecosystems and, therefore, Sr isotopes have been considered in order to understand the behaviour of Ca (Elias et al. 1982; Åberg et al. 1989; Åberg 1995; Bailey et al. 1996; Drouet et al. 2005; Wiegand and Schwendenmann 2013; Bedel et al. 2016). However, studies using Sr/Ca elemental ratios in forested ecosystems point to a potential decoupling between both elements due to biological processes which occur within trees along the transpiration stream, that fractionate these elements from each other (Baes and Bloom 1988; Poszwa et al. 2000; Bullen and Bailey 2005; Drouet et al. 2005; Berger et al. 2006). The recent developments of stable Ca isotopic measurements confirmed this decoupling between Sr and Ca isotopic systems. Ca does not only allow to trace sources like Sr isotopes but also to decipher processes within the biosphere, such as internal recycling of Ca between litter, organic soil horizons and vegetation (Schmitt et al. 2003; Schmitt and Stille 2005; Cenki-Tok et al. 2009; Holmden and Bélanger 2010), or abiotic processes such as secondary mineral precipitations (Tipper et al. 2006; Ewing et al. 2008; Jacobson and Holmden 2008; Hindshaw et al. 2011, 2013; Nielsen and DePaolo 2013; Jacobson et al. 2015).

In this section the relative importance of the four main Ca reservoirs (water-soil-vegetation-atmosphere) will be discussed from the local to global scales, and we will try to identify the interactions between all of them. In order to depict the fractionation processes (water-rock interactions and biological (re)cycling) governing the Earth-surface Ca isotopic system, we will distinguish between forested and non-forested catchments. We will then have a look at the influence of these local signatures on the global signatures of Ca carried by the world rivers. This finally allows to recognize the potential of Ca isotopes applied to Earth-surface processes.

2 $\delta^{44/40}$Ca Fractionations Related to Continental Weathering Processes

2.1 Range of $\delta^{44/40}$Ca Variations in Earth-Surface Processes

The global range of $\delta^{44/40}$Ca variations in Earth-surface processes has been reported Fig. 1. Vegetation presents one of the widest ranges of variation (~ 4.1 ‰) on Earth, with values comprised between -2.2 and 1.76 ‰. This amplitude is similar to that observed in mammals and fishes (~ 5 ‰) (see Chapter "Biominerals and Biomaterials"; Russell et al. 1978; Skulan et al. 1997; Skulan and DePaolo 1999; Clementz et al. 2003; Chu et al. 2006; Reynard et al. 2010; Heuser et al. 2011), but is huge compared to the variations (~ 2 ‰) caused by abiotic calcium carbonate precipitations (see Chapter "Calcium Isotope Fractionation During Mineral Precipitation from Aqueous Solution"; Skulan et al. 1997; Zhu and MacDougall 1998; De La Rocha and DePaolo 2000; Gussone et al. 2003; Lemarchand et al. 2004; Marriott et al. 2004; Blättler et al. 2015; Fantle and DePaolo 2005; Gussone et al. 2005; Kasemann et al. 2005; Sime et al. 2005; Steuber and Buhl 2006; Fantle and DePaolo 2007; Amini et al. 2008; Jacobson and Holmden 2008; Ewing et al. 2008; Teichert et al. 2009; Payne et al. 2010; Griffith et al. 2011; Gussone et al. 2011; Reynard

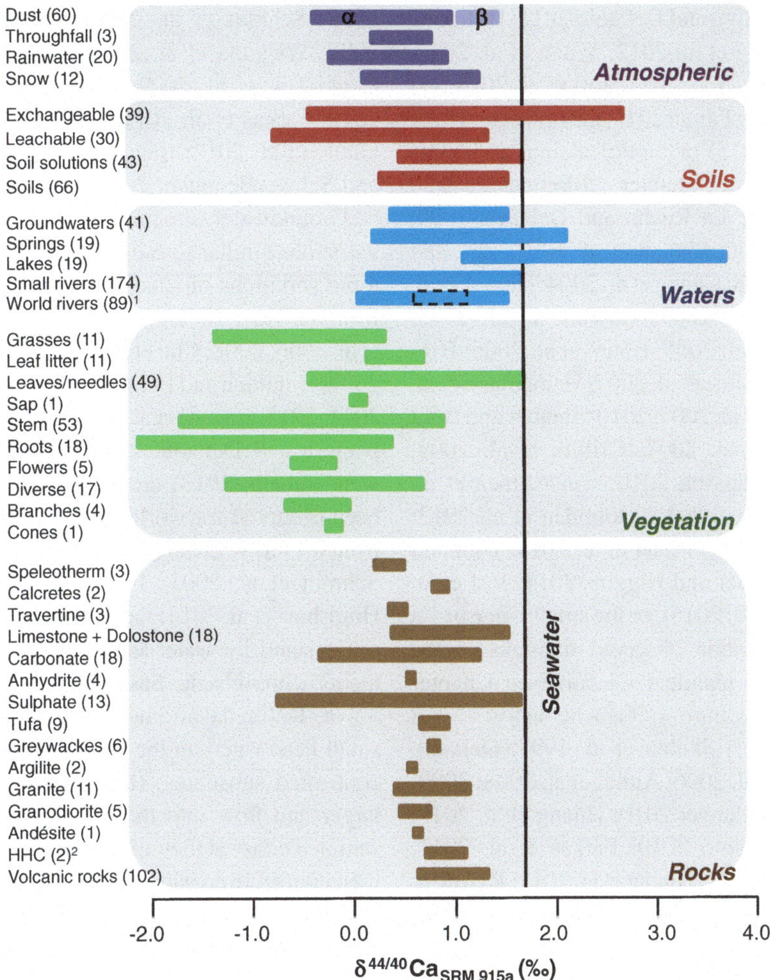

Fig. 1 Range of measured $\delta^{44/40}$Ca values in main Earth-surface reservoirs. All values are expressed against NIST SMR 915a standard. Atmospheric deposits (Schmitt et al. 2003; Schmitt and Stille 2005; Wiegand et al. 2005; Ewing et al. 2008; Cenki-Tok et al. 2009; Holmden and Bélanger 2010; Farkaš et al. 2011; Hindshaw et al. 2011; Fantle et al. 2012; Bagard et al. 2013; Wiegand and Schwendenmann 2013), soils (Schmitt et al. 2003; Wiegand et al. 2005; Perakis et al. 2006; Tipper et al. 2006; Page et al. 2008; Cenki-Tok et al. 2009; Holmden and Bélanger 2010; Farkaš et al. 2011; Hindshaw et al. 2011; Bagard et al. 2013; Moore et al. 2013; Gangloff et al. 2014), waters (Zhu and MacDougall 1998; Schmitt et al. 2003; Chu et al. 2006; Jacobson and Holmden 2008; Tipper et al. 2006, 2008, 2010; Cenki-Tok et al. 2009; Holmden and Bélanger 2010; Hindshaw et al. 2011; Bagard et al. 2013; Hindshaw et al. 2013; Moore et al. 2013; Nielsen and DePaolo 2013; Wiegand and Schwendenmann 2013; Oehlerich et al. 2015), vegetation (Skulan and DePaolo 1999; Schmitt et al. 2003; Bullen et al. 2004; Wiegand et al. 2005; Chu et al. 2006; Page et al. 2008; Cenki-Tok et al. 2009; Holmden and Bélanger 2010; Farkaš et al. 2011; Hindshaw et al. 2011, 2012; Bagard et al. 2013; Hindshaw et al. 2013; Moore et al. 2013), and rocks (Skulan et al. 1997; Halicz et al. 1999; DePaolo 2004; Tipper et al. 2006; Ewing et al. 2008; Jacobson and Holmden 2008; Amini et al. 2009; Holmden 2009; Hindshaw et al. 2011; Holmden and Bélanger 2010; Huang et al. 2010; 2011; Simon and DePaolo 2010; Farkas et al. 2011; Ryu et al. 2011; Bagard et al. 2013; Moore et al. 2013; Nielsen and DePaolo 2013; Valdes et al. 2014; Jacobson et al. 2015; Oehlerich et al. 2015). *1* The *dotted rectangle* stands for rivers sampled at their mouth. *2* HHC refers to High Himalayan Cristalline. *α* acid soluble, residue and bulk; *β* water soluble. The *number in parenthesis* refer to the number of samples from each category. *N.B.* Only field sample values have been represented, no experimentally obtained value are presented

et al. 2011; Turchyn and DePaolo 2011; Holmden et al. 2012; Tang et al. 2012; Wang et al. 2012, 2013; Kasemann et al. 2014; Jost et al. 2014; Du Vivier et al. 2015; Fantle 2015; Griffith et al. 2015; Husson et al. 2015) or biotic calcium carbonate precipitations (see Chapter "Biominerals and Biomaterial"; De La Rocha and DePaolo 2000; Nägler et al. 2000; Skulan et al. 1997; Zhu and MacDougall 1998; Chang et al. 2004; Heuser et al. 2005; Sime et al. 2005; Gussone et al. 2005; Immenhauser et al. 2005; Böhm et al. 2006; Hippler et al. 2006; Sime et al. 2007; Heinemann et al. 2008; Gussone et al. 2009, 2010; Steuber and Buhl 2006; Farkaš et al. 2007; Griffith et al. 2008; Gussone and Filipsson 2010; von Allmen et al. 2010; Blättler et al. 2012; Holmden et al. 2012; Hinojosa et al. 2012; Pretet et al. 2013; Ullmann et al. 2013; Blättler and Higgins 2014; Jost et al. 2014; Brazier et al. 2015), or the small range of Ca isotopic fractionation observed in rocks of the crust and upper mantle (~ 1 ‰) (see Chapter "High Temperature Geochemistry and Cosmochemistry"; Skulan et al. 1997; DePaolo 2004; Tipper et al. 2006; Amini et al. 2008; 2009; Holmden and Bélanger 2010; Huang et al. 2010; Simon and DePaolo 2010; Farkaš et al. 2011; Hindshaw et al. 2011; Huang et al. 2011; Ryu et al. 2011; Bagard et al. 2013; Valdes et al. 2014; Jacobson et al. 2015).

Compared to the vegetation reservoir, soil solutions, bulk and leachable soil fractions are the second most variable continental reservoirs, enriched in the heavy ^{44}Ca isotope. The soil reservoirs span a global $\delta^{44/40}$Ca variability of 3.51 ‰, with values ranging between -0.74 and 2.77 ‰ (Schmitt et al. 2003; Wiegand et al. 2005; Perakis et al. 2006; Tipper et al. 2006; Page et al. 2008; Cenki-Tok et al. 2009; Holmden and Bélanger 2010; Farkaš et al. 2011; Hindshaw et al. 2011; Bagard et al. 2013; Moore et al. 2013; Gangloff et al. 2014).

Atmospheric deposits (rainwater, snow, dust and throughfall) display positive and similar $\delta^{44/40}$Ca values comprised between 0.27 and 1.36 ‰, with a global variability of 1.09 ‰, except for one dust sample (-0.42 ± 0.20 ‰, Wiegand et al. 2005) and one rainwater sample (-0.34 ± 0.20 ‰, Wiegand and Schwendenmann 2013; Schmitt et al. 2003; Schmitt and Stille 2005; Wiegand et al. 2005; Ewing et al. 2008; Cenki-Tok et al. 2009; Holmden and Bélanger 2010; Farkaš et al. 2011; Hindshaw et al. 2011; Fantle et al. 2012; Bagard et al. 2013; Wiegand and Schwendenmann 2013).

Groundwater, streams and world rivers display variations similar to rainwater and snow. Ca isotopic variations of small rivers (1.48 ‰; ranging from 0.27 to 1.70 ‰) (Schmitt et al. 2003; Tipper et al. 2006, 2008; Chu et al. 2006; Cenki-Tok et al. 2009; Holmden and Bélanger 2010; Hindsaw et al. 2011, 2013; Bagard et al. 2013; Moore et al. 2013; Nielsen and DePaolo 2013; Wiegand and Schwendenmann 2013) are very close to what has been observed for world rivers (1.38 ‰; ranging from 0.17 to 1.55 ‰) (Zhu and MacDougall 1998; Schmitt et al. 2003; Tipper et al. 2008, 2010; Hindshaw et al. 2013; Jacobson et al. 2015). Small rivers stand for water samples from small catchments (Strengbach, Saskatchewan, Damma, La Selva, Kulingdakan and Kochechum as well as small headwaters in the Tibetan plateau draining contrasted substrates) (Fig. 2). World rivers are larger and flow into the ocean; they have been sampled either at their mouth or somewhere along their course. In considering only samples collected at the river mouth, the $\delta^{44/40}$Ca variation of large rivers is reduced to 0.58 ‰ and ranges between 0.67 and 1.25 ‰ (Zhu and MacDougall 1998; Schmitt et al. 2003; Tipper et al. 2010), with an average equal to 0.86 ‰ (see Chapter "Global Ca Cycles: Coupling of Continental and Oceanic Processes"). Groundwaters present a variability slightly higher to that of small rivers (1.91 ‰; ranging from 0.17 to 2.08 ‰) (Schmitt et al. 2003; Jacobson and Holmden 2008; Tipper et al. 2008; Cenki-Tok et al. 2009; Holmden and Bélanger 2010; Hindshaw et al. 2011, 2013; Holmden et al. 2012; Nielsen and DePaolo 2013; Wiegand and Schwendenmann 2013; Jacobson et al. 2015), and similar to spring waters (1.90 ‰; ranging from 0.17 to 2.08 ‰) (Cenki-Tok et al. 2009; Nielsen and DePaolo 2013). Lake waters are the most ^{44}Ca enriched natural waters and correspond to the third most fractionated continental reservoir (2.47 ‰; ranging from 1.3 to 3.77 ‰) (Nielsen and DePaolo 2013; Oehlerich et al. 2013).

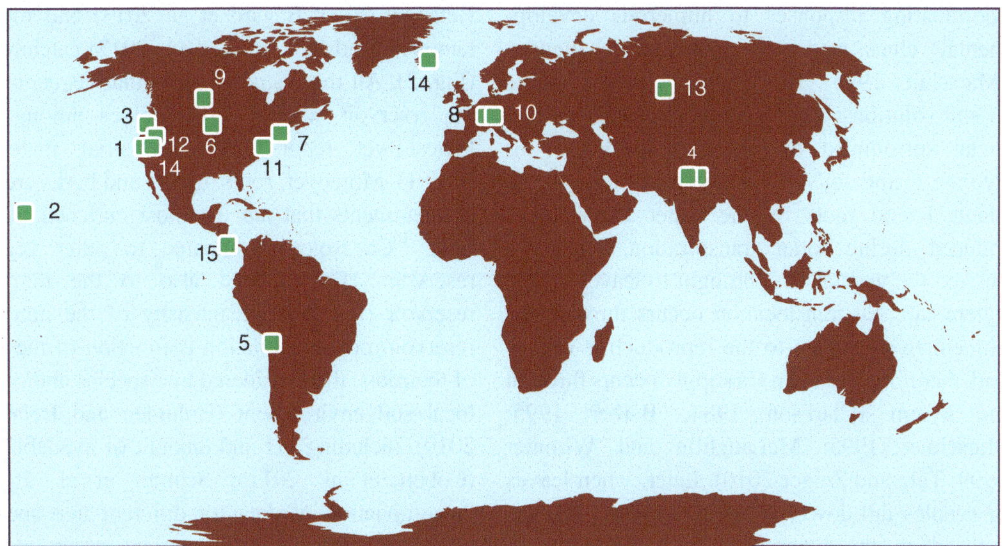

Fig. 2 Map showing the location of the sites studied using Ca isotopes in Earth-surface processes. *1* Santa Cruz, USA, Bullen et al. (2004); *2* Hawaï, USA, Wiegand et al. (2005); *3* Oregon Coast Range, USA, Perakis et al. (2006); *4* Marsyandi River, Nepal and Bhote Kosi, Southern Tibet, Tipper et al. (2006, 2008); *5* Atacama desert, Chile, Ewing et al. (2008); *6* Madison aquifer, South Dakota, USA, Jacobsen and Holmden (2008); *7* Arbutus Lake Watershed, USA, Page et al. (2008); *8* Strengbach catchment, France, Cenki-Tok et al. (2009), Gangloff et al. (2014); *9* La Ronge, Saskatchewan, Canada, Holmden and Bélanger (2010); *10* Damma glacier, Switzerland, Hindshaw et al. (2011, 2012); *11* Mt Wachusett, USA, Farkaš et al. (2011); *12* Black Rock Desert Nevada, USA, Fantle et al. (2012); *13* Kulingdakan catchment, Russia, Bagard et al. (2013); *14* Iceland, Hindshaw et al. (2013), Jacobson et al. (2015); *14* Mono Lake, California, USA, Nielsen and DePaolo (2013); *15* La Selva catchment, Costa Rica, Wiegand and Schwendenmann (2013)

Although a large quantity of oceanic sediments has been studied (see chapter "Global Ca Cycles: Coupling of Continental and Oceanic processes"), few is known about continental rocks. From the small number of available data we can infer that continental sedimentary rocks vary largely (2.5 ‰ variability, spanning from −0.77 to 1.73 ‰), whereas igneous rocks are more constant (volcanic rocks: 0.81 ‰ variability, spanning from 0.61 to 1.34 ‰; plutonic rocks: 0.82 ‰ variability, spanning from 0.31 to 1.13 ‰) (Skulan et al. 1997; Halicz et al. 1999; DePaolo 2004; Tipper et al. 2006; Ewing et al. 2008; Jacobson and Holmden 2008; Amini et al. 2009; Holmden 2009; Hindshaw et al. 2011; Holmden and Bélanger 2010; Huang et al. 2010, 2011; Simon and DePaolo 2010; Farkaš et al. 2011; Ryu et al. 2011; Bagard et al. 2013; Moore et al. 2013; Nielsen and DePaolo 2013; Valdes et al. 2014; Jacobson et al. 2015; Oehlerich et al. 2015).

2.2 Forested Ecosystems

2.2.1 Ca Cycling Through the Vegetation

Recent studies have shown that plants may play an important role in the biogeochemical cycling of calcium (Berner and Berner 1996; Poszwa et al. 2000; Schmitt et al. 2003; Schmitt and Stille 2005; Wiegand et al. 2005; Perakis et al. 2006; Page et al. 2008; Cenki-Tok et al. 2009; Holmden and Bélanger 2010; Farkaš et al. 2011; Hindshaw et al. 2011, 2012; Bagard et al. 2013). Ca is an essential nutrient for plants that is mainly involved in the cell stabilization of walls and membranes (Marschner 1995; Taiz and Zeiger 2010). It acts as a counter-ion for inorganic and organic anions in the vacuole (organelle in the plant cell), and the cytosolic (liquid phase in which the organelles are bathed) Ca concentration is an obligate intracellular messenger coordinating responses to numerous developmental clues and environmental challenges (Marschner 1995). This divalent cation is stable in soil solutions: it is encountered either as free metal surrounded by water molecules or as hydrate in the soils. Ca is taken up through the plants lateral roots by the water mass fluxes induced during foliar transpiration. Once Ca entered the vegetal, it is brought to leaves by the xylem sap. No translocation occurs through the phloem from foliage to the reproductive organs and, therefore, the main Ca supply occurs through the xylem (Clarkson 1984; Barber 1995; Marschner 1995; McLaughlin and Wimmer 1999; Taiz and Zeiger 2010). Later, when leaves or needles fall down, nutrient elements are mineralised in the litter and Ca becomes again available in soil solutions for a new uptake through the fine lateral roots in the organic-rich forest floor and underlying mineral soils of forested catchments (Likens et al. 1998). This Ca recycling by the vegetation constitutes an essential way of nutrient element supply for forests, especially in catchments where low buffer capacities of soils are recorded, linked to reduced atmospheric deposition of base cations due to acid rains, or to the presence of silicate soils that present only low cation contents and are thus sensitive to acid depositions or to ecosystem export caused by forest harvesting (Hedin et al. 1994; Lesack and Melack 1996; Likens et al. 1998; Aubert et al. 2001). Indeed, temperate forest canopies buffer part of the incoming acid depositions by exchanging Ca^{2+}, K^+, Mg^{2+} for H^+ and NH_4^+ (Potter 1991; Lovett and Schaefer 1992; Draaijers et al. 1997). Similarly, under tropical high rainfall conditions, weathering processes cause rapid breakdown of minerals.

2.2.2 $\delta^{44/40}$Ca Variations Within the Vegetation

Several recent studies manifest the importance of calcium isotopes for a better understanding of the control of the Ca budget and cycling by vegetation in small forested temperate (Schmitt et al. 2003; Schmitt and Stille 2005; Page et al. 2008; Cenki-Tok et al. 2009; Farkaš et al. 2011), tropical (Wiegand et al. 2005), boreal (Holmden and Bélanger 2010; Bagard et al. 2013) and mountainous (Hindshaw et al. 2011, 2012) catchments (Fig. 2). All these studies show that vegetation is the reservoir with $\delta^{44/40}$Ca values among the lowest yet reported for terrestrial materials (Fig. 1). Moreover, roots, trunks and barks are the compartments that are the most enriched in the light ^{40}Ca isotope compared to other vegetal reservoirs (Fig. 1), and also to the nutritive reservoir (Fig. 1). The intensity of the nutritive reservoir/root fractionation is function of the size of the roots, the considered tree species and/or the local soil environment (Holmden and Bélanger 2010), including pH and amount of available Ca (Cobert et al. 2011a; Schmitt et al. 2013). A compilation of data for different tree species points to variable soil nutrient reservoir/root fractionation (Table 1). Some of the variability can also result from the variability of considered soil nutrient reservoirs (soil solutions, porewater, exchangeable fraction) and tree root size. Anyway, in every case the tree roots get enriched in the light isotope compared to the nutritive solution and the weakest difference is recorded for a larch tree from Siberia, whereas the most important difference is reported for American beech and sugar maple from the Adidonrack mountains (Table 1). Tree roots present also variable $\delta^{44/40}$Ca values with finest roots enriched in ^{44}Ca compared to larger ones (Cenki-Tok et al. 2009; Holmden and Bélanger 2010). Moreover, by studying woody vegetation (*Rhododendron ferrugineum*), Hindshaw et al. (2012) suggested that the presence of mycorrhyza may shift above-ground biomass $\delta^{44/40}$Ca values towards lower $\delta^{44/40}$Ca values, whereas the percentage of mycorrhyza infection is without effect. Further investigations are necessary to understand precisely the role of fungi on the acquisition of $\delta^{44/40}$Ca values in roots. A preliminary *in vitro* experiment pointed to the fact that in presence of soil bacteria nutrients become directly available to roots, so that the Ca isotopic fractionation of the nutritive solution is ruled out (Cobert et al. 2011b). This has recently been confirmed by a field study (Gangloff et al. 2014). The precise mechanisms are, however, yet not well understood and request further investigations.

Table 1 Soil liquid pool/roots fractionation factors for several watersheds worldwide

Watershed	Bedrock	Soil plot	Tree species	$\Delta_{\text{average soil solution/roots}}$ (‰)	Reference
Strengbach	Grantitic and metamorphic (Paleozoïc)	BP	European Beech (*Fagus sylvatica*)	1.47	Cenki-Tok et al. (2009)
		SP	Norway Spruce (*Picea Abies*)	0.73	
La Ronge	Granitic (Precambrian shield)	Plot 1.1	Jack Pine (*Pinus banksiana*)	0.67	Holmden and Bélanger (2010)
			Trembling Aspen (*Populus tremuloides*)	1.28	
		Plot 1.2	Trembling Aspen (*Populus tremuloides*)	1.16	
			Black Spruce (*Picea mariana*)	1.24	
		Plot 2.1	Black Spruce (*Picea mariana*)	1.16	
		Plot 2.2	Jack Pine (*Pinus banksiana*)	0.74	
			Black Spruce (*Picea mariana*)	1.40	
Kulingdakan	Basaltic (Permo-Triassic flood basalts)	South-facing slope	Larch (*Larix gmelinii*) Rhododendron C	0.21	Bagard et al. (2013)
Damma glacier	Granitic and polymetamorphic	110 years	(*Rhododendron ferrugineum*)	0.63[a]	Hindshaw et al. (2011, 2012)
Arbutus lake	Granitic gneiss and metasedimentary rocks	Catchment 14	American beech (*Fagus Grandifolia*)	2.79[b]	Page et al. (2008)
			Sugar maple (*Acer saccharum*)	2.88[b]	
		Catchment 15	American beech (*Fagus Grandifolia*)	1.38[b]	
			Sugar maple (*Acer saccharum*)	2.29[b]	
Hawaï	Tholeiitic basalt	0.3 ka	*Metrosideros* tree	0.93[c]	Wiegand et al. (2005)
	Post-shield alkalic basalt	4100 ka	*Metrosideros* tree	1.22[c]	

$\delta^{44/40}$Ca values are normalized to NIST SRM 915a

[a] $\Delta_{\text{porewater/roots}}$

[b] $\Delta_{\text{exchangeable at 50 cm depth/roots}}$

[c] $\Delta_{\text{exchangeable/roots}}$

Within the presently studied plant species, fractionation takes also place along the transpiration stream, with enrichment in the heavy ^{44}Ca isotope from roots to stemwood and up to leaves or needles (Platzner and Degani 1990; Wiegand et al. 2005; Page et al. 2008; Cenki-Tok et al. 2009; Holmden and Bélanger 2010; Bagard et al. 2013) (Fig. 3). This implies a preferential ^{40}Ca partitioning in wood, and a relative enrichment in ^{44}Ca in leaves and needles compared to wood. Consequently, leaves, needles and wood contribute in function of their speed of decomposition to relative ^{40}Ca enrichments in the leaf litter. The overall recorded $\delta^{44/40}$Ca variation from bottom to top of the different species of trees is about 0.80 ‰, with no difference between the species (Cenki-Tok et al. 2009; Holmden and Bélanger 2010; Bagard et al. 2013). No $\delta^{44/40}$Ca differences between foliage and stemwood were recorded for old (85–110 years) or young (10–40 years) trees (Holmden and Bélanger 2010). In contrast, ^{44}Ca enrichments were recorded between basal (0.24 ± 0.15 ‰) and apical (0.64 ± 0.21 ‰) European beach leaves and between young (0.28 ± 0.15 ‰) and old (0.80 ± 0.12 ‰) Norway spruce needles (Cenki-Tok et al. 2009).

Deciphering causes of isotopic fractionation within trees in catchments is difficult because, in addition to biological processes, the isotopic signals might be overprinted by abiotic fractionations. Complementary studies using experimental approaches have been performed to specify the mechanisms controlling the Ca uptake by plants. Holmden and Bélanger (2010) were the first who have undergone a preliminary in vivo experiment with aspen seedling grown on crushed basalt. The Ca isotopic fractionations were limited. They confirmed that roots were enriched in ^{40}Ca compared to the basalt, but the stem showed $\delta^{44/40}$Ca values similar to roots within error bars and only the leaves were enriched in the light ^{40}Ca. Two other hydroponic experiments point to at least three different $\delta^{44/40}$Ca fractionation steps (Cobert et al. 2011a; Schmitt et al. 2013). The first one takes place at the nutritive solution/root interface before the Ca enters the lateral roots, following a closed-system equilibrium fractionation with a fractionation factor ($\alpha_{\text{bean plant/nutritive solution}}$) of 0.9988 suggesting that Ca forms exchangeable bonds with the root surface, during Ca adsorption on cation-exchange binding sites on the external surface of the cell wall (Schmitt et al. 2013). The

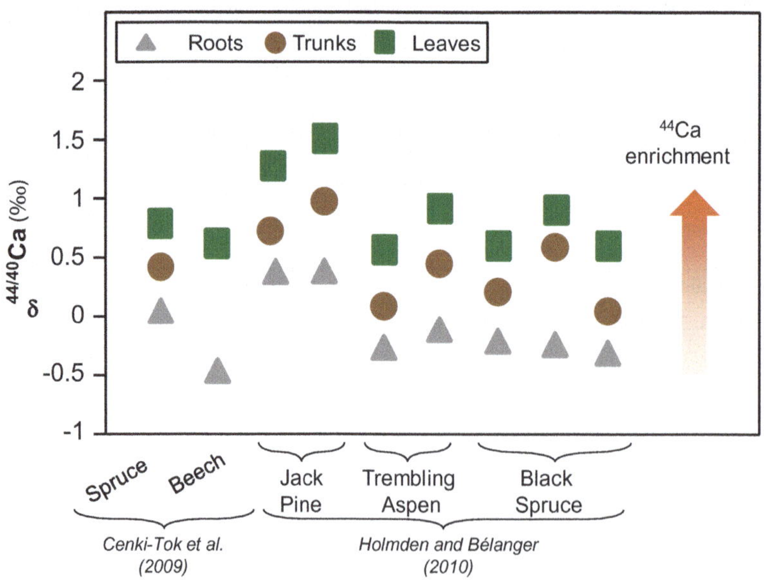

Fig. 3 $\delta^{44/40}$Ca variability during translocation from roots to shoot of trees (Cenki-Tok et al. 2009; Holmden and Bélanger 2010). All values are normalized to NIST SMR 915a standard

second occurs along the transpiration stream, linked to ion-exchange reactions with the pectins in the middle lamella of the conducting xylem. The third takes place when Ca enters the reproductive organs, that are enriched in pectins compared to other organs and seem, consequently, less subject to acidification of the nutritive medium than other plant organs. The intensity of the Ca isotopic fractionation within the plants seems also to be dependent on the Ca content and pH of the nutritive solution (Cobert et al. 2011a). Moreover, the Ca isotopic composition of the average plant is dependent on the $\delta^{44/40}$Ca value of the nutritive solution, which points to a control of the external nutritive medium on the Ca isotopic signature within plants (Schmitt et al. 2013).

When trying to transpose the hydroponic results to the only field study that has undergone a sampling of several organs of a larch tree (Bagard et al. 2013), one recognizes the three previously described fractionation levels, with some local differences. For instance, the flow of the xylem sap is greater in the trunk than in the branches (Bresinsky et al. 2008; Hacke and Sperry 2001; Sperry 2011). Therefore, similarly to ion-exchange processes, the ^{40}Ca enrichment of pectins from the branches (−0.45 ± 0.12 and −0.76 ± 0.17 ‰) is greater than that of the trunk (0.12 ± 0.12‰) due to a larger number of equivalent theoretical plates, i.e. of possible exchangeable sites (Russell and Papanastassiou 1978). Moreover, we can explain the Ca isotopic compositions of thin branches (<4 mm; −0.45 ± 0.12 ‰), thicker ones (∼1 cm; −0.76 ± 0.12 ‰) and needles (0.40 ± 0.12 ‰ and 0.66 ± 0.08 ‰) by the xylem flux feeding the branches, which gets enriched in ^{44}Ca when ^{40}Ca binds to pectins. By analogy with the experiments in hydroponic conditions, the xylem cannot be anymore considered as a non-limiting Ca reservoir, at least in the lateral branches. As the xylem travels more way to reach the thin branches than the thicker ones, directly attached to the trunk, it makes sense to obtain isotopic compositions enriched in ^{44}Ca in the fine branches. The needles being located at the end of the path of the xylem are also enriched in ^{44}Ca relative to the thin branches. Finally, the isotopic composition of the Ca in larch cones (−0.08 ± 0.12 ‰), enriched in ^{40}Ca in needles, is compatible with an enrichment of Ca oxalates in these organs, as well as the low $\delta^{44/40}$Ca value of the bark (−0.96 ± 0.08 ‰). As a result, the Ca speciation within the tree seems to be the main factor controlling its isotopic composition (free-Ca, Ca linked to pectins or to oxalates).

2.2.3 δ⁴⁴/⁴⁰Ca Variations Within the Soil Pool

The soil pool comprises soil solutions, porewater soil extracts as well as the bulk soil minerals themselves. The absence of common soil pools to different studies makes comparison often difficult. Some studies also complicate the comparison by reporting only leaching experiments using different reactants. Two successive leaching experiments are usually performed in order to have access to the nutrient pools that are available to fine roots in soils: the first obtained fraction aims to evaluate the soil's exchangeable cation isotopic signature, using 1 N ammonium acetate or barium chloride, whereas the second one corresponds to the acid-leachable fraction, using 1 N HNO$_3$ (Bullen et al. 2004; Wiegand et al. 2005; Perakis et al. 2006; Page et al. 2008; Holmden and Bélanger 2010; Farkaš et al. 2011). The exchangeable ions in soils are considered as the most "plant available" ions (Suarez 1996), whereas the acid-leachable fraction is thought to be representative of cations available for release by weathering (Blum et al. 2008). Various processes occurring in soils have been reported with Ca isotope fractionation depending on the nature of the bedrock (granite or basalt), the weathering stage (early or more developed), the vegetation cover (forests or sparse vegetation), the climate… These studies have especially shown that the degree of "^{44}Ca enrichments in soils is function of the calcium uptake rate into trees, the magnitude of the fractionation factor and the supply rate of ^{40}Ca-enriched input fluxes from litterfall, atmospheric deposition and soil mineral weathering" (Holmden and Bélanger 2010). If these latter fluxes are more important than the uptake by the vegetation, the Ca soil reservoirs can be considered as non-limiting with regard to Ca, and the

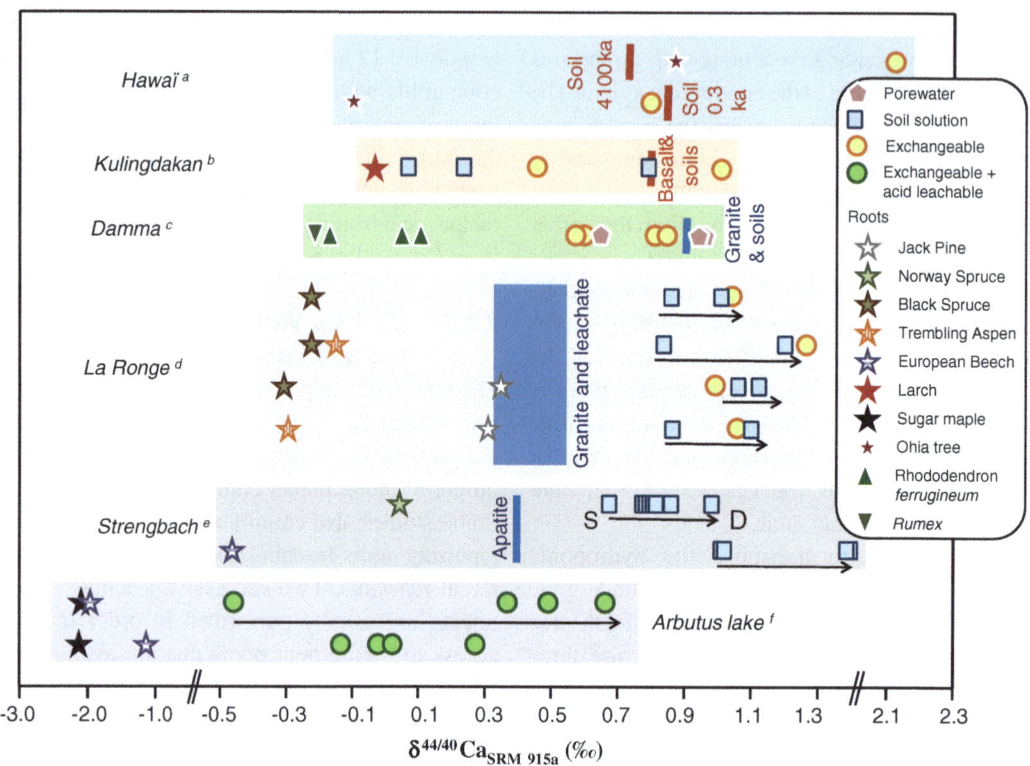

Fig. 4 Ca isotopic amplitude of fractionation between soil liquids (including porewaters, soil solutions and exchangeable fraction) and roots (2 mm when available) for the presently studied watersheds. All values are normalized to NIST SMR 915a standard. *a* Wiegand et al. (2005); *b* Bagard et al. (2013); *c* Hindshaw et al. (2011); *d* Holmden and Bélanger (2010); *e* Cenki-Tok et al. (2009); *f* Page et al. (2008). *Arrows* from *left* (surface) to *right* (depth). A typical value from Wiegand and Schwendenmann (2013) has not been reported

vegetation uptake, which is enriched in ^{40}Ca, will almost be without any major effect on the Ca soil pool reservoir as has been observed in Hawaï for young soils or in Damma glacier subject to recent weathering processes (Wiegand et al. 2005; Hindshaw et al. 2011; Fig. 4). In contrast, in Hawaii, for older soils (4100 ka) impoverished in nutrients, soil reservoirs can no longer be considered as non-limiting and the ^{44}Ca exchangeable pool increase can be explained either by the ^{40}Ca uptake by the vegetation or by marine aerosol inputs (Wiegand et al. 2005). In forested catchments grown on granites, such as Strengbach and Saskatchewan ones, the main process causing Ca isotope fractionation is vegetation uptake, with ^{40}Ca enrichments in roots and subsequent ^{44}Ca enrichment in soil solutions compared to average soils or rocks (Cenki-Tok et al.

2009; Holmden and Bélanger 2010; Fig. 4). In permafrost environments organic matter degradation rather than vegetation uptake, as well as the contribution from another pool, enriched in ^{40}Ca such as soil colloids seem to control the Ca isotopic signature of soil solutions (Bagard et al. 2013) (Fig. 4). Using simple numerical models, Fantle and Tipper (2014) observed that the exchangeable Ca fraction may be enriched or depleted in Ca in accordance with the intensity of the recycling flux compared to the uptake flux in slowly weathering granite catchments: when the uptake (recycling) is more important than the recycling (uptake), then the exchangeable pool is driven to heavier (lower) $\delta^{44/40}$Ca values.

$\delta^{44/40}$Ca variations of soil solutions and/or Ca leachable fractions along the depth profiles have also been recorded by several authors. In this context, Bullen et al. (2004) observed a Ca isotopic disequilibrium between porewaters and Ca labile soil sites, and they suggested that cation-exchange processes are non-instantaneous in their soils. Wiegand et al. (2005) evidenced an increase in $\delta^{44/40}$Ca of exchangeable soil pool along with increasing age of their four million years old Hawaiian soil sequence, which was attributable to changes in the Ca sources (from primary minerals to atmospheric deposition with increasing basalt age) and to Ca uptake by vegetation. Perakis et al. (2006) have shown that the exchangeable and acid leachable pools at 0–5 cm soil depth are 1 ‰ lighter relative to Ca in HF digested residues. They explained this observation by Ca recycling through sparingly soluble Ca precipitated soil oxalates in the upper soil floor as well as by recycling of isotopically light Ca derived from decomposed leaf litter. Page et al. (2008) have for their part shown that $\delta^{44/40}$Ca of exchangeable and acid leachable Ca fractions increased with increasing depth through the soil profile. These ^{40}Ca enriched values in surficial soil profiles can be attributed to leaf litter and woody biomass decomposition. Holmden and Bélanger (2010) have also reported that $\delta^{44/40}$Ca of soil waters are lower in the forest floor compared to the mineral soil pools. They suggested that trees take up preferentially ^{40}Ca in the forest floor, whereas below the thin root zone, the percolated Ca isotopic signature is influenced by other Ca sources, coming from the upper mineral soils. If cation-exchange processes during Ca percolation in the soil below the roots level can be neglected, $\delta^{44/40}$Ca values will behave as conservative tracers of Ca transport in natural waters (soil-, ground- and surface waters). In this context, Cenki-Tok et al. (2009) also showed a $\delta^{44/40}$Ca increase of soil solution with depth below beech (*Fagus sylvatica* L) trees, whereas the values remained rather constant below spruces (*Picea* sp.) (Fig. 4). The ^{44}Ca enrichment below beeches points to the dissolution of another soil pool, linked to alteration and/or water-rock interactions, such as for instance secondary mineral phases, whose role has not yet been elaborated in this catchment. Similarly, Bagard et al. (2013) suggest the

presence of organic and organo-mineral colloids that preferentially scavenge the light ^{40}Ca isotope. However, Holmden and Bélanger (2010) have excluded any formation of such Ca-bearing phases in the acidic soils of their boreal catchment. Thus, further investigations are needed to confirm or to fully exclude this hypothesis. Finally, Farkaš et al. (2011) have shown that in the base-poor Wasuchett Mountain catchment (NE USA) tree roots are able to bypass the exchangeable cation pool of deep mineral soils and take up their nutrients from wet atmospheric deposition, organically-complexed Ca in the forest floor and topsoil and from dissolution of apatite and Ca-rich plagioclase.

2.2.4 $\delta^{44/40}$Ca Variations Within Atmospheric Deposits

Previous studies in forested catchments have tried to evaluate the influence of Ca isotopic compositions of atmospheric deposits on the geochemical budget of the catchment. Atmospheric deposits include incident wet and dry incomes. Wet depositions that meet the forest cover attain the soil as throughfall or stemflow. Throughfall represents the part of precipitation that interacts with the canopy whereas stemflow corresponds to the part that funnels down the trunk (Crockford and Richardson 2000; Park and Cameron 2008 and refs). Page et al. (2008) pointed out that retention and accumulation of atmospherically derived Ca in the upper soil profile probably had little influence on the Ca isotopic composition of the soils they studied in the Arbutus Lake watershed (NE USA). Similarly, Bagard et al. (2013) observed that two snow samples were within the range of basalt and soil values, suggesting that Ca atmospheric inputs would not impact local $\delta^{44/40}$Ca values (Fig. 5).

In contrast, Hindshaw et al. (2011) have shown that the Ca isotopic composition of precipitations, although close to those of the silicate rocks, varied throughout the year, with 2 % contribution on the Ca budget in the winter and 35 % in the summer. Perakis et al. (2006) estimated for their part that Ca atmospheric depositions supply about 12 % of plant demands in Oregon. Wiegand et al. (2005) observed a

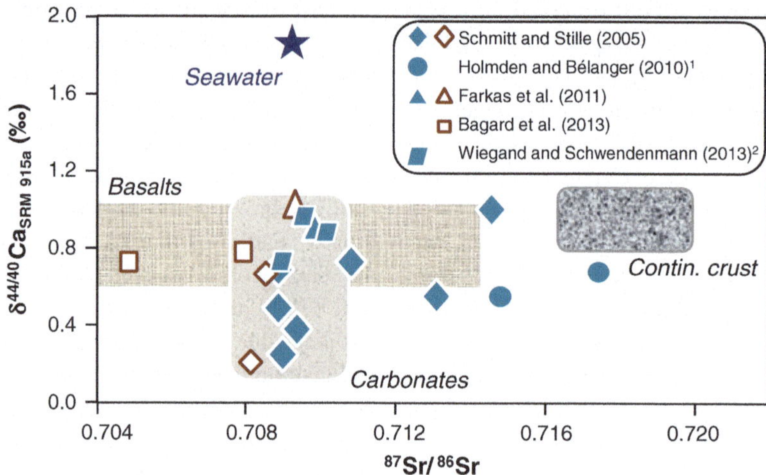

Fig. 5 $\delta^{44/40}$Ca versus ^{87}Sr/^{86}Sr of presently published rain and snow waters. $\delta^{44/40}$Ca values are normalized to NIST SRM 915a. *1* Sr isotopic data from Bélanger and Holmden (2010). *2* Highly fractionated $\delta^{44/40}$Ca value from Wiegand and Schwendenmann (2013) was not represented Carbonate end-member: Ca isotopic data from Farkaš et al. (2007) and Gussone et al. (2005); Sr isotopic data from Veizer et al. (1999). Continental crust end-member: Ca isotopic data from Bullen et al. (2004) and Huang et al. (2010, 2011); Sr isotopic data from Goldstein and Jacobsen (1988) and Hofmann (1997) Modern seawater: Hippler et al. (2003). *White symbols* Snow samples. *Colored symbols* Rain samples

dominance of marine-aerosol derived Ca (with $\delta^{44/40}$Ca ~ -0.4 ‰) in small soil exchangeable pools of older soils compared to younger ones. Based on Sr isotopes, Bélanger and Holmden (2010) evidenced that large amounts (40–95 %) of Ca present in their studied boreal catchment in Saskatchevan (Canada) had atmospheric origins (Fig. 5). They have also shown that Ca fluxes of dry deposition are up to four times higher than those of the wet deposition, which implies to consider both atmospheric compartments (dry and wet) in further studies. Finally, previous Ca atmospheric budgets performed in the temperate Strengbach catchment (France) have shown that, based on rainwaters alone, only 15 % of the Ca catchment budget is atmospherically-derived, whereas it can be comprised between 20 and 40 % when considering throughfall samples (Schmitt and Stille 2005; Chabaux et al. 2005; Cenki-Tok et al. 2009). Indeed, throughfall samples that are slightly enriched in ^{40}Ca compared to wet precipitation correspond to a mixing of dry and wet atmospheric deposits with recycled Ca from leaf and needle exudations (Lovett et al. 1996; Dambrine et al. 1998; Mougougou

et al. 1998). A simple mixing budget calculation shows that at least 70 % from throughfall has an incident atmospheric origin (Schmitt and Stille 2005). This suggests that the atmospheric Ca budget can be strongly affected by biological activity and vegetation excretions. Consequently, determination of the atmospheric Ca contribution to catchment surface waters using mass balance budget calculations based only on annual Ca fluxes of rainwater and/or throughfall must be taken with caution (Cenki-Tok et al. 2009). The determination of Ca isotopic data for dry and wet atmospheric deposits that may contribute substantially to the Ca budget of the vegetation is therefore required to constrain the global Ca cycle. Fantle et al. (2012) examined for their part surface sediments (<5 mm depth) from a hyper-arid dust producer environment (Black Rock desert in northern Nevada) in order to constrain dust Ca isotope signatures. They especially noticed that acid soluble and dust residue fractions have similar $\delta^{44/40}$Ca values, also similar to other published dust samples (Ewing et al. 2008), and correspond to average river waters. This implies that the Ca isotopic composition of dust is strongly linked to weathering and has little leverage to change the ocean's isotopic composition. In contrast, the water soluble dust fraction is enriched in the heavy ^{44}Ca compared to previous fractions, which could be due to Ca adsorption on clays (see below).

2.3 Non-forested Ecosystems

2.3.1 Abiotic Processes

Bulk rock modifications occur through physical disintegration, chemical alteration and biological activity. The weathering front generally increases with depth and is partly controlled by meteoric- and ground-waters. Indeed, meteoric water for instance is acid and is an oxidant, properties acquired during its contact with the atmosphere. Thus, the disequilibrium resulting from the meeting of acid and oxidant water and basic and/or reductive rocks induce chemical alteration. Physical weathering (rock disintegration) occurs through changes in physical properties of the rock (porosity, temperature, moisture...). The different rock minerals have not the same weathering stability. Basic rocks such as basalts, but also sedimentary rocks such as carbonates, are more easily weatherable than acid rocks such as granites or sandstones. Similarly, the susceptibility of silicate minerals to weathering is also function of Bowen's reaction series: the first minerals to crystallize correspond to the most weatherable (Langmuir 1997). In soils derived from silicate rocks, the most easily weatherable Ca-bearing minerals are apatite or plagioclase. Besides primary rocks and/or minerals, secondary minerals, that is chemical weathering products (clays, carbonates, sulfates, gypsum...), can also precipitate/dissolve in soils when soil water conditions are suitable (saturation indexes, oxidative properties, kinetics...). Ca precipitation in oxalate crystals is another possible sink for Ca. Oxalic acid is indeed one of the more common and abundant low-molecular weight organic acids in forest soils where it accumulates mainly under the action of mycorrhizal hyphae in the upper soil horizons as salts of Ca oxalates (Graustein et al.

1977; Certini et al. 2000). Seasonal precipitation and dissolution of secondary minerals have also to be loocked for (Hindshaw et al. 2011).

Calcium carbonates in soils can be considered as carbon reservoirs. Their study has long been neglected in contrast to silicate weathering, because their budget is considered to be equilibrated on the long term ($>10^6$ years). This reservoir could, however, be considered as a carbon sink at middle-term (10^2–10^6 years), in particular when the response of the CO_2 budget to anthropogenic influences is seeked. Carbonate rocks react faster than silicates, so that carbonate weathering will respond more easily to climatic changes and thus to the modification of the Earth System to anthropogenic processes (Retallack 1990; Capo and Chadwick 1999; Chiquet et al. 2000; Hamidi et al. 2001; Durand et al. 2010). Moreover, with increasing temperatures, the ocean storage capacity will become less effective, so that other sinks such as secondary precipitations in soils should become more important.

Precipitation and dissolution of secondary minerals are susceptible to fractionate Ca isotopic signatures (DePaolo 2011). It is commonly accepted for stable isotopes that kinetic isotope fractionation processes favour the light isotope, whereas mass-dependent equilibrium fractionation mechanisms favour heavy isotopes on molecular sites (Schauble et al. 2006; Schauble 2007). Previous in vitro calcium carbonate precipitations have indeed shown that heavier Ca isotopes occur preferentially in the dissolved phase compared to the solid one (See Chap. III; Gussone et al. 2003, 2005; Lemarchand et al. 2004; Tang et al. 2008; Reynard et al. 2011). The studies have shown that Ca isotopic fractionation factors are dependent of the precipitation-rate: rapid calcite precipitation induces small Ca isotopic fractionations, whereas the opposite is observed for low precipitation rates. More recently, analyses of pore fluids have shown an absence of $\delta^{44/40}$Ca variations during marine calcite precipitation under chemical equilibrium conditions (Fantle and DePaolo 2007). Ion-exchange mechanisms are also likely to take place in soils and groundwater and could also fractionate $\delta^{44/40}$Ca values, similarly to precipitation of clay phases during incongruent weathering, with ^{40}Ca entering the solid phase and ^{44}Ca staying in the liquid one. Nothing is presently known about Ca isotopic fractionation associated with Ca adsorption on clays in the weathering environment. However, based on a study performed in the marine environment, lighter Ca isotopes are expected to adsorb preferentially on the solid phase (Ockert et al. 2013).

2.3.2 $\delta^{44/40}$Ca Fractionations Caused by Water-Rock Interactions

Experimental weathering of a granodiorite in a plug flow reactor demonstrated that congruent low-temperature dissolution of the granodiorite minerals does not fractionate Ca isotopes (Ryu et al. 2011). Similarly, another study suggests that congruent dissolution of apatite performed under controlled conditions, using mineral (supposed to reflect water-rock interactions) and organic acids (supposed to reflect root exudates or soil micro-organisms) as weathering agents do not fractionate the Ca isotopic compositions of the solution (Cobert et al. 2011b). In contrast, $\delta^{44/40}$Ca mass-dependent isotope fractionation occurs during igneous differentiation and crystallization (Amini et al. 2009; Huang et al. 2010).

In the field, very few studies tried to evaluate the influence of water-rock interactions on the Ca isotopic compositions of soils and waters in catchments where the biological cycling is not the preponderant process (Tipper et al. 2006; Ewing et al. 2008; Jacobson and Holmden 2008; Hindshaw et al. 2011, 2013; Nielsen and DePaolo 2013; Moore et al. 2013; Jacobson et al. 2015).

Hindshaw et al. (2011) were the first studying an early stage of weathering of primary silicate minerals from an Alpine soil profile (Damma glacier, Central Swiss Alps). Their main result is that young soils are isotopically identical in Ca isotopic composition to the rock from which they are derived, suggesting that primary mineral dissolution of bulk rocks should not strongly fractionate Ca isotopes. Consequently, fractionations are likely to occur where the soils are no longer buffered by primary mineral dissolution, and secondary processes such as biological cycling and/or secondary mineral precipitation become dominant.

$\delta^{44/40}$Ca fractionations have indeed been observed during precipitation of secondary minerals (Tipper et al. 2006; Ewing et al. 2008; Jacobson and Holmden 2008; Hindshaw et al. 2013; Jacobson et al. 2015). Tipper et al. (2006) studied Ca isotopic fractionations in soils from the Southern Tibetan Plateau with negligible vegetation cover. They have identified a ^{40}Ca-enriched reservoir corresponding to travertines. As a result they have proposed a control of the $\delta^{44/40}$Ca by secondary carbonate precipitation, counterbalanced by groundwater enriched in the heavy isotope. In contrast, Jacobson and Holmden (2008) studied a 236 km long flow path in an aquifer (Madison aquifer, spanning over 15 ky, South Dakota) and observed no isotopic fractionation when Ca precipitated under saturated conditions. They proposed that a combination of long-timescales of interactions between water and rocks and slow rates of calcite precipitation allow fluids to chemically and isotopically equilibrate with calcite. They explain these observations by kinetic versus equilibrium fractionations controlling calcite precipitation. The steady-state 1D reactive-transport model they have applied to the aquifer suggests that besides calcite precipitation, Ca-for-Na ion exchange mechanisms could fractionate Ca isotopes during the initial and final stages of water transport. However, due to an insufficient dataset they could not prove such cationic exchange process in the field. At the same time Ewing et al. (2008) studied a 2 My hyper arid soil of the Atacama desert where no vascular plants and only minimal microbial activity are recorded. The sulphate and carbonate variations observed along the studied soil profile have been affected by changes in the source (terrestrial or marine) of the atmospheric deposits. They extracted and analysed the Ca from sulfates and carbonates using ultrapure water and 1 M acetic acid, respectively. They observed in-depth $\delta^{44/40}$Ca increases, which they explained by a transport model. Shallow horizons that are enriched in ^{40}Ca were subject to repeated dissolution and reprecipitation events described by Rayleigh fractionation. This system is supposed to reach isotopic steady state after 100 ky. As a result, deeper horizons are subject to downward transport of ^{44}Ca during small rainfall events over millions of years. More recently Nielsen and DePaolo (2013) studied precipitation of calcium carbonate minerals from aqueous systems. It results from their study that lake waters can be out of steady-state and that $\delta^{44/40}$Ca of its carbonate sediments do not reflect variations between weathering and carbonate precipitation to and from the lake through time.

3 Change in $\delta^{44/40}$Ca Signature During Downstream Transportation into the Ocean

3.1 Importance of the $\delta^{44/40}$Ca Weathering Flux to the Oceans

The oceanic mass balance of Ca is a key challenge in Earth Sciences since the ocean plays an important role in moderating atmospheric CO_2 and climate. Riverine Ca fluxes are a key component for the Ca mass balance in the oceans and are thought to control short-term fluctuations of the oceans isotopic compositions, since they constitute the predominant source of Ca to the oceans (Schmitt et al. 2003; Farkaš et al. 2007). Precise understanding of past fluxes requires to ascertain the modern control of the riverine $\delta^{44/40}$Ca flux. At the global scale, two thirds of the riverine Ca originates from the weathering of carbonates rocks, what has implications for the global carbon cycle over short timescales (Meybeck 1987; Berner and Berner 1996; Gaillardet et al. 1999; Edmond and Huh 2003; Calmels 2007; Li and Zhang 2008). In contrast, long-term removal of CO_2 from the atmosphere is linked to silicate weathering processes (Walker and Kastings 1981; Berner et al. 1983; Gaillardet et al. 1999).

The riverine signal consists of mixtures of waters with various compositions, inherited both from the drained lithological signature (igneous and sedimentary rocks) and the isotopic fractionations occurring during a series of processes associated with weathering reactions. As can be inferred from previous paragraphs and summed up in Fig. 6, two main fractionation processes occur on continents:

plants and secondary mineral phases (soils carbonates, travertines, clays) are enriched in the light ^{40}Ca. As a result the corresponding soil waters or associated river waters are enriched in the heavy ^{44}Ca isotope (Fig. 6). It is thus important to evaluate the variation of the $\delta^{44/40}$Ca record of the dissolved Ca in rivers from upstream to downstream until the ocean.

To attain these objectives, two scales of studies are usually proposed (Fig. 2): small scale catchments, which allow to study rivers draining one rock type under a given climate, and large scale catchments, which allow to integrate the informations over larger portions of continental crust from different climatic regions (Meybeck 1979; Amiotte-Suchet and Probst 1993, 1995; Gaillardet et al. 1999). Current $\delta^{44/40}$Ca characterisation of the flux into the ocean is extremely limited and only a few studies dealt with natural waters at local or regional scales (Zhu and MacDougall 1998; Schmitt et al. 2003; Tipper et al. 2006, 2008, 2010; Cenki-Tok et al. 2009; Hindshaw et al. 2011; Bagard et al. 2013; Wiegand and Schwendenmann 2013).

3.2 Small Scale Catchments

A detailed $\delta^{44/40}$Ca study of several spring, brook and stream waters performed at the local scale of the granitic and forested Strengbach catchment shows that Ca isotopic compositions vary in function of several parameters: season, depth of the water sources, and dry/humid periods (Fig. 7) (Cenki-Tok et al. 2009). Low $\delta^{44/40}$Ca values have been reported in winter, for deep soil compartments and dry periods with low water-flow rate. This has been linked to a lithological signature dominated by intense interactions between waters and Ca-rich primary minerals (apatite and feldspar). In contrast, higher $\delta^{44/40}$Ca values have been measured in spring, summer and autumn for surface soil compartments during wet periods and high flow rates. The authors explained this difference in Ca isotopic compositions by the influence of vegetation on slightly mineralized waters: the uppermost soil horizons are ^{40}Ca depleted due to

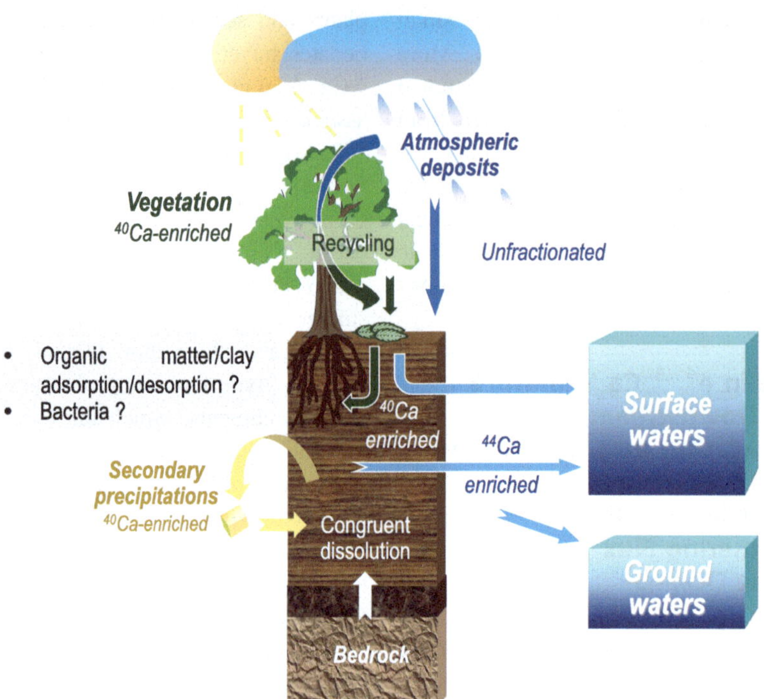

Fig. 6 Summary of identified and potential mechanisms leading to Ca isotopic fractionations, as well as Ca end-members at the scale of small watersheds

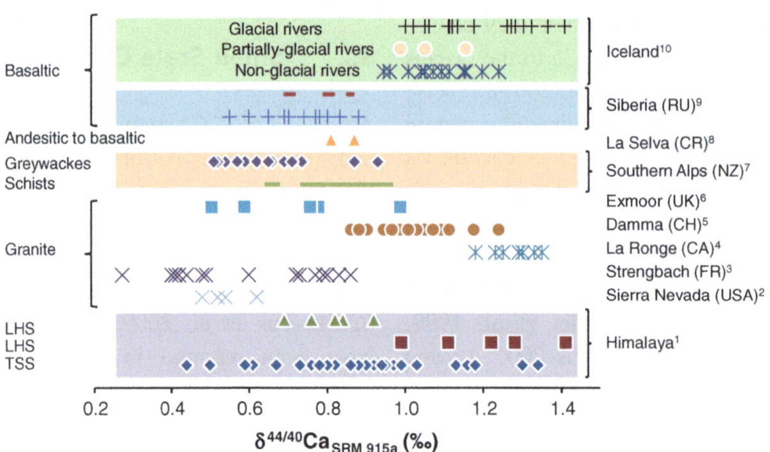

Fig. 7 Ca isotopic variation in riverwaters. All values are normalized to NIST SMR 915a standard. *1* Tipper et al. (2008); *2* Nielsen and DePaolo (2013); *3* Schmitt et al. (2003), Cenki-Tok et al. (2009); *4* Holmden and Bélanger (2010); *5* Hindshaw et al. (2011); *6* Chu et al. (2006); *7* Moore et al. (2013); *8* Wiegand and Schwendenmann (2013); *9* Bagard et al. (2013); *10* Hindshaw et al. (2013), Jacobson et al. (2015)

vegetation uptake and recycling, and residual waters are thereby enriched in ^{44}Ca (Fig. 4). Although no correlation was recorded between $\delta^{44/40}$Ca values and discharge rates, $\delta^{44/40}$Ca values of snow are enriched in ^{44}Ca in the Strengbach catchment (Cenki-Tok et al. 2009) and in the Damma catchment (Hindshaw et al. 2011). Hindshaw et al. (2011) proposed that in the winter, not frozen, isolated, water pockets promotes sorption of ^{40}Ca (and/or exchange mechanisms) on soil components, leaving the corresponding water enriched in ^{44}Ca. Consequently, in the spring, increased snowmelt flushes out these ^{44}Ca enriched waters into the streamwaters.

Tipper et al. (2010) have observed similar $\delta^{44/40}$Ca variations for wet and dry seasons in the Tibetan Plateau, which they attributed to secondary calcite or calcrete precipitation in soils. Indeed, in Himalayan rivers, up to 70 % of the total, initially dissolved, Ca are reported to be removed by secondary calcite precipitation (Galy et al. 1999; Jacobson et al. 2002; Bickle et al. 2005; Tipper et al. 2006). Alternatively, they proposed that this Ca isotopic variation could be linked to the inflow of groundwater during low water stand and to shallow soil runoff at high water stand. This is consistent with the results of Tipper et al. (2006). They have indeed suggested that the downstream $\delta^{44/40}$Ca evolution of the Bhote Kosi river results from mixing between the isotopic signal of upstream waters controlled by limestone dissolution (enriched in ^{40}Ca) and of downstream waters dominated by fractionated groundwater inputs (enriched in ^{44}Ca). In contrast, Jacobson and Holmden (2008), who studied $\delta^{44/40}$Ca evolution in an aquifer, concluded that the Ca isotopic signatures reflect mixing of water masses controlled by anhydrite and dolomite dissolution, with an absence of fractionation linked to calcite precipitation under saturated conditions. They explained this by "long timescale of water-rock interaction and slow rate of calcite precipitation". Another study performed in a forested catchment from Costa Rica (Wiegand and Schwendenmann 2013) showed that Ca isotopic signatures of river waters are characterized by source-specific signatures (groundwater versus rainwater).

Tipper et al. (2006) explained their non-fractionated and bedrock-like $\delta^{44/40}$Ca values in surface runoff waters by short reaction times between water and rocks that do not allow precipitation of secondary phases to take place. Tipper et al. (2008) concluded that with the exception of waters draining mainly dolostone

rocks with high $\delta^{44/40}$Ca values, the other Himalayan river waters they studied are not rock controlled. Therefore, Tipper et al. (2010) suggested that except for dolostones that undergo congruent dissolution, rivers draining silicate and limestone rocks are generally only marginally offset from the $\delta^{44/40}$Ca value of the rock itself by fractionation processes (biological activity and/or secondary precipitations; Fig. 7). In accord with this, Siberian Kulingdakan small-scale stream waters reflect only a lithology contribution from the basaltic substratum (Fig. 7). At a larger scale, the Kochechum river waters are for their part explained by deep brine-related underground waters during winter low-water periods and by surface waters influenced by basalt weathering. In order to reconcile the respective influences of vegetation and abiotic processes on small-scale river waters, Hindshaw et al. (2012) proposed that vegetation activity impacts river waters only in low weathering rate areas. More recently, Fantle and Tipper (2014) suggested that a biosphere off steady state can significantly impact both the Ca isotopic compositions of soils and rivers.

3.3 Global Scale Catchments

The Ca isotopic composition of continental runoff has an average $\delta^{44/40}$Ca value of 0.86 ± 0.06 ‰ (see chapter "Global Ca Cycles: Coupling of Continental and Oceanic Processes"). In contrast to small scale rivers (1.48 ‰ variability), large rivers sampled at their mouth show small $\delta^{44/40}$Ca variations of 0.56 ‰, ranging between 0.69 and 1.25 ‰ (Figs. 1 and 8). This is certainly linked to the fact that they integrate their $\delta^{44/40}$Ca signature over huge continental spatial scales. As a result, small and variable Ca reservoirs isotopic signatures are homogenized during upstream to downstream flow, and no lithology-dependence can a priori be identified (Schmitt et al. 2003; Tipper et al. 2010). Schmitt et al. (2003) and Tipper et al. (2010) have also suggested that process-related Ca isotopic fractionation due to vegetal uptake or secondary calcite precipitations seem to exert only a second order in the global $\delta^{44/40}$Ca riverine

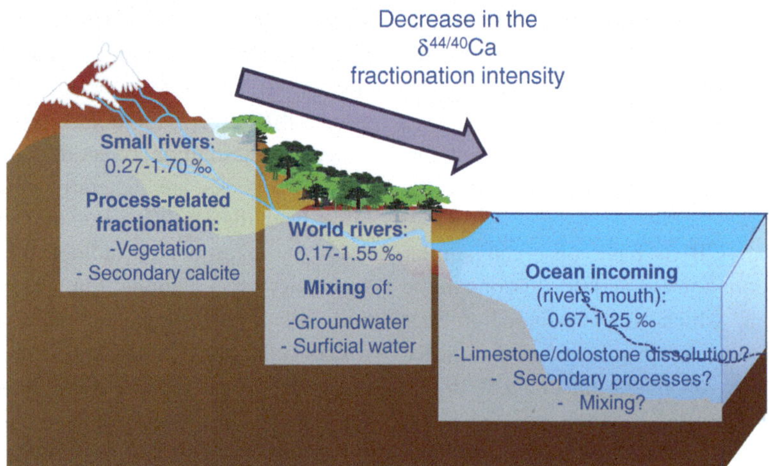

Fig. 8 Summary of downstream $\delta^{44/40}$Ca variations and corresponding mechanisms. All values are normalized to NIST SMR 915a standard (Zhu and MacDougall 1998; Schmitt et al. 2003; Tipper et al. 2006, 2008, 2010; Chu et al. 2006; Cenki-Tok et al. 2009; Holmden and Bélanger 2010; Hindshaw et al. 2011, 2013; Moore et al. 2013; Nielsen and DePaolo 2013; Wiegand and Schwendenmann 2013; Jacobson et al. 2015)

control. Similarly, Farkaš et al. (2011) suggested that the absence of recordable vegetation signal in $\delta^{44/40}$Ca values could be due to the fact that organically-complexed Ca in the topsoil horizon is tightly bound to the ion-exchange sites, inducing limited mobility of this complex. It has also to be noted that Himalayan Rivers (Ganges and Brahmaputra) show ^{40}Ca excesses associated with high ^{87}Sr/^{86}Sr values that can be due to the contribution of radiogenic metasediments (Caro et al. 2010). More recently, Hindshaw et al. (2013) observed that for rivers draining Icelandic catchments with less than 10 % glacial cover, $\delta^{44/40}$Ca of waters reflect a mixing between basalt-derived, hydrothermal- and meltwaters. In contrast, when more than 25 % of the studied catchments were ice-covered, precipitation of Ca-bearing secondary precipitated minerals and/or adsorption or ion exchange of Ca onto mineral surfaces (with a fractionation factor α equal to 0.9999) has to be invoked. Such difference in the control of the isotopic signature of Ca was attributed to differences in weathering regimes. Indeed, the glacial catchments drain sediments containing altered basaltic glass, whereas the non-glacial catchments drain crystalline basalts which are less weatherable (Hindshaw et al. 2013). For their part Jacobson et al. (2015) explained the higher Icelandic

riverine Ca isotopic compositions compared to basalt by conservative mixing of Ca released during the weathering of basalt and hydrothermally precipitated calcite. If their conclusions are confirmed by further studies, this would imply that inputs from carbonate weathering to the rivers' budget is not negligible (up to 90 %). As a result, contribution of basalt weathering to long-term CO_2 consumption and climate regulation may not be that important as previously suggested, especially if carbonates precipitated from hydrochemical and not from atmospheric sources (Dessert et al. 2003).

On the other hand, by studying around 60 rivers covering various substratum worldwide, Tipper et al. (2010) identified a predominant control on large rivers Ca isotopic ratios by limestone/dolostone lithologies, by comparing the $\delta^{44/40}$Ca values of limestone/dolostone with those of limestone/dolostone draining river waters. This is consistent with the hypothesis developed by several authors who suggested that carbonate weathering governed two-third of global riverine Ca (Meybeck 1987; Berner and Berner 1996; Gaillardet et al. 1999). In a more recent study performed in the Southern Alps of New Zealand, Moore et al. (2013) confirmed that riverine Ca mainly originated from carbonate weathering, as a function of tectonic activity. About 50–60 % of Ca originated from carbonates in regions where uplift rates were the lowest (0.1 mm/year) whereas >90 % of Ca came from carbonates when the uplift rates were maximum (10 mm/year). Only for glacial rivers more Ca came from silicate weathering, mainly due to non-stoechiometric release of silicate Ca. However, when extrapolated to other similar regions, such as the Himalaya, the authors concluded that the spatial extent of mountainous glaciation was not spread enough to have a global impact.

Farkaš et al. (2007) reported oceanic limestone secular variations over the Phanerozoïc, which $\delta^{44/40}$Ca reflects the dominant mineralogy in marine carbonate deposits (calcite versus aragonite). Consequently, Tipper et al. (2010) suggested that $\delta^{44/40}$Ca variations in world major rivers were due to dissolution of limestone/dolostone rocks of different ages and having variable $\delta^{44/40}$Ca isotopic compositions. Further data from direct comparison of waters and corresponding bedrocks are therefore necessary to understand the precise mechanisms governing the $\delta^{44/40}$Ca values of large rivers.

A recent review paper challenges these interpretations. Fantle and Tipper (2014) have indeed calculated an average carbonate value equal to 0.60 ‰, whereas it reaches 0.94 ‰ for silicates. When considering an average contribution of silicate weathering to global Ca riverine flux comprised between 10 and 26 % (Meybeck 1987; Berner and Berner 1996; Gaillardet et al. 1999), one should attain a Ca isotopic composition of 0.63–0.69 ‰ for the riverine flux to the ocean (and not 0.86 ‰). Thus, Fantle and Tipper (2014) have invoked secondary processes (Ca uptake by plants, secondary mineral formations or adsorption onto clays) to explain this discrepancy. These authors demonstrated indeed by using simple numerical models that a steady state biosphere can only impact the Ca isotopic composition of rivers when $\delta^{44/40}$Ca from the vegetation cycling flux is isotopically distinct from the uptake one. In contrast, a biosphere off steady-state can impact riverine Ca isotopic composition over thousand-year timescales, though a modest extent (on the order of tens of a

permil). As a result, they concluded their study by the suggestion that either present-day riverine value reflects a terrestrial Ca cycle that is currently in a transient state (i.e. global net growth of the biosphere) or there remain fractionation mechanisms that need to be characterized in the weathering environment. More recently Jacobson et al. (2015) questioned this interpretation by suggesting that mixing processes rather than fractionation mechanisms could take place at global scale.

4 Potential of Ca Isotopes Applied to Earth-Surface Processes

Few studies presently focused on Ca isotopes applied in Earth-Surface processes. Nevertheless, they identified several research directions to be taken in the near future and that will be described hereafter.

4.1 Internal Ca Cycling Processes Within the Tree

Although understanding of the internal Ca partition between vegetation organs has been highly improved during past years (von Blanckenburg et al. 2009; Cobert et al. 2011a; Schmitt et al. 2013), increasing $\delta^{44/40}$Ca values from roots to foliage in trees grown in the field point to internal fractionation processes that are not well understood yet. Holmden and Bélanger (2010) proposed that tree rings cellulose may record secular variations of Ca isotope ratios due to natural and anthropogenic disturbances of the forest Ca cycle. Two recent preliminary studies show the potential of $\delta^{44/40}$Ca variations in tree rings for the reconstruction of environmental conditions through time (Nielsen et al. 2011; Stille et al. 2012).

One of the studies was performed on tree rings spanning $\sim$50 years of *Tamarix aphylla*, an exotic perennial and evergreen tree in Death Valley (USA), where spring water is the only Ca source (Nielsen et al. 2011). Only small $\delta^{44/40}$Ca variations (0.40 ‰) were recorded with an increase of heavy values during drought conditions. The authors suggested that when the tree is stressed either the soil pool is limited with regard to its Ca content during drought conditions and thus gets enriched in ^{44}Ca whereas the vegetal takes up the ^{40}Ca, or the Ca is held more tightly in the small pores of the soil. Another study performed on spruces from the nutrient-depleted Strengbach catchment in Vosges mountains (France), where the main Ca source is the atmosphere, indicates that the biologically recycled Ca does not allow to detect any $\delta^{44/40}$Ca variations in the trees growth rings over past century (Stille et al. 2012). This could be due to the low weathering flux of Ca in this catchment that forced the spruces' root system to pump the Ca mainly from the organic matter rich top-soil over the past century. As a consequence, the Strengbach catchment has certainly been depleted in nutrient elements already since the beginning of the 20th century. Applying Ca isotopes to tree rings could thus bring information on past environmental conditions on the Ca available reservoirs in soils and on the functioning of the terrestrial biosphere over time, assuming a constant fractionation factor during uptake.

4.2 Ca Uptake Mechanisms Within the Rhizosphere

All field work and in vitro studies, performed up to now, reported systematic root ^{40}Ca enrichments compared to the soil solutions (Table 1) (Platzner and Degani 1990; Schmitt et al. 2003, 2013; Wiegand et al. 2005; Page et al. 2008; Cenki-Tok et al. 2009; Holmden and Bélanger 2010; Cobert et al. 2011a; Bagard et al. 2013). Thus, Ca isotopes could be a powerfull tool to distinguish between Ca uptake mechanisms within the rhizosphere. The rhizosphere constitutes indeed a narrow interface between soil and roots that is directly influenced by root secretions and associated soil micro-organisms. Consequently, small roots and fungal associates are able to derive nutrients directly from minerals, using chemical exudates that are more aggressive than typical salt extracts (van Breemen et al.

2000; Wallander et al. 2001). Thus, they may enhance soil nutrient mobilization and uptake around the roots (Robert and Berthelin 1986; Blum et al. 2002; Courty et al. 2010). The precise study of Ca isotopes in the rhizosphere could give information on the role of soil micro-organisms on the Ca uptake by trees and induced fractionation mechanisms. It could also help to identify the depth from which Ca is taken up by the roots or help to understand nutrient provenance.

4.3 Ca Recycling by the Vegetation

In acidic ecosystems an important part of Ca is provided by atmosphere deposits. If wet deposits have sometimes been considered in catchment studies using Ca isotopes, few is known concerning dry deposits or processes occurring in the canopy. Indeed, a part of the incident rainfall passes through the forest canopy, directly or by interacting with the vegetation: the throughfall (Lloyd and Marques 1988; Bruijnzeel 1989). Another part is funnelled down the trunk of the trees: the stemflow (Jordan 1978; Herwitz 1986). The difference between the sum of these two fluxes and the incident rainfall corresponds to the canopy interception. Foliar nutrient leaching during acid rain events has also been reported (Fink 1991; Hambuckers and Remacle 1993; Draaijers et al. 1997; Dambrine et al. 1998; Mougougou et al. 1998). These components need to be precisely characterised by Ca isotopes. Indeed, their contribution to mineral nutrient fluxes to the soil surface may be not negligible since these fluxes are more rapid than those coming from the litter decomposition. Thus, the Ca ions are in dissolved inorganic form which can be taken up immediately by the trees. Similarly to what has been observed for Rare Earth Elements (Stille et al. 2009), the natural waters that percolate through the litter will interact with it and contribute to the Ca recycling flux. The resulting $\delta^{44/40}$Ca values of the percolated waters and degraded litter are then enriched in ^{40}Ca, as was suggested by Perakis et al. (2006) and Bagard et al. (2013). This also suggests that Ca isotopes might become a tracer of vegetation uptake (waters enriched in ^{44}Ca) versus decomposing litter leaching (waters enriched in ^{40}Ca). Quantifying these fluxes is fundamental for calculation of Ca budgets at the scale of small forested catchments. The Ca cycle may be perturbed by forest harvesting, which removes significant amounts of light Ca from forest ecosystems (Federer et al. 1989; Johnson et al. 1992; Bailey et al. 2003). The age of forests might also have to be taken into account since forest growth rates are typically highest during their middle stage of development (Miller 1995), which could have an impact on the Ca uptake rates by trees. Moreover, since local plant biomass is growing in spring, the resulting storehouse of biologically processed Ca (^{40}Ca-enriched) will also increase. All these parameters have to be taken into account by further investigations at the scale of small forested catchments.

4.4 Time Integrated Vegetal Turnover Marker

A study performed in a small forested Siberian catchment subjected to permafrost shows that Ca isotopic variations are very different from one hillslope to the other (Bagard et al. 2013). The slope presenting the most important $\delta^{44/40}$Ca variation and ^{40}Ca enrichments in soil solutions and in soil surface acetic acid extracts corresponds to the slope where vegetation and the decomposition rate of needle litter are the more important. In this way, preserved Ca isotopic compositions in paleosol exchangeable and acid leachable phases might bring relevant information on the evolution of biological activity at the catchment scale.

This has to be linked to the results obtained for sedimentary records of Californian marine terrasses and Hawaiian soil chronosequences (Bullen et al. 2004; Wiegand et al. 2005). The comparison of $\delta^{44/40}$Ca values in 300 years and 4.1 Ma years old soils suggests that early forests preferentially took up ^{40}Ca from primary minerals of the fresh bulk rocks. Later on, when the soil thickens, the ecosystem develops and plants take up preferentially their ^{40}Ca from recycling litter and decomposing wood (Bullen and Bailey 2005). As a result, one could propose in accord with von Blankenburg et al. (2009) that Ca isotopes might become time-integrated turnover markers. "Furthermore, these markers may be preserved in the geological record, for example, in the form of precipitated calcites or oxides in paleosols and late sediments. Using these indicators the impact of plants on isotope fractionation can be traced as vegetation changes with climate."

4.5 Hydrological Tracer

Understanding hydrological processes such as water origin, mixing history or distribution between surficial and deep water masses is necessary to allow sustainable management of water resources. Stable (O, C, H, N, S), radioactive (U–Th–Ra) and radiogenic (Sr) isotopes have proved to be useful tracers of hydrologic pathways, mixing of waters and solutes… (Riotte and Chabaux 1999; Tricca et al. 1999; Aubert et al. 2002; Durand et al. 2005; Reddy et al. 2006; Chabaux et al. 2008; Elsbury et al. 2009; Kendall et al. 2010). However, there is a lack of understanding of the physical processes and chemical reactions influencing isotopic fractionation mechanisms. In this context, Ca isotopes might become a complementary tracer in order to distinguish between deep-seated reservoirs (weathering) and water reservoirs situated close to the soil surface within the trees root system (Cenki-Tok et al. 2009). Thus $\delta^{44/40}$Ca values in stream and source waters carry a dual signal and their analysis might become a relevant and helpful approach to constrain, at a catchment scale, the inter-relationships between vegetation and weathering processes (Cenki-Tok et al. 2009). Moreover, at the scale of forested catchments where the $\delta^{44/40}$Ca of vegetation is not so important (Wiegand and Schwendenmann 2013) or at larger scale catchments where vegetation is not the main controlling parameter (Moore et al. 2013; Jacobson et al. 2015), Ca isotopes could also be controlled by mixing processes. This highlights the potential of Ca isotopes to become a tracer of hydrological processes.

5 Conclusion

Current understanding of Ca isotopic behaviour in continental scale weathering system is extremely limited. Recent studies performed in small scale catchments have shown that natural waters (surface, ground, soil solutions) are impacted by fractionation mechanisms (biological or abiotic) or mixing processes between different end-members. Congruent bedrock and primary mineral dissolution should a priori not fractionate Ca isotopic compositions. In contrast, nothing is known about their potential incongruent dissolution.

In forested catchments, atmospheric deposits (wet, dry and throughfall) play an important role in providing nutrients to acid ecosystems. Nutrients from the atmosphere, the soil exchangeable and acid leachable fractions as well as nutrients originating from litter degradation are taken up again by the roots. The roots are systematically enriched in ^{40}Ca compared to the nutritive solution (that is enriched in ^{44}Ca). Soil micro-organisms may also have an impact on Ca isotopic signatures that need to be investigated. Degrading litter leached by percolating waters constitutes the recycling flux that enriches surface waters in ^{40}Ca. Within the vegetation itself, organs get enriched in ^{44}Ca from roots to shoots, but remain enriched in ^{40}Ca compared to the nutritive solution, and the intensity of fractionation within the vegetation is function of the Ca content of the nutritive solution and its pH.

In non-forested catchments, secondary mineral precipitations are also enriched in ^{40}Ca, and the resulting waters are thus enriched in ^{44}Ca. Few is actually known about sorption/desorption effects on clays and colloids, cation exchange processes, role of organic matter or Ca precipitation in oxalate on Ca isotopic signatures. Seasonal precipitation and dissolution of secondary minerals have also to be looked for. These processes might also occur in forested catchments, where they seem to be obscured by the biological effects. Consequently, fractionation mechanisms during natural processes are not yet fully understood and request further investigations, which could also allow to quantify the respective proportions of biological or mineral processes in forested catchments and the interactions between them due to micro-organisms.

In contrast, at the scale of global catchments, the lithological and fractionation mechanisms are obscured and $\delta^{44/40}$Ca values of river water do not vary much, although the precise mechanisms have not been identified yet. This observation requires more systematic studies on rivers draining limestones and silicates in order to improve current understanding of the origin of Ca and the mechanisms occurring from upstream to downstream flow of large rivers, as well as the relative proportion of carbonate *versus* silicate weathering processes.

This review of Earth-Surface Ca isotopic fractionations has pointed to several future applications of Ca isotopes going from plant physiology to chemical interactions at the soil-water-plant level, such as: internal Ca cycling processes within trees, Ca uptake mechanisms within the rhizosphere, Ca recycling by the vegetation, time-integrated vegetal turnover marker, and hydrological tracer. To attain these objectives, coupled small and global scale catchments and in vitro experiments are requested. The field studies have to be performed at settings that allow identifying the mechanisms controlling Ca isotope variations (e.g. by varying one parameter at each time: lithology or vegetal or secondary precipitations). In order to identify mixing versus fractionation mechanisms, multi-proxy approaches, including $\delta^{44/40}$Ca, ^{87}Sr/^{86}Sr isotopes and Sr/Ca elemental ratios are welcome.

References

Åberg G (1995) The use of natural strontium isotopes as tracers in environmental studies. Water Air Soil Poll 79:309–322

Åberg G et al (1989) Weathering rates and ^{87}Sr/^{86}Sr ratios: an isotopic approach. J Hydrol 109:65–78

Åberg G et al (1990) Strontium isotopes in trees as an indicator for calcium availability. Catena 17:1–11

Amini M et al (2008) Calcium isotope ($\delta^{44/40}$Ca) fractionation along hydrothermal pathways, Logatchev Field (Mid-Atlantic Ridge, 14°45′N). Geochim Cosmochim Acta 72:4107–4122

Amini M et al (2009) Calcium isotopes ($\delta^{44/40}$Ca) in MPI-DING reference glasses, USGS rock powders and various rocks: Evidence for Ca isotope fractionation in terrestrial silicates. Geost Geoanal Res 33:231–247

Amiotte-Suchet P, Probst JL (1993) Modelling of atmospheric CO_2 consumption by chemical weathering of rocks: application to the Garonne, Congo and Amazon basins. Chem Geol 107:205–210

Amiotte-Suchet P, Probst JL (1995) A global model for present day atmospheric/soil CO_2 consumption by chemical erosion of continental rocks (GEM-CO_2). Tellus 47B:273–280

Aubert D et al (2001) REE fractionation during granite weathering and removal by waters and suspended loads: Sr and Nd isotopic evidence. Geochim Cosmochim Acta 65:387–406

Aubert D et al (2002) Evidence of hydrological control of Sr behaviour in stream water (Stengbach catchment, Vosges mountains, France). Appl Geochem 17:285–300

Baes AU, Bloom PR (1988) Exchange of alkaline earth cations in soil organic matter. Soil Sci 146:6–14

Bagard M-L et al (2013) Biogeochemistry of stable Ca and radiogenic Sr isotopes in a larch-covered permafrost-dominated watershed of Central Siberia. Geochim Cosmochim Acta 114:169–187

Bailey SW et al (1996) Calcium inputs and transport in a base-poor forest ecosystem as interpreted by Sr isotopes. Water Resour Res 32:707–719

Bailey SW et al (2003) Implications of sodium mass balance for interpreting the calcium cycle of a forested ecosystem. Ecology 84:471–484

Barber SA (1995) Soil nutrient bioavailability; a mechanistic approach, 2nd edn. Wiley, New York

Bedel L et al (2016) Unexpected calcium sources in deep soil layers in low-fertility forest soils identified by strontium isotopes (Lorraine plateau, eastern France). Geoderma 264:103–116

Bélanger N, Holmden C (2010) Influence of landscape on the apportionment of Ca nutrition in a Boreal Shield forest of Saskatchewan (Canada) using $^{87}Sr/^{86}Sr$ as a tracer. Can J Soil Sci 90:267–288

Berger TW et al (2006) The role of calcium uptake from deep soils for spruce (Picea abies) and beech (Fagus sylvatica). Forest Ecol Manag 229:234–246

Berner EK, Berner RA (1996) Global environment: water, air and geochemical cycles. Prentice Hall, Englewood Cliffs

Berner RA et al (1983) The carbonate-silicate geochemical cycle and its effect on atmospheric carbon dioxide over the past 100 million years. Am J Sci 283:641–683

Bickle MJ et al (2005) Relative contributions of silicate and carbonate rocks to riverine Sr fluxes in the headwaters of the Ganges. Geochim Cosmochim Acta 69:2221–2240

Blättler CL, Higgins JA (2014) Calcium isotopes in evaporites record variations in Phanerozoic seawater SO_4 and Ca. Geology. doi:10.1130/G35721.1

Blättler CL et al (2012) Explaining the Phanerozoic Ca isotope history of seawater. Geology 40:843–846

Blättler CL et al (2015) Mg and Ca isotope signatures of authigenic dolomite in siliceous deep-sea sediments. Earth Planet Sci Lett 419:32–42

Blum JD et al (2002) Mycorrhizal weathering of apatite as an important calcium source in base-poor forest ecosystems. Nature 417:729–731

Blum JD et al (2008) Use of foliar Ca/Sr discrimination and $^{87}Sr/^{86}Sr$ ratios to determine soil Ca sources to sugar maple foliage in a northern hardwood forest. Biogeochemistry 87:287–296

Böhm F et al (2006) Calcium isotope fractionation in modern scleractinian corals. Geochim Cosmochim Acta 70:4452–4462

Brazier J-M et al (2015) Calcium isotopic evidence for dramatic increase of continental weathering during the Toarcian oceanic anoxic event (Early Jurassic). Earth Planet Sci Lett 411:164–176

Bresinsky A et al (2008) Strasburger Lehrbuch der Botanik. Spektrum Akademischer Verlag

Bruijnzeel LA (1989) (De)forestation and dry season flow in the tropics: a closer look. J Trop For Sci 1:229–243

Bullen TD, Bailey SW (2005) Identifying calcium sources at an acid deposition impacted spruce forest: a strontium isotope, alkaline earth element multi-tracer approach. Biogeochem 74:63–99

Bullen TD et al (2004) Calcium stable isotope evidence for three soil calcium pools at a granitoid chrono-sequence. In: Wanty RB, Seal RR (eds) Water-rock interaction. Proceedings of the 11th international symposium on water-rock interaction, vol 1, Saratoga Springs, New York, July 2004. Taylor & Francis, London, pp 813–817

Calmels D (2007) Chemical weathering of carbonates: the role of differing weathering agents for global budgets. PhD thesis, University of Paris 7

Capo RC, Chadwick OA (1999) Sources of strontium and calcium in desert soil and calcrete. Earth Planet Sci Lett 170:61–72

Capo RC et al (1998) Strontium isotopes as tracers of ecosystem processes: theory and methods. Geoderma 82:197–225

Caro G et al (2010) ^{40}K–^{40}Ca isotopic constraints on the oceanic calcium cycle. Earth Planet Sci Lett 296:124–132

Cenki-Tok B et al (2009) The impact of water-rock interaction and vegetation on calcium isotope fractionation in soil- and stream waters of a small, forested catchment (the Strengbach case). Geochim Cosmochim Acta 73:2215–2228

Certini G et al (2000) Vertical trends of oxalate concentration in two soils under Abies alba from Tuscany (Italy). J Plant Nutr Soil Sci 163:173–177

Chabaux F et al (2008) U-series geochemistry in weathering profiles, river waters and lakes. Radioactiv Env 13:49–104

Chabaux F, Riotte J, Schmitt AD et al (2005) Variations of U and Sr isotope ratios in Alsace and Luxembourg rainwatrs: hydrodynamic implications. Comptes Rendus Géosci 337:1447–1456

Chadwick OA et al (1999) Changing sources of nutrients during four million years of ecosystem development. Nature 397:491–497

Chang VT-C et al (2004) Mg and Ca isotope fractionation during CaCO$_3$ biomineralisation. Biochem Biophys Res Comm 323:79–85

Chiquet A et al (2000) Chemical mass balance of calcrete genesis on the Toledo granite (Spain). Chem Geol 170:19–35

Chu N-C et al (2006) Establishing the potential of Ca isotopes as proxy for consumption of dairy products. Appl Geochem 21:1656–1667

Clarkson DT (1984) Calcium transport between tissues and its distribution in the plant. Plant Cell Environ 7:449–456

Clementz MT et al (2003) Are calcium isotopes a reliable monitor of trophic level in marine settings? Int J Osteoarchaeol 13:29–36

Cobert F et al (2011a) Experimental identification of Ca isotopic fractionations in higher plants. Geochim Cosmochim Acta 75:5467–5482

Cobert F et al (2011b) Biotic and abiotic experimental identification of bacterial influence on calcium isotopic signatures. Rapid Comm Mass Spectrom 25:2760–2768

Courty PE et al (2010) The role of ectomycorrhizal communities in forest ecosystem processes: new perspectives and emerging concepts. Soil Biol Biochem 42:679–698

Crockford RH, Richardson DP (2000) Partitioning of rainfall into throughfall, stemflow and interception: effect of forest type, ground cover and climate. Hydr Proc 14:2903–2970

Dambrine E et al (1998) Evidence of current soil acidification in spruce stands (Strengbach catchment, Vosges mountains, North-Eastern France). Water Air Soil Poll 105:43–52

De La Rocha CL, DePaolo DJ (2000) Isotopic evidence for variations in the marine calcium cycle over the Cenozoic. Science 289:1176–1178

DePaolo DJ (2004) Calcium isotope variations produced by biological, kinetic, radiogenic and nucleosynthetic processes. In: Johnson C, Beard B, Albarede F (eds) Reviews in mineralogy and geochemistry: geochemistry of the nontraditional stable isotopes, 52. Mineralogical Society of America, Washington, pp 255–288

DePaolo DJ (2011) Surface kinetic model for isotopic and trace element fractionation during precipitation of calcite from aqueous solutions. Geochim Cosmochim Acta 75:1039–1056

Dessert C et al (2003) Basalt weathering laws and the impact of basalt weathring on the global carbon cycle. Chem Geol 202:257–273

Dijkstra FA (2003) Calcium mineralization in the forest floor and surface soil beneath different tree species in the northeastern US. For Ecol Manage 175:185–194

Dijkstra FA, Smits MM (2002) Tree species effects on calcium cycling: the role of calcium uptake in deep soils. Ecosystems 5:385–398

Draaijers GPJ et al (1997) The impact of canopy exchange on differences observed between atmospheric deposition and throughfall fluxes. Atmos Environ 31:387–397

Drouet T et al (2005) Long-term records of strontium isotopic composition in tree rings suggest changes in forest calcium sources in the early 20th century. Global Change Biol 11:1926–1940

Durand S et al (2005) U isotope ratios as tracers of roundwater inputs into surface waters: example of the Upper Rhine hydrosystem. Chem Geol 220:1–19

Durand N et al (2010) Calcium carbonate features. In: Stoops G, Marcelino V, Mees F (eds) Interpretation of micromorphological features of soils and regoliths, 1st edn. Elsevier, Amsterdam

Du Vivier ADC et al (2015) Ca isotope stratigraphy across the Cenomanian-Turonian OAE 2: links between volcanism, seawater geochemistry, and the carbonate fractionation factor. Earth Planet Sci Lett 416:121–131

Edmond JM, Huh Y (2003) Non-steady state carbonate recycling and implications for the evolution of atmospheric pCO$_2$. Earth Planet Sci Lett 216:125–139

Elias RW et al (1982) The circumvention of the natural biopurification of calcium along nutrient pathways by atmospheric inputs of industrial lead. Geochim Cosmochim Acta 46:2561–2580

Elsbury KE et al (2009) Using oxygen isotopes of phosphate to trace phosphorus sources and cycling in Lake Erie. Env Sci Technol 43:3108–3114

Ewing S et al (2008) Non-biological fractionation of stable Ca isotopes in soils of the Atacama Desert, Chile. Geochim Cosmochim Acta 72:1096–1110

Fantle MS (2015) Calcium isotopic evidence for rapid recrystallization of bulk marine carbonates and implications for geochemical proxies. Geochim Cosmochim Acta 148:378–401

Fantle MS, DePaolo DJ (2005) Variations in the marine Ca cycle over the past 20 million years. Earth Planet Sci Lett 237:102–117

Fantle MS, DePaolo DJ (2007) Ca isotopes in carbonate sediment and pore fluid from ODP Site 807A: the Ca^{2+}(aq)-calcite equilibrium fractionation factor and calcite recrystallization rates in Pleistocene sediments. Geochim Comsochim Acta 71:2524–2546

Fantle MS, Tipper ET (2014) Calcium isotopes in the global biogeochemical Ca cycle: implications for development of a Ca isotope proxy. Earth Sci Rev 129:148–177

Fantle MSA et al (2012) The Ca isotopic composition of dust-producing regions: measurements of surface sediments in the Black Rock Desert, Nevada. Geochim Cosmochim Acta 87:178–193

Farkaš J et al (2007) Evolution of the oceanic calcium cycle during late Mesozoic: evidence from δ$^{44/40}$Ca of marine skeletak carbonates. Earth Planet Sci Lett 253:96–111

Farkaš J et al (2011) Calcium isotope constraints on the uptake and sources of Ca^{2+} in a base-poor forest: a new concept of combining stable (δ$^{44/42}$Ca) and radiogenic (εCa) signals. Geochim Cosmochim Acta 75:7031–7046

Federer CA et al (1989) Long-term depletion of calcium and other nutrients in eastern US forests. Environ Manag 13:593–601

Fink S (1991) The micromorphological distribution of bound calcium in needles of Norway spruces (*Picea abies* L.). New Phytol 119:33–40

Gaillardet J et al (1999) Global silicate weathering and CO_2 consumption rates deduced from the chemistry of large rivers. Chem Geol 159:3–30

Galy A et al (1999) The strontium isotopic budget of Himalayan Rivers in Nepal and Bangladesh. Geochim Cosmochim Acta 63:1905–1925

Gangloff S et al (2014) Impact of bacterial activity on Sr and Ca isotopic compositions (^{87}S/^{86}Sr and $\delta^{44/40}$Ca) in soil solutions (the Strengbach CZO). Procedia Earth Planet Sci 10:109–113

Goldstein SJ, Jacobsen SB (1988) Nd and Sr isotopic systematics of river water suspended material. Implications for crustal evolution. Earth Planet Sci Lett 87:249–265

Graustein WC, Armstrong RL (1983) The use of strontium-87/strontium-86 ratios to measure transport into forested watersheds. Science 219:289–292

Graustein WC et al (1977) Calcium oxalate: occurrence in soils and effect on nutrient and geochemical cycles. Science 198:1252–1254

Griffith EM et al (2008) Influences on the fractionation of calcium isotopes in planktonic foraminifera. Earth Planet Sci Lett 268:124–136

Griffith EM et al (2011) Seawater calcium isotope ratios across the Eocene-Oligocene transition. Geology 39:683–686

Griffith EM et al (2015) Effects of ocean acidification on the marine calcium isotope record at the Paleocene-Eocene Thermal Maximum. Earth Planet Sci Lett 419:81–92

Gussone N, Filipsson HL (2010) Calcium isotope ratios in calcitic tests of benthic foraminifers. Earth Planet Sci Lett 290:108–117

Gussone N et al (2003) Model for kinetic effects on calcium isotope fractionation (δ^{44}Ca) in inorganic aragonite and cultured planktonic foraminifera. Geochim Cosmochim Acta 67:1375–1382

Gussone N et al (2005) Calcium isotope fractionation in calcite and aragonite. Geochim Cosmochim Acta 69:4485–4494

Gussone N et al (2009) A critical evaluation of calcium isotope ratios in tests of planktonic foraminifers. Geochim Cosmochim Acta 73:7241–7255

Gussone N et al (2010) Minor element and Ca isotope composition of calcareous dinoflagellate cysts of cultured *Thoracosphaera heimii*. Earth Planet Sci Lett 289:180–188

Gussone N et al (2011) Calcium isotope fractionation in ikaite and vaterite. Chem Geol 285:194–202

Hacke U, Sperry J (2001) Functional and ecological xylem anatomy. Perspect Plant Ecol Evol Syst 4:97–115

Halicz L et al (1999) High-precision measurement of calcium isotopes in carbonates and related materials by multiple collector inductively coupled plasma mass spectrometry (MC-ICP-MS). J Anal At Spectrom 14:1835–1838

Hambuckers A, Remacle J (1993) Relative importance of factors controlling the leaching and uptake of inorganic-ions in the canopy of a Spruce Forest. Biogeochemistry 23:99–117

Hamidi EM et al (2001) Isotopic tracers of the origin of Ca in a carbonate crust from the Middle Atlas, Morocco. Chem Geol 176:93–104

Hedin RS et al (1994) Passive treatment of acid mine drainage with limestone. J Env Qual 23:1338–1345

Heinemann A et al (2008) Modification of Ca isotope and trace metal composition of the major matrices involved in shell formation of Mytilus edulis. Geochem Geophys Geosyst 9:Q01006. doi:10.1029/2007GC001777

Herwitz S (1986) Infiltration excess caused by stemflow in a cyclone-prone tropical rainforest. Earth Surf Proc Land 11:401–412

Heuser A et al (2005) Calcium Isotope ($\delta^{44/40}$Ca) variations of Neogene planktonic foraminifera. Paleoceanogr 20:PA2013. doi:10.1029/2004PA001048

Heuser A et al (2011) Calcium isotopes in fossil bones and teeth—diagenetic versus biogenic origin. Geochim Cosmochim Acta 75:3419–3433

Hindshaw RS et al (2011) Calcium isotopes in a proglacial weathering environment: Damma glacier, Switzerland. Geochim Cosmochim Acta 75:106–118

Hindshaw RS et al (2012) Calcium isotope fractionation in alpine plants. Biogeochem. doi:10.1007/s10533-012-9732-1

Hindshaw RS et al (2013) The stable calcium isotopic composition of rivers draining basaltic catchments in Iceland. Earth Planet Sci Lett 374:173–184

Hinojosa JL et al (2012) Evidence for end-Permian ocean acidification from calcium isotopes in biogenic apatite. Geology 40:743–746

Hippler D et al (2006) Tropical Atlantic SST history inferred from Ca isotope thermometry over the last 140ka. Geochim Cosmochim Acta 70:90–100

Hippler D, Schmitt AD, Gussone N et al (2003) Calcium isotopic composition of various reference materials and seawater. Geostand Geoanal Res 27:13–19

Hofmann AW (1997) Mantle geochemistry: the message from oceanic volcanism. Nature 385:219–229

Holmden C (2009) Ca isotope study of Ordovician dolomite, limestone and anhydrite in the Williston basin: implications for subsurface dolomitization and local Ca cycling. Chem Geol 268:180–188

Holmden C, Bélanger N (2010) Ca isotope cycling in a forested ecosystem. Geochim Cosmochim Acta 74:995–1015

Holmden C et al (2012) $\delta^{44/40}$Ca variability in shallow water carbonates and the impact of submarine groundwater discharge on Ca-cycling in marine environments. Geochim Cosmochim Acta 83:179–194

Huang S et al (2010) Calcium isotopic fractionation between clinopyroxene and orthopyroxene from mantle peridotites. Earth Planet Sci Lett 292:337–344

Huang S et al (2011) Stable calcium isotopic compositions of Hawaiian shield lavas: evidence for recycling of ancient marine carbonates into the mantle. Geochim Cosmochim Acta 75:4987–4997

Husson JM et al (2015) Ca and Mg isotope constraints on the origin of earth's deepest $\delta^{13}C$ excursion. Geochim Cosmochim Acta 160:243–266

Immenhauser A et al (2005) A critical assessment of mollusk $^{18}O/^{16}O$, Mg/Ca, and $^{44}Ca/^{40}Ca$ ratios as proxies for Cretaceous seawater temperature seasonality. Palaeogeogr Palaeoclim Palaeoecol 215:221–237

Jacobson AD, Holmden C (2008) $\delta^{44}Ca$ evolution in a carbonate aquifer and its bearing on the equilibrium isotope fractionation factor for calcite. Earth Planet Sci Lett 270:349–353

Jacobson AD, Blum JD, Walter LM (2002) Reconciling the elemental and Sr isotope composition of Himalayan weathering fluxes: insights from the carbonate geochemistry of stream waters. Geochim Cosmochim Acta 66:3417–3429

Jacobson AD, Andrews MG, Lehn GO et al (2015) Silicate versus carbonate weathering in Iceland: new insights from Ca isotopes. Earth Planet Sci Lett 416:132–142

Johnson AH, McLaughlin SB, Adams MB et al (1992) Synthesis and conclusions from epidemiological and mechanistic studies of red spruce decline. In: Eager C (ed) Ecology and decline of red spruce in the eastern U.S. Springer, Berlin, pp 385–411

Jordan CF (1978) Stem flow and nutrient transfer in a tropical rain forest. Oikos 31:257–263

Jost AB, Mundil R, He B et al (2014) Constraining the cause of the end-Guadalupian extinction with coupled records of carbon and calcium isotopes. Earth Planet Sci Lett. 396:201–212

Kasemann SA, Hawkesworth CJ, Prave AR et al (2005) Boron and calcium isotope composition in Neoproterozoic carbonate rocks from Namibia: evidence for extreme environmental change. Earth Planet Sci Lett 231:73–86

Kasemann SA, Pogge von Strandmann PAE, Prave AR et al (2014) Continental weathering following a Cryogenian glaciation: evidence from calcium and magnesium isotopes. Earth Planet Sci Lett 396:66–77

Kendall C, Young MB, Silva SR (2010) Applications of stable isotopes for regional to national-scale water quality and environmental monitoring programs. In: West JB, Bowen GJ, Dawson TE, Tu KP (eds) Isoscapes: understanding movement, pattern and processes on earth through isotope mapping, pp 89–111

Kennedy MJ, Hedin LO, Derry LA (2002) Decoupling of unpolluted temperate forests from rock nutrient sources revealed by natural $^{87}Sr/^{86}Sr$ and ^{84}Sr tracer addition. PNAS 99:9639–9644

Langmuir D (1997) Aqueous environmental geochemistry. Prentice-Hall, Englewood Cliffs

Lasaga A, Berner RA, Garrels RM (1985) An improved geochemical model of atmospheric CO_2 fluctuations over the past 100 million years. In: Sundquist ET, Broecker WS (eds) The carbon cycle and atmospheric CO_2, natural variations archean to present. American Geophysical Union Monograph, vol 32, pp 397–411

Lemarchand D, Wasserburg GJ, Papanastassiou DA (2004) Rate-controlled calcium isotope fractionation in synthetic calcite. Geochim Cosmochim Acta 68:4665–4678

Lesack LFW, Melack JM (1996) Mass balance of major solutes in a rainforest catchment in the Central Amazon: implication for nutrient budgets in tropical rainforests. Biogeochim 32:115–142

Li S, Zhang Q (2008) Geochemistry of the upper Han River basin, China, 1: spatial distribution of major ion compositions and their controlling factors. Appl Geochem 23:3535–3544

Likens GE, Driscoll CT, Buso DC et al (1998) The biogeochemistry of calcium at Hubbard Brook. Biogeochemistry 41:89–193

Lloyd CR, Marques AD (1988) Spatial variability of throughfall and stemflow measurements in Amazonian rainforest. Agri For Meteorol 42:63–73

Lovett GM, Schaefer DA (1992) Canopy interactions of Ca, Mg and K. In: Johnson DW, Lindberg SE (eds) Atmospheric deposition and forest nutrient cycling. Springer, New York, pp 253–275

Lovett GM, Nolan SS, Driscoll CT et al (1996) Factors regulating throughfall flux in New Hampshire forested landscape. Can J For Res 26:2134–2144

Marriott CS, Henderson GM, Belshaw NS et al (2004) Temperature dependence of δ^7Li, $\delta^{44}Ca$ and Li/Ca during growth of calcium carbonate. Earth Planet Sci Lett 222:615–624

Marschner H (1995) Mineral nutrition of higher plants, 2nd edn. Academic Press, London

McLaughlin SB, Wimmer R (1999) Calcium physiology and terrestrial ecosystem processes. New Phytol 142:373–417

Meybeck M (1979) Concentrations des eaux fluviales en 6lements majeurs et apports en solution aux oceans. Rev Geol Dyn Geogr Phys 21:215–246

Meybeck M (1987) Global chemical weathering of surficial rocks estimated from river dissolved loads. Am J Sci 287:401–428

Miller HG (1995) The influence of stand development on nutrient demand, growth and allocation. Plant Soil 168–169:225–232

Miller EK, Blum JD, Friedland AJ (1993) Determination of soil exchangeable-cation loss and weathering rates using Sr isotopes. Nature 362:438–441

Moore J, Jacobson AD, Holmden C et al (2013) Tracking the relationship between mountain uplift, silicate weathering, and long-term CO_2 consumption with Ca isotopes: Southern Alps, New Zealand. Chem Geol 341:110–127

Mougougou A, Trémolières M, Sánchez-Pérez JM et al (1998) Réalité de l'éxcrétion-récrétion foliaire en milieu forestier alluvial chez deux espèces ligneuses de la sous-trate arborescente. CR Acad Sci Paris, Sciences de la vie 321:915–922

Nägler Th, Eisenhauer A, Müller A et al (2000) The $\delta^{44}Ca$-temperature calibration on fossil and cultured Globigerinoides sacculifer: New tool for reconstruction of past sea surface temperatures. Geochem Geophys Geosyst 1. doi:10.1029/2000GC000091

Nielsen LC, DePaolo DJ (2013) Ca isotope fractionation in a high-alkalinity lake system: Mono Lake, California. Geochim Cosmochim Acta 118:276–294

Nielsen LC, Druhan JL, Yang W et al (2011) Calcium isotopes as tracers of biogeochemical processes. In: Baskaran M (ed) Handbook of environmental isotope geochemistry, advances in isotope geochemistry. Springer, Berlin, pp 105–124

Ockert C, Gussone N, Kaufhold S et al (2013) Isotopic fractionation during Ca exchange on clay minerals in a marine environment. Geochim Cosmochim Acta 112:374–388

Oehlerich M, Mayr C, Griesshaber E et al (2013) Ikaite precipitation in a lacustrine environment - implications for palaeoclimatic studies using carbonates from Laguna Potrak Aike (Patagonia, Argentina). Quat Sci Rev 71:46–53

Oehlerich M, Mayr C, Gussone N et al (2015) Late Glacial and Holocene climatic changes in south-eastern Patagonia inferred from carbonate isotope records of Laguna Potrok Aike (Argentina). Quat Sci Rev 114:189–202

Page B, Bullen T, Mitchell M (2008) Influences of calcium availability and tree species on Ca isotope fractionation in soil and vegetation. Biogeochem 88:1–13

Park A, Cameron JL (2008) The influence of canopy traits on throughfall and stemflow in five tropical trees growing in a Panamanian plantation. For Ecol Manag 255:1915–1925

Payne JL, Turchyn AV, Paytan A et al (2010) Calcium isotope constraints on the end-Permian mass extinction. Proc Natl Acad Sci 107:8543–8548

Perakis SS, Maguire DA, Bullen TD et al (2006) Coupled nitrogen and calcium cycles in forests of the Oregon Coast range. Ecosystems 9:63–74

Platzner I, Degani N (1990) Fractionation of stable calcium isotopes in tissues of date palm trees. Biomed Env Mass Spectrom 19:822–824

Poszwa A, Dambrine E, Pollier B et al (2000) A comparison between Ca and Sr cycling in forest ecosystems. Plant Soil 225:299–310

Potter CS (1991) Nutrient leaching from Acer rubrum leaves by experimental acid rainfall. Can J For Res 21:222–229

Pretet C, Samankassou E, Felis T et al (2013) Constraining calcium isotope fractionation ($\delta^{44/40}$Ca) in modern and fossil scleractinian coral skeleton. Chem Geol 340:49–58

Raymo ME, Ruddiman WF, Froelich PN (1988) Influence of late Cenozoic mountain building on ocean geochemical cycles. Geol 16:649–653

Reddy MM, Schuster PF, Kendall C et al (2006) Characterization of surface and ground water δ^{18}O seasonal variation and its use for estimating groundwater residence times. Hydrol Proc 20:1753–1772

Retallack GJ (1990) Soils of the past: an introduction to paleopedology. Unwin-Hyman, London

Reynard LM, Henderson GM, Hedges REM (2010) Calcium isotope ratios in animal and human bone. Geochim Cosmochim Acta 74:3735–3750

Reynard LM, Day CC, Henderson GM (2011) Large fractionation of calcium isotopes during cave-analogue calcium carbonate growth. Geochim Cosmochim Acta 75:3726–3740

Riotte J, Chabaux F (1999) ^{243}U/^{238}U) Activity ratios in freshwaters as tracers of hydrological processes: the Strengbach watershed (Vosges, France. Geochim Cosmochim Acta 63:1263–1275

Robert M, Berthelin J (1986) Role of biological and biochemical factors in soil mineral weathering. In: Huang PM, Schnitzer M (eds) Intern actions of soil minerals with natural organics and microbes. Soil Science Society of America, Special Publication No. 17, Madison, Wisconsin, U.S.A., pp 453–495

Russell W, Papanastassiou D (1978) Calcium isotope fractionation in ion-exchange chromatography. Anal Chem 50:1151–1154

Russell WA, Papanastassiou DA, Tombrello TA (1978) Ca isotope fractionation on the Earth and other solar system materials. Geochim Cosmochim Acta 42:1075–1090

Ryu JS, Jacobson AD, Holmden C et al (2011) The major ion, δ44/40Ca, δ44/42Ca, and δ26/24 Mg geochemistry of granite weathering at pH = 1 and T = 25 °C: power-law processes and the relative reactivity of minerals. Geochim Cosmochim Acta 75:6004–6026

Schauble EA (2007) Role of nuclear volume in driving equilibrium stable isotope fractionation of mercury, thallium, and other very heavy elements. Geochim Cosmochim Acta 71:2170–2189

Schauble EA, Ghosh P, Eiler JM (2006) Preferential formation of ^{13}C–^{18}O bonds in carbonate minerals, estimated using first-principles lattice dynamics. Geochim Cosmochim Acta 70:2510–2529

Schmitt AD, Stille P (2005) The source of calcium in wet atmospheric deposits: Ca-Sr isotope evidence. Geochim Cosmochim Acta 69.3463–3468

Schmitt AD, Chabaux F, Stille P (2003) The calcium riverine and hydrothermal isotopic fluxes and the oceanic calcium mass balance. Earth Planet Sci Lett 213:503–518

Schmitt AD, Cobert F, Bourgeade P et al (2013) Calcium isotope fractionation during plant growth under a limiting nutrient supply. Geochim Cosmochim Acta 110:70–83

Sime NG, De La Rocha CL, Galy A (2005) Negligible temperature dependence of calcium isotope fractionation in 12 species of planktonic foraminifera. Earth Planet Sci Lett 232:51–66

Sime NG, De La Rocha CL, Tipper ET et al (2007) Interpreting the Ca isotope record of marine biogenic carbonates. Geochim Cosmochim Acta 71:3979–3989

Simon JI, DePaolo DJ (2010) Stable calcium isotopic composition of meteorites and rocky planets. Earth Planet Sci Lett 289:457–466

Skulan J, DePaolo DJ (1999) Calcium isotope fractionation between soft and mineralized tissues as a monitor of calcium use in vertebrates. Proc Nat Acad Sci 96:13709–13713

Skulan J, DePaolo DJ, Owens TL (1997) Biological control of calcium isotopic abundances in the global calcium cycle. Geochim Cosmochim Acta 61:2505–2510

Sperry J (2011) Hydraulics of vascular water transport. In: Wojtaszek P (ed) Mechanical integration of plant cells and plants. Springer, Berlin, pp 303–327

Steuber T, Buhl D (2006) Calcium-isotope fractionation in selected modern and ancient marine carbonates. Geochim Cosmochim Acta 70:5507–5521

Stille P, Pierret MC, Steinmann M et al (2009) Impact of atmospheric deposition, biogeochemical cycling and water-mineral interaction on REE fractionation in acidic surface soils and soil water (the Strengbach case). Chem Geol 264:173–186

Stille P, Schmitt AD, Labolle F et al (2012) The suitability of annual tree growth rings as environmental archives: evidence from Sr, Nd, Pb and Ca isotopes in spruce growth rings from the Strengbach watershed. CR Géosci 344:297–311

Suarez DL (1996) Beryllium, Magnesium, Calcium, Strontium, and Barium. In: Sparks DL (ed) Methods of soil analysis: chemical methods, Part 3. Soil Science Society of America, Madison, WI, pp 575–601

Taiz L, Zeiger E (2010) Plant Physiology, 5th edn. Sinauer Associates Inc

Tang J, Dietzel M, Böhm F et al (2008) Sr^{2+} /Ca^{2+} and $^{44}Ca/^{40}Ca$ fractionation during inorganic calcite formation: II. Ca isotopes. Geochim Cosmochim Acta 72:3733–3745

Tang J, Niedermayr A, Köhler SJ et al (2012) Sr^{2+}/Ca^{2+} and $^{44}Ca/^{40}Ca$ fractionation during inorganic calcite formation: III. Impact of salinity/ionic strength. Geochim Cosmochim Acta 77:432–443

Teichert BMA, Gussone N, Torres ME (2009) Controls on calcium isotope fractionation in sedimentary pore-waters. Earth Planet Sci Lett 279:373–382

Tipper ET, Galy A, Bickle MJ (2006) Riverine evidence for a fractionated reservoir of Ca and Mg on the continents: Implications for the oceanic Ca cycle. Earth Planet Sci Lett 247:267–279

Tipper ET, Galy A, Bickle MJ (2008) Calcium and magnesium systematics in rivers draining the Himalaya-Tibetan-Plateau region: Lithological or fractionation control? Geochim Cosmochim Acta 72:1057–1075

Tipper ET, Gaillardet J, Galy A et al (2010) Calcium isotope ratios in the world's largest rivers: a constraint on the maximum imbalance of oceanic calcium fluxes. Global Biogeochem Cycles 24:GB3019. doi:10.1029/2009GB003574

Tricca A, Stille P, Steinmann M et al (1999) Rare earth elements and Sr and Nd isotopic composition of dissolved and suspended loads from small river systems in the Vosges mountains (France), the river Rhine and groundwater. Chem Geol 160:139–158

Turchyn AV, DePaolo DJ (2011) Calcium isotope evidence for suppression of carbonate dissolution in carbonate-bearing organic-rich sediments. Geochim Cosmochim Acta 75:7081–7098

Ullmann CV, Böhm F, Rickaby REM et al (2013) The Giant Pacific Oyster (Crassostrea gigas) as a modern analog for fossil ostreoids: Isotopic (Ca, O, C) and elemental (Mg/Ca, Sr/Ca, Mn/Ca) proxies. Geochem Geophys Geosyst 14:4109–4120. doi:10.1002/ggge.20257

Valdes MC, Moreira M, Foriel J et al (2014) The nature of Earth's building blocks as revealed by calcium isotopes. Earth Planet Sci Lett 394:135–145

van Breemen N, Finlay R, Lundström U et al (2000) Mycorrhizal weathering: a true case of mineral plant nutrition? Biogeochemistry 49:53–67

Veizer J, Ala D, Azmy K et al (1999) $^{87}Sr/^{86}Sr$, $\delta^{13}C$ and $\delta^{18}O$ evolution of Phanerozoïc seawater. Chem Geol 161:59–88

von Allmen K, Nägler ThF, Pettke Th et al (2010) Stable isotope profiles (Ca, O, C) through modern brachiopod shells of *T. septentrionalis* and *G. vitreus*: implications for calcium isotope paleo-ocean chemistry. Chem Geol 269:210–219

von Blanckenburg F, von Wirén N, Guelke M et al (2009) Fractionation of metal stable isotopes by higher plants. Elements 5:375–380

Walker JCG, Kasting JF (1981) A negative feedback mechanisms for the long-term stabilization of Earth's surface temperature. J Geophys Res 86:9776–9782

Wallander H, Nilsson LO, Hagerberg D et al (2001) Estimation of the biomass and seasonal growth of external mycelium of ectomycorrhizal fungi in the field. New Phytol 151:753–760

Wang SH, Yan W, Magalhaes HV et al (2012) Calcium isotope fractionation and its controlling factors over authigenic carbonates in the cold seeps of the northern South China Sea. Chin Sci Bull 57:1325–1332

Wang S, Yan W, Magalhães HV et al (2013) Factors influencing methane-derived authigenic carbonate formation at cold seep from southwestern Dongsha area in the northern South China. Env Earth Sci. doi:10.1007/s12665-013-2611-9

Wiegand BA, Schwendenmann L (2013) Determination of Sr and Ca sources in small tropical catchments (La Selva, Costa Rica)—a comparison of Sr and Ca isotopes. J Hydrol 488:110–117

Wiegand BA, Chadwick OA, Vitousek PM et al (2005) Ca cycling and isotopic fluxes in forested exosystems in Hawaii. Geophys Res Lett 32:L11404. doi:10.1029/2005GL022746

Zhu P, MacDougall J (1998) Calcium isotopes in the marine environment and the oceanic calcium cycle. Geochim Cosmochim Acta 62:1691–1698

Global Ca Cycles: Coupling of Continental and Oceanic Processes

Edward T. Tipper, Anne-Désirée Schmitt and Nikolaus Gussone

Abstract

Calcium is one of the most important mobile metals that can migrate easily between major geochemical reservoirs at the Earth's surface; the hydrosphere and the biosphere and crust. In doing so calcium plays a key role in regulating climate over million year time-scales, transferring carbon from the atmosphere and storing it as calcium carbonate. Calcium isotopes potentially provide a way of tracing the mobility of calcium within the Earth surface environment. This chapter reviews the steps where calcium isotopes are fractionated in the weathering and ocean environments, and how these fractionations can be used to constrain mass transfers on both the continents and in the oceans.

Keywords

Calcium · Stable calcium isotopes · Global Ca cycle · Chemical weathering · Marine Ca cycle · Ca continental cycle

E.T. Tipper
Department of Earth Sciences, University of Cambridge, Cambridge, UK
e-mail: ett20@cam.ac.uk

A.-D. Schmitt
LHyGeS/EOST, Université de Strasbourg, Strasbourg, France
e-mail: adschmitt@unistra.fr

A.-D. Schmitt
Laboratoire Chrono-Environnement, Université Bourgogne Franche-Comté, Besançon, France

N. Gussone (✉)
Institut für Mineralogie, Universität Münster, Münster, Germany
e-mail: nguss_01@uni-muenster.de

1 Introduction

Calcium (Ca) is a metal that is mobile in aqueous fluids and can move easily between major geochemical reservoirs at the Earth's surface, such as the hydrosphere, crust and the biosphere. It is the most abundant alkaline earth metal and the fifth most abundant element in the Earth's crust overall (Rudnick and Gao 2003). With only one redox state its chemistry is simpler than many other elements that are cycled at the Earth's surface by redox processes. Calcium is an essential nutrient in the marine and terrestrial

© Springer-Verlag Berlin Heidelberg 2016
N. Gussone et al., *Calcium Stable Isotope Geochemistry*,
Advances in Isotope Geochemistry, DOI 10.1007/978-3-540-68953-9_6

"

biospheres, it is a major component in several important mineral groups in natural systems in both marine and terrestrial environments (calcite, dolomite, phosphate, gypsum) Via calcium carbonate, calcium provides the hard skeletal framework for a variety of marine organisms with protective shells in the aquatic environment. These calcium rich shells and skeletons provide the main calcium and carbon sink from the oceans over geological time-scales, and serve as a recorder of ocean chemical evolution through time. Through the weathering of primary Ca-bearing silicate minerals and the precipitation of calcium carbonate, the element Ca has played a major role in the regulation of the long-term carbon cycle and provides a critical negative feedback that has maintained Earth's climate equable over geological history (Berner and Kothavala 2001; Walker et al. 1981). Calcium carbonate in limestone is by far the largest reservoir of carbon at the Earth's surface. Calcium is, therefore, a pivotal element that links the lithosphere, hydrosphere, biosphere, and atmosphere via a major global biogeochemical cycle and is pivotal to some of the most important Earth surface processes.

The global biogeochemical cycle of Ca has been a topic of critical importance to geoscientists for decades because of it's significance to the carbon cycle and nutrient supply. Much effort has been put into developing isotopic proxies to trace Ca cycling, such as $^{87}Sr/^{86}Sr$, over a variety of temporal and spatial scales, from changes in seawater chemistry over tens of millions of years, to diurnal changes in small rivers. However, the routine measurement of Ca isotope ratios (expressed as $\delta^{44/40}Ca$) over the past 15 years has provided a possibility to trace both the sources and processes that affect Ca directly, using Ca isotopes and this chapter will review much of the advances that have been made in trying to understand the linkages between the continental and marine cycles of Ca.

In its most simplistic form, the continental Ca budget or cycle is composed of the loss of the Ca from the continents via carbonate and silicate mineral weathering. Ca is transferred mainly via rivers and groundwaters to the oceans where the Ca becomes part of the marine Ca cycle (Fig. 1). Quantifying the fluxes and attendant isotopic compositions of continental Ca transfers has proved challenging even when the Ca cycle is

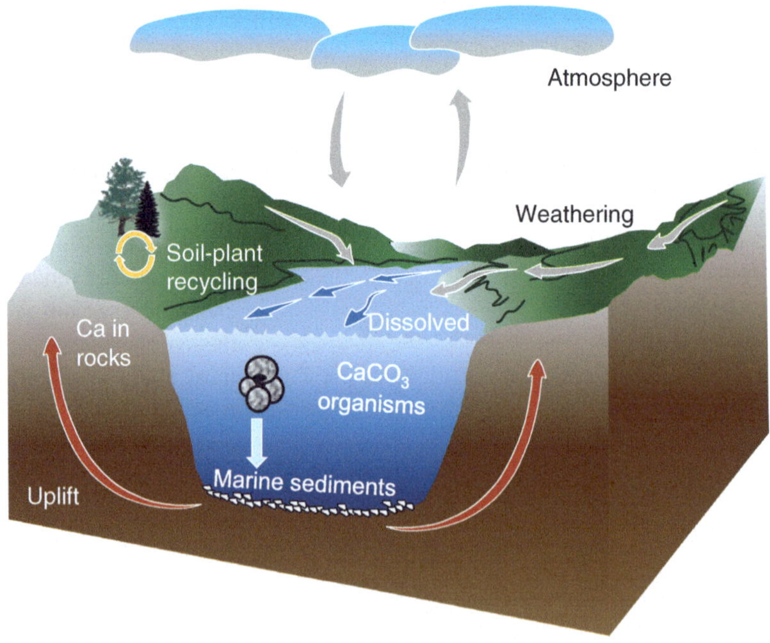

Fig. 1 Global biogeochemical cycle of Ca involving the continent, the ocean and the atmosphere and major processes involving Ca transfer on Earth are shown (adapted for Ca from Mackenzie 2003)

considered in its most simplistic form, and the state of knowledge is probably less advanced than the marine Ca cycle, largely because there is less data. Between 60 and 90 % of solute Ca is derived from the weathering of carbonate mineral which have much faster dissolution kinetics than silicate minerals (Berner and Berner 1996; Gaillardet et al. 1999b; Meybeck 1987; Tipper et al. 2006a), but many studies of Ca isotopes have focused on regions of the world where silicate catchments dominate, and both soils and rivers are depleted in Ca (Fantle and Tipper 2014). In more detail, the processes which release and transfer Ca from the continents to the oceans are complex. The mineral pool of Ca is by far the largest on the planet, though the fraction of Ca residing in sedimentary rocks versus crystalline rocks is poorly known, despite the critical balance of the proportion of carbonate and silicate rocks for the climatic evolution of the planet. Ca is released to a series of reactive pools in the critical zone, the outer veneer of planet described as the fragile skin of the planet defined from the outer extent of vegetation down to the lower limits of groundwater (Brantley et al. 2005) where Ca can reside in aqueous solutions, soil mineral or organic reservoirs or the biosphere. Dissolution reactions can occur both abiotically and through catalysis by organisms, providing nutrients and energy for the sustenance of terrestrial ecosystems (e.g., Brantley et al. 2007). The current rapid evolution of the Earth's surface conditions (temperature and hydrology, IPCC 2014; Raymond et al. 2008), coupled for a need of sustainable agriculture has created a need to understand supply of nutrients in soils and water quality in rivers (Likens et al. 1998).

This has provided the stimulus for a wealth of work to understand the critical zone and to identify mechanisms scavenging elements such as Ca and providing nutrients for the subsistence of terrestrial ecosystems. These studies are based on multidisciplinary approaches, and an array of "non-traditional" stable isotopes have played an important role in trying to understand mass transfers and chemical reactions within the critical zone (e.g., White et al. 2015). Ca isotopes have been at the fore-front of this work (e.g., Schmitt et al. 2003a; Wiegand et al. 2005; Page et al. 2008; Tipper et al. 2006b, 2008a, 2010a; Cenki-Tok et al. 2009; Holmden and Bélanger 2010; Farkaš et al. 2011; Hindshaw et al. 2011, 2013; Fantle and Tipper 2014; Bagard et al. 2013; Wiegand and Schwendenmann 2013; Gangloff et al. 2014; Jacobson et al. 2015) firstly because Ca is one of the major elements by mass, and secondly it is intimately linked to the global carbon cycle on a range of time-scales (e.g., Berner et al. 1983).

There are two sources of Ca inputs to the critical zone; mineral weathering and atmospheric inputs (dust and rain waters) (Likens et al. 1998). Ca can either be removed from the catchment or soil, by groundwater or river water, or be returned to a mineral or organic reservoir in the catchment, either as (1) a secondary mineral such as pedogenic calcite in calcrete soils, (2) adsorption on clay minerals and/or Fe–Al oxy-hydroxydes and/or humic acids in soils (loosely referred to as the exchangeable pool) or (3) Ca present in biomass (either living or decaying). The total pool of Ca in the critical zone is controlled by the balance of "external" inputs via mineral weathering or from the atmosphere, and outputs via erosion (minerals and organically bound Ca) and export of solutes via rivers/groundwaters. The decay of organic matter and attendant loss of organically bound Ca does not change the total pool of Ca in the critical zone, but moves it from the biomass pool to the solute or exchangeable pool. A large exchangeable pool might permit rapid growth rate of the biosphere, but the total amount of growth possible will be limited by the supply of nutrients via mineral weathering. A significant body of work relating to Ca transfers in forest ecosystems and soils has been conducted. Forests require nutrients to develop their annual biomass but soil reserves can be limited, especially in environments where (1) weathering rates are low (Likens et al. 1967; Likens and Bormann 1995), (2) soils are poor in nutrients such as acid tropical soils (Jordan and Herrera 1981; Nykvist 2000), (3) soil acidification due to acid rain takes place (Likens et al. 1996), or (4) intensive forestry and forest tree growth (Raulund-Rasmussen et al. 2011) as well as forest harvesting (Thiffault et al. 2006) take place. All these factors lower Ca availability and thus alter tree nutrition, health and growth rates. In addition, Ca can be lost from the catchment due to run-off. Consequently, Ca cycling and tree nutrition mechanisms need to be identified and characterized in order to manage soil and forest nutrient sources (Bélanger et al. 2012). To date, few catchments have been intensively studied specifically for Ca; one notable exception is the experimental Hubbard Brook catchment (New Hampshire, USA) (Likens et al. 1998).

In the oceanic domain, Ca isotopes have been suggested as a tool for elucidating variations in Ca cycling in the geological past (e.g., De La Rocha and DePaolo 2000; DePaolo 2004; Fantle 2010; Fantle and DePaolo 2005; Farkaš et al. 2007; Griffith et al. 2008a; Heuser et al. 2005; Sime et al. 2007; Blättler and Higgins 2014; Vivier et al. 2015; Griffith et al. 2015). Such an application of Ca isotopes has great potential for the reconstruction of the Ca concentration of the ocean through time, which is essential for understanding marine carbonate equilibria, the pH of the past ocean, and the carbon cycle. Fantle (2010) recently suggested that the Ca isotopic composition of marine authigenic minerals might contain information regarding the partitioning of Ca output fluxes between the deep and shallow ocean, and/or changes in the global fractionation factor associated with Ca removal from the ocean. Further, Ca isotopes have been proposed as a proxy for paleo-sea surface temperatures (Nägler et al. 2000; Gussone et al. 2004; Immenhauser et al. 2005; Hippler et al. 2006) whilst Böhm et al. (2006) and Langer et al. (2007) concluded that the temperature sensitivity was too small for Ca isotopes to be useful in this manner. Recent experimental work suggests that Ca isotopic fractionation during $CaCO_3$ formation is complicated and may be controlled by solution chemistry and/or reactions at the mineral surface (Gussone et al. 2003, 2005, 2011; Lemarchand et al. 2004; Fantle and DePaolo 2007; Tang et al. 2008; DePaolo 2011).

However before it is possible to characterize the global Ca flux to the ocean, it is necessary to identify and characterize the different reservoirs containing Ca, and how and when they liberate it. In the subsequent sections we will review the existing state of knowledge and present selected results from the literature and discuss the key aspects of the global Ca cycle.

2 Principal Ca Reservoirs at the Earth's Surface: Estimates of $\delta^{44/40}Ca$

2.1 Rocks

The total global mass of Ca on the continents varies widely depending on how the estimates have been made (Rudnick and Gao 2003; Taylor and McLennan 1985). Hartmann and Moosdorf (2012) estimated that 64 % of the Earth is covered by sedimentary rocks, a third of these being carbonates. The other 13, 7 and 6 % are covered by metamorphic, plutonic and volcanic rocks, respectively. The remaining 10 % represent water and ice. The total global mass of sediments is thought to be $\sim 2.7 \times 10^{18}$ t (Veizer and MacKenzie 2003). Of this mass, approximately 73 % are located on continents (52 % are situated within orogenic belts, 21 % on platforms), 13 % at passive margin basins, 6 % at active marine basins and 8 % cover the ocean floor (Ronov 1982; Gregor 1985; Veizer and Jansen 1985; Veizer and MacKenzie 2003). Hence to a first approximation we estimate that the total mass of sediments present on the continent amount to $\sim 1.97 \times 10^{18}$ t. Using these masses of sedimentary and igneous rocks and the average Ca concentration of the different rocks (Tables 1 and 2), it is possible to estimate to the first order the average mass of Ca in sedimentary and igneous rocks, as $\sim 108 \times 10^{15}$ t and $\sim 4.48 \times 10^{15}$ t, respectively. Whilst these numbers are subject to considerable uncertainty, they provide the basis for discussion.

Table 1 Proportion of atmospherically-derived Ca in different watersheds worldwide

Country	Locality	% of atmospherically-derived Ca	Reference
USA	Hubbard Brook	20	Likens et al. (1998)
	New York	53	Miller et al. (1993)
	New Hampshire	32	Bailey et al. (1996)
	New Hampshire	5–25	Blum et al. (2002)
	Oregon coast	>97	Perakis et al. (2006)
	Hawaï	5–90	Kennedy et al. (1998)
	Hawaï	<20–30	Vitousek et al. (1999)
		42–58	Lovett et al. (1985)
Canada	Saskatchewan	31–98	Bélanger and Holmden (2010)
Chili	Andes	>80	Kennedy et al. (2002)
Costa Rica		<10	Bern et al. (2005)
France	French Guyana	~100	Poszwa et al. (2009)
	Stengbach	<15	Cenki-Tok et al. (2009)
Swiss	Damma	2–38	Hindshaw et al. (2011)
Belgium	Central and high	71–82	Drouet et al. (2005)

2.1.1 Continental Silicates

In terms of the Ca isotopic composition of continental rocks, the largest dataset is for volcanic rocks. Their variability ranges between 0.61 and 1.34, with an average equal to 0.89 ± 0.03 ‰ ($2\sigma_{mean}$; N = 102, Fig. 2; Table 2) (Skulan et al. 1997; DePaolo 2004; Amini et al. 2009; Hindshaw et al. 2011; Holmden and Bélanger 2010; Huang et al. 2010, 2011; Simon and DePaolo 2010; Bagard et al. 2013; Valdes et al. 2014; Jacobson et al. 2015).

To date there is relatively little data for plutonic and metamorphic rocks, with plutonic Ca isotope compositions spanning 0.31–1.13 ‰, with an average equal to 0.75 ± 0.10 ‰ ($2\sigma_{mean}$; N = 19, Fig. 2; Table 2) (Tipper et al. 2006b; Hindshaw et al. 2011; Holmden and Bélanger 2010; Ryu et al. 2011; Farkaš et al. 2011; Valdes et al. 2014). Metamorphic rocks vary between 0.57 and 1.32 ‰, with an average value equal to 0.81 ± 0.17 ‰ ($2\sigma_{mean}$; N = 11, Fig. 2; Table 2) (Tipper et al. 2006b; Amini et al. 2009; Moore et al. 2013). When considering that metamorphic and plutonic represent respectively 65 and 35 % of igneous rocks on Earth (Hartmann and Moosdorf 2012) it is possible to calculate an igneous-weighted rock average equal to 0.80

$\pm$ 0.20 ‰. This is slightly lower than the arithmetic mean of all silicate rocks and minerals (Fantle and Tipper 2014) of 0.94 ‰, but within uncertainty. The slight difference between these two estimates is a consequence of different literature compilations and different weighting. There is a particular lack of Ca isotope data for plutonic (N = 9) igneous rocks and metamorphic (N = 11) rocks despite the fact that these represent 80 % of all crystalline rocks (Hartmann and Moosdorf 2012), with a far greater abundance of data for volcanic rocks (N = 102).

2.1.2 Continental Sediments

Over Earth's history, carbonate rocks have become the largest reservoir of carbon at the Earth's surface, containing about 4×10^7–5×10^7 PgC. A significant mass of this carbon is bound to Ca in limestone. The Ca isotopic composition of sedimentary rocks varies from -0.77 to 1.73 ‰, with an average value equal to 0.64 ± 0.09 ($2\sigma_{mean}$; N = 78) (Fig. 2; Table 2) (Halicz et al. 1999; Tipper et al. 2006b, 2008a; Ewing et al. 2008; Jacobson and Holmden 2008; Holmden 2009; Moore et al. 2013), identical to the average of carbonate rocks (0.63 ± 0.09, $2\sigma_{mean}$, N = 57, Table 2). For the present

Table 2 Distribution of Ca in main sedimentary and igneous rocks

	Sedimentary rocks				Igneous rocks			Water + ice	Total	References
	Sandstone/sand	Schists	Carbonates	Sum	Shield (plutonic + metam.)	Volcanic	Sum			
%	25.8^{β}	25^{β}	13.2^{β}	64^{α}	20	6	26^{α}	10^{α}	100*	$^{\alpha}$Hartmann and Moosdorf (2012), $^{\beta}$Amiotte-Suchet et al. (2003)
Mass ($*10^{15}$ t)				1970			800*	308*	3078*	Veizer and Mackenzie (2003)
Ca (%)	3.91	2.21	30.2	5.55*	0.51	7.6	0.56*		6.11*	Turekian and Wedepohl (1961)

*Calculated by using the proportion of crystalline rocks from Amiotte-Suchet et al. (2003) and global mass of sedimentary continental rocks from Veizer and Mackenzie (2003)

**Calculated by adding global continental masses of sedimentary and igneous rocks

estimate of the $\delta^{44/40}$Ca of sedimentary rocks, marine carbonates have not been considered (in an attempt to allow for additional diagenetic modification during continental emplacement). Instead, the $\delta^{44/40}$Ca of sedimentary rocks has been based on published continental clays, speleotherms, calcretes, travertine and tufa, dolostone, anhydrite, limestone, soil carbonates and sulphates. This value is identical within uncertainty to the average value of carbonate rocks by considering mean carbonate Ca isotopic composition over geologic time (0.61 ‰) (Fantle and Tipper 2014).

2.2   Hydrosphere

There are three major parts of the hydrosphere in terms of Ca reservoirs; oceans, rivers and groundwaters. Rivers and groundwaters transfer Ca in its ionic and particulate forms from the continents to the oceans. Berner et al. (1983) estimated the global dissolved Ca weathering flux from the continents to the oceans as 0.53×10^9 t/year. Of this weathering flux ~ 85 % (0.45×10^9 t/year) is thought to be provided by dissolution of sedimentary rocks and 0.076×10^9 t/year (~ 15 %) by crystalline rocks. In addition rivers transport $\sim 333 \times 10^6$ t/year of Ca as suspended particulate matter (Berner and Berner 1996) (Table 3). Most suspended sediment is thought to be deposited in coastal water, with little being transported to deeper than the carbonate compensation depth, so that particulates do not influence the carbonate system, maybe except during sea level low stands (Milliman 1993). However, there are clearly exceptions to this, in particular in areas of large sediment supply such as the Bengal Fan.

2.2.1   Oceans

Ca is the 5th most concentrated element in the oceans, at 412 mg/l (10.5 mmol/l, Berner and Berner 1996). The total mass of Ca in the oceans is approximately 0.577×10^{15} t (Berner and Berner 1996) (Table 2). The concentration of Ca and isotopic composition is homogenous

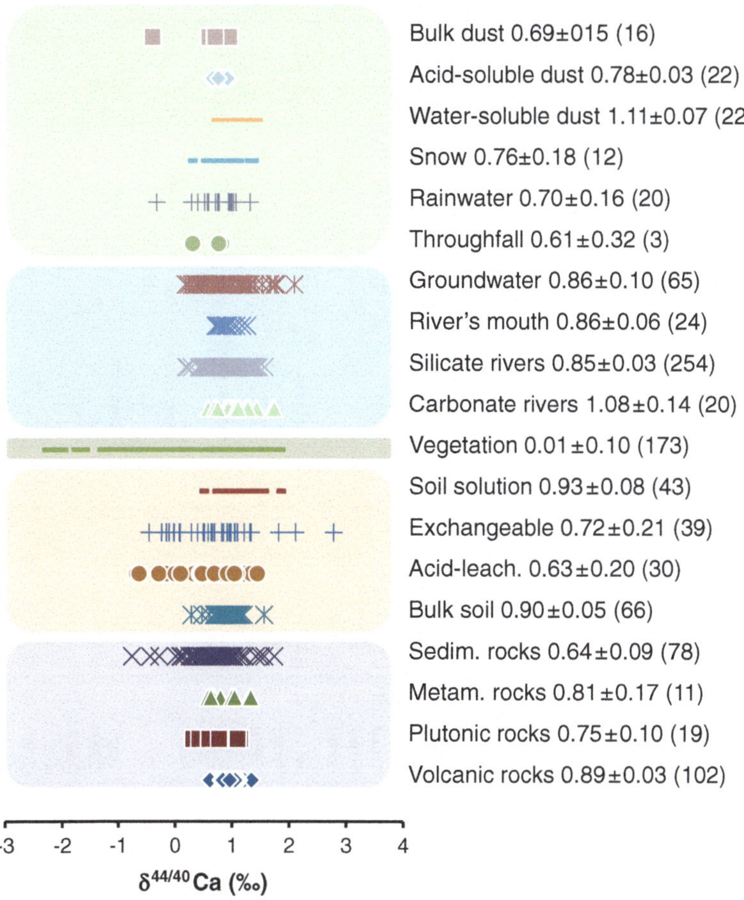

Fig. 2 Variability of Ca isotopic compositions in continental reservoirs. Continental rocks (Skulan et al. 1997; Halicz et al. 1999; DePaolo 2004; Tipper et al. 2006b; Ewing et al. 2008; Jacobson and Holmden 2008; Amini et al. 2009; Holmden 2009; Hindshaw et al. 2011; Holmden and Bélanger 2010; Huang et al. 2010, 2011; Simon and DePaolo 2010; Farkaš et al. 2011; Ryu et al. 2011; Bagard et al. 2013; Moore et al. 2013; Nielsen and DePaolo 2013; Valdes et al. 2014; Jacobson et al. 2015; Oehlerich et al. 2015), soils (Schmitt et al. 2003a; Wiegand et al. 2005; Perakis et al. 2006; Tipper et al. 2006b; Page et al. 2008; Cenki-Tok et al. 2009; Holmden and Bélanger 2010; Farkaš et al. 2011; Hindshaw et al. 2011; Bagard et al. 2013; Moore et al. 2013; Gangloff et al. 2014), vegetation (Skulan and DePaolo 1999; Schmitt et al. 2003a; Bullen et al. 2004; Wiegand et al. 2005; Chu et al. 2006; Page et al. 2008; Cenki-Tok et al. 2009; Holmden and Bélanger 2010; Farkaš et al. 2011; Hindshaw et al. 2011, 2012; Bagard et al. 2013; Hindshaw et al. 2013; Moore et al. 2013), water (Zhu and MacDougall 1998; Schmitt et al. 2003a; Chu et al. 2006; Jacobson and Holmden 2008; Tipper et al. 2006b, 2008a; Cenki-Tok et al. 2009; Holmden and Bélanger 2010; Tipper et al. 2010a; Hindshaw et al. 2011; Bagard et al. 2013; Hindshaw et al. 2013; Moore et al. 2013; Nielsen and DePaolo 2013; Wiegand and Schwendenmann 2013; Oehlerich et al. 2015) and atmospheric (Schmitt et al. 2003a; Schmitt and Stille 2005; Wiegand et al. 2005; Ewing et al. 2008; Cenki-Tok et al. 2009; Holmden and Bélanger 2010; Farkaš et al. 2011; Hindshaw et al. 2011; Fantle et al. 2012; Bagard et al. 2013; Wiegand and Schwendenmann 2013) samples are shown. All literature data has been corrected for interlaboratory bias, ^{40}Ca contribution and converted to $\delta^{44/40}$Ca$_{SRM915a}$ (‰). For each sample type, the average Ca isotopic composition, corresponding 2SE and the number of samples are given. Only field samples are presented

Table 3 Masses of the different reservoirs of Ca

Reservoir	Compartment	Size (t)	Reference	Comment	Ca concentration	Reference	Mass of Ca (t)	Reference	Type	$\delta^{44/40}$Ca range of variation (‰)	Average $\delta^{44/40}$Ca (‰)	2SE	N
Biosphere	Vegetation	1.8×10^{12}	Faurie (2011)		0.5 % dry weight	Taiz and Zeiger (2010)	9.0×10^{9}		Biosphere	−2.2 to 1.76	0.01	0.10	173
Soil	$81–5000 \times 10^{9}$			By assuming 1.5×10^{8} km^2 soil coverage, an average thickness of 0.05–3 m, and an average density of 1.3 g/cm^3	3.50 %	Berner and Berner (1996)	$2.8–175 \times 10^{9}$		Bulk	0.21 to 1.55	0.90	0.05	66
									Acid-leachable	−0.74 to 1.32	0.63	0.20	30
									Exchangeable	−0.34 to 2.77	0.72	0.21	39
									Soil solutions	0.42 to 1.77	0.93	0.08	43
Hydrosphere	Ocean	1400×10^{15}	Berner and Berner (1996)		412 mg/l	Berner and Berner (1996)	0.577×10^{15}		Ocean	1.77 to 2.23	1.89	0.02	62
	Rivers	1.7×10^{12}	Berner and Berner (1996)		13.4 mg/l	Berner and Berner (1996)	0.023×10^{9}		Carbonate rivers	0.57 to 1.70	1.08	0.14	20
									Silicate rivers	0.17 to 1.55	0.85	0.03	254
									Large rivers	0.67 to 1.25	0.86	0.06	24
	Groundwater	15.3×10^{15}	Berner and Berner (1996)		0.01–450 mg/l	Chapelle (2003), Dowling et al. (2003)	$0.153–6885 \times 10^{9}$		Groundwater	0.17 to 2.08	0.86	0.10	65

(continued)

Table 3 (continued)

Reservoir	Compartment	Size (t)	Reference	Comment	Ca concentration	Reference	Mass of Ca (t)	Reference	Type	$\delta^{44/40}$Ca range of variation (‰)	Average $\delta^{44/40}$Ca (‰)	2SE	N
Rocks	Sedimentary	1.97×10^{18}	Veizer and Mackenzie (2003)		6.55 %	Turekian and Wedepohl (1961)	129×10^{15}		Sedimentary	−0.77 to 1.73	0.64	0.09	78
									Carbonates		0.63	0.09	57
	Igneous	0.8×10^{18}	Veizer and Mackenzie (2003)		0.56 %	Turekian and Wedepohl (1961)	4.48×10^{15}		Plutonic	0.31 to 1.13	0.75	0.10	19
									Metamorphic	0.57 to 1.32	0.81	0.17	11
									Volcanic	0.61 to 1.34	0.89	0.03	102
									Average igneous		0.80	0.20	
Ocean	CaCO$_3$ sedimentation						34×10^{15}	Wilkinson and Algeo (1989)	Phanerozoïc	−1.09 to 1.70	0.59	0.02	1979
									Modern	−0.18 to 1.70	0.76	0.03	538

Table 4 Fluxes between the different reservoirs of Ca

Reservoir	Compartment	Sub-compartment	Mass flux of water (t/year)	Reference	Ca concentration	Comment	Reference	Ca flux (t/year)	Reference	Type	$\delta^{44/40}Ca$ range of variation (‰)	Average $\delta^{44/40}Ca$ (‰)	2SE	N
Atmosphere	Wet deposition	Continental	0.107×10^{15}	Berner and Berner (1996)	0.1–3.0 mg/l	–	Berner and Berner (1996)	$0.011–0.321 \times 10^{9}$		Rain	−0.34 to 1.30	0.70	0.16	20
		Marine and coastal	0.398×10^{15}	Berner and Berner (1996)	0.2–1.5 mg/l	–	Berner and Berner (1996)	$0.080–0.597 \times 10^{9}$		Snow	0.22 to 1.29	0.76	0.18	12
										Throughfall	0.29 to 0.80	0.61	0.32	3
	Dust		$1–3 \times 10^{9}$	Hinds (1999)	2.56 %	Assumed similar to average continental crust	Rudnick and Gao (2003)	$26–77 \times 10^{6}$		Bulk	−0.42 to 0.93	0.69	0.15	16
										Water-soluble	0.77 to 1.36	1.11	0.07	22
										Acid-soluble	0.64 to 0.92	0.78	0.03	22
	Sea salt particles		$1–10 \times 10^{9}$	Hinds (1999)	0.412 g/kg	Assumed similar to seawater	Berner and Berner (1996)	$0.412–4.12 \times 10^{6}$						
	Evaporation from the ocean		0.434×10^{15}	Berner and Berner (1996)	0.412 g/kg	Assumed similar to seawater	Berner and Berner (1996)	0.179×10^{12}		Seawater		1.89	0.02	62
	Evaporation from the continent		0.071×10^{15}	Berner and Berner (1996)	13.4 mg/l	Assumed similar to average river water	Meybeck (1979)	0.95×10^{9}						
	Seawater vapour transported to the continent		0.036×10^{15}	Berner and Berner (1996)	0.412 g/kg	Assumed similar to seawater	Berner and Berner (1996)	0.015×10^{12}						

(continued)

Table 4 (continued)

Reservoir	Compartment	Sub-compartment	Mass flux of water (t/year)	Reference	Ca concentration	Comment	Reference	Ca flux (t/year)	Reference	Type	$\delta^{44/40}$Ca range of variation (‰)	Average $\delta^{44/40}$Ca (‰)	2SE	N
Hydrosphere	Rivers to ocean									Large rivers (mouth)	0.67 to 1.25	0.86	0.06	24
		Dissolved	0.53×10^9	Berner et al. (1983)				0.528×10^9	Berner and Berner (1996)	Weathering flux				
		Particulate load	333×10^6	Berner et al. (1983)				0.333×10^9	Berner and Berner (1996)	Particulate load				
		Sedimentary rocks	0.45×10^9	Berner et al. (1983)				0.452×10^9	Berner and Berner (1996)	Sedimentary rocks				
		Crystalline rocks	0.076×10^9	Berner et al. (1983)				0.076×10^9	Berner and Berner (1996)	Igneous rocks				
	Groundwater to ocean							$23–53 \times 10^6$	Chaudhuri and Clauer (1986), Zekster and Dzhamalov (1988), Taniguschi et al. (2002)					
Ocean	Cation exchange	Clay minerals						$37–38 \times 10^6$	Berner and Berner (1996), Drever (1988)					
	Hydrothermal							$0.08–0.8 \times 10^9$	Drever (1988), Wilkinson and Algeo (1989), Milliman (1993), de Villiers (1998)		0.79 to 1.86	1.44	0.14	22
	Diagenesis	From marine sediments						144×10^6	Drever (1988)					
	Sedimentation	production flux						2.1×10^9	Milliman (1993)					
		Precipitation						1.3×10^9	Milliman (1993)					
		Dissolution						0.84×10^9	Milliman (1993)					
	Burial of pore water							0.8×10^6	Drever (1997)	Pore waters	0.19 to 1.92	1.31	0.07	103

throughout the entire ocean, hence they are never limiting to the growth of organisms (mainly corals, foraminifera and coccolithophorids) which use Ca to produce shells and hard parts (De La Rocha et al. 2008). The Ca isotopic composition of seawater is probably one of the best determined reservoirs on the planet, with a measured $\delta^{44/40}$Ca value of 1.89 ± 0.02, $(2\sigma_{mean}$, N = 62; see Table 4), and as such is the isotopically heaviest reservoir of Ca on Earth, and has the most homogeneous composition compared with all the other data in the present chapter.

2.2.2 Rivers

The total reservoir of riverine solute Ca has been estimated at $\sim 0.23 \times 10^9$ t by combining the weight of river water on Earth (1.7×10^{12} t) and the world average river water Ca concentration (13.4 mg/l) (Table 2, Berner and Berner 1996). Whilst this reservoir is relatively small, its significance lies in the fact that riverine and groundwaters link the continental and oceanic cycles of Ca. In particular, whether the Ca was originally derived from the dissolution of silicate or carbonate minerals has a significant role in Earth's carbon budget, and in maintaining an equable climate over million year timescales (Sect. 1). In many natural waters Ca is one of the principal cations (in it's ionic form as Ca^{2+}). For example, in the Yangtze River, one of the world's largest, Ca accounts for 70 % of the total annual cation flux (Li et al. 2011). There is a significant body of work that has attempted to (1) determine the mineral source of Ca (principally trying to partition carbonate and silicate inputs as this has an impact on long-term climate), (2) determine the level of recycling of Ca within a catchment, by for example storage of Ca in the biosphere or secondary phases, and (3) characterise the size and composition of sediment or fine particles exported from catchments.

Partitioning Mineral Inputs of Ca to the Critical Zone

Partitioning silicate and carbonate derived Ca has been a fundamental goal of many river chemistry studies. There are several methods for partitioning carbonate and silicate Ca, based on forward models, inverse models and using Sr isotopes as a proxy for Ca. Ca isotopes therefore potentially hold the promise of tracing silicate and carbonate Ca directly (e.g., Jacobson et al. 2015). A particular novelty in this respect is the use of radiogenic ^{40}Ca as a direct tracer of lithology (Davenport et al. 2014). In the Yangtze River, it is estimated (based mainly on the sodium budget) that 83 % of the Ca is derived from the dissolution of carbonate minerals, and 17 % from the dissolution of silicate minerals (Li et al. 2011). Indeed, at a global scale it is thought that Ca is mainly derived from carbonate weathering, with several estimates suggesting that silicate weathering only accounts for 10–33 % of total Ca fluxes (e.g., Gaillardet et al. 1999b; Meybeck 1987; Berner and Berner 1996; Tipper et al. 2010a; Moore et al. 2013; Moon et al. 2014). However, most of this work has relied on indirect tracers of silicate Ca, such as sodium content in the river, combined with a Ca/Na ratio in the silicate rock being weathered (e.g., Galy and France-Lanord 1999), such that:

$$Ca_{sil} = Na \cdot (Ca/Na)_{rock} \qquad (1)$$

and

$$Ca_{carb} = Ca_{river} - Ca_{sil} \qquad (2)$$

or based on Sr isotopes, which have proven useful at partitioning carbonate and silicate inputs (e.g., Bickle et al. 2015).

Rivers Draining Silicate Rocks

A significant body of work has made measurements of $\delta^{44/40}$Ca in silicate small and larger-scale catchments, yielding a Ca isotopic variability of waters ranging from 0.17–1.55 ‰ and an average equal to 0.85 ± 0.03 ‰ $(2\sigma_{mean}$; N = 254, Fig. 2; Table 2) (Schmitt et al. 2003a; Chu et al. 2006; Tipper et al. 2006b, 2008a, 2010a; Cenki-Tok et al. 2009; Holmden and Bélanger 2010; Hindshaw et al. 2011, 2013; Bagard et al. 2013; Moore et al. 2013; Nielsen and DePaolo 2013; Wiegand and Schwendenmann 2013; Jacobson et al. 2015). Some very dilute first order streams from the west of

Scotland have $\delta^{44/40}$Ca values that are unusually high (mean is 1.28 ‰) compared to most other rivers draining silicate rock. Although it is tempting to speculate that the Ca in these rivers has been influenced by rain or sea-salt (seawater has $\delta^{44/40}$Ca of 1.89 ‰, see Sect. 2.2.1), the Sr isotope ratios of the same samples argue strongly against this (Tipper et al. 2010a) making it plausible that these rivers are significantly influenced by fractionation processes such as precipitation of secondary phases or uptake by plants during weathering. However, elucidating any fractionation control on riverine $\delta^{44/40}$Ca in rivers draining silicate rocks has so far been difficult (e.g., Hindshaw et al. 2011; Bagard et al. 2013)

Rivers Draining Carbonate Rocks

As discussed previously, it is likely that most of riverine Ca is derived from the weathering of sedimentary rocks, mainly carbonates. Fewer studies have analysed rivers from catchments that are dominated by carbonate rocks but Tipper et al. (2006b, 2008a, 2010a) analysed small rivers draining small catchments where carbonate weathering is significant. In catchments draining mainly carbonate rocks, Ca isotopic variability ranges from 0.57 to 1.70 ‰ with an average equal to 1.08 ± 0.14 ($2\sigma_{mean}$; N = 20, Fig. 2; Table 2). (Schmitt et al. 2003a; Tipper et al. 2006b, 2008a, 2010a). When small rivers draining limestone are compared to limestone rock there is a hint that the rivers have $\delta^{44/40}$Ca values that are offset from limestone rocks (Tipper et al. 2006b, 2008a, 2010a). In Himalayan rivers, Tipper et al. (2006b) proposed that the isotopically heavy values of the rivers could be reconciled by ubiquitous secondary carbonate precipitation (with lower Ca isotope ratios than the waters), a phenomenon known to occur in Himalayan rivers (Galy et al. 1999; Jacobson et al. 2002; Bickle et al. 2005). Ca isotope data from small rivers draining limestones from southern and Eastern France, show a slightly more complex behaviour (Tipper et al. 2010a). Small rivers draining the Jura mountains from Eastern France show variable $\delta^{44/40}$Ca (range is 0.6 ‰). The

stream, Choranche, sampled as it exited a karstic network of the Vercors in the South of France, was observed to be precipitating secondary calcite very close to the sampling site, and indeed has an elevated $\delta^{44/40}$Ca of 1.2 ‰. In contrast, the Huveane River in Provence, was also sampled close to abundant travertine precipitation, but has a low $\delta^{44/40}$Ca value of 0.6 ‰. One difficulty with interpreting these variable spot samples results is that the bedrock $\delta^{44/40}$Ca values are not known for the specific localities, which are potentially variable. Tipper et al. (2010a) pointed out that there is no simple relationship between $\delta^{44/40}$Ca values and the calcite saturation index, for most rivers. Moore et al. (2013) suggested that there is a link between carbonate weathering and tectonic activity. Indeed, they showed that 50–60 % of Ca originates from carbonates in regions where uplift rates are the lowest (0.1 mm/year) whereas >90 % of Ca comes from carbonates when the uplift rates are maximal (10 mm/year). Exceptions are glacial rivers where the Ca can be derived non-stoichiometrically from silicates. However, when extrapolated to other, similar regions, such as the Himalaya, the authors conclude that the spatial extent of mountainous glaciation is not widespread enough to have a global impact. Further systematic work on Ca isotopes on rivers draining limestone is required. There are even fewer $\delta^{44/40}$Ca data of rivers draining dolostone catchments but Tipper et al. (2008a) showed that the Ca isotope compositions of these rivers was consistent with the Ca isotope data of the rock.

Large Rivers

In large rivers sampled at their mouth, Ca isotopes vary from 0.67 to 1.25 ‰ with an average equal to 0.86 ± 0.06 ‰ ($2\sigma_{mean}$; N = 24, Fig. 2; Table 2, Schmitt et al. 2003a; Tipper et al. 2006b, 2008a, 2010a; Zhu and Macdougall 1998). It results from Sects. 2.1.1 and 2.1.2 that carbonate and silicate rocks are isotopically significantly different (0.63 ± 0.19 ‰ and 0.80 ± 0.20 ‰, respectively). Given that the majority of Ca at a global scale is thought to derive from carbonate weathering, this implies a theoretical Ca-flux weighted value of 0.63–0.69 ‰ for the average river, significantly lower than the measured value of $\delta^{44/40}Ca$ in rivers of 0.86 ± 0.06 ‰ making the Ca isotopic data at a global scale difficult to reconcile with the partitioning of carbonate and silicate weathering based on conventional methods (Fantle and Tipper 2014; Tipper et al. 2010a). The Ca isotope data could be used to suggest that silicate isotopic compositions dominate the Ca isotopic signature of river flux. This discrepancy of ∼0.20 ‰ is unlikely to be caused by a fundamental error in the partitioning of carbonate and silicate weathering based on other methods. Potentially the data are not representative, but is more likely caused by a continental Ca cycle that is not at steady state. For example, a global net growth of the biosphere could theoretically reconcile the differences or alternatively there remain fractionation mechanisms that need to be characterized in the weathering environment (Fantle and Tipper 2014). More recently Jacobson et al. (2015) suggested that mixing processes rather than fractionation mechanisms could take place at global scale, implying that more Ca is derived from silicate rocks than previously thought.

At present there are three potential causes of this difference between the calculated global riverine $\delta^{44/40}Ca$ and the measured value, none of which are mutually exclusive, and all of which act in the same direction in terms of isotopic fractionation:

1. **Growth or decay of the biosphere**. While the Ca isotopic composition of rivers is relatively restricted, and there is minimal evidence for Ca isotopic fractionation during chemical weathering, the Ca isotopic composition of plants is highly variable (3 ‰ range). The mean value of natural plant material measured to date is 0 ‰, or ∼1 ‰ lower than mean silicate rock and ∼0.6 ‰ lower than mean carbonate rock (the mineral sources of Ca to rivers). Plants are enriched in ^{40}Ca ($\Delta_{plant-source} \sim -0.7$ ‰) relative to nutrient solutions or soil pore fluids (e.g. Schmitt et al. 2003a; Bullen et al. 2004; Wiegand et al. 2005; Page et al. 2008;

Cenki-Tok et al. 2009; Holmden 2009; Cobert et al. 2011a, b; Farkaš et al. 2011; Hindshaw et al. 2012). This is, potentially, reflected in the $\delta^{44/40}Ca$ of soil pore fluids, the average of which is 0.96 (or ∼+1 ‰ relative to mean vegetation). Fantle and Tipper (2014) explored the impact of changes in the biospheric reservoir of Ca with some simple box model calculations and demonstrated that in principle, growth of the biosphere could drive the mean river $\delta^{44/40}Ca$ to values that are higher than expected.

2. **Growth of the pedogenic soil reservoir: pedogenic calcite**. The formation of secondary carbonates such as pedogenic carbonates, travertines, and calcretes could lead to a fractionated output from the continents (Tipper et al. 2006b). It has been fairly well established through experimental work that Ca isotopes are fractionated during abiotic $CaCO_3$ precipitation, favoring the preferential incorporation of the light nuclide (Gussone et al. 2003; Lemarchand et al. 2004; Marriott et al. 2004; Gussone et al. 2005; Tang et al. 2008; Gussone et al. 2011; Tang et al. 2012). Thus, it is feasible that $CaCO_3$ precipitation in catchments can sequester isotopically light Ca, driving exported Ca to heavier values and potentially creating an isotopically light, relatively soluble mineral reservoir that can be susbsequently mobilized. If Ca isotopes can be used to trace secondary carbonate formation in the terrestrial realm, then this would represent an important tool for elucidating the terrestrial carbon cycle, specifically the impact of terrestrial carbonates on the silicate weathering cycle and climate.

Such a scenario is potentially reflected in the Ca isotopic composition of some Himalayan catchments (referred to above), where the formation of secondary carbonates (i.e., travertines and calcretes) sequesters more than twice the mass of the dissolved Ca export flux on an annual basis (Jacobson et al. 2002; Bickle et al. 2005; Tipper et al. 2006b, 2008a). Tipper et al. (2006b) reported travertines enriched in the light nuclide compared to surrounding limestone and local river waters. Despite the very small range in riverine $\delta^{44/40}$Ca values in large rivers, Tipper et al. (2010a) showed that several rivers show the same systematic trend between high and low water stand for both Ca concentrations, Mg/Ca ratios and $\delta^{44/40}$Ca values. At high water stand, the Chiang-Jiang, Huanghe and Ganges Rivers show lower Ca concentrations (between a factor of 0.3 and 0.8 lower). $\delta^{44/40}$Ca values are also lower in the wet season in each of these rivers, implying that the Ca has a different source between the wet and dry season, and is not merely diluted by high runoff. When seasonal changes in $\delta^{44/40}$Ca are compared to Mg/Ca ratios (which normalises out any dilution effects due to higher runoff in the wet season), a correlation was observed for several rivers (Tipper et al. 2010a). Both Mg/Ca ratios and $\delta^{44/40}$Ca values are higher in the dry season. Such trends would be consistent with secondary calcite precipitation in soils or calcretes as has been observed in Himalayan rivers (Jacobson et al. 2002; Tipper et al. 2006c, 2008a), where up to 70 % of Ca removal fractionates both Mg/Ca ratios and Ca isotope ratios. Quantitive interpretations are complicated by the rate dependence of fractionation factor, and to a lesser extent temperature and/or solution chemistry, yet highlights the importance of constraining the kinetics of mineral precipitation in natural systems. The seasonal response in the processes causing fractionation could have implications for how $\delta^{44/40}$Ca may respond to changing climate over longer time-scales such as glacial-interglacial periods where the hydrological cycle is significantly re-organised. Continental storage of Ca (and carbon) is thought to be 35–40 % higher in interglacial periods (Adams and Post 1999) and could influence the Ca isotope composition of rivers. However, although very small seasonal variations are observed in some rivers, the overall homogeneity of the data makes the average $\delta^{44/40}$Ca of the Ca flux to

the oceans from weathering well constrained, and is of particular use in bringing new constraints to the oceanic budget of Ca.

3. **Changes in the size of the soil exchangeable reservoir.** Although little documented there is evidence that suggests that the exchange pool of Ca fractionates the isotopes, with the exchange pool having an affinity for the light isotope (Ockert et al. 2013). Growth in the exchange pool associated with the growth of soils could potentially drive rivers to heavier values than expected. However, in the continental environment, the $\delta^{44/40}$Ca of the exchange pool depends on both the source of the Ca as well any mass dependent fractionation. The size of the exchange pool is essentially controlled by the cation exchange capacity which is largely controlled by clay minerals and organic surfaces.

River Sediments and Colloids

Ca can be transported in dissolved, particulate or colloidal form. Particulates (0.4 μm–0.5 mm) and dissolved forms (<0.4 μm) are the two phases that have been mostly considered. It should be noted that what is generally meant by dissolved flux includes what is sensu stricto dissolved (<1 kDa) as well as the colloidal phase (1 kDa–0.4 μm). Dupré et al. (1996) have shown that in rivers from the Congo catchment, Ca is partitioned between dissolved (80–90 %) and particulate phases. Similarly, Braun et al. (2005) showed for a small Cameroon catchment, that Ca is essentially transferred in the dissolved form, but both of these catchments are supply limited (West et al. 2005). In contrast, at a global scale, although particulate fluxes dominate dissolved fluxes for most elements, the Ca budget is almost equally partitioned by dissolved and particulate fluxes (Oelkers et al. 2011). This is likely because of the high carbonate content of these sediments, and the reactivity of carbonates. This is reflected in the suspended sediment composition of the world's largest rivers which are commonly depleted in Ca relative to the upper continental crust (Gaillardet et al. 1999a).

Similarly, Ca is mainly found as dissolved cation in river waters, although significant contents (between 16 and 30 %) of colloidal Ca has been observed particularly in boreal regions (Dahlqvist et al. 2004; Bagard et al. 2011; Pokrovsky et al. 2010, 2012) where there is a high organic content in the rivers. Ca can be adsorbed on negatively charged surfaces of colloids or bound by fibrillar polysaccharides extracted by algae and bacteria (Dahlqvist et al. 2004). This work emphasized the importance of colloidal Ca as a control of Ca-mobility in rivers, since up to 16 % of Ca is linked to colloidal phases in rivers like the Amazon or the boreal Kalix river. Pokrovsky et al. (2010) reported higher colloidal Ca contents for boreal river water in summer than in winter, due to the fact that dissolved organic matter is more pronounced during the spring flood than in winter. This has been confirmed recently by using Ca isotopes in the Kulingdakan watershed: Bagard et al. (2013) showed that the Ca biogeochemical cycle is controlled by the degradation of the biomass and by the presence of colloids.

2.2.3 Groundwaters

Whilst there is considerable knowledge about the Ca flux from rivers, and their attendant Ca isotopic compositions, far less is known about groundwaters. Groundwater dissolved Ca concentrations can be higher than in surface waters, such that a small input of groundwater to the oceans could be the source of significant Ca (Milliman 1993). A recent review article attempting to estimate the submarine groundwater discharge pointed to the lack of data from coastal zones in almost all parts of the world (Taniguchi et al. 2002). The total global mass of groundwater is thought to be 15.3×10^{15} t (Berner and Berner 1996), but its average Ca concentration is difficult to estimate due to the chemical heterogeneity of most aquifers. The concentration of Ca in water depends on the residence time of the water and the lithology. A very large range in Ca concentrations have been reported in the literature (from 0.01 to 450 mg/l (Table 2) Chapelle 2003; Dowling et al. 2003).

By using the end-member Ca concentration values, it is possible to estimate the Ca masses for the groundwater reservoir ranging between 0.153 and 6890×10^9 t. There is very little isotopic data available. Ground and spring water $\delta^{44/40}$Ca values range from 0.17 to 2.08 ‰, with an average $\delta^{44/40}$Ca equal to 0.86 ± 0.10 ‰ ($2\sigma_{mean}$; N = 65) (Fig. 2; Table 2) (Schmitt et al. 2003a; Jacobson and Holmden 2008; Tipper et al. 2008a; Cenki-Tok et al. 2009; Holmden et al. 2012; Holmden and Bélanger 2010; Hindshaw et al. 2011, 2013; Wiegand and Schwendenmann 2013; Nielsen and DePaolo 2013; Jacobson et al. 2015).

The groundwater contribution of Ca to seawater is controversial. The groundwater flux has often been considered as a variable in models, and adjusted such that the present-day ocean is in steady-state for Ca content and Ca isotopes (Holmden et al. 2012). It has thus been considered either negligible (Mackenzie and Morse 1992), or as high as 5–10 % of the mass carried by river waters (Chaudhuri and Clauer 1986; Zekster and Dzhamalov 1988; Taniguchi et al. 2002), which implies a Ca flux to the ocean equal to $26–53 \times 10^6$ t/year. Holmden et al. (2012) calculated a submarine groundwater $\delta^{44/40}$Ca value equal to 0.65 ± 0.23 ($2\sigma_{mean}$) on a global scale (Fig. 4).

2.3 Biospheric Cycling of Ca

The mass of Ca in the biosphere depends on how it is estimated. The total mass of Ca in the global biomass can be estimated from the average weight of dry biomass of 1.8×10^{12} t (Faurie et al. 2011) and assuming that Ca represents 0.5 % of this dry weight (Taiz and Zeiger 2010), yielding a mass of 9.2×10^9 t Ca. This value is lower than the one obtained by Fantle and Tipper (2014) who considered fresh biomass. Isotopically vegetation displays the largest variation on Earth (about 4.1 ‰), ranging from -2.2 to 1.76 ‰, with an average value equal to 0.01 ± 0.10 ‰ ($2\sigma_{mean}$; N = 173) (Fig. 2; Table 2) (Skulan and DePaolo 1999; Schmitt et al. 2003a; Wiegand et al. 2005; Chu et al. 2006; Page et al. 2008; Cenki-Tok et al. 2009; Holmden and Bélanger 2010; Farkaš et al. 2011; Hindshaw et al. 2011; Bagard et al. 2013; Moore et al. 2013; Bullen et al. 2004; Hindshaw et al. 2012).

When the substrate is poor in Ca and atmospheric inputs are small, recycling by the biosphere (or decay) plays an important role in providing nutrients to new vegetation that are required for growth. There is thus a balance between the relative proportion of the removal of Ca by plants and the recycling of Ca via the degradation of the litter. This delicate balance can be significant for forest management when harvesting trees for maintaining the fertility of forest soils. After leaf and needle fall, Ca is re-mineralized in the litter, composed of leaves, needles, branches, twigs, fruits, and migrates into soil solutions where some of it becomes bio-available (Likens et al. 1998; Dijkstra 2003; McLaughlin and Wimmer 1999; Taiz and Zeiger 2010). The litter is decomposed through the action of macro-fauna (worms, insects, spiders, snails) which fragments or digests it; increasing the reactive surface area for the colonisation by microorganisms (bacteria, fungi).

The rate of litter decomposition and nutrient release can regulate the energy flow, the primary productivity and the nutrient cycling in forest ecosystems (Bray and Gorham 1964; Liao et al. 2006; Olson 1963). It is mainly influenced by climatic parameters (temperature, humidity, rainfall, soil moisture), the type of litter, the nature and abundance of the decomposing organisms, the vegetal species (conifer or hardwood, type of hardwood) (Bernhard-Reversat 1972; Coûteaux et al. 1995; Zeller and Martin 1998). Most rapid litter decomposition occurs for species having maximal contents in ash and nitrogen, and minimal content in lignin. For instance, leaves disappear much faster than branches or twigs (John 1973; Rochow 1974), the half-lives of leaves and barks being equal to 0.9 and 5.4 years, respectively, in a temperate climate forest from Uruguay (Hernández et al. 2009). This has an influence of the $\delta^{44/40}$Ca of recycling since leaves and needles are enriched in ^{44}Ca compared to the trunk or the roots (see Chapter "Earth-Surface Ca Isotopic Fractionations").

2.4   Soils

Soils represent an important pool of the critical zone (NRC 2001). They result from the weathering of rocks by physical, chemical and biological processes and represent weathering gradients from deep mineral soil layers that have compositions similar to the bedrock towards the surface which can be highly altered, the transition depth being ill-defined and differing from one environment to another (Richter and Markewitz 1995). The amount of Ca released from the weathering of rocks and minerals is dependent on the nature of the parent rock (Fichter et al. 1998; Dijkstra and Smits 2002; Dijkstra 2003). In surficial soils there are significant masses of decomposing biomass through the action of micro-organisms (bacteria and fungi). During this process, Ca is solubilized and is released to soil solutions. External sources (precipitation, throughfall, supercortical flow, and dust) also contribute to the Ca content of the soil (Fig. 3).

It is estimated that 29 % (or 1.5×10^8 km^2) of the Earth's surface is covered by soils, based on knowledge of global land areas. At a global scale soil depth varies with topography, bedrock, climate, biological, physical and chemical processes that are involved (Summerfield 1997; Pelletier and Rasmussen 2009; Nicótina et al. 2011), but average depths have been mapped and is it possible to distinguish shallow soils (<10 cm) from deep ones >3 m (FAO 2007). Bulk soil density mainly depends on the mineral content of the considered soil and its degree of compaction. Most mineral soils, have bulk densities between 1.1 and 1.6 g/cm^3 (Manrique and Jones 1991). Some organic soils have bulk densities below 1.0 g/cm^3 (Schaetzl 2005). By assuming a thickness comprised between 0.05 and 3 m and an average density of 1.3 g/cm^3, it is possible to estimate a global soil inventory of between 81 and 5000×10^9 t. By assuming an average Ca content of 3.5 % (Berner and Berner 1996), an estimated size of the soil Ca pool is between 2.8–175×10^9 t, approximately four to six orders of magnitude smaller than that contained in rocks.

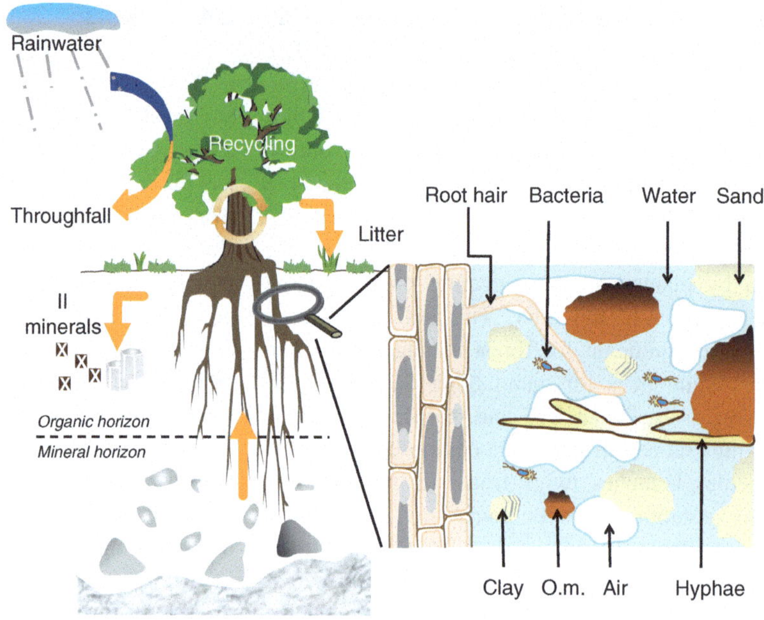

Fig. 3 Reservoirs involved in the Ca biogeochemical cycle at the water-soil-plant interface. Soils are made up of organic and mineral matter, water and gaseous phases that are derived from the chemical reactions and biological activities in soil, as well as from the biological decay of organic matter

Bulk soils show a significant range in $\delta^{44/40}$Ca, from 0.21 to 1.55 ‰, with an average of 0.90 ± 0.05 ‰ ($2\sigma_{mean}$; N = 66) (Fig. 2; Table 2) (Wiegand et al. 2005; Tipper et al. 2006b; Farkaš et al. 2011; Hindshaw et al. 2011; Bagard et al. 2013). In contrast, 1 N HNO$_3$ leachable fractions are enriched in the light ^{40}Ca isotope with a variability comprised between −0.74 and 1.32 ‰, and an average value equal to 0.63 ± 0.20 ‰ ($2\sigma_{mean}$; N = 30) (Fig. 2) (Perakis et al. 2006; Page et al. 2008; Holmden and Bélanger 2010; Farkaš et al. 2011; Hindshaw et al. 2011; Bagard et al. 2013). Soil exchangeable fractions vary from −0.34 to 2.77 ‰, with an average value equal to 0.72 ± 0.21 ‰ ($2\sigma_{mean}$; N = 39, Fig. 2; Table 2) (Wiegand et al. 2005; Perakis et al. 2006; Page et al. 2008; Holmden and Bélanger 2010; Farkaš et al. 2011; Hindshaw et al. 2011; Bagard et al. 2013; Moore et al. 2013). Soil solutions (pore waters) have also been collected and analyzed by several studies. They range from 0.42 to 1.77 ‰, with an average value equal to 0.93 ± 0.08 ‰ ($2\sigma_{mean}$; N = 43, Fig. 2; Table 2, Schmitt et al. 2003a; Cenki-Tok et al. 2009; Holmden and Bélanger 2010; Hindshaw et al. 2011; Gangloff et al. 2014).

Soils are complex and can be compartmentalised into distinct zones, where either there are distinct pools of Ca, or are associated with particular processes. These are considered separately in the subsequent sections.

2.4.1 Weathering of Primary Minerals

Carbonate rocks cover only about 10–14 % of the continents (Dürr et al. 2005; Ford and Williams 2007; Hartmann and Moosdorf 2012) but chemical weathering of carbonate rocks provides 45–60 % of the total solutes provided by rock weathering (Meybeck 1987; Gaillardet et al. 1999b; Moon et al. 2014). This feature is due to the high solubility of Ca–Mg carbonates, their faster kinetics of dissolution compared to silicates (Tipper et al. 2006a) and the congruent nature of carbonate weathering.

In silicate dominated soils plagioclase, apatite and calcite are the main Ca-contributors to soils (e.g., Likens et al. 1998; Probst et al. 2000; White et al. 2008; Aubert et al. 2001; Blum et al. 2002; Bakker et al. 2005). In granites, accessory phases such as calcite and apatite ($\sim$1 % of the total volume of rocks) can contribute to more than 90 % of the Ca export (e.g., White et al.

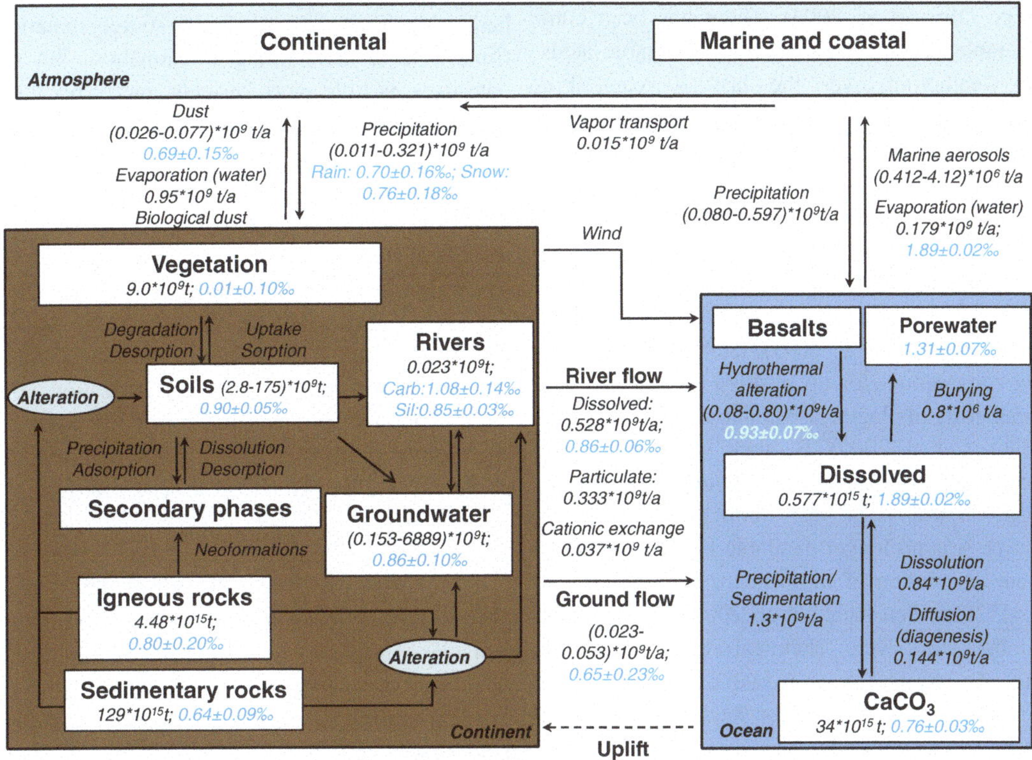

Fig. 4 Variability of Ca isotopic compositions in the ocean. Oceanic fluids (Zhu and MacDougall 1998; De La Rocha and De Paolo 2000; Schmitt et al. 2001; Hippler et al. 2003; Schmitt et al. 2003b; Farkaš et al. 2007; Gussone et al. 2007; Fantle and DePaolo 2007; Amini et al. 2008, 2009; Schmitt et al. 2009; Teichert et al. 2009; Huang et al. 2010; Turchyn and DePaolo 2011; Du Vivier et al. 2015; Fantle 2015; Jacobson et al. 2015), non-carbonated (Schmitt et al. 2003b; Soudry et al. 2004; Teichert et al. 2005, Amini et al. 2008; Komiya et al. 2008; Griffith et al. 2008a, c; Arning et al. 2009; Holmden et al. 2009; Du Vivier et al. 2015) and carbonated (Skulan et al. 1997; Zhu and MacDougall 1998; De La Rocha and DePaolo 2000; Fantle and Depaolo 2005; Kasemann et al. 2005; Sime et al. 2005; Steuber and Buhl 2006; Fantle and DePaolo 2007; Amini et al. 2008; Teichert et al. 2009; Payne et al. 2010; Blättler et al. 2011; Griffith et al. 2011; Gussone et al. 2011; Turchyn et al. 2011; Wang et al. 2012; Holmden et al. 2012; Jost et al. 2014; Kasemann et al. 2014; Du Vivier et al. 2015; Fantle 2015; Griffith et al. 2015; Husson et al. 2015) sediments, as well as biological taxa, classified by their phylum (Skulan et al. 1997; Zhu and MacDougall 1998; Skulan and DePaolo 1999; Nägler et al. 2000; Schmitt et al. 2003b; Chang et al. 2004; Fantle and DePaolo 2005; Gussone et al. 2005; Heuser et al. 2005; Immenhauser et al. 2005; Sime et al. 2005; Böhm et al. 2006; Hippler et al. 2006; Steuber and Buhl 2006; Farkaš et al. 2007; Sime et al. 2007; Griffith et al. 2008b; Heinemann et al. 2008; Gussone et al. 2009; Gussone and Filipsson 2010; Gussone et al. 2010; von Allmen et al. 2010; Blättler et al. 2012; Hinojosa et al. 2012; Holmden et al. 2012; Hippler et al. 2013; Pretet et al. 2013; Ullmann et al. 2013; Blättler and Higgins 2014; Jost et al. 2014; Brazier et al. 2015) are shown. Fluids include seawater, hydrothermal vents and pore fluids. Bulk carbonates include carbonate mud, limestone, chalk, dolostone and ooze samples. Mollusca contain bivalve, blue mussel, chiton, oyster, snail, limpet, belemnites and pteropods. Chordata represent conodont. Protozoa regroup benthic and planktonic foraminifera. Brachiopoda, Haptophyta, Porifera and Cnidaria correspond to brachiopods, coccolithophores, schlerosponges, and corals, respectively. Chlorophyta and Rhodophyta represent green and red algae respectively. Finally, Echinodermata contain sea urchin, echinoid spine and starfish samples. Every single literature value has been corrected for interlaboratory bias and converted to $\delta^{44/40}$ Ca$_{SRM915a}$ (‰). For each sample type, the average Ca isotopic composition, corresponding 2SE and the number of samples are given. Only field samples are presented

2005; Oliva et al. 2003). There has been considerable interest in the effects of organic acids on weathering over the last 20 years. For instance, biological weathering was shown to enhance dissolution in Ca-bearing minerals such as carbonates, phosphates or phyllosilicates, and to translocate metal cations from the mineral to the organic horizon, driven by evapotranspiration processes (Drever 1994; Hinsinger et al. 1995; Hinsinger and Gilkes 1996, 1997; Courty et al. 2010; Turpault et al. 2009). Biological weathering can account for up to 40 % of the total Ca weathering of plagioclase under acidic conditions (Bakker et al. 2005). Bedrock will provide the main source of Ca to the soil (and biosphere) for many thousands of years until the bedrock Ca source was no longer available because of complete dissolution of the primary minerals (Bern et al. 2005; Pett-Ridge et al. 2009), for example in the case were chemical weathering rate exceeds physical erosion rate meaning that no new material is exposed for weathering. In such cases where the upper reaches of soils become very depleted in primary minerals (due to intense leaching in tropical soils for example) the main source of Ca can become mineral inputs from the atmosphere (Jordan 1985; Kennedy et al. 1998; Chadwick et al. 1999) (Table 5). It is however worth noting that even in the absence of the weathering of primary minerals, the exchangeable pool can sustain the biosphere for several thousands of years (Bullen et al. 1997).

The $\delta^{44/40}$Ca derived from mineral weathering will be a function of both the $\delta^{44/40}$Ca of the minerals that are being dissolved, and any mass dependent processes that accompany weathering. Experimental and field studies have indicated that congruent low-temperature mineral dissolution does not fractionate Ca isotopes (e.g., Ryu et al. 2011; Cobert et al. 2011b; Hindshaw et al. 2011). Mass dependent fractionation does however occur during mineral formation both at high and low temperatures (for example carbonate and clay precipitation), though there is not much data for this in soils. Tipper et al. (2006b, 2008a) showed that secondary carbonates (travertines) had a lower $\delta^{44/40}$Ca than the source material (limestone) demonstrating fractionation, but to date there is little work on clay, partly because the Ca content of clay is low. However, no fractionation during dissolution combined with fractionation during the formation of secondary phases means that Ca isotopes could be used as a tracer of soil processes. Complications arise because primary minerals show some variation as discussed above either because of mass-dependent isotope fractionation associated with igneous differentiation and crystallization, or incorporation of a range of Ca isotopic compositions into the source materials of melts (e.g., Amini et al. 2009; Huang et al. 2010, 2011).

2.4.2 Soil Carbonate Biomineralization

Soil carbonate forms under arid to sub-humid climatic conditions. In general, it is found in relatively dry soils where grasses or mixed grasses and shrubs are the dominant vegetation. Under these conditions soil pH is generally 7 or above, in contrast to forested soils where pH is below 6. Authigenic soil carbonate is common in soils where mean annual rainfall is less than 75 cm, while it is rarely found in soils receiving more than 100 cm precipitation per year (Cerling 1984). There is a contrasting climatic response between soil carbonate and soil organic carbon. While organic carbon storage on land appears to have been much less than present during the cold, dry glacial maximum, calcrete, soil carbonate carbon storage would have been greater (Adams and Post 1999). The formation of secondary carbonates plays a vital role in the building of many soils and can be microbial to a greater or lesser extent (e.g., Verrecchia and Dumont 1996; Milliere et al. 2014). Many secondary carbonate fractions can thus be found in soil, as $CaCO_3$: shell fragments (gastropods, birds), indurated nests (coleoptera), excrements (termites), bio-spheroids (earthworms), calcite needles (fungi), rhizolites (calcifications). The bacteria also play a role in the precipitation of $CaCO_3$ by changing the local soil conditions. Finally, the combined action of plant-fungi

Table 5 Survey of worldwide seawater values reported in the literature[a]

Reference	Location	Depth (m)	$\delta^{44/40}Ca_{NIST\ SRM\ 915a}$ (‰)	2SD	Reference
SE England	NE Atlantic	0	1.92	0.10	Schmitt et al. (2001)
SE England	NE Atlantic	0	1.89	0.15	Hippler et al. (2003)
SE England	NE Atlantic		1.89	0.08	Schmitt et al. (2003a)
SE England	NE Atlantic	0	1.86	0.20	Schmitt et al. (2009)
SE England	NE Atlantic	0	1.91	0.10	Cobert et al. (2011a)
SE England	NE Atlantic	0	1.77	0.17	Bagard et al. (2013)
17 CTD-6 (CTD/rosette)	Mid-Atlantic	3500	1.81	0.17	Amini et al. (2008)
17 CTD-16 (CTD rosette)	Mid-Atlantic	2500	1.94	0.17	Amini et al. (2008)
217 CTD (CTD/rosette)	Mid-Atlantic	2660	1.80	0.01	Amini et al. (2008)
231 CTD-8 (CTD/rosette)	Mid-Atlantic	3038	1.89		Amini et al. (2008)
Mauritania	E Atlantic	0	1.86	0.10	Schmitt et al. (2001)
Sargasso Sea	NW Atlantic (31° 40′ N, 64° 10′ W)	300	1.82	0.19	De La Rocha and DePaolo (2000)
	Subantarctic Atlantic (46° 57′ S, 6° 15′ E)	4066	1.88	0.21	De La Rocha and DePaolo (2000)
Coast of Wales	NE Atlantic		1.88	0.12	Gussone et al. (2007)
NASS-2	N Atlantic (32° 10′ N, 64° 30′ W)	1300	1.78	0.15	Zhu and MacDougall (1998)
Florida reef tract	NW Atlantic		1.81	0.08	Holmden et al. (2012)
IAPSO	Modern Atlantic seawater	0	2.01	0.19	Hippler et al. (2003)
IAPSO	Modern Atlantic seawater	0	1.89	0.26	Chang et al. (2004)
IAPSO	Modern Atlantic seawater	0	1.80	0.30	Wieser et al. (2004)
IAPSO	Modern Atlantic seawater	0	1.83	0.54	Böhm et al. (2006)
IAPSO	Modern Atlantic seawater	0	1.85		Steuber and Buhl (2006)
IAPSO	Modern Atlantic seawater	0	1.86	0.30	Amini et al. (2008)
IAPSO	Modern Atlantic seawater	0	1.88	0.08	Jacobson and Holmden (2008)
IAPSO	Modern Atlantic seawater	0	1.96	0.12	Kasemann et al. (2008)
IAPSO	Modern Atlantic seawater	0	1.78	0.16	Amini et al. (2009)
IAPSO	Modern Atlantic seawater	0	1.77	0.10	Teichert el al. (2009)
IAPSO	Modern Atlantic seawater	0	1.86	0.20	Farkaš et al. (2007)
IAPSO	Modern Atlantic seawater	0	1.90	0.12	Huang et al. (2010)
IAPSO	Modern Atlantic seawater	0	1.90	0.11	Farkaš et al. (2011)
IAPSO	Modern Atlantic seawater	0	1.95	0.26	Müller et al. (2008)
IAPSO	Modern Atlantic seawater	0	1.82	0.16	Hippler et al. (2013)
IAPSO	Modern Atlantic seawater	0	1.84	0.09	Harouaka et al. (2014)
IAPSO	Modern Atlantic seawater	0	1.85	0.05	Jacobson et al. (2015)
IAPSO	Modern Atlantic seawater	0	1.86	0.04	Du Vivier et al. (2015)

(continued)

Table 5 (continued)

Reference	Location	Depth (m)	$\delta^{44/40}Ca_{\text{NIST SRM 915a}}$ (‰)	2SD	Reference
	Mediterranean seawater	0	2.02	0.14	Sime et al. (2005)
	Mediterranean seawater		1.78	0.40	Böhm et al. (2006)
	Central North Pacific (31° 28′ N, 136° 6′ W)	500	1.87	0.09	De La Rocha and DePaolo (2000)
Santa Barbara Basin	NE Pacific (34° 15′ N, 119° 55′ W)	0	1.89	0.26	De La Rocha and DePaolo (2000)
San Diego	NE Pacific (32° 52′ N, 117° 16′ W)	0	1.94	0.18	Skulan et al. (1997)
52 CTD	Pacific (North rift zone)	1637	1.92	0.15	Hippler et al. (2003)
sw#2	N Pacific (22° 31′ N, 122° 12′ W)	67	1.97	0.06	Zhu and MacDougall (1998)
sw#3	N Pacific (22° 31′ N, 122° 12′ W)	2184	1.90	0.13	Zhu and MacDougall (1998)
sw#4	S Pacific (58° S, 174° W)	63	1.88	0.07	Zhu and MacDougall (1998)
24 CTD	Indian	Deep sea	1.99	0.12	Hippler et al. (2003)
sw#5	Indian (6° 9′ S, 50° 54′ E)	43	1.87	0.06	Zhu and MacDougall (1998)
NASS-5	Unknown		1.95		Gussone et al. (2003)
	Unknown		1.95	0.10	Fantle and DePaolo (2005)
	Unknown		1.96	0.12	Kasemann et al. (2005)
	Unknown		2.23	0.14	Tipper et al. (2006b)
	Unknown		2.05	0.07	Fantle and DePaolo (2007)
	Unknown		2.02	0.08	Fantle and DePaolo (2007)
	Unknown		1.88	0.08	Jacobson and Holmden (2008)
	Unknown		2.01	0.18	Page et al. (2008)
	Unknown		1.93		Tipper et al. (2008a)
	Unknown		1.86	0.07	Holmden (2009)
	Unknown		1.87	0.33	Wombacher et al. (2009)
	Unknown		1.86	0.05	Holmden and Bélanger (2010)
	Unknown		1.97		Reynard et al. (2010)
	Unknown		1.88		Gussone et al. (2011)
	Unknown		1.97		Reynard et al. (2011)
	Unknown		1.86	0.07	Ruy et al. (2011)
	Unknown		1.81	0.08	Holmden et al. (2012)

[a]Literature data where converted to SRM 915a, if expressed against another standard (see Chapter "Analytical Methods"); we also employed $\delta^{44/40}Ca = 2.05 \times \delta^{44/42}Ca$ when necessary

interactions, bacteria in tropical soils leads to the formation of Ca oxalate (e.g., Verrecchia and Dumont 1996; Milliere et al. 2014; Martignier and Verrecchia 2013). Ca isotope fractionations have been recorded during precipitation of secondary minerals, such as travertine, sulfates or carbonates, which are enriched in ^{40}Ca compared to their source of Ca (Tipper et al. 2006b; Ewing et al. 2008), the same direction of fractionation that is observed in marine carbonates.

2.4.3 The Rhizosphere

The rhizosphere (the narrow region of soil that is directly influenced by root secretions and associated soil microorganisms) constitutes an important component in the soil for nutrient uptake. It is a micro-environment that corresponds to the volume of soil around living plant roots which is influenced by the root activity. It corresponds to a zone of high microbial activity (Hiltner 1904; Hinsinger 1998; Hinsinger et al. 2005, 2006; Högberg and Read 2006; Smith et al. 2002). Although it occupies a small volume of the soil, it plays a central role in the maintenance of the soil-plant system and influences the biogeochemistry of forest ecosystems (Leyval and Berthelin 1991; Gobran et al. 1998). However, the precise knowledge of the functioning of nutrients, such as Ca, in this part of the soil is still lacking. Hindshaw et al. (2012) suggested that the presence of mycorrhiza may shift the $\delta^{44/40}Ca$ of plant roots towards higher values, although the total percentage of mycorrhiza infections in the soil does not appear to influence $\delta^{44/40}Ca$. Similarly, a preliminary in vitro experiment pointed to differential uptake mechanisms and Ca isotopic fractionations by roots in presence or absence of soil bacteria (Cobert et al. 2011a, b). More recently, this was confirmed in the field (Gangloff et al. 2014). The presence of bacteria might facilitate the access of nutrients to the plant, suppressing a possible reservoir effect of the nutritive medium (e.g. it becomes non Ca-limited). Further work will be necessary to fully understand the function of mycorrhiza and bacteria in the Ca uptake process by roots.

2.4.4 Humic Substances

Ca can be bonded to organic complexes such as humic substances by a number of different mechanisms. They represent supramolecular aggregates negatively charged and formed by high molecular weight polymers, such as humic and fulvic acids with a hydrophobic interior and a hydrophilic exterior. Ca can bond with the organic functional groups (with non-covalent interactions such as van der Waal forces), which can protect the humic substances from degradation (Simpson et al. 2002). Previous studies have shown that many Ca-bridges exist at a pH of about 8; and Ca has a greater affinity for humic substances in fertile compared to nutrient poor soils (Clarholm and Skyllberg 2013). Batch sorption experiments in the laboratory have also shown that sorption of humic substances onto minerals can be modified in presence of Ca^{2+} and Mg^{2+} ions (Tipping and Hurley 1992). Ca can also make bonds between negatively-charged humic substances and clays during complexolyse reactions, making the clay-humus complex very resistant to chemical reactions. Finally, Ali and Dzombak (1996) have shown that sorption of Ca^{2+} increases with increasing pH, whereas an increase of ionic strength slightly decreases Ca^{2+} sorption. To date there is little or no data about how isotopes behave in many of these processes.

2.4.5 Soil Exchangeable Pool

As a consequence of the breakdown of primary minerals and loss of solutes the predominant neoformed minerals are clays, Fe–Al oxy-hydroxydes, pedogenic carbonates or Ca-oxalate biominerals (Nagy 1995; Jackson 1956; Sposito 1980). Many clays have permanent negative charges that result from isomorphic substitutions occurring between the constituent atoms of tetrahedral or octahedral clay layers. These charges are compensated by incorporation of cations within the interlayer sites of the clay, or on mineral surfaces (Mulder and Cresser 1994). Ca ions can also be exchanged within the interlayer space of clay minerals following reaction 3 (Mooney et al. 1952; Bonnot-Courtois 1982; Chiou and Rutherford 1997):

$$2K^+ + Clay - Ca \Rightarrow Ca^{2+} + Clay - 2K^+ \quad (3)$$

The OH$^-$ groups present on the edge of the clay also induce exchange capacity and adsorption of charged species. These are however pH dependent. Presently nothing is known about Ca isotopic fractionation associated with Ca adsorption on clays in the weathering environment. However, based on a study performed in the marine environment, lighter Ca isotopes are expected to adsorb preferentially and the intensity of fractionation is dependent on the clay mineral in question (Ockert et al. 2013). However, it is interesting to note that there is field evidence for isotopically light and heavy exchangeable reservoirs (e.g., Bullen et al. 2004; Moore et al. 2013). Bullen et al. (2004) suggested the existence of an isotopically light Ca "organically-complexed pool, derived from biological cycling" in Santa Cruz soil pore fluids whilst the ^{44}Ca in the older soils from the Santa Cruz chronosequence reflected addition of an external input of Ca such as rain with the same isotopic composition as seawater [consistent with the interpretation of Sr and Mg isotope data (White et al. 2009; Tipper et al. 2010b)]. Simple modelling work has suggested that transient changes in the biosphere could influence the bio-available exchangeable reservoir (Fantle and Tipper 2014) demonstrating that the isotopically light nature of the Santa Cruz exchangeable Ca is difficult to obtain and maintain over large spatial scales without non-steady state behaviour and/or previously unrecognized fractionation during Ca cycling/recycling (preferential release of light Ca).

2.4.6 Soil Porewaters

Ca is the predominant charged ion held in its hydrated form in soils and is abundant in soil solutions. This solute Ca can either leave the system via lateral flow of water, act as a nutrient for plants or Ca supply for adsorption onto solid phases. The Ca content of soil pore waters can be highly variable. Marques et al. (1997) have for example reported that soil solution Ca concentrations are high in surficial soil waters, whereas they diminish as drainage water percolates through the mineral soil, either due to root uptake or retention on the soil exchange complex. Unsurprisingly given the complexity of this environment, the Ca isotope data to date is quite variable for soil pore-waters.

Cenki-Tok et al. (2009) showed that $\delta^{44/40}$Ca in soil solutions from the Strengbach experimental catchment increased with depth below beech trees, whereas the $\delta^{44/40}$Ca remained constant below spruces. Bagard et al. (2013) suggested the presence of organic and organomineral colloids that preferentially scavenge the light ^{40}Ca isotope. Alternatively, this ^{44}Ca enrichment below beeches could be caused by the dissolution of another soil pool, linked to weathering and/or water-rock interactions, such as for instance secondary mineral phases, whose role has not yet been elaborated in this watershed. Fantle and Tipper (2014) commented that such enrichments in the heavy isotope are consistent with uptake of the light isotope by the biosphere, leaving a residual pool that is enriched in the heavy isotopes, and developed some simple mass balance models to explore this idea.

2.5 Atmospheric Ca in Dust and Rain Waters

Atmospheric Ca is mainly produced from sea salts and soil dust particles and is deposited again on land and in the ocean through wet and dry deposits (Hinds 1999) When deposited on the continents it can interact with the canopy to become throughfall and stemflow incomes in forested ecosystems. In some catchments where rates of bedrock weathering are low, atmospheric deposits can constitute a major source of inputs to both the soil and solutes (Table 5). Below, dry and wet atmospheric deposition are discussed separately.

2.5.1 Dry Deposition and Dust

Dry deposition consists of fine (<1 μm) and coarse (>1 μm) particles as well as gases (Cappellato and Peters 1995). These deposits have a mineral (volcanic particles, clays, silts, sands, salts), organic (pollen, vegetal or animal fibres), natural or anthropogenic origin. Hundreds of millions of tons of dust are emitted from deserts and volcanoes into the atmosphere each year, that are transported by winds before being deposited on the surface of the Earth (e.g., Derry and Chadwick 2007), with the Sahara Desert being the largest source of atmospheric dust on the planet. The paleo-lakes or playas are also important sources of atmospheric dust (e.g., Fantle et al. 2012). These dusts participate in the general sedimentation in both oceans and on the continents. It is also known that they have a strong impact on the Earth's radiation budget and that they participate in the fertilisation of the surface ocean (Grousset and Biscaye 2005). These wind inputs are particularly important in areas downwind of major arid and semi-arid lands (i.e. Eastern Tropical Atlantic, Northwest Pacific, Arabian Sea and the Mediterranean).

The annual flux of soil dust can be estimated by considering the soil dust flux comprised between $1–3 \times 10^9$ t/year (Hinds 1999). To a first approximation, the average Ca content can be considered equal to that of the upper continental crust [CaO ~ 3.59 % wt; Rudnick and Gao (2003)], implying that the Ca soil dust flux to the atmosphere ranges from $0.03–0.8 \times 10^9$ t/year. This range is consistent with the recently estimated mineral dust flux at $0.0026–0.077 \times 10^9$ t/year Ca flux (Lawrence and Neff 2009). Mahowald et al. (2006) estimated that the atmospheric flux during the last glacial maximum was of the order of 2–3 times the current one, but these estimates have substantial uncertainty associated with them.

Similarly, the flux from sea salt particles is estimated to be between $1–10 \times 10^9$ t/year (Hinds 1999). Assuming that this atmospheric contribution has a Ca concentration equal to that of seawater, the sea-salt Ca flux ranges between $0.412–4.12 \times 10^6$ t/year, a small percentage of the total Ca flux to the oceans.

There are few Ca isotopic data of atmospheric dust but their $\delta^{44/40}$Ca varies from 0.58 to 0.93 ‰ for bulk/residue samples, with an average value equal to 0.69 ± 0.15 ($2\sigma_{mean}$; N = 16) (Ewing et al. 2008; Fantle et al. 2012; Wiegand et al. 2005); from 0.77 to 1.36 ‰ for the water-soluble fraction, with an average value equal to 1.11 ± 0.07 ‰ ($2\sigma_{mean}$; N = 22) and from 0.64 to 0.92 ‰ for the acid-soluble fraction (Fantle et al. 2012), with an average equal to 0.78 ± 0.03 ‰ ($2\sigma_{mean}$; N = 22) (Fig. 2; Table 2).

2.5.2 Wet Deposition

A significant flux of water is evaporated from the ocean (0.424×10^{15} t/year, Berner and Berner 1996). By assuming an average Ca concentration similar to that of seawater (Berner and Berner 1996), an upper limit (because of the distillation process) for the flux of Ca *from* seawater evaporation can be estimated as 0.18×10^9 t/year. Some of this water vapour gets transported to the continents, estimated at 0.036×10^{15} t/year (Berner and Berner 1996), with an attendant Ca flux of 0.015×10^{12} t/year. However, these are likely overestimates because marine rainwaters are much more dilute than seawater. Similarly, about 66 % of continental waters are evaporated during their transport to the oceans corresponding to a water flux of 0.071×10^{15} t/year (Berner and Berner 1996). An upper limit for the Ca flux can be estimated in the same way as for seawater by assuming that the evaporated water has a Ca content similar to that of average river waters (13.4 mg/l Meybeck 1979), yielding a Ca flux of 0.95×10^9 t/year (Table 3).

For wet deposition, both the concentration of Ca, and the total rainfall is highly variable by region. Berner and Berner (1996) distinguished between continental rain (with 0.1–3.0 mg/l Ca and 0.107×10^{15} t/year water) and marine and coastal rain, (with 0.2–1.5 mg/l Ca and 0.398×10^{15} t/year water). This yields a wet deposition flux of Ca of $0.01–0.33 \times 10^9$ t/year over the continents and of $0.08–0.60 \times 10^9$ t/year for marine and coastal fluxes (Table 3).

Isotopically, wet atmospheric deposits can be divided into rainwater deposits and snow. Although much of the Ca is derived from the evaporation of seawater, it is not correct to assume that the $\delta^{44/40}$Ca is the same as seawater. Measured $\delta^{44/40}$Ca values of rain vary from −0.34 to 1.30 ‰ with an average value equal to

0.70 ± 0.16 ‰ $(2\sigma_{mean};\quad N = 20)$ (Fig. 2; Table 2) (Schmitt et al. 2003a; Schmitt and Stille 2005; Cenki-Tok et al. 2009; Holmden and Bélanger 2010; Farkaš et al. 2011; Hindshaw et al. 2011; White et al. 2012; Wiegand and Schwendenmann 2013). This is very similar to the mean $\delta^{44/40}$Ca of carbonates (Sect. 2.1.2), consistent with the dissolution of carbonate dust in the atmosphere. This has been observed for Sr isotopes based on many more measurements (e.g., Capo and Chadwick 1999), and has also been observed for Mg isotopes for Alpine rain (Tipper et al. 2012). The $\delta^{44/40}$Ca of snow varies from 0.22 to 1.29 ‰, with an average value equal to 0.76 ± 0.18 ‰ $(2\sigma_{mean};\quad N = 12)$ (Fig. 3) (Schmitt and Stille 2005; Cenki-Tok et al. 2009; Holmden and Bélanger 2010; Farkaš et al. 2011; Hindshaw et al. 2011; Bagard et al. 2013).

The significance of wet atmospheric deposition to catchment inputs is variable and several studies have quantified the proportion of atmospheric inputs of Ca at catchment scale. Out of the total atmospheric deposition Ca constitutes a significant part. Some forests appear to derive most of their Ca from atmospheric inputs, while at other places the part from weathering of soil minerals is predominant (Table 5). Ca isotopes have added to these mass flux constraints. For instance Bagard et al. (2013) have shown that the contribution of wet atmospheric Ca deposits have a negligible impact to Ca budgets at the catchment scale. In contrast, Hindshaw et al. (2011) have highlighted seasonal variations in the proportion of Ca derived from the atmosphere with a 2 % contribution in winter, but up to 38 % in summer at the Damma glacier CZO (Swiss Alps). Wiegand et al. (2005) demonstrated the dominance of marine-aerosol derived Ca in older soils compared to younger ones at Hawaii.

2.5.3 Interaction Between the Biosphere and Atmospheric Deposition

In forests, incident precipitation is modified by the interaction with the trees and reach the soil surface as throughfall or stemflow. Throughfall represents the part of precipitation that meets the forest canopy whereas stemflow represents the part of precipitations funnelled down from side branches of the trees to the main stem and then to the ground (Crockford and Richardson 2000; Park and Cameron 2008). Throughfall and stemflow volumes are influenced by structural parameters of the canopy (volume, depth, roughness, branching patterns of the leaves and branches, leave angles), as well as the tree age, shape and size (Park and Cameron 2008). Usually the Ca contribution from stemflow is negligible compared to that of throughfall (Bellot et al. 1999; Parker 1983; Dezzeo and Chacón 2006). However, it can be important for the nutrient inputs to soils that are impoverished in nutrient elements such as in tropical rainforests (Herwitz 1986). During rain circulation through the canopy, nutrients may be leached from or taken up by the tree tissues making the throughfall enriched in nutrients compared to incoming precipitation (Parker 1983). This Ca enrichment may be due to (1) foliar leaching that occurs passively or by ion-exchange processes with transport of the soluble ions in the free spaces between cells (apoplastic transport), and (2) washing off of dry deposited solutes (Parker 1983; Lovett and Lindberg 1984; Cappellato and Peters 1995; Lovett et al. 1996). Probst et al. (1992) have shown that net atmospheric input of Ca represent 85 % of throughfall inputs in the Strengbach experimental catchment (Vosges mountains, NE France). This is consistent with a more recent Ca isotope study, showing for the same catchment that at least 70 % of the Ca present in the throughfall has an incident atmospheric origin (Schmitt and Stille 2005). Dezzeo and Chacón (2006) have shown that annual inputs of Ca are 5–8 times higher in throughfall than in incident rainfall for a forest from Southern Venezuela. This enrichment in base cations in throughfall may neutralize some of the acidity in bulk precipitation, and may buffer the acidity of rainfall (Zeng et al. 2005; Shen et al. 2013). This may be particularly important for the neutralisation of sulphuric acid in rainfall generated from industrial sulphur emissions by the following schematic reaction:

$$Cell\ wall\ \mathrm{Ca}^{2+} + 2\mathrm{H}^+ + \mathrm{SO_4}^{2-}$$
$$\Rightarrow Cell\ wall\ 2\mathrm{H}^+ + \mathrm{Ca}^{2+} + \mathrm{SO_4}^{2-} \qquad (4)$$

The difference in nutrient fluxes between deposition (including wet and dry deposition) and the sum of throughfall and stemflow is termed net canopy exchange and allows the amount of nutrients released by the canopy to be quantified (Lovett and Lindberg 1984; Lovett et al. 1996):

$$Net\ canopy\ exchange = Total\ deposition\ (wet + dry)$$
$$- (Throughfall + Stemflow)$$
$$(5)$$

There are only a small number of $\delta^{44/40}\mathrm{Ca}$ data for throughfall samples, ranging from 0.29 to 0.80 ‰, with an average value equal to 0.61 ± 0.32 ‰ ($2\sigma_{mean}$; N = 3, Fig. 2; Table 3) (Schmitt et al. 2003a; Cenki-Tok et al. 2009). Ca isotope mass balance constraints have demonstrated that calculations of atmospheric Ca contribution to catchment surface waters using only mass budget calculations based on yearly Ca fluxes of rainwater must be treated with caution. For example, at the Strengbach catchment scale, when considering only incident wet deposition-derived Ca, a contribution of 15 % to the catchment's Ca budget has been calculated, whereas it increases up to 20–40 % when including throughfall samples (Schmitt and Stille 2005; Cenki-Tok et al. 2009; Chabaux et al. 2005).

Although throughfall nutrient transfer to the soil is much more rapid than litterfall decomposition because it occurs in dissolved inorganic form (Chuyong et al. 2004; Parker 1983), a much greater mass flux is from the later process (Dezzeo and Chacón 2006; Blum et al. 2008).

3 Modern Global Budgets of Ca

The size of the main surficial Ca reservoirs on Earth, their average Ca isotopic compositions and a quantification of the fluxes between these reservoirs have been summarized in Fig. 5. The larger reservoir of Ca in the lithosphere makes the residence time of Ca in the lithosphere at least two orders of magnitude longer than in the surficial reservoirs, 10^6 years versus 10^8 years (Van Cappellen 2003). Consequently, on millennial time-scales only small fractions of Ca can be transferred from the lithosphere, and redistributed among the atmosphere, hydrosphere and biosphere. In contrast, when considering longer timescales (>10^4 years), interactions between the lithosphere and surficial reservoirs can no longer be neglected (Van Cappellen 2003). Testament to this is that it is estimated that more Ca is present in limestones than crystalline rocks, all of which must have transited through surficial reservoirs.

By far the most work to date has concentrated on the oceanic Ca cycle, but the continental Ca cycle is an integral part of this, not least because it supplies the Ca to the oceans. The biomass reservoir of Ca is about 400 times larger than the instantaneous river water reservoir, with soils intermediate in size (Fig. 4), highlighting the importance of the biosphere-soil-water interactions for the Ca biogeochemical cycle. Moreover, adding Ca isotopes to elemental flux studies has made major advances in the last decade, in the identification and quantification of mechanisms involved in the mass transfer of Ca between different reservoirs.

3.1 The Continental Cycle of Ca

The continental Ca cycle is complex, not least because there are many different reservoirs of Ca as described above and a series of different processes. Depending on the spatial and temporal time-scales, there are several areas where Ca isotopes have already made a major contribution. For example at a forest scale over decadal time-scales, Farkaš et al. (2011) showed that in a base poor forest in Massachusetts mineral weathering (of biotite) in particular only contributed a maximum of ∼5 % Ca to the red-oak trees growing at that locality, highlighting the recycling of Ca between the biosphere, soil exchangeable fractions and bulk soil. Simple box models have begun to incorporate Ca isotope data to provide a better constraint (Fantle and Tipper 2014), but these models are in a nascent

Fig. 5 Present-day Ca quantitative reservoirs and fluxes, including Ca isotopic variations (in *blue*) (expressed as $\delta^{44/40}$ Ca$_{SRM915a}$ (‰), the errors correspond to the 2 σ_{mean})

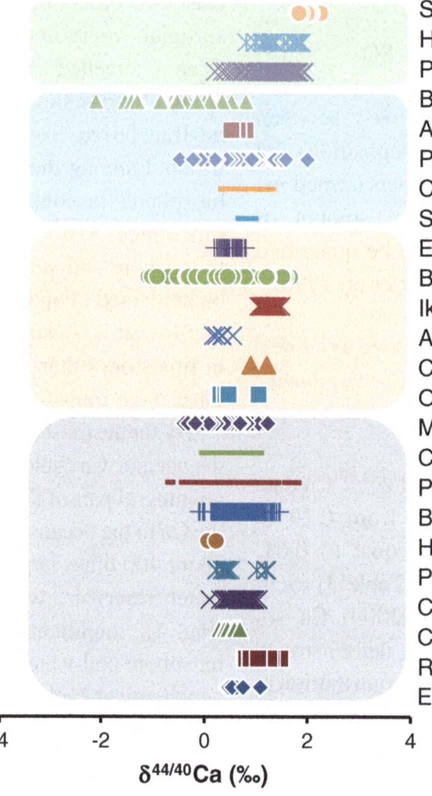

state and additional data from well constrained catchments is required.

At global scales, an important observation is that there is an offset between the calculated and measured riverine $\delta^{44/40}$Ca. Whilst the explanation for this is not well known, it must be caused by a process relating to the continental Ca cycle being in a transient state at the continental scale. To date this is not well constrained, but it demonstrates the potential of high precision Ca isotope ratios to contribute to our understanding of Ca cycling at both a catchment, regional and global scale. This will require a thorough study of continental processes in order to identify and characterize the processes that control the Ca isotopic signature arriving to the ocean, to extrapolate to past and future variations, and to develop simple and testable models of both the continental and oceanic budgets. Consequently, more detailed study at the soil-water-plant interface and within the soil itself need to be undertaken to decipher the details of the processes occurring in these environments. To characterize precisely the continental Ca biogeochemical cycle, multi-disciplinary approaches (pedologs, geochemists, mineralogists, biologists) and tools (combination of isotopes, elemental concentrations, flux calculations) are now necessary (e.g., Bernasconi et al. 2008, 2011; White et al. 2015).

3.2 The Oceanic Cycle of Ca

The oceanic Ca cycle has been more extensively studied in comparison with the continental budget, mainly because oceanic calcium carbonate sediments represent the largest planetary sink of carbon, and thus strongly influence the global carbon cycle. The main sources of Ca to the oceans are rivers, groundwater (Sects. 2.2.2 and 2.2.3) and hydrothermal inputs. The estimated hydrothermal fluxes vary from 80–204 × 10⁶ t/year (Wilkinson and Algeo 1989; Milliman 1993; Drever 1988). More recently de

Villiers (1998) calculated the hydrothermal flux as equal to $400–800 \times 10^6$ t/year assuming that the modern ocean is at steady state. Isotopically, hydrothermal fluids range from 0.79 to 1.86 ‰, with an average value equal to 1.44 ± 0.14 ‰ ($2\sigma_{mean}$, N = 23) (Table 3) (Zhu and Macdougall 1998; Schmitt et al. 2003a; Amini et al. 2008).

Additional smaller but poorly quantified fluxes of Ca to the ocean are (1) Cation exchange or the removal or replacement of Ca loosely bound to surfaces or inter-layer sites of clay minerals (France-Lanord and Derry 1997; Cerling et al. 1989), estimated as $37–38 \times 10^6$ t/year (Berner and Berner 1996; Drever 1988) (Table 3) and (2) a diffusional diagenesis flux from marine sediments has been estimated at 144×10^6 t/year (Drever 1988).

The main Ca sink from the oceans corresponds to the sedimentation of calcified organisms (Fig. 5), either on the slope-rise or on the bottom of the ocean. Wilkinson and Algeo (1989) estimated the total mass of Phanerozoic oceanic carbonate as 84×10^{15} t, equivalent to a mass of Ca of 34×10^{15} t. Deep-sea environments account for 55 % of $CaCO_3$ accumulation (planktic foraminifera and coccoliths), the remaining 45 % originate from shallow-water environments (reef corals, coraline red algae, benthic foraminifera, ooids) (Milliman 1993; Neumann and Land 1975; Wefer 1980; Hubbard et al. 1990). Aragonite is thought to account for 70 % of the modern shallow-water sink, which is about 30 % of total Ca oceanic removal Blättler et al. (2012). There is a difference in mineralogy dependent on the depth and the type of organisms, with carbonate production dominated by benthic aragonite and magnesium calcite in deep water and planktonic calcite in shallow water. Corals represent about 20 % of the whole oceanic carbonate sedimentation and foraminifera between 20 and 60 % (Milliman and Droxler 1996; Milliman 1993). Production rates are one to three orders of magnitude greater in shallow water but global shallow water accumulation approximates that from deep water (Milliman 1993). Milliman (1993) estimated a whole ocean accumulation rate of 3.2×10^9 t/year, taking into account a dissolution rate of 2.1×10^9 t/year.

Isotopically, Phanerozoic $CaCO_3$ sedimentation ranges from -1.09 to 1.70 ‰, with an average equal to 0.59 ± 0.02 ($2\sigma_{mean}$, N = 1979). In contrast, modern $CaCO_3$ samples are slightly enriched in ^{44}Ca with a range of variation from -0.18 to 1.70 ‰ and an average value equal to 0.76 ± 0.03 ($2\sigma_{mean}$, N = 538) (Table 2) (Skulan et al. 1997; Zhu and Macdougall 1998; Skulan and DePaolo 1999; De La Rocha and DePaolo 2000; Nägler et al. 2000; Schmitt et al. 2003a; Chang et al. 2004; Fantle and DePaolo 2005; Gussone et al. 2005, 2009, 2010, 2011; Heuser et al. 2005; Immenhauser et al. 2005; Kasemann et al. 2005; Teichert et al. 2005, 2009; Sime et al. 2005; Böhm et al. 2006; Hippler et al. 2006; Steuber and Buhl 2006; Fantle and DePaolo 2007; Farkaš et al. 2007, 2006; Sime et al. 2007; Amini et al. 2008; Griffith et al. 2008a, 2011; Heinemann et al. 2008; Gussone and Filipsson 2010; Payne et al. 2010; von Allmen et al. 2010; Blättler et al. 2011; Turchyn and DePaolo 2011; Blättler et al. 2012; Holmden et al. 2012; Wang et al. 2012, 2013; Hippler et al. 2013; Pretet et al. 2013; Ullmann et al. 2013; Blättler and Higgins 2014; Hinojosa et al. 2012; Jost et al. 2014; Kasemann et al. 2014; Fantle 2015; Brazier et al. 2015; Vivier et al. 2015; Husson et al. 2015).

Burial of pore water, with 0.8×10^6 t/year can also constitute a minor Ca sink flux (Drever 1997). Isotopically, pore water ranges from 0.19 to 1.92 ‰ with an average equal to 1.31 ± 0.07 ‰ ($2\sigma_{mean}$, N = 103, Table 3, Teichert et al. 2005, 2009; Henderson et al. 2006; Fantle and DePaolo 2007; Turchyn and DePaolo 2011; Fantle 2015).

Ultimately, the concentration of Ca in seawater is governed by the balance between the inputs and outputs of Ca. When the inputs and outputs are in balance, the ocean is at steady state. Mathematically the rate of change of the Ca content of seawater ($N_{\mathrm{Ca}_{sw}}$) is generally described by:

$$\frac{dN_{\mathrm{Ca}_{sw}}}{dt} = \sum_{inputs,i} (F_i - F_{sed}) \qquad (6)$$

where F_i is the mass flux of input flux i. Simplifying this equation to consider only the most significant fluxes the change in Ca content of seawater can be reduced to:

$$\frac{dN_{\mathrm{Ca}_{sw}}}{dt} = F_{riv} + F_{hyd} - F_{sed} \qquad (7)$$

where F_{riv} is the riverine flux of Ca to seawater, F_{hyd} is the hydrothermal flux of Ca to seawater, and F_{sed} is the sedimentary flux of Ca from seawater.

There have been several attempts to quantify the balance of inputs and outputs of Ca to and from the oceans, but the mass budget is difficult to constrain because the fluxes of Ca to and from the oceans are not straightforward to quantify (Opdyke and Walker 1992; Mackenzie and Morse 1992; Milliman 1993; Sabine and Mackenzie 1995; Ridgwell and Zeebe 2005). Milliman (1993) concluded that the modern marine Ca budget is not at steady state or alternatively suggesting that either the outputs have been overestimated or the inputs underestimated. The isotope ratios of Ca provide another dimension to this problem, and add a key mass balance constraint, modifying Eq. 7 to become:

$$N_{\mathrm{Ca}_{sw}} \cdot \frac{d\delta_{sw}}{dt} = \sum_{inputs,i} F_i \cdot \delta_i - \sum_{outputs,j} F_j \cdot \delta_j \qquad (8)$$

where δ_i, j refers to the $\delta^{44/40}\mathrm{Ca}$ of the input i or output j flux, and F_i is the mass flux of input i or output j. Because the output flux of Ca is derived from the precipitation of seawater, its composition is related directly to that of seawater. Each output has its own fractionation factor (Δ_j) because of the differing mineralogies and type of bioclast that removes Ca over the full depth range of the oceans.

Many studies have assumed that the output from the oceans is "bulk" carbonate, with an average fractionation factor defined as the mean weighted average isotopic composition of the output flux relative to seawater (Fantle and Tipper 2014):

$$\Delta_{sed}^{global} = \frac{1}{F_{sed}^{global}} \sum_{outputs,j} F_j \cdot \Delta_j \qquad (9)$$

Equation 8 can then be specifically related to seawater as:

$$N_{\mathrm{Ca}_{sw}} \cdot \frac{d\delta_{sw}}{dt} = \sum_{inputs,i} F_i \cdot \delta_i - \sum_{outputs,j} F_j \cdot (\delta_{sw} - \Delta_j)$$
$$(10)$$

where δ_{sw} is equal to the isotopic composition of seawater as a function of time.

Reducing Eq. 10 to the principal terms it becomes:

$$N_{\mathrm{Ca}_{sw}} \cdot \frac{d\delta_{sw}}{dt} = \delta_{riv} \cdot F_{riv} + \delta_{hyd} \cdot F_{hyd} - (\delta_{sw}$$
$$- \Delta_{sed}^{global}) \cdot F_{sed}^{global}$$
$$(11)$$

where $\delta_{riv}, \delta_{hyd}$ are the Ca isotopic compositions of the riverine input and hydrothermal respectively.

There is Ca isotope data for most of the different oceanic reservoirs of Ca. Present-day seawater has been shown to be constant, whatever the location or the depth, so that it yields only a narrow variation, from 1.77 to 2.23 ‰, with an average equal to 1.89 ± 0.02 ‰ ($2\sigma_{mean}$; N = 62) (Table 4).

The main input of Ca to seawater, rivers have an average $\delta^{44/40}\mathrm{Ca}$ of 0.86 ± 0.06 ‰ (Sect. 2.2.2) and hydrothermal fluids vary from 0.79 to 1.86 ‰, with an average value equal to 1.44 ± 0.14 ‰ $2\sigma_{mean}$; N = 23 (Fig. 4; Table 3, Zhu and Macdougall 1998; Schmitt et al. 2003a; Amini et al. 2008). From this hydrothermal flux variability, Amini et al. (2008) calculated a hydrothermal end-member value equal to 0.93 ± 0.07 ‰. Out of the minor fluxes to seawater, pore fluids vary from 0.19 to 1.92 ‰, with an average value equal to 1.31 ± 0.07 ‰ ($2\sigma_{mean}$; N = 103) (Fig. 4; Table 3) (Fantle and DePaolo 2007; Teichert et al. 2009; Turchyn and DePaolo 2011; Fantle 2015) and groundwaters range from 0.17 to 2.08 ‰, with an average equal to 0.86 ± 0.10 ‰ ($2\sigma_{mean}$; N = 65) (Fig. 4; Table 2) (Schmitt et al. 2003a; Jacobson and Holmden 2008; Tipper et al. 2008a; Cenki-Tok et al. 2009; Holmden et al. 2012; Holmden and Bélanger 2010; Hindshaw et al. 2011, 2013; Wiegand and Schwendenmann 2013; Nielsen and DePaolo 2013; Jacobson et al. 2015).

Modern (i.e. Holocene) biotic and abiotic oceanic $CaCO_3$ yield a variability from −0.18 to 1.70 ‰, with an average value equal to 0.76 ± 0.03 ‰ ($2\sigma_{mean}$; N = 457, Table 2) (Skulan et al. 1997; Zhu and Macdougall 1998; Nägler et al. 2000; Schmitt et al. 2003a; Chang et al. 2004; Gussone et al. 2005; Heuser et al. 2005; Sime et al. 2005, 2007; Böhm et al. 2006; Farkaš et al. 2007; Amini et al. 2008; Griffith et al. 2008b; Gussone and Filipsson 2010; von Allmen et al. 2010; Hippler et al. 2013; Blättler et al. 2012; Blättler and Higgins 2014; Holmden et al. 2012; Gussone et al. 2009, 2011; Pretet et al. 2013; Heinemann et al. 2008; Skulan and DePaolo 1999; Steuber and Buhl 2006; Ullmann et al. 2013). This value is very similar to those calculated in the compilation of Fantle and Tipper (2014) (0.77 ‰), and the calculated weighted Ca-ouptut flux into $CaCO_3$ of 0.76 ± 0.11 ‰ ($2\sigma_{mean}$) from Holmden et al. (2012). Several studies have attempted to construct a present-day oceanic Ca budget based on Eqs. 6–10 in the modern day (Skulan et al. 1997; De La Rocha and DePaolo 2000; Schmitt et al. 2003a; Tipper et al. 2010a). Most of this work assumed that present-day seawater is close to steady state. Tipper et al. (2010a) argued that given apparent similarity of the input and output $\delta^{44/40}Ca$ values, the maximum deviation from steady state could only be ~ 0.2 ‰, which was used to demonstrate that the input and output fluxes of Ca must be within 15 % of each other on time-scales similar to the residence time of Ca in the oceans (~ 1 Ma), otherwise there should be a greater isotopic difference between the input and output $\delta^{44/40}Ca$ values.

4 Global Ca Cycling in Earth's History

Changes in the Ca concentration of ocean water is a fundamental variable in the evolution of Earth's climate, intimately linked to the global carbon cycle, through the saturation state and precipitation of calcium carbonate minerals. While the fractionation between seawater and marine (biogenic) carbonates (Eq. 9) leads to the relation between imbalances in the Ca input and output and $\delta^{44/40}Ca_{SW}$ (Eq. 10), it complicates the reconstruction of $\delta^{44/40}Ca_{SW}$, due to biomineralisation related species-specific fractionation patterns. For example, a switch from calcite to aragonite seas will likely lead to a different bulk fractionation factor $\left(\Delta_{sed}^{global}\right)$ between marine biogenic carbonate and seawater. In turn, this will permanently modify $\delta^{44/40}Ca_{SW}$ (Blättler et al. 2012), but only transiently modify the $\delta^{44/40}Ca$ of the bulk carbonate output (Sime et al. 2007; Fantle 2010; Tipper et al. 2010a; Fantle and Tipper 2014). Therefore a tracer of seawater $\delta^{44/40}Ca$ will yield a different result to a tracer of the bulk $\delta^{44/40}Ca$' of the carbonate output from the ocean. An important consideration therefore is whether the proxy material being used to reconstruct δ_{sw} is a "passive" tracer of seawater, (mimicking the isotopic composition of seawater, whilst being offset from the actual composition of seawater by an isotopic fractionation factor) without being a sizeable component of the global output flux or reflecting the isotopic composition of the global output flux (Fantle 2010; Fantle and Tipper 2014).

Consequently, archives need to be thoroughly calibrated, to determine (1) the influence of environmental parameters and species specific vital effects on the Ca isotope fractionation, which can potentially bias the records and (2) understand whether the recorder only removes a small fraction of the total Ca from seawater, and so is a passive tracer.

To date several different archives have been proposed for the reconstruction of $\delta^{44/40}Ca_{SW}$. In this section we briefly review the different archives that have been published, show their contributions and application limits. We also discuss past Ca isotope seawater records and the processes that affect the Ca isotopic variation through time.

4.1 Archives of $\delta^{44/40}Ca_{SW}$

Finding appropriate archives is crucial for the reliable reconstruction of $\delta^{44/40}Ca_{SW}$ throughout Earth's history. A significant part of the Ca isotope work on abiotic and biogenic minerals has been motivated by the need to find a sample set that can record $\delta^{44/40}Ca_{SW}$ with fidelity (Fig. 6). While detailed descriptions on fractionation characteristics of inorganic Ca bearing minerals and biominerals are provided in Chapters "Calcium Isotope Fractionation During Mineral Precipitation from Aqueous Solution" and "Biominerals and Biomaterial" respectively, this section briefly discusses the advantages and limitations of archives that have been applied to reconstruct $\delta^{44/40}Ca_{SW}$, in order to explain potential discrepancies between records based on different archives.

4.1.1 Bulk Carbonate

The bulk carbonate fraction of sediments is used in several studies to reconstruct $\delta^{44/40}Ca_{SW}$, though there are some significant differences in methodology. Some studies used the micritic fraction with and without visible bioclasts and cements (Blättler et al. 2011; Brazier et al. 2015). Some studies have used bulk nannofossil oozes, including microfossilrich intervals (e.g., De La Rocha and DePaolo 2000; Fantle and DePaolo 2005, 2007; Vivier et al. 2015) or bulk carbonate including skeletal carbonates (e.g., De La Rocha and DePaolo 2000; Steuber and Buhl 2006; Blättler et al. 2012, 2011; Brazier et al. 2015).

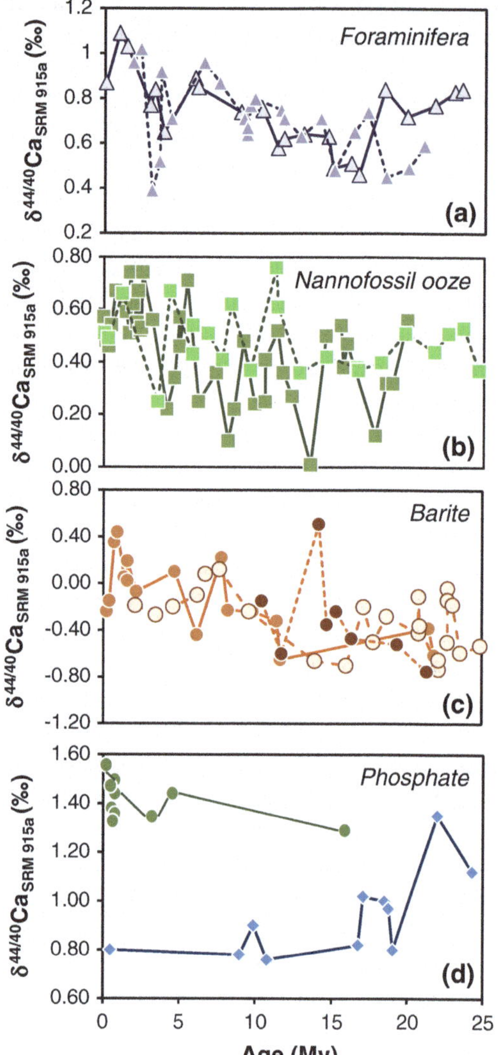

Fig. 6 Ca isotopic composition $\delta^{44/40}$ $Ca_{SRM915a}$ (‰). Since the Miocene recorded by different proxy minerals: **a** Marine foraminifera: Globigerinella spp from ODP 144-871 and 144-872 and G. bulloides from ODP 183-1138 (Heuser et al. 2005); **b** Marine nannofossil carbonate ooze from DSDP site 590 and ODP site 80717 (Fantle and DePaolo 2005, 2007); **c** Marine barite from DSDP 572-573, DSDP 574 and DSDP 575 (Griffith et al. 2008a); **d** Marine phosphates (Schmitt et al. 2003b; Arning et al. 2009)

The main obvious advantages of using bulk carbonate is that this sample material is not limited, and that no special time consuming sample selection or knowledge on taxonomy, e.g. identification and picking of certain species is required. Secondly, it potentially represents the bulk output of Ca from seawater, which can have advantages over a "passive tracer" such as foraminifera, barite or phosphates discussed below. However, there are complications with using bulk carbonate as a recorder of $\delta^{44/40}Ca_{SW}$ because of a lack of control on faunal and floral composition and thus species specific fractionation. The risk is that the bulk $\delta^{44/40}Ca$ is controlled by a mixture between different bioclasts with differing fractionation factors from seawater (Sime et al. 2007). If this is not carefully controlled, a bulk carbonate record might represent neither the $\delta^{44/40}Ca$ of seawater, nor the bulk carbonate output. In addition potential post depositional alteration can be more difficult to ascertain for bulk samples rather than specific tests. Carbonate cements formed in marine sediments exhibit a different fractionation compared to biogenic carbonates and were suggested by Steuber and Buhl (2006) as potential archive for $\delta^{44/40}Ca_{SW}$, but there is only limited data available on cements. Because of this, there is still uncertainty in the use of bulk carbonates for the reconstruction of $\delta^{44/40}Ca_{SW}$. For these reasons, a considerable body of work has been invested in developing alternative proxy records, with constant fractionation factors from seawater.

4.1.2 Abiotic Records of $\delta^{44/40}Ca_{SW}$

Barite Mineral Separates

Barite occurs in different depositional environments, such as in cold vents and in evaporates and is deposited through a large number of processes including biogenic and hydrothermal processes. Barites from cold vent and hydrothermal settings show different ranges in Ca isotope fractionation relative to seawater (Griffith et al. 2008a). Barite only removes a minor proportion of the total Ca from seawater, and as such is a passive tracer of seawater. It therefore will respond to changes in $\delta^{44/40}Ca_{SW}$, and will differ from changes in the bulk $\delta^{44/40}Ca$ output of Ca from the oceans. For $\delta^{44/40}Ca_{SW}$ reconstructions only marine pelagic barite has been used so far (Griffith et al. 2008a,

2011). This is due to the fact that present day calibrations of pelagic barite show no significant influence of environmental parameters like temperature, water column carbonate and dissolves organic carbon concentrations, barite saturation, sedimentation and barite accumulation rates (see also Chapter "Calcium Isotope Fractionation during Mineral Precipitation from Aqueous Solution" for Ca isotope fractionation in natural and synthetic barite). Potential complications might arise from the observation that the fractionation found in pelagic barite is not reproduced in lab experiments, as synthetic barite is offset from pelagic barite. In addition, the exact reaction pathways in the microenvironments in which pelagic barite forms are not yet completely understood. This might be relevant for long term $\delta^{44/40}Ca_{SW}$ records, as ocean chemistry changes on long and short terms with respect to Ba/SO$_4$, Ba/Ca ratios, which may influence the fractionation of Ca isotopes in barite (Griffith et al. 2008a, c) One possible source of uncertainty that needs to be considered is barite Ca isotope fractionation related to dissolution-reprecipitation, caused by organotrophic sulfate reduction in organic rich sediments. In addition it is important that only pure pelagic barite is sampled, which is not influenced by hydrothermal or cold seep activity. Nevertheless, Griffith et al. (2008a) generated a record of $\delta^{44/40}Ca_{SW}$ based on barite, and proposed a dynamic marine Ca cycle.

Gypsum and Anhydrite

Anhydrite and gypsum have been suggested as archives for the reconstruction of Ca isotope evolution of marginal seas, e.g. during formation of evaporate sequences (Hensley 2006; Blättler and Higgins 2014; Harouaka et al. 2014). Due to Rayleigh type fractionation effects during the precipitation of evaporates, the $\delta^{44/40}Ca_{SW}$ of such Ca sulphate deposits is not representative for the open ocean, but may provide important insights into the evolution of evaporate sequences of marginal seas. While anhydrite precipitation experiments show little variability in the Ca isotope fractionation (Hensley 2006), the range of fractionation factors for gypsum spans a relatively large range (Harouaka et al. 2014). The fractionation characteristics of inorganic Ca sulfate minerals are discussed in more detail in Chapter "Calcium Isotope Fractionation during Mineral Precipitation from Aqueous Solution".

Apatite

Carbonate fluor-apatite (CFA) formed in marine organic rich sediments is used as archive for $\delta^{44/40}Ca_{SW}$ reconstructions, because it forms near the sediment-water interface from a seawater dominated fluid, associated with suboxic decay of organic matter (Jarvis et al. 1994). As with barite discussed above, it is a minor constituent of the output flux of Ca, and as such is a passive tracer. Carbonate fluor-apatite is in principle suited for long-term reconstructions, as P burial rates via carbonate fluor-apatite (CFA) formation are constant throughout geologic time (Schmitt et al. 2003b; Soudry et al. 2004, 2006). In addition, coupled $\delta^{44/40}Ca$ and $\delta^{18}O$ analyses of CFA suggest no resolvable dependence on temperature (Schmitt et al. 2003b) and there is general agreement between $\delta^{44/40}Ca_{SW}$ records generated from phosphate, bulk carbonate and foraminifers (e.g., Schmitt et al. 2003b) demonstrating the potential of phosphates. However, some uncertainties may arise because of the poor control on Ca fractionation during CFA formation in terms of growth conditions, e.g. precipitation rates (Schmitt et al. 2003b; Soudry et al. 2004). This potential restriction is confirmed by the observation that different types of CFA exhibit different $\delta^{44/40}Ca_{SW}$, e.g. phosphate crusts are generally enriched in ^{44}Ca compared to peloidal phosphates (Schmitt et al. 2003b; Arning et al. 2009). This difference is likely related to the different microenvironment, as phosphate crusts form within the sediment. Consequently, the reduced fractionation compared to the peloidal phosphate might be related to Rayleigh type fractionation, due to precipitation within a restricted microenvironment, or due to differences in precipitation rate or precipitation pathway (via an amorphous Ca phosphate phase, Arning et al. 2009). Incorporation of different Ca sources in CFA (skeletal Ca^{2+} versus seawater Ca^{2+}) with probably different Ca isotopic compositions may also complicate the interpretation of CFA $\delta^{44/40}Ca_{SW}$ variations (Soudry et al. 2004). While CFA has a potential to record $\delta^{44/40}Ca_{SW}$, it is important to distinguish between different types of CFA, to avoid artefact in the $\delta^{44/40}Ca_{SW}$ reconstruction.

4.1.3 Taxon-Specific Records of $\delta^{44/40}Ca_{SW}$

Reconstructions of $\delta^{44/40}Ca_{SW}$ of past seawater based on specific taxa is in general more laborious compared to records based on bulk material because (1) a large effort is required to identify and separate taxa from bulk sample material, and (2) taxon-specific calibrations are required to verify that the Ca isotope fractionation characteristics are suitable for $\delta^{44/40}Ca_{SW}$ reconstructions. Potential unknown taxon-specific anomalies in Ca isotope fractionation (e.g. large temperature sensitivities and vital effects, etc.) would directly translate into artefacts when reconstructing $\delta^{44/40}Ca_{SW}$. In principle, such artefacts could be corrected for, just as vital effects are sometimes corrected for in other isotopic systems such as $\delta^{18}O$ records in foraminifera. Such artefacts are much harder to correct for in records of $\delta^{44/40}Ca_{SW}$ from bulk carbonates made from many different components because there are more than one effect to correct for.

Foraminifera

Most planktic foraminifer species exhibit a similar range in Ca isotope composition and small dependence on environmental parameters like temperature, carbonate chemistry and salinity (see Chapter "Biominerals and Biomaterial" for discussion and references). Their long tradition and frequent usage as geochemical proxy archives makes a direct comparison to other tracers possible. Foraminifera reach back in time as far as the earlier Cambrian (Pawlowski et al. 2003), but in general it is necessary to construct long-term records from multiple species, splicing them together in a bootstrap fashion because of the evolution of foraminifer taxa or the abundance in any particular core. The Ca isotope fractionation systematics are still not fully understood for some species. For example, *G. sacculifer* shows a bimodal temperature dependence, and *N. pachyderma* shows reduced Ca isotope fractionation at low temperatures and /or carbonate ion concentrations (Hippler et al. 2009; Gussone et al. 2009). Similar atypical behaviour has been observed in benthic species at low temperatures, but because of the small temperature dependence of $\delta^{44/40}$Ca in the temperature range above 5 °C, benthic foraminifers could make an excellent recorder $\delta^{44/40}$Ca$_{SW}$ over geological time. Their better preservation, constant habitat/calcification depth and slower evolution make benthic foraminifers ideal for archives of $\delta^{44/40}$Ca$_{SW}$ if the artefacts caused by the low temperature anomaly can be excluded (Gussone and Filipsson 2010). Foraminifer based records seem to reflect seawater changes accurately, as multiple records based on planktic foraminifers have thus far yielded consistent results (Heuser et al. 2005; Sime et al. 2005, 2007) and fit well to the Miocene barite record (Griffith et al. 2008a).

Brachiopods

Brachiopods have been used for the reconstruction of $\delta^{44/40}$Ca$_{SW}$ (Farkaš et al. 2007; Brazier et al. 2015), because initial results on recent and fossil samples indicate a minor dependence of $\delta^{44/40}$Ca on temperature (Farkaš et al. 2007; Gussone et al. 2005), and see Chapter "Biominerals and Biomaterial"). In addition, their benthonic habitat, restricted mobility and past application for the use of environmental reconstructions using stable isotopes and relatively large amounts of sample material provided by a shell, are promising preconditions for a $\delta^{44/40}$Ca$_{SW}$ archive. Recent temperature calibrations of modern brachiopods show however a more variable Ca isotope fractionation (von Allmen et al. 2010) and complications related to shell heterogeneities and inter-specific variability. Nevertheless, brachiopods form a key part of the Phanerozoic record of seawater $\delta^{44/40}$Ca.

Cephalopods

Belemnite rostra are abundant in Mesozoic sediments and they have been used as archives for light stable isotopes and radiogenic isotopes for a long time. Belemnites provide relatively large amounts of sample material and a good stratigraphic coverage. As they are extinct, knowledge on Ca isotope fractionation characteristics are obtained from cross calibration with other archives (Farkaš et al. 2006; Blättler et al. 2012). The best estimate for a fractionation factor for belemnites is derived by cross calibration with brachiopods (Farkaš et al. 2007), but may require future calibration as datasets expand and a greater understanding of both fractionation factors and mechanisms is developed. Potential additional complications might arise from belemnites mobility in the water column, and eventually changes in Ca isotope fractionation related to trophic level effects (see also Chapter "Biominerals and Biomaterial").

Bivalves

Calibration studies on recent bivalves and palaeo $\delta^{44/40}$Ca$_{SW}$ values inferred from bivalves suggest a general suitability for $\delta^{44/40}$Ca$_{SW}$ reconstructions, but there are also some questions that need to be solved (see also this chapter for more detailed discussion). A small dependence of $\delta^{44/40}$Ca$_{SW}$ on temperature and other environmental factors is found for *Arctica islandica, mytilus edulis* and (Hippler et al. 2013; Hiebenthal 2009; Ullmann et al. 2013). Discrepant observations exist for the temperature dependence in rudists, extinct reef forming bilvales for $\delta^{44/40}$Ca$_{SW}$. While Immenhauser et al. (2005) found evidence for a substantial temperature sensitivity based on δ^{18}O-$\delta^{44/40}$Ca$_{SW}$ cross calibration, data from Steuber and Buhl (2006) suggest a minor temperature dependence. Nevertheless, in general rudists and oysters seem to reflect long-term seawater changes in agreement to other records (Steuber and Buhl 2006; Farkaš et al. 2007).

Biogenic Apatite

Hydroxyapatite conodont microfossils from the Permian-Triassic boundary have been analyzed by Hinojosa et al. (2012). They have observed a negative $\delta^{44/40}$Ca excursion similar to the one obtained by carbonate rocks from the same period (Payne et al. 2010). This suggests that $\delta^{44/40}$Ca variations recorded from biogenic apatites reflect changes in seawater Ca isotopic composition, rather than in carbonate mineralogy. More generally, the use of bones and teeth as a recorder of past seawater variations could be complicated by isotopic differences between kinds of hard-tissues (e.g. bone, dentin, enamel), bone remodeling, changes in Ca metabolisms (Skulan and DePaolo 1999), and trophic level effects (see Chapter "Biomedical Application of Ca Stable Isotopes"). For conodonts (a worm-like organism with teeth, phylum Chordata), the main disadvantage resides in the fact that these species are extinct, meaning that it is not possible to calibrate fractionation factors and to test the effects of environmental factors such as temperature and pH.

Other Taxa

Studies on other taxa including coccolithophores and scleractinian corals revealed Ca isotope fractionation patterns, which are suitable for $\delta^{44/40}$Ca$_{SW}$ reconstructions. However, their applicability is restricted by either the small size (few μm diameter) of coccolithophores and related difficulties of obtaining pure monospecific samples or the tendency of aragonitic corals to recrystallize.

4.2 Past Changes in $\delta^{44/40}$Ca$_{SW}$

The need to determine past changes in $\delta^{44/40}$Ca$_{SW}$ has largely been motivated by the need to quantify past changes in the Ca content of seawater, and as such changes in the relative balance between the inputs (weathering) and outputs (carbonate sedimentation) discussed in more detail in Fantle (2010) and Fantle and Tipper (2014).

The reconstruction of the Ca mass of seawater using $\delta^{44/40}$Ca$_{SW}$ (from Eq. 10) relies on the assumption that the isotopic compositions of the inputs and outputs has remained relatively constant over geological time (De La Rocha and DePaolo 2000; DePaolo 2004; Fantle and DePaolo 2005; Griffith et al. 2008a; Fantle 2010). The data discussed in the present review has demonstrated considerable range in the main reservoirs in the oceanic Ca cycle, though it is not clear how this small scale variability translates into heterogeneity at a global scale. Many questions remain concerning the temporal variability of rivers and whether or not they are at steady state on a range of time-scales (ranging from annual to millenial). When δ_{sw} is driven by imbalances between the input and output mass fluxes (or δ_{input}^{global}), the evolution of $\delta^{44/40}$Ca$_{SW}$ is mirrored by the δ^{44}Ca of the global output flux. If changes in $\delta^{44/40}$Ca$_{SW}$ are driven by changes in the input value of $\delta^{44/40}$Ca from rivers, groundwaters and hydrothermal fluids, then the δ^{44}Ca of the global output flux and $\delta^{44/40}$Ca$_{SW}$ will both respond in the same way. However, if changes in δ^{44}Ca are driven by changes in Δ_{output}^{global}, then the δ^{44}Ca of the global output flux will remain constant on a timescale longer than the residence time of Ca in the oceans, whereas $\delta^{44/40}$Ca$_{SW}$ will adjust to the change in bulk fractionation factor. This highlights the difference between a passive tracer, versus a tracer of the bulk output. A fundamental issue with interpreting Ca isotopes in terms of $F_{input}^{global}/F_{output}^{global}$ or δ_{input}^{global}, is hence how to obtain a reasonable representation of δ_{output}^{global}. However, whilst the interpretation of proxy records is still incomplete, a number of records from the geological past have emerged over the last 15 years, based on the proxy archives discussed above.

4.2.1 Neogene Seawater Records of Ca Isotopes

There is broad agreement between proxy records of foraminifera, nannofossil ooze, barite and phosphates for the past 25 Myr though the interpretation of these data have been different (Fig. 6). Fantle and DePaolo (2005) and De La

Rocha and DePaolo (2000) suggested there were potentially large variations in the marine cycle of Ca based on reconstruction of $\delta^{44/40}\mathrm{Ca}_{SW}$ from bulk carbonates, whereas a record of $\delta^{44/40}\mathrm{Ca}_{SW}$ based on pelagic barite (Griffith et al. 2008a) shows a pronounced increase (of 0.3 ‰) around 13 million years ago. Griffith et al. (2008a) argue that this likely corresponds to a significant change in Ca concentration, coincident with a climatic transition and global change in the carbon cycle.

Sime et al. (2007) on the other hand generated a foraminifera record, consistent with that of Heuser et al. (2005) that was remarkably constant, suggesting in turn that the $\delta^{44/40}\mathrm{Ca}$ of seawater was relatively constant. These workers suggested that discrepancies between the foraminifera and bulk carbonate record could be reconciled by changes in the fractionation factor.

At present there is no definitive record of $\delta^{44/40}\mathrm{Ca}$ in seawater for the Neogene. Future work will need to reconcile the differences between the existing data sets. Part of the discrepancies may arise from differences between calibrated fractionation factors or because some records are passive tracers, whilst others are bulk and so don't record the same thing.

4.2.2 Phanerozoic and Deeper Time Seawater Records of Ca Isotopes

Further back in time the data is still relatively sparse given the noise in the data, with only three studies reconstructing a Phanerozic record (Farkaš et al. 2006, 2007; Blättler et al. 2012). Farkaš et al. (2007) published the first composite seawater record of Phanerozoic $\delta^{44/40}\mathrm{Ca}$ based on brachiopods, belemnites, rudists and foraminifera and phosphates for the Tertiary. Most of the Phanerozoic $\delta^{44/40}\mathrm{Ca}_{SW}$ is within the range 1–2 ‰ though there is considerable scatter in the data compared to any changes in the data. Some of this scatter almost certainly relates to the factors discussed above. However, there does appear to be a significant rise in $\delta^{44/40}\mathrm{Ca}_{SW}$ during the late Devonian/Carboniferous, and (Farkaš et al.

2007) noted that their record showed an increase from ~ 1.3 ‰ at the beginning of the Ordovician to ~ 2 ‰ at present. Initial modelling showed that such changes could not be reconciled by changes between the input/output fluxes, but rather must be caused by changes in the isotopic composition of the input/output fluxes (Farkaš et al. 2007). By mass balance, this makes sense because a consistent increase in $\delta^{44/40}\mathrm{Ca}_{SW}$ would imply a continual change in the Ca content of seawater, if the isotopic compositions of the outputs and inputs remained constant. As discussed above there are many factors that control the weathering input Ca isotope compositions that are only just beginning to be elucidated, but a change in the $\delta^{44/40}\mathrm{Ca}$ of seawater might reflect a major change in continental processes.

Blättler et al. (2012) commented that there were more abrupt changes in $\delta^{44/40}\mathrm{Ca}_{SW}$ that are coincident with changes between calcite and aragonite seas. They suggest that large-scale changes in the Ca isotope ratio of seawater, such as those in the late Carboniferous, were no longer possible after Jurassic time because of the generation of a deep-sea calcite sink expressed by deposition of foraminiferal-coccolith ooze across the world ocean. However, this assertion needs to be reconciled with the need for mass balance between the inputs and outputs of Ca to the oceans. Ultimately, on long time-scales, the bulk output flux of Ca, whether calcite or aragonite, must mirror the bulk input flux, and such changes in the fractionation factor between the bulk output of Ca and seawater should only be preserved with a passive tracer of $\delta^{44/40}\mathrm{Ca}_{SW}$.

Farkaš et al. (2007) remarked that there are several short-term, mostly negative, oscillations and several studies have focused on major perturbations to the carbon cycle such as at the end Permian (Payne et al. 2010). An abrupt change in sedimentation style occurs across the end-Permian extinction, thought to represent a major change in carbon cycle dynamics across this boundary. In carbonate strata from both the Tethys and Panthalassa oceans, microbialites and oolites overlie fossiliferous limestones of the latest Permian age, corresponding to a mass extinction and a large negative excursion in the carbon isotope $\delta^{13}C$ composition of carbonate minerals. The Permian-Triassic boundary from marine limestone in south China shows a significant but transient negative excursion of -0.3 ‰, that lasts several hundred thousand years (Payne et al. 2010). The causes of the change in sediment styles and attendant carbon isotope excursions have been debated for many years, but the plausible causes of this transient shift in $\delta^{44/40}Ca$ of seawater are more restricted. They could be a transient shift caused by (1) a change from calcite to aragonite seas, (2) a change in the weathering input isotopic composition or (3) a decrease in the amount of carbonate sedimentation caused by ocean acidification (Payne et al. 2010). Coupling Ca, C and B isotope records has enabled the ocean acidification hypothesis to be tested (Clarkson et al. 2015) suggesting that at least part of the Ca isotope excursion can be explained by ocean acidification.

The Mesozoic ocean anoxia events have also been the subject of several studies (e.g., Blättler et al. 2011; Brazier et al. 2015). These events correspond to geological periods where it is thought that there was significant environmental change. The early Toarcian is thought to have been punctuated by pulses of carbon release to the ocean-atmosphere system, that in turn are thought to have driven increased continental discharge, nutrient input, marine anoxia, seawater acidification and species extinctions. It has been suggested that these periods of environmental change are accompanied by increased carbonate and silicate weathering on the continents supplying increased Ca to seawater. Ca isotopes in the marine realm clearly have the potential to make a considerable contribution to furthering our understanding of the past marine Ca cycle (Blättler et al. 2011; Brazier et al. 2015). The extent of environmental change in the early Toarcian may be revealed by Ca isotopes on bulk carbonate from a section in Portugal (Brazier et al. 2015) which show a -0.8 ‰ decrease in $\delta^{44/40}Ca$, though coeval data on brachiopods do not appear to reveal the same trend. A key issue with bulk carbonates is demonstrating that there are no diagenetic effects, but unlike for trace elements it is more difficult to reset Ca isotopes diagenetically. Models of the data have been used to suggest that the data may reflect an increase in the continental weathering input of up to 400 %, a very significant environment change.

Similarly Cretaceous carbonate-rich sedimentary sections deposited during Oceanic Anoxic Events 1a (Early Aptian) and 2 (Cenomanian-Turonian) show small but significant negative Ca isotope excursions which recover to their pre-excursion values over several million years. These negative excursions have been used to infer a threefold increase in weathering fluxes to the oceans (Blättler et al. 2011).

Kasemann et al. (2005) investigated Ca isotope systematics in deeper geological time, across a major Neoproterozoic glaciation in Namibia. The signal to noise ratio of the data is relatively large but there is a tantalising hint of a

Table 6 Simplified modern ocean Ca budget

Budget	Ca flux (10^9 t/year)	Input mass (%)	$\delta^{44/40}Ca_{NIST\ SRM\ 915a}$ (‰)	2SD
Inputs				
Rivers	0.528	38.6–82.3	0.86	0.06
Hydrothermal vents	0.08–0.8	12.6–14.6	0.93	0.07
Groundwater	0.026–0.53	4.1–46.8	0.65	0.23
Input weighted average			0.78–0.88	
Outputs				
CaCO$_3$ sediment	0.84		0.76	0.03

negative Ca isotope excursion associated with this event, that coupled to other isotope systems has helped reconstruct the sequence of events attendant to one of the planets most profound climatic perturbations.

5 Conclusions

Calcium is one of the most important mobile metals that can migrate easily between major geochemical reservoirs at the Earth's surface; the hydrosphere and the biosphere and crust. It is the fifth most abundant element in the Earth's crust and the most abundant alkaline earth metal. It's global biogeochemical cycle has shaped the way the surface of the Earth has evolved over Earth's 4.5 Ga history. The above discussion has attempted to summarise the main processes that occur on both the continents and oceans that are relevant to Ca, and where available discuss Ca isotope evidence which has made significant advances in our understanding, or to suggest ways where Ca isotopes might provide a way forward. It is impossible to offer an exhaustive treatment of Ca in one chapter, but the above does summarise most of the key processes, that are likely to form part of research investigations over the next 10–20 years of research. Key to this chapter is that oceanic and continental cycles of Ca are coupled through time and space, and an understanding of both is required to constrain either one. Ca is supplied to the oceans through processes that occur on the continents, and Ca is restored to the continents tectonically, via the emplacement of limestones onto the continents. Ca isotopes are providing a major way of understanding both the continental and oceanic cycles (Table 6).

References

Adams JM, Post WM (1999) A preliminary estimate of changing calcrete carbon storage on land since the last glacial maximum. Global Planet Change 20(4):243–256

Ali MA, Dzombak DA (1996) Effects of simple organic acids on sorption of Cu^{2+} and Ca^{2+} on goethite. Geochim Cosmochim Acta 60(2):291–304. doi:http://dx.doi.org/10.1016/0016-7037(95)00385-1

Amini M, Eisenhauer A, Bohm F, Fietzke J, Bach W, Garbe-Schonberg D, Rosner M, Bock B, Lackschewitz KS, Hauff F (2008) Calcium Isotope ($\delta^{44/40}$Ca) fractionation along hydrothermal pathways, Logatchev field (Mid-Atlantic Ridge, 14°45'N). Geochim Cosmochim Act 72(16):4107–4122. doi:10.1016/j.gca.2008.05.055

Amini M, Eisenhauer A, Böhm F, Holmden C, Kreissig K, Hauff F, Jochum KP (2009) Calcium isotopes ($\delta^{44/40}$ Ca) in MPI-DING reference glasses, USGS rock powders and various rocks: evidence for Ca isotope fractionation in terrestrial silicates. Geostandards and Geoanalytical Research 33(2):231–247

Amiotte Suchet P, Probst JL, Ludwig W (2003) Worldwide distribution of continental rock lithology: implications for the atmospheric/soil CO_2 uptake by continental weathering and alkalinity river transport to the oceans. Global Biogeochem Cycles 17(2):1038. doi:10.1029/2002GB001891

Arning E, Lückge A, Breuer C, Gussone N, Birgel D, Peckmann J (2009) Genesis of phosphorite crusts off Peru. Marine Geology 262(1–4):68–81. doi:http://dx.doi.org/10.1016/j.margeo.2009.03.006

Aubert D, Stille P, Probst A (2001) REE fractionation during granite weathering and removal by waters and suspended loads: Sr and Nd isotopic evidence. Geochim Cosmochim Act 65(3):387–406. doi:http://dx.doi.org/10.1016/S0016-7037(00)00546-9

Bagard ML, Chabaux F, Pokrovsky OS, Viers J, Prokushkin AS, Stille P, Rihs S, Schmitt AD, Dupré B (2011) Seasonal variability of element fluxes in two Central Siberian rivers draining high latitude permafrost dominated areas. Geochim Cosmochim Acta 75 (12):3335–3357. doi:http://dx.doi.org/10.1016/j.gca.2011.03.024

Bagard ML, Schmitt AD, Chabaux F, Pokrovsky OS, Viers J, Stille P, Labolle F, Prokushkin AS (2013) Biogeochemistry of stable Ca and radiogenic Sr isotopes in a larch-covered permafrost-dominated watershed of Central Siberia. Geochim Cosmochim Acta 114(0):169–187. doi:http://dx.doi.org/10.1016/j.gca.2013.03.038

Bailey SW, Hornbeck JW, Driscoll CT, Gaudette HE (1996) Calcium inputs and transport in a base-poor forest ecosystem as interpreted by Sr isotopes. Water Resour Res 32(3):707–719

Bakker MR, George E, Turpault MP, Zhang JL, Zeller B (2005) Impact of douglas-fir and scots pine seedlings on plagioclase weathering under acidic conditions 266 (1–2):247–259. doi:10.1007/s11104-005-1153-7

Bélanger N, Holmden C (2010) Influence of landscape on the apportionment of Ca nutrition in a Boreal Shield forest of Saskatchewan (Canada) using $^{87}Sr/^{86}Sr$ as a tracer. Can J Soil Sci 90(2):267–288. doi:10.4141/CJSS09079, http://dx.doi.org/10.4141/CJSS09079

Bélanger N, Holmden C, Courchesne F, Cote B, Hendershot WH (2012) Constraining soil mineral weathering $^{87}Sr/^{86}Sr$ for calcium apportionment studies of a deciduous forest growing on soils developed from granitoid igneous rocks. Geoderma 185–186:84–96

Bellot J, Roda F, Retana J, Gracia CA (eds) (1999) Ecology of Mediterranean evergreen oak forests, ecological studies. Springer, Berlin, vol 137

Bern CR, Townsend AR, Farmer GL (2005) Unexpected dominance of parent-material strontium in a tropical forest on highly weathered soils. Ecology 86(3):626–632. doi:10.1890/03-0766

Bernasconi SM, Christi I, Hajdas I, Zimmermann S, Hagedom F, Smittenberg RH, Furrer G, Zeyer J, Brunner I, Frey B, Plötze M, Lapanje A, Edwards P, Venterink HO, Göransson H, Frossard E, Bünemann E, Jansa J, Tamburini F, Welc M, Mitchell E, Bourdon B, Kretzschmar R, Reynolds B, Lemarchand E, Wiederhold J, Tipper E, Kiczka M, Hindshaw R, Stähli M, Jonas T, Magnusson J, Bander A, Farinotti D, Huss M, Wacker L, Abbaspour K (2008) Weathering, soil formation and initial ecosystem evolution on a glacier forefield: a case study from the damma glacier, switzerland. Min Mag 72(1):19–22

Bernasconi SM, Bauder A, Bourdon B, Brunner I, Bünemann E, Christl I, Derungs N, Edwards P, Farinotti D, Frey B, Frossard E, Furrer G, Gierga M, Göransson H, Gülland K, Hagedorn F, Hajdas I, Hindshaw R, Ivy-Ochs S, Jansa J, Jonas T, Kiczka M, Kretzschmar R, Lemarchand E, Luster J, Magnusson J, Mitchell EA, Venterink HO, Plötze M, Reynolds B, Smittenberg RH, Stähli M, Tamburini F, Tipper ET, Wacker L, Welc M, Wiederhold JG, Zeyer J, Zimmermann S, Zumsteg A (2011) Chemical and biological gradients along the damma glacier soil chronosequence, switzerland. Vadose Zone J 10:867–883. doi:10.2136/vzj2010.0129

Berner EK, Berner RA (1996) global environment: water, air, and geochemical cycles. Prentice Hall, Upper Saddle River, New Jersey

Berner RA, Kothavala Z (2001) Geocarb III: a revised model of atmospheric CO_2 over Phanerozoic time. Am J Sci 301:182–204

Berner RA, Lasaga AC, Garrels RM (1983) The carbonate-silicate geochemical cycle and its effect on atmospheric carbon dioxide over the past 100 million years. Am J Sci 283:641–683

Bernhard-Reversat F (1972) Décomposition de la litière de feuilles en forêt ombrophile de basse Côte d'Ivoire. Oecologia Plantarum 7:279–300

Bickle MJ, Chapman HJ, Bunbury J, Harris NBW, Fairchild IJ, Ahmad T, Pomiès C (2005) Relative contributions of silicate and carbonate rocks to riverine Sr fluxes in the headwaters of the Ganges. Geochim Cosmochim Act 69(9):2221–2240. doi:10.1016/j.gca.2004.11.019

Bickle MJ, Tipper ET, Galy A, Chapman H, Harris N (2015) On discrimination between carbonate and silicate inputs to Himalayan rivers. Am J Sci 315:120–166

Blättler CL, Higgins JA (2014) Calcium isotopes in evaporites record variations in phanerozoic seawater SO_4 and Ca. Geology 42(8):711–714

Blättler CL, Jenkyns HC, Reynard LM, Henderson GM (2011) Significant increases in global weathering during Oceanic Anoxic Events 1a and 2 indicated by calcium isotopes. Earth Planet Sci Lett 309(1–2):77–88

Blättler CL, Henderson GM, Jenkyns HC (2012) Explaining the Phanerozoic Ca isotope history of seawater. Geology 40(9):843–846

Blum JD, Klaue A, Nezat CA, Driscoll CT, Johnson CE, Siccama TG, Eagar C, Fahey TJ, Likens GE (2002) Mycorrhizal weathering of apatite as an important calcium source in base-poor forest ecosystems. Nature 417(6890):729–731

Blum JD, Dasch AA, Hamburg SP, Yanai RD, Arthur MA (2008) Use of foliar Ca/Sr discrimination and $^{87}Sr/^{86}Sr$ ratios to determine soil Ca sources to sugar maple foliage in a northern hardwood forest. Biogeochemistry 87(3):287–296. doi:10.2307/40343550

Böhm F, Gussone N, Eisenhauer A, Dullo WC, Reynaud S, Paytan A (2006) Calcium isotope fractionation in modern scleractinian corals. Geochim Cosmochim Act 70(17):4452–4462. doi:10.1016/j.gca.2006.06.1546

Bonnot-Courtois CJRN (1982) Etude des échanges entre terres rares et cations interfoliaires de deux argiles. Clay Miner 17:409–420

Brantley S, White TS, White AF, Sparks D, Richter D, Pregitzer K, Derry L, Chorover J, Chadwick O, April R, Anderson S, Amundson R (2005) Frontiers in exploration of the critical zone: report of a workshop sponsored by the national science foundation (nsf)

Brantley SL, Goldhaber MB, Ragnarsdottir KV (2007) Crossing disciplines and scales to understand the critical zone. Elements 3(5):307

Braun JJ, Ngoupayou JRN, Viers J, Dupre B, Bedimo JPB, Boeglin JL, Robain H, Nyeck B, Freydier R, Nkamdjou LS, Rouiller J, Muller JP (2005) Present weathering rates in a humid tropical watershed: Nsimi, south cameroon. Geochim Cosmochim Acta 69(2):357–387. doi:http://dx.doi.org/10.1016/j.gca.2004.06.022

Bray J, Gorham E (1964) Litter production in forests of the world. Adv Ecol Res 2:101–157. doi: 10.1016/S0065-2504(08)60331-1

Brazier JM, Suan G, Tacail T, Simon L, Martin JE, Mattioli E, Balter V (2015) Calcium isotope evidence for dramatic increase of continental weathering during the Toarcian oceanic anoxic event (Early Jurassic). Earth Planet Sci Lett 411(0):164–176. doi:http://dx.doi.org/10.1016/j.epsl.2014.11.028

Bullen T, White A, Blum A, Harden J, Schulz M (1997) Chemical weathering of a soil chronosequence on granitoid alluvium: II. Mineralogic and isotopic constraints on the behavior of strontium. Geochim Cosmochim Acta 61(2):291–306

Bullen T, Fitzpatrick J, White A, Schulz M, Vivit D (2004) Calcium stable isotope evidence for three soil calcium pools at a granitoid chronosequence. In: Bullen T, Wang Y (eds) Proceedings of the 12th international symposium on water-rock interaction, Kunming, China, p 1734

Capo RC, Chadwick OA (1999) Sources of strontium and calcium in desert soil and calcrete. Earth Planet Sci Lett 170(1–2):61–72

Cappellato R, Peters NE (1995) Dry deposition and canopy leaching rates in deciduous and coniferous forests of the Georgia Piedmont: an assessment of a regression model. J Hydrol 169(1–4):131–150. doi: http://dx.doi.org/10.1016/0022-1694(94)02653-S

Cenki-Tok B, Chabaux F, Lemarchand D, Schmitt AD, Pierret MC, Viville D, Bagard ML, Stille P (2009) The impact of water-rock interaction and vegetation on calcium isotope fractionation in soil- and stream waters of a small, forested catchment (the Strengbach case). Geochim Cosmochim Act 73(8):2215–2228

Cerling TE (1984) The stable isotopic composition of modern soil carbonate and its relationship to climate. Earth Planet Sci Lett 71:229–240

Cerling TE, Pederson BL, Von Damm KL (1989) Sodium-calcium ion exchange in the weathering of shales: implications for global weathering budgets. Geol 17(6):552–554

Chabaux F, Riotte J, Schmitt AD, Carignan J, Herckes P, Pierret MC, Wortham H (2005) Variations of U and Sr isotope ratios in Alsace and Luxembourg rain waters: origin and hydrogeochemical implications. C R Geosci 337(16):1447–1456. doi:http://dx.doi.org/10.1016/j.crte.2005.07.008

Chadwick OA, Derry LA, Vitousek PM, Huebert BJ, Hedin LO (1999) Changing sources of nutrients during four million years of ecosystem development. Nature 397(6719):491–497

Chang VTC, Williams R, Makishima A, Belshaw NS, O'Nions RK (2004) Mg and Ca isotope fractionation during CaCO3 biomineralisation. Biochem Biophys Res Commun 323:79–85

Chapelle F (2003) 5.14—Geochemistry of groundwater. In: Turekian KK, Holland HD (eds) Treatise on geochemistry. Pergamon, Oxford, pp 425–449. doi: http://dx.doi.org/10.1016/B0-08-043751-6/05167-7

Chaudhuri S, Clauer N (1986) Fluctuations of isotopic composition of strontium in seawater during the phanerozoic eon. Chem Geol 59:293–303

Chiou CT, Rutherford DW (1997) Effects of exchanged cation and layer charge on the sorption of water and EGME vapors on montmorillonite clays 45(6):867–880

Chu NC, Henderson GM, Belshaw NS, Hedges REM (2006) Establishing the potential of ca isotopes as proxy for consumption of dairy products. Appl Geochem 21(10):1656–1667

Chuyong GB, Newbery DM, Songwe NC (2004) Rainfall input, throughfall and stemflow of nutrients in a central African rain forest dominated by ectomycorrhizal trees 67(1):73–91. doi: 10.1023/B:BIOG.0000015316.90198.cf

Clarholm M, Skyllberg U (2013) Translocation of metals by trees and fungi regulates pH, soil organic matter turnover and nitrogen availability in acidic forest soils. Soil Biol Biochem 63(0):142–153. doi:http://dx.doi.org/10.1016/j.soilbio.2013.03.019

Clarkson MO, Kasemann SA, Wood RA, Lenton TM, Daines SJ, Richoz S, Ohnemueller F, Meixner A, Poulton SW, Tipper ET (2015) Ocean acidification and the Permo-Triassic mass extinction. Science 348 (6231):229–232. doi:10.1126/science.aaa0193, http://www.sciencemag.org/content/348/6231/229.full.pdf

Cobert F, Schmitt AD, Bourgeade P, Labolle F, Badot PM, Chabaux F, Stille P (2011a) Experimental identification of Ca isotopic fractionations in higher plants. Geochim Cosmochim Act 75(19):5467–5482

Cobert F, Schmitt AD, Calvaruso C, Turpault MP, Lemarchand D, Collignon C, Chabaux F, Stille P (2011b) Biotic and abiotic experimental identification of bacterial influence on calcium isotopic signatures. Rapid Commun Mass Spectrom 25(19):2760–2768

Courty PE, Buée M, Diedhiou AG, Frey-Klett P, Tacon FL, Rineau F, Turpault MP, Uroz S, Garbaye J (2010) The role of ectomycorrhizal communities in forest ecosystem processes: new perspectives and emerging concepts. Soil Biol Biochem 42(5):679–698. doi:http://dx.doi.org/10.1016/j.soilbio.2009.12.006

Coûteaux MM, Bottner P, Berg B (1995) Litter decomposition, climate and liter quality. Trends Ecol Evol 10 (2):63–66. doi:http://dx.doi.org/10.1016/S0169-5347(00)88978-8

Crockford RH, Richardson DP (2000) Partitioning of rainfall into throughfall, stemflow and interception: effect of forest type, ground cover and climate. Hydrol Process 14 (16–17):2903–2920. doi:10.1002/1099-1085(200011/12)14:16/17<2903::AID-HYP126>3.0.CO;2-6

Dahlqvist R, Benedetti MF, Andersson K, Turner D, Larsson T, Stolpe B, Ingri J (2004) Association of calcium with colloidal particles and speciation of calcium in the Kalix and Amazon rivers. Geochim Cosmochim Acta 68(20):4059–4075. doi:http://dx.doi.org/10.1016/j.gca.2004.04.007

Davenport J, Caro G, France-Lanord C (2014) Tracing silicate weathering in the himalaya using the 40 K-40Ca system: a reconnaissance study. Procedia Earth Planet Sci 10(0):238–242. doi:http://dx.doi.org/10.1016/j.proeps.2014.08.030

De La Rocha CL, DePaolo DJ (2000) Isotopic evidence for variations in the marine calcium cycle over the Cenozoic. Science 289:1176–1178. doi:10.1126/science.289.5482.1176

De La Rocha CL, Hoff C, Bryce J (2008) Calcium cycle. In: Fath SEJD (ed) Encyclopedia of ecology. Academic Press, Oxford, pp 507–513. doi: http://dx.doi.org/10.1016/B978-008045405-4.00569-3

DePaolo DJ (2004) Calcium isotopic variatons produced by biological, kinetic, radiogenic and nucleosytheitic processes. Rev Min Geochem 55:255–288

DePaolo DJ (2011) Surface kinetic model for isotopic and trace element fractionation during precipitation of calcite from aqueous solutions. Geochim Cosmochim Act 75(4):1039–1056. doi:10.1016/j.gca.2010.11.020

de Villiers S (1998) Excess dissolved Ca in the deep ocean: a hydrothermal hypothesis. Earth Planet Sci Lett 164(3–4):627–641

Derry LA, Chadwick OA (2007) Contributions from earth's atmosphere to soil. Elements 3(5):333

Dezzeo N, Chacón N (2006) Nutrient fluxes in incident rainfall, throughfall, and stemflow in adjacent primary and secondary forests of the Gran Sabana, southern Venezuela. For Ecol Manage 234(1–3):218–226. doi: http://dx.doi.org/10.1016/j.foreco.2006.07.003

Dijkstra FA (2003) Calcium mineralization in the forest floor and surface soil beneath different tree species in the northeastern {US}. For Ecol Manage 175(1–3):185–194. doi:http://dx.doi.org/10.1016/S0378-1127(02)00128-7

Dijkstra FA, Smits MM (2002) Tree species effects on calcium cycling: the role of calcium uptake in deep soils. Ecosystems 5(4):385–398. doi:10.1007/s10021-001-0082-4

Dowling CB, Poreda RJ, Basu AR (2003) The groundwater geochemistry of the Bengal Basin: weathering, chemsorption, and trace metal flux to the oceans. Geochim Cosmochim Acta 67(12):2117–2136

Drever JI (1994) The effect of land plants on weathering rates of silicate minerals. Geochim Cosmochim Acta 58(10):2325–2332

Drever JI et al (1988) Geochemical cycles: the continental crust and the oceans. In: Chemical cycles in the evolution of the earth. A Wiley Interscience publication, Hoboken

Drever JI (1997) The geochemisrty of Natural waters, 3rd edn. Prentice-Hall Inc., Upper Saddle River

Drouet T, Herbauts J, Demaiffe D (2005) Long-term records of strontium isotopic composition in tree rings suggest changes in forest calcium sources in the early 20th century. Global Change Biol 11(11):1926–1940. doi:10.1111/j.1365-2486.2005.01034.x

Dupré B, Gaillardet J, Rousseau D, Allègre CJ (1996) Major and trace elements of river-borne material: the Congo Basin. Geochim Cosmochim Acta 60:1301–1321

Dürr HH, Meybeck M, Dürr SH (2005) Lithologic composition of the earth's continental surfaces derived from a new digital map emphasizing riverine material transfer. Global Biogeochem Cycles 19(4):n/a–n/a. doi:10.1029/2005GB002515, gB4S10

Ewing SA, Yang W, DePaolo DJ, Michalski G, Kendall C, Stewart BW, Thiemens M, Amundson R (2008) Non-biological fractionation of stable Ca isotopes in soils of the Atacama Desert, Chile. Geochim Cosmochim Act 72:1096–1110. doi:10.1016/j.gca.2007.10.029

Fantle MS (2010) Evaluating the Ca isotope proxy. Am J Sci 310(3):194–230

Fantle MS (2015) Calcium isotopic evidence for rapid recrystallization of bulk marine carbonates and implications for geochemical proxies. Geochim Cosmochim Acta 148:378–401. doi:http://dx.doi.org/10.1016/j.gca.2014.10.005

Fantle MS, DePaolo DJ (2005) Variations in the marine Ca cycle over the past 20 million years. Earth Planet Sci Lett 237(1–2):102–117. doi:10.1016/j.epsl.2005.06.024

Fantle MS, DePaolo DJ (2007) Ca isotopes in carbonate sediment and pore fluid from ODP site 807A: The Ca^{2+} (aq)–calcite equilibrium fractionation factor and calcite recrystallization rates in Pleistocene sediments. Geochim Cosmochim Act 71(10):2524–2546. doi:10.1016/j.gca.2007.03.006

Fantle MS, Tipper ET (2014) The global calcium cycle: a review. Earth-Sci Rev 129:148–177

Fantle MS, Tollerud H, Eisenhauer A, Holmden C (2012) The Ca isotopic composition of dust-producing regions: measurements of surface sediments in the Black Rock Desert, Nevada. Geochim Cosmochim Act 87:178–193

FAO (2007) The state of food and agriculture

Farkaš J, Buhl D, Blenkinsop J, Veizer J (2006) Evolution of the oceanic calcium cycle during the late Mesozoic: Evidence from $\delta^{44/40}$Ca of marine skeletal carbonates. Earth Planet Sci Lett 253(1–2):96–111. doi:10.1016/j.epsl.2006.10.015

Farkaš J, Böhm F, Wallmann K, Blenkinsop J, Eisenhauer A, van Geldern R, Munnecke A, Voigt S, Veizer J (2007) Calcium isotope record of phanerozoic oceans: Implications for chemical evolution of seawater and its causative mechanisms. Geochim Cosmochim Act 71(21):5117–5134. doi:10.1016/j.gca.2007.09.004

Farkaš J, Déjeant A, Novák M, Jacobsen SB (2011) Calcium isotope constraints on the uptake and sources of Ca^{2+} in a base-poor forest: a new concept of combining stable ($\delta^{44/42}$Ca) and radiogenic (ϵCa) signals. Geochim Cosmochim Act 75(22):7031–7046. doi:http://dx.doi.org/10.1016/j.gca.2011.09.021

Faurie C, Ferra C, Medori P, Devaux J, Hemptinne J (2011) Ecologie—approche scientifique et pratique, 6th edn. Lavoisier, Paris

Fichter J, Turpault MP, Dambrine E, Ranger J (1998) Localization of base cations in particle size fractions of acid forest soils (Vosges Mountains, N-E France). Geoderma 82(4):295–314. doi:http://dx.doi.org/10.1016/S0016-7061(97)00106-7

Ford D, Williams PD (2007) Karst hydrogeology and geomorphology. Wiley, Hoboken

France-Lanord C, Derry LA (1997) Organic carbon burial forcing of the carbon cycle from Himalayan Erosion. Nature 390:65–67

Gaillardet J, Dupré B, Allègre CJ (1999a) Geochemistry of large river suspended sediments: silicate weathering or recycling tracer? Geochim Cosmochim Act 63(23–24):1037–4051. doi:10.1016/S0016-7037(99)00307-5

Gaillardet J, Dupré B, Louvat P, Allègre CJ (1999b) Global silicate weathering and CO_2 consumption rates deduced from the chemistry of large rivers. Chem Geol 159:3–30

Galy A, France-Lanord C (1999) Weathering processes in the Ganges-Brahmaputra basin and the riverine alkalinity budget. Chem Geol 159(1–4):31–60. doi:10.1016/S0009-2541(99)00033-9

Galy A, France-Lanord C, Derry LA (1999) The strontium isotopic budget of Himalayan Rivers in Nepal and Bangladesh. Geochim Cosmochim Act 63(13–14):1905–1925. doi:10.1016/S0016-7037(99)00081-2

Gangloff S, Stille P, Pierret MC, Weber T, Chabaux F (2014) Characterization and evolution of dissolved organic matter in acidic forest soil and its impact on the mobility of major and trace elements (case of the Strengbach watershed). Geochim Cosmochim Acta 130:21–41. doi:http://dx.doi.org/10.1016/j.gca.2013.12.033

Gobran GR, Clegg S, Courchesne F (1998) Rhizospheric processes influencing the biogeochemistry of forest ecosystems 42(1–2):107–120. doi:10.1023/A:1005967203053

Gregor CB (1985) The mass-age distribution of Phanerozoic sediments. Geological Society, London, Memoirs 10(1):284–289

Griffith EM, Paytan A, Caldeira K, Bullen TD, Thomas E (2008a) A dynamic marine calcium cycle during the past 28 million years. Science 322(5908):1671–1674

Griffith EM, Paytan A, Kozdon R, Eisenhauer A, Ravelo AC (2008b) Influences on the fractionation of calcium isotopes in planktonic foraminifera. Earth Planet Sci Lett 268(12):124–136

Griffith EM, Schauble EA, Bullen TD, Paytan A (2008c) Characterization of calcium isotopes in natural and synthetic barite. Geochim Cosmochim Act 72 (23):5641–5658

Griffith EM, Paytan A, Eisenhauer A, Bullen TD, Thomas E (2011) Seawater calcium isotope ratios across the eocene-oligocene transition. Geology 39:683–686

Griffith EM, Fantle MS, Eisenhauer A, Paytan A, Bullen TD (2015) Effects of ocean acidification on the marine calcium isotope record at the Paleocene–Eocene thermal maximum. Earth Planet Sci Lett 419:81–92. doi:http://dx.doi.org/10.1016/j.epsl.2015.03.010

Grousset FE, Biscaye PE (2005) Tracing dust sources and transport patterns using Sr, Nd and Pb isotopes. Chem Geol 222(3–4):149–167. doi:http://dx.doi.org/10.1016/j.chemgeo.2005.05.006

Gussone N, Filipsson HL (2010) Calcium isotope ratios in calcitic tests of benthic foraminifers. Earth Planet Sci Lett 290(12):108–117

Gussone N, Eisenhauer A, Heuser A, Dietzel M, Bock B, Böhm F, Spero HJ, Lea DW, Bijma J, Nägler TF (2003) Model for kinetic effects on calcium isotope fractionation (δ^{44} Ca) in inorganic aragonite and cultured planktonic foraminifera. Geochim Cosmochim Acta 67(7):1375–1382

Gussone N, Eisenhauer A, Tiedemann R, Haug GH, Heuser A, Bock B, Ngler TF, Müller A (2004) Reconstruction of caribbean sea surface temperature and salinity fluctuations in response to the pliocene closure of the central american gateway and radiative forcing, using $\delta^{44/40}$Ca, $\delta^{18}O$ and Mg/Ca ratios. Earth Planet Sci Lett 227(3–4):201–214

Gussone N, Böhm F, Eisenhauer A, Dietzel M, Heuser A, Teichert BMA, Reitner J, Wörheide G, Dullo WC (2005) Calcium isotope fractionation in calcite and aragonite. Geochim Cosmochim Act 69(18):4485–4494. doi:10.1016/S0016-7037(02)01296-6

Gussone N, Langer G, Geisen M, Steel BA, Riebesell U (2007) Calcium isotope fractionation in coccoliths of cultured calcidiscus leptoporus, helicosphaera carteri, syracosphaera pulchra and umbilicosphaera foliosa. Earth Planet Sci Lett 260(3–4):505–515

Gussone N, Hönisch B, Heuser A, Eisenhauer A, Spindler M, Hemleben C (2009) A critical evaluation of calcium isotope ratios in tests of planktonic foraminifers. Geochim Cosmochim Acta 73(24):7241–7255. doi:http://dx.doi.org/10.1016/j.gca.2009.08.035

Gussone N, Zonneveld K, Kuhnert H (2010) Minor element and Ca isotope composition of calcareous dinoflagellate cysts of cultured Thoracosphaera heimii. Earth Planet Sci Lett 289(1–2):180–188. doi:http://dx.doi.org/10.1016/j.epsl.2009.11.006

Gussone N, Nehrke G, Teichert BMA (2011) Calcium isotope fractionation in ikaite and vaterite. Chem Geol 285(14):194–202

Halicz L, Galy A, Belshaw NS, O'Nions RK (1999) High-precision measuremeomt of calcium isotopes in carbonates and related materials by multiple collector inductively coupled plasma mass spectrometry (MC-ICP-MS). J Anal At Spectrosc 14:1835–1838

Harouaka K, Eisenhauer A, Fantle MS (2014) Experimental investigation of Ca isotopic fractionation during abiotic gypsum precipitation. Geochim Cosmochim Acta 129:157–176. doi:http://dx.doi.org/10.1016/j.gca.2013.12.004

Hartmann J, Moosdorf N (2012) The new global lithological map database glim: a representation of rock properties at the earth surface. Geochemistry, Geophysics, Geosystems 13(12):n/a–n/a. doi:10.1029/2012GC004370

Heinemann A, Fietzke J, Eisenhauer A, Zumholz K (2008) Modification of Ca isotope and trace metal composition of the major matrices involved in shell formation of Mytilus edulis. Geochemistry, Geophysics, Geosystems 9(1):n/a–n/a. doi:10.1029/2007GC001777

Henderson G, Bayon G, Benoit M, Chu NC (2006) $\delta^{44/42}$ Ca in gas hydrates, porewaters and authigenic carbonates from Niger Delta sediments. Geochim Cosmochim Act 70(18):A244

Hensley TM (2006) Calcium isotopic variation in marine evaporites and carbonates: applications to late miocene mediterranean brine chemistry and late cenozoic calcium cycling in the oceans. PhD thesis, Scripps Institution of Oceanography

Hernández J, del Pino A, Salvo L, Arrarte G (2009) Nutrient export and harvest residue decomposition patterns of a Eucalyptus dunnii Maiden plantation in temperate climate of Uruguay. For Ecol Manage 258 (2):92–99. doi:http://dx.doi.org/10.1016/j.foreco.2009.03.050

Herwitz SR (1986) Infiltration-excess caused by Stemflow in a cyclone-prone tropical rainforest. Earth Surf Proc Land 11(4):401–412. doi:10.1002/esp.3290110406

Heuser A, Eisenhauer A, Böhm F, Wallmann K, Gussone N, Pearson PN, Nägler TF, Dullo WC (2005) Calcium isotope ($\delta^{44/40}$ Ca) variations of Neogene planktonic foraminifera. Paleoceanography 20: PA2013. doi:10.1029/2004PA001048

Hiebenthal C (2009) Sensitivity of A. islandica and M. edulis towards environmental changes: a threat to the bivalves—an opportunity for palaeo-climatology?

Hiltner L (1904) Über neuere Erfahrungen und Probleme auf dem Gebiet der Bodenbakteriologie und unter besonderer Berücksichtigung der Gründüngung und Brache. Arbeiten der Deutschen Landwirtschafts Gesellschaft 1904(98):59–78

Hinds WC (1999) Aerosol technology: properties, behavior, and measurement of airborne particles, 2nd edn. Wiley-Interscience, Hoboken

Hindshaw RS, Reynolds BC, Wiederhold JG, Kretzschmar R, Bourdon B (2011) Calcium isotopes in a proglacial weathering environment: Damma glacier, Switzerland. Geochim Cosmochim Act 75(1):106–118

Hindshaw R, Reynolds B, Wiederhold J, Kiczka M, Kretzschmar R, Bourdon B (2012) Calcium isotope fractionation in alpine plants. Biogeochemistry 1–16. doi:10.1007/s10533-012-9732-1

Hindshaw RS, Bourdon B, Pogge von Strandmann PAE, Vigier N, Burton KW (2013) The stable calcium isotopic composition of rivers draining basaltic catchments in iceland. Earth Planet Sci Lett 374:173–184. doi:http://dx.doi.org/10.1016/j.epsl.2013.05.038

Hinojosa JL, Brown ST, Chen J, DePaolo DJ, Paytan A, Shen Sz, Payne JL (2012) Evidence for end-Permian ocean acidification from calcium isotopes in biogenic apatite. Geology 40(8):743–746. doi:10.1130/G33048.1, http://geology.gsapubs.org/content/40/8/743.full.pdf+html

Hinsinger P (1998) How do plant roots acquire mineral nutrients? Chemical processes involved in the rhizosphere. Adv Agr 64:225–265

Hinsinger P, Gilkes RJ (1996) Mobilization of phosphate from phosphate rock and alumina–sorbed phosphate by the roots of ryegrass and clover as related to rhizosphere pH. Eur J Soil Sci 47(4):533–544. doi:10.1111/j.1365-2389.1996.tb01853.x

Hinsinger P, Gilkes R (1997) Dissolution of phosphate rock in the rhizosphere of five plant species grown in an acid, P-fixing mineral substrate. Geoderma 75(3–4):231–249. doi:http://dx.doi.org/10.1016/S0016-7061(96)00094-8

Hinsinger P, Bolland M, Gilkes R (1995) Silicate rock powder: effect on selected chemical properties of a range of soils from Western Australia and on plant growth as assessed in a glasshouse experiment. Fertilizer Res 45(1):69–79

Hinsinger P, Gobran GR, Gregory PJ, Wenzel WW (2005) Rhizosphere geometry and heterogeneity arising from root-mediated physical and chemical processes. New Phytologist 168(2):293–303. doi:10.1111/j.1469-8137.2005.01512.x

Hinsinger P, Plassard C, Jaillard B (2006) Rhizosphere: a new frontier for soil biogeochemistry. J Geochem Explor 88(1):210–213

Hippler D, Schmitt A, Gussone N, Heuser A, Stille PAE, Nägler T (2003) Ca isotopic composition of various standards and seawater. Geostandard Newslett 27:13–19

Hippler D, Eisenhauer A, Nägler TF (2006) Tropical atlantic SST history inferred from Ca isotope thermometry over the last 140 ka. Geochim Cosmochim Act 70(1):90–100. doi:10.1016/j.gca.2005.07.022

Hippler D, Buhl D, Witbaard R, Richter DK, Immenhauser A (2009) Towards a better understanding of magnesium-isotope ratios from marine skeletal carbonates. Geochim Cosmochim Act 73(20):6134–6146. doi:10.1016/j.gca.2009.07.031

Hippler D, Witbaard R, van Aken HM, Buhl D, Immenhauser A (2013) Exploring the calcium isotope signature of Arctica islandica as an environmental proxy using laboratory- and field-cultured specimens. Palaeogeogr Palaeoclimatol Palaeoecol 373:75–87

Högberg P, Read DJ (2006) Towards a more plant physiological perspective on soil ecology. Trends Ecol Evol 21(10):548–554. doi:http://dx.doi.org/10.1016/j.tree.2006.06.004

Holmden C (2009) Ca isotope study of ordovician dolomite, limestone, and anhydrite in the williston basin: implications for subsurface dolomitization and local ca cycling. Chem Geol 268:180–188

Holmden C, Bélanger N (2010) Ca isotope cycling in a forested ecosystem. Geochim Cosmochim Act 74 (3):995–1015

Holmden C, Papanastassiou DA, Blanchon P, Evans S (2012) $\delta^{44/40}$ ca variability in shallow water carbonates and the impact of submarine groundwater discharge on ca-cycling in marine environments. Geochim Cosmochim Act 83:179–194. doi:10.1016/j.gca.2011.12.031

Huang S, Farkaš J, Jacobsen SB (2010) Calcium isotopic fractionation between clinopyroxene and orthopyroxene from mantle peridotites. Earth Planet Sci Lett 292 (3–4):337–344

Huang S, Farkaš J, Jacobsen SB (2011) Stable calcium isotopic compositions of hawaiian shield lavas: evidence for recycling of ancient marine carbonates into the mantle. Geochim Cosmochim Acta 75(17):4987–4997

Hubbard DK, Miller AI, Scaturo D (1990) Production and cycling of calcium carbonate in a shelf-edge reef system (St. Croix, US Virgin Islands): applications to the nature of reef systems in the fossil record. J Sediment Res 60(3)

Husson J, Higgins J, Maloof A et al (2015) Ca and mg isotope constraints on the origin of earth's deepest excursion. Geochim Cosmochim Act 160:243–266

Immenhauser A, Nägler TF, Steuber T, Hippler D (2005) A critical assessment of mollusk $^{18}O/^{16}O$, Mg/Ca, and $^{44}Ca/^{40}Ca$ ratios as proxies for Cretaceous seawater temperature seasonality. Palaeogeogr Palaeoclimatol Palaeoecol 215(3–4):221–237. doi:http://dx.doi.org/10.1016/j.palaeo.2004.09.005

IPCC (2014) Climate Change 2014. CUP, NY

Jackson M (1956) Soil chemical analysis—advanced course. Department of Soils University of Wisconsin, Madison

Jacobson AD, Holmden C (2008) δ^{40} Ca evolution in a carbonate aquifer and its bearing on the equilibrium isotope fractionation factor for calcite. Earth Planet Sci Lett 270(3–4):349–353. doi:10.1016/j.epsl.2008.03. 039

Jacobson AD, Blum JD, Walter LM (2002) Reconciling the elemental and Sr isotope composition of Himalayan weathering fluxes: insights from the carbonate geochemistry of stream waters. Geochim Cosmochim Act 66(19):3417–3429

Jacobson AD, Grace Andrews M, Lehn GO, Holmden C (2015) Silicate versus carbonate weathering in iceland: new insights from ca isotopes. Earth Planet Sci Lett 416:132–142. doi:http://dx.doi.org/10.1016/j.epsl. 2015.01.030

Jarvis I, Burnett W, Nathan Y, Almbaydin F, Attia AKM, Castro LN, Flicoteaux R, Himly ME, Husain V, Outawnah AA, Serjani A, Zanin YN (1994) Phosphorite geochemistry: state-of-the-art and environmental concerns. In concepts and controversies in phosphogenesis. Eclogae Geol Helv 87(3):643–700

John DM (1973) Accumulation and decay of litter and net production of forest in tropical West Africa. Oikos 24 (3):430–435. doi:10.2307/3543819

Jordan CF (1985) Nutrient cycling in tropical forest ecosystems: principles and their application in management and conservation. Wiley, Chichester, P019 1UD

Jordan CF, Herrera R (1981) Tropical rain forests: are nutrients really critical? Am Nat 117(2):167–180. doi:10.2307/2460498

Jost AB, Mundil R, He B, Brown ST, Altiner D, Sun Y, DePaolo DJ, Payne JL (2014) Constraining the cause of the end-Guadalupian extinction with coupled records of carbon and calcium isotopes. Earth Planet Sci Lett 396:201–212. doi:http://dx.doi.org/10.1016/j. epsl.2014.04.014

Kasemann SA, Hawkesworth C, Prave AR, Fallick AE, Pearson PN (2005) Boron and calcium isotope composition in Neoproterozoic carbonate rocks from Namibia: evidence for extreme environmental change. Earth Planet Sci Lett 231(1–2):73–86

Kasemann SA, Schmidt DN, Pearson PN, Hawkesworth CJ (2008) Biological and ecological insights into Ca isotopes in planktic foraminifers as a palaeotemperature proxy. Earth Planet Sci Lett 271(1):292–302

Kasemann SA, Pogge von Strandmann PAE, Prave AR, Fallick AE, Elliott T, Hoffmann KH (2014) Continental weathering following a cryogenian glaciation: evidence from calcium and magnesium isotopes. Earth Planet Sci Lett 396:66–77. doi:http://dx.doi.org/10. 1016/j.epsl.2014.03.048

Kennedy MJ, Chadwick O, Vitousek P, Derry LA, Hendricks D (1998) Replacement of weathering with atmospheric sources of base cations during ecosystem development, Hawaiian Islands. Geol 26:1015–1018

Kennedy MJ, Hedin LO, Derry LA (2002) Decoupling of unpolluted temperate forests from rock nutrient sources revealed by natural ^{87}Sr/^{86}Sr and ^{84}Sr tracer addition. Proc Natl Acad Sci 99(15):9639–9644

Komiya T, Suga A, Ohno T, Han J, Guo J, Yamamoto S, Hirata T, Li Y (2008) Ca isotopic compositions of dolomite, phosphorite and the oldest animal embryo fossils from the Neoproterozoic in Weng'an, South China. Gondwana Research 14(1–2):209–218

Langer G, Gussone N, Nehrke G, Riebesell U, Eisenhauer A, Thoms S (2007) Calcium isotope fractionation during coccolith formation in Emiliania huxleyi: independence of growth and calcification rate. Geochem Geophys Geosyst 8(Q05007). doi:10.1029/ 2006GC001,422

Lawrence CR, Neff JC (2009) The contemporary physical and chemical flux of aeolian dust: a synthesis of direct measurements of dust deposition. Chem Geol 267(1–2):46–63. doi:http://dx.doi.org/10.1016/j.chemgeo. 2009.02.005

Lemarchand D, Wasserburg G, Papanastassiou D (2004) Rate-controlled calcium isotope fractionation in synthetic calcite. Geochim Cosmochim Acta 68 (22):4665–4678

Leyval C, Berthelin J (1991) Weathering of a mica by roots and rhizospheric microorganisms of pine 55:1009–1016. doi:10.2136/sssaj1991. 03615995005500040020x

Li S, Lu XX, He M, Zhou Y, Bei R, Li L, Ziegler AD (2011) Major element chemistry in the upper Yangtze River: a case study of the Longchuanjiang River. Geomorphology 129(1–2):29–42. doi:http://dx.doi. org/10.1016/j.geomorph.2011.01.010

Liao JH, Wang HH, Tsai CC, Hseu ZY (2006) Litter production, decomposition and nutrient return of uplifted coral reef tropical forest. For Ecol Manage 235(1–3):174–185. doi:http://dx.doi.org/10.1016/j. foreco.2006.08.010

Likens GE, Bormann F (1995) Biogeochemistry of a forest ecosystem, 2nd edn. Springer, Berlin

Likens GE, Bormann FH, Johnson NM, Pierce RS (1967) The calcium, magnesium, potassium, and sodium budgets for a small forested ecosystem. Ecology 48 (5):772–785. doi:10.2307/1933735

Likens GE, Driscoll CT, Buso DC (1996) Long-term effects of acid rain: response and recovery of a forest ecosystem. Science 272(5259):244–246

Likens GE, Driscoll CT, Buso DC, Siccama TG, Johnson CE, Lovett GM, Fahey TJ, Reiners WA, Ryan DF, Martin CW, Bailey SW (1998) The biogeochemistry of calcium at Hubbard Brook. Biogeochemistry 41 (2):89–173

Lovett GM, Lindberg SE (1984) Dry deposition and canopy exchange in a mixed oak forest as determined by analysis of throughfall. J Appl Ecol 21(3):1013–1027. doi:10.2307/2405064

Lovett GM, Nolan SS, Driscoll CT, Fahey TJ (1996) Factors regulating throughfall flux in a New Hampshire forested landscape. Can J For Res 26(12):2134–

2144. doi:10.1139/x26-242, http://dx.doi.org/10.1139/x26-242

Mackenzie FT, Morse JW (1992) Sedimentary carbonates through phanerozoic time. Geochim Cosmochim Acta 56:3281–3295

Mackenzie FT (2003) Our Changing Planet: An Introduction to Earth System Science and Global Environmental Change. Prentice-Hall

Mahowald NM, Muhs DR, Levis S, Rasch PJ, Yoshioka M, Zender CS, Luo C (2006) Change in atmospheric mineral aerosols in response to climate: last glacial period, preindustrial, modern, and doubled carbon dioxide climates. J Geophys Res: Atmos 111 (D10):n/a–n/a. doi:10.1029/2005JD006653

Manrique LA, Jones CA (1991) Bulk density of soils in relation to soil physical and chemical properties. Soil Sci Soc Am J 55:476–481. doi:10.2136/sssaj1991.03615995005500020030x

Marques R, Ranger J, Villette S, Granier A (1997) Nutrient dynamics in a chronosequence of Douglas-fir (Pseudotsuga menziesii (Mirb). Franco) stands on the Beaujolais Mounts (France). 2. Quantitative approach. For Ecol Manage 92(1):167–197. doi:10.1016/S0378-1127(96)03913-8

Marriott CS, Henderson GM, Belshaw NS, Tudhope AW (2004) Temperature dependence of δ^7 Li, δ^{44} Ca and Li/Ca during growth of calcium carbonate. Earth Planet Sci Lett 222:615–624

Martignier L, Verrecchia EP (2013) Weathering processes in superficial deposits (regolith) and their influence on pedogenesis: a case study in the Swiss Jura Mountains. Geomorphology 189:26–40

McLaughlin SB, Wimmer R (1999) Tansley review no. 104. New Phytol 142(3):373–417. doi:10.1046/j.1469-8137.1999.00420.x

Meybeck M (1979) Concentration des eaux fluviales en éléments majeurs et apports en solutions aux océans. Rev Géologie Dynamique et Géographie Physique 21 (3):215–246

Meybeck M (1987) Global chemical weathering of surficial rocks estimated from river dissolved load. Am J Sci 287:401–428

Miller EK, Blum JD, Friedland AJ (1993) Determination of soil exchangeable-cation loss and weathering rates using Sr isotopes. Nature 362(6419):438–441

Milliere L, Verrecchia E, Gussone N (2014) Stable calcium isotope composition of a pedogenic carbonate in forested ecosystem: the case of the needle fibre calcite (NFC). In: EGU General Assembly Conference Abstracts, EGU General Assembly Conference Abstracts, vol 16, p 5812

Milliman J (1993) Production and accumulation of calcium carbonate in the ocean: budget of a nonsteady state. Global Biogeochem Cycles 7:927–957

Milliman JD, Droxler AW (1996) Neritic and pelagic carbonate sedimentation in the marine environment: ignorance is not bliss. Geol Rundsch 85:496–504

Moon S, Chamberlain CP, Hilley GE (2014) New estimates of silicate weathering rates and their uncertainties in global rivers. Geochim Cosmochim Act 134:257–274. doi:http://dx.doi.org/10.1016/j.gca.2014.02.033

Mooney RW, Keenan AG, Wood LA (1952) Adsorption of Water Vapor by Montmorillonite. II. Effect of Exchangeable Ions and Lattice Swelling as Measured by X-Ray Diffraction. Journal of the American Chemical Society 74(6):1371–1374. doi:10.1021/ja01126a002, http://dx.doi.org/10.1021/ja01126a002

Moore J, Jacobson AD, Holmden C, Craw D (2013) Tracking the relationship between mountain uplift, silicate weathering, and long-term co2 consumption with ca isotopes: Southern alps, new zealand. Chem Geol 341:110–127. doi:http://dx.doi.org/10.1016/j.chemgeo.2013.01.005

Mulder J, Cresser M (1994) Soil and soil solution chemistry. In: Biogeochemistry of small catchments: a tool for environmental research, Wiley, Hoboken pp 107–131

Müller RD, Sdrolias M, Gaina C, Steinberger S, Heine C (2008) Long-term sea-level fluctuations driven by ocean basin dynamics. Science 319:1357–1362

Nägler T, Eisenhauer A, Muller A, Hemleben C, Kramers J (2000) The δ^{44} ca-temperature calibration on fossil and cultured globigerinoides sacculifer: new tool for reconstruction of past sea surface temperatures. Geochem Geophys Geosyst 1 (9):2000GC000,091

Nagy KL (1995) Dissolution and precipitation kinetics of sheet silicates. Rev Mineral Geochem 31(1):173–233

Neumann AC, Land LS (1975) Lime mud deposition and calcareous algae in the Bight of Abaco, Bahamas: a budget. J Sediment Res 45(4)

Nicótina L, Tarboton DG, Tesfa TK, Rinaldo A (2011) Hydrologic controls on equilibrium soil depths. Water Resour Res 47(4):n/a–n/a. doi:10.1029/2010WR009538

Nielsen LC, DePaolo DJ (2013) Ca isotope fractionation in a high-alkalinity lake system: Mono Lake, California. Geochim Cosmochim Acta 118:276–294. doi:http://dx.doi.org/10.1016/j.gca.2013.05.007

NRC (2001) Basic research opportunities in earth science. National Academy Press, Washington DC, USA

Nykvist N (2000) Tropical forests can suffer from a serious deficiency of calcium after logging. AMBIO: J Hum Environ 29(6):310–313. doi:10.1579/0044-7447-29.6.310

Ockert C, Gussone N, Kaufhold S, Teichert BMA (2013) Isotope fractionation during ca exchange on clay minerals in a marine environment. Geochim Cosmochim Acta 112:374–388. doi:http://dx.doi.org/10.1016/j.gca.2012.09.041

Oehlerich M, Mayr C, Gussone N, Hahn A, Hölzl S, Lücke A, Ohlendorf C, Rummel S, Teichert BMA, Zolitschka B (2015) Lateglacial and Holocene climatic changes in south-eastern Patagonia inferred from carbonate isotope records of La- guna Potrok Aike (Argentina). Quat Sci Rev 114:189–202. doi: http://dx.doi.org/10.1016/j.quascirev.2015.02.006

Oelkers EH, Gislason SR, Eiriksdottir ES, Jones M, Pearce CR, Jeandel C (2011) The role of riverine particulate material on the global cycles of the elements. Appl Geochem 26 Supplement(0):S365–S369. doi:http://dx.doi.org/10.1016/j.apgeochem.2011.03.062, ninth International Symposium on the Geochemistry of the Earth's Surface (GES-9)

Oliva Viers, Duprà Oliva P, Viers J, Dupré B (2003) Chemical weathering in granitic environments. Chem Geol 202(3–4):225–256

Olson JS (1963) Energy storage and the balance of producers and decomposers in ecological systems. Ecology 44(2):322–331. doi:10.2307/1932179

Opdyke BN, Walker JC (1992) Return of the coral reef hypothesis: basin to shelf partitioning of $CaCO_3$ and its effect on atmospheric CO_2. Geology 20(8):733–736, nASA: Grant numbers: NAGW-176

Page BD, Bullen TD, Mitchell MJ (2008) Influences of calcium availability and tree species on Ca isotope fractionation in soil and vegetation. Biogeochem doi:10.1007/s10533-008-9188-5

Park A, Cameron JL (2008) The influence of canopy traits on throughfall and stemflow in five tropical trees growing in a Panamanian plantation. For Ecol Manage 255(5–6):1915–1925. doi:http://dx.doi.org/10.1016/j.foreco.2007.12.025

Parker GG (1983) Throughfall and stemflow in the forest nutrient cycle. Adv Ecol Res 13:57–133

Pawlowski J, Holzmann M, Berney C, Fahrni J, Gooday AJ, Cedhagen T, Habura A, Bowser SS (2003) The evolution of early Foraminifera. Proc Nat Acad Sci 100(20):11494–11498. doi:10.1073/pnas.2035132100, http://www.pnas.org/content/100/20/11494.full.pdf

Payne JL, Turchyn AV, Paytan A, DePaolo DJ, Lehrmann DJ, Yu M, Wei J (2010) Calcium isotope constraints on the end-permian mass extinction. Proc Natl Acad Sci 107(19):8543–8548

Pelletier JD, Rasmussen C (2009) Geomorphically based predictive mapping of soil thickness in upland watersheds. Water Resour Res 45(9):n/a–n/a. doi:10.1029/2008WR007319

Perakis SS, Maguire DA, Bullen TD, Cromack K, Waring RH, Boyle JR (2006) Coupled nitrogen and calcium cycles in forests of the Oregon Coast Range. Ecosystems 9(1):63–74. doi:10.1007/s10021-004-0039-5

Pett-Ridge JC, Derry LA, Kurtz AC (2009) Sr isotopes as a tracer of weathering processes and dust inputs in a tropical granitoid watershed, luquillo mountains, puerto rico. Geochim Cosmochim Acta 73(1):25–43

Pokrovsky OS, Viers J, Shirokova LS, Shevchenko VP, Filipov AS, Duprà B, Pokrovsky O, Viers J, Shirokova L, Shevchenko V, Filipov A, Dupré B (2010) Dissolved, suspended, and colloidal fluxes of organic carbon, major and trace elements in the Severnaya Dvina River and its tributary. Chem Geol 273(1–2):136–149. doi:http://dx.doi.org/10.1016/j.chemgeo.2010.02.018

Pokrovsky OS, Viers J, Dupré B, Chabaux F, Gaillardet J, Audry S, Prokushkin AS, Shirokova LS, Kirpotin SN, Lapitsky SA, Shevchenko VP (2012) Biogeochemistry of carbon, major and trace elements in watersheds of northern Eurasia drained to the Arctic Ocean: the change of fluxes, sources and mechanisms under the climate warming prospective. C R Geosci 344(11–12):663–677. doi:http://dx.doi.org/10.1016/j.crte.2012.08.003, erosion–Alteration: from fundamental mechanisms to geodynamic consequences (Ebelmen's Symposium)

Pretet C, Samankassou E, Felis T, Reynaud S, Böhm F, Eisenhauer A, Ferrier-Pagès C, Gattuso JP, Camoin G (2013) Constraining calcium isotope fractionation ($\delta^{44}/^{40}$ Ca) in modern and fossil scleractinian coral skeleton. Chem Geol 340:49–58. doi:http://dx.doi.org/10.1016/j.chemgeo.2012.12.006

Probst A, Viville D, Fritz B, Ambroise B, Dambrine E (1992) Hydrochemical budgets of a small forested granitic catchment exposed to acid deposition: the Strengbach catchment case study (Vosges massif, France) 62(3–4):337–347. doi:10.1007/BF00480265

Probst A, El Gh'mari A, Aubert D, Fritz B, McNutt R (2000) Strontium as a tracer of weathering processes in a silicate catchment polluted by acid atmospheric inputs, Strengbach, France. Chem Geol 170(1–4):203–219. doi:http://dx.doi.org/10.1016/S0009-2541(99)00248-X

Raulund-Rasmussen K, De Jong J, Humphrey J, Smith M, Ravn H, Katzensteiner K, Klimo E, Szukics U, Delaney C, Hansen K, Stupak I, Ring E, Gundersen P, Loustau D (2011) Papers on impacts of forest management on environmental services. Tech. rep., EFI Technical Report 57p

Raymond PA, Oh NH, Turner RE, Broussard W (2008) Anthropogenically enhanced fluxes of water and carbon from the mississippi river. Nature 451(7177):449–452

Reynard LM, Henderson GM, Hedges REM (2010) Calcium isotope ratios in animal and human bone. Geochim et Cosmochim Act 74(13):3735–3750

Reynard LM, Day CC, Henderson GM (2011) Large fractionation of calcium isotopes during cave-analogue calcium carbonate growth. Geochim Cosmochim Act 75(13):3726–3740

Richter DD, Markewitz D (1995) How deep is soil? BioScience 45(9):600–609. doi:10.2307/1312764

Ridgwell A, Zeebe RE (2005) The role of the global carbonate cycle in the regulation and evolution of the Earth system. Earth Planet Sci Lett 234(3–4):299–315. doi:http://dx.doi.org/10.1016/j.epsl.2005.03.006

Rochow JJ (1974) Litter fall relations in a missouri forest. Oikos 25(1):80–85. doi:10.2307/3543548

Ronov AB (1982) The Earth's sedimentary shell (quantitative patterns of its structure, compositions, and evolution). Int Geol Rev 24(11):1313–1363. doi:10.1080/00206818209451075

Rudnick R, Gao S (2003) Treatise on geochemistry, vol 3, Elsevier Science, Oxford, chap Composition of the Continental Crust, pp 1–64

Ryu JS, Jacobson AD, Holmden C, Lundstrom C, Zhang Z (2011) The major ion, $\delta^{44/40}$ Ca, $\delta^{44/42}$ Ca, and δ^{26} Mg geochemistry of granite weathering at pH = 1 and T = 25°C: power-law processes and the relative reactivity of minerals. Geochim Cosmochim Act 75(20):6004–6026. doi:10.1016/j.gca.2011.07.025

Sabine CL, Mackenzie FT (1995) Bank-derived carbonate sediment transport and dissolution in the Hawaiian Archipelago. Aquat Geochem 1(2):189–230. doi:10.1007/BF00702891

Schaetzl R (2005) Soils: genesis and geomorphology. Cambridge University Press, Cambridge

Schmitt AD, Bracke G, Stille P, Kiefel B (2001) The calcium isotope composition of modern seawater determined by thermal ionisation mass spectrometry. Geostandard Newslett 25(2-3):267–275

Schmitt AD, Stille P (2005) The source of calcium in wet atmospheric deposits: Ca-Sr isotope evidence. Geochim Cosmochim Act 69(14):3463–3468. doi: 10.1016/j.gca.2004.11.010

Schmitt AD, Chabaux F, Stille P (2003a) The calcium riverine and hydrothermal isotopic fluxes and the oceanic calcium mass balance. Earth Planet Sci Lett 213(3–4):503–518. doi:10.1016/S0012-821X(03)00341-8

Schmitt AD, Stille P, Vennemann T (2003b) Variations of the ^{44}Ca/^{40}Ca ratio in seawater during the past 24 million years: evidence from δ^{44} Ca and δ^{18} O values of Miocene phosphates. Geochim Cosmochim Act 67(14):2607–2614. doi:10.1016/S0016-7037(03)00100-5

Schmitt AD, Gangloff S, Cobert F, Lemarchand D, Stille P, Chabaux F (2009) High performance automated ion chromatography separation for Ca isotope measurements in geological and biological samples. J Anal Atom Spect 24(8):1089–1097

Shen W, Ren H, Jenerette GD, Hui D, Ren H (2013) Atmospheric deposition and canopy exchange of anions and cations in two plantation forests under acid rain influence. Atmos Environ 64(0):242 – 250. doi:http://dx.doi.org/10.1016/j.atmosenv.2012.10.015

Sime NG, De La Rocha CL, Galy A (2005) Negligible temperature dependence of calcium isotope fractionation in twelve species of planktonic foraminifera. Earth Planet Sci Lett 232(1–2):51–66. doi:10.1016/j.epsl.2005.01.011

Sime NG, De La Rocha CL, Tipper ET, Tripati A, Galy A, Bickle MJ (2007) Interpreting the Ca Isotope record of marine biogenic carbonates. Geochim Cosmochim Act 71(16):3979–3989. doi:10.1016/j.gca.2007.06.009

Simon JI, DePaolo DJ (2010) Stable calcium isotopic composition of meteorites and rocky planets. Earth Planet Sci Lett 289(3):457–466

Simpson A, Kingery W, Hayes M, Spraul M, Humpfer E, Dvortsak P, Kerssebaum R, Godejohann M, Hofmann M (2002) Molecular structures and associations of humic substances in the terrestrial environment. Naturwissenschaften 89(2):84–88. doi:10.1007/s00114-001-0293-8

Skulan J, DePaolo DJ (1999) Calcium isotope fractionation between soft and mineralised tissues as a monitor of calcium use in vertabrates. Proc Nat Acad Sci USA 96(24):13709–13713

Skulan J, DePaolo DJ, Owens TL (1997) Biological control of calcium isotope abundances in the global calcium cycle. Geochim Cosmochim Act 61(12):2505–2510. doi:10.1016/S0016-7037(97)00047-1

Smith D, Prithiviraj B, Zhang F (2002) Nitrogen fixation: global perspectives. Wiley, CABI Publishing, Wallingford, Oxon, UK, pp 327–330

Soudry D, Segal I, Nathan Y, Glenn CR, Halicz L, Lewy Z, VonderHaar DL (2004) ^{44}Ca/^{42}Ca and ^{143}Nd/^{144}Nd isotope variations in Cretaceous-Eocene Tethyan francolites and their bearing on phosphogenesis in the southern Tethys. Geology 32(5):389–392. doi:10.1130/G20438.1

Soudry D, Glenn C, Nathan Y, Segal I, VonderHaar D (2006) Evolution of Tethyan phosphogenesis along the northern edges of the Arabian–African shield during the Cretaceous–Eocene as deduced from temporal variations of Ca and Nd isotopes and rates of P accumulation. Earth Sci Rev 78(1–2):27–57. doi:http://dx.doi.org/10.1016/j.earscirev.2006.03.005

Sposito G (1980) Derivation of the Freundlich equation for Ion Exchange Reactions in soils 44:652–654. doi:10.2136/sssaj1980.03615995004400030045x

Steuber T, Buhl D (2006) Calcium-isotope fractionation in selected modern and ancient marine carbonates. Geochim Cosmochim Act 70(22):5507–5521. doi:10.1016/j.gca.2006.08.028

Summerfield M (1997) Supercontinent break-up and landscape development. Geogr Rev 10:36–40

Taiz L, Zeiger E (2010) Plant physiology, 5th edn. Sinauer Assoc, Sunderland

Tang J, Dietzel M, Böhm F, Köhler SJ, Eisenhauer A (2008) Sr^{2+}/Ca^{2+} and ^{44}Ca/^{40}Ca fractionation during inorganic calcite formation: II. Ca isotopes. Geochim Cosmochim Act 72(15):3733–3745

Tang J, Niedermayr A, Köhler SJ, Böhm F, KÄsakürek B, Eisenhauer A, Dietzel M (2012) Sr^{2+}/Ca^{2+} and ^{44}Ca/^{40}Ca fractionation during inorganic calcite formation: III. Impact of salinity/ionic strength. Geochim Cosmochim Acta 77:432–443. doi:http://dx.doi.org/10.1016/j.gca.2011.10.039

Taniguchi M, Burnett WC, Cable JE, Turner JV (2002) Investigation of submarine groundwater discharge. Hydrol Process 16:2115–2129

Taylor SR, McLennan SM (1985) The continental crust. Its evolution and composition. Blackwell Science, Oxford

Teichert B, Gussone N, Eisenhauer A, Bohrmann G (2005) Clathrites: archives of near-seafloor pore-fluid evolution (delta Ca-44/40, delta C-13, delta O-18) in gas hydrate environments. Geology 33(3):213–216

Teichert BMA, Gussone N, Torres ME (2009) Controls on calcium isotope fractionation in sedimentary porewaters. Earth Planet Sci Lett 279(34):373–382

Thiffault E, Paré D, Bélanger N, Munson A, Marquis F (2006) Harvesting intensity at clear-felling in the boreal forest Soil Sci Soc Am J 70:691–701. doi:10.2136/sssaj2005.0155

Tipper ET, Bickle MJ, Galy A, West AJ, Pomiès C, Chapman HJ (2006a) The short term climatic sensitivity of carbonate and silicate weathering fluxes: insight from seasonal variations in river chemistry. Geochim Cosmochim Act 70(11):2737–2754. doi:10.1016/j.gca.2006.03.005

Tipper ET, Galy A, Bickle MJ (2006b) Riverine evidence for a fractionated reservoir of Ca and Mg on the continents: implications for the oceanic Ca cycle. Earth Planet Sci Lett 247(3–4):267–279. doi:10.1016/j.epsl.2006.04.033

Tipper ET, Galy A, Gaillardet J, Bickle MJ, Elderfield H, Carder EA (2006c) The Mg isotope budget of the modern ocean: constraints from riverine Mg isotope ratios. Earth Planet Sci Lett 250:241–253. doi:10.1016/j.epsl.2006.07.037

Tipper ET, Galy A, Bickle MJ (2008a) Calcium and magnesium isotope systematics in rivers draining the Himalaya-Tibetan-Plateau region: lithological or fractionation control? Geochim Cosmochim Act 72(4):1057–1075. doi:10.1016/j.gca.2007.11.029

Tipper ET, Gaillardet J, Galy A, Louvat P, Bickle MJ (2010a) Calcium isotope ratios in the world's largest rivers: a constraint on the maximum imbalance of oceanic calcium fluxes. Global Biogeochem Cycles 24:GB3019. doi:10.1029/2009GB003574

Tipper ET, Gaillardet J, Louvat P, Capmas F, White AF (2010b) Mg isotope constraints on soil pore-fluid chemistry: evidence from Santa Cruz, California. Geochim Cosmochim Act 74:3883–3896. doi:10.1016/j.gca.2010.04.021

Tipper ET, Lemarchand E, Hindshaw R, Reynolds BC, Bourdon B (2012) Seasonal sensitivity of weathering processes: hints from magnesium isotopes in a glacial stream. Chem Geol 312–313:80–92. doi:10.1016/j.chemgeo.2012.04.002

Tipping E, Hurley MA (1992) A unifying model of cation binding by humic substances. Geochim Cosmochim Acta 56(10):3627–3641. doi:http://dx.doi.org/10.1016/0016-7037(92)90158-F

Turchyn AV, DePaolo DJ (2011) Calcium isotope evidence for suppression of carbonate dissolution in carbonate-bearing organic-rich sediments. Geochim Cosmochim Acta 75(22):7081–7098

Turekian KK, Wedepohl KH (1961) Distribution of the elements in some major units of the earth's crust. Geol Soc Am Bull 72(2):175–192

Turpault MP, Nys C, Calvaruso C (2009) Rhizosphere impact on the dissolution of test minerals in a forest ecosystem. Geoderma 153(1–2):147–154. doi:http://dx.doi.org/10.1016/j.geoderma.2009.07.023

Ullmann CV, Böhm F, Rickaby REM, Wiechert U, Korte C (2013) The Giant Pacific Oyster (Crassostrea gigas) as a modern analog for fossil ostreoids: isotopic (Ca, O, C) and elemental (Mg/Ca, Sr/Ca, Mn/Ca) proxies. Geochem Geophys Geosyst 14(10):4109–4120. doi:10.1002/ggge.20257

Valdes MC, Moreira M, Foriel J, Moynier F (2014) The nature of Earth's building blocks as revealed by calcium isotopes. Earth Planet Sci Lett 394:135–145. doi:http://dx.doi.org/10.1016/j.epsl.2014.02.052

Van Cappellen P (2003) Biomineralization and global biogeochemical cycles. Rev Mineral Geochem 54(1):357–381

Veizer J, Jansen SL (1985) Basement and sedimentary recycling-2: time dimension to global tectonics. J Geol 93:625–643. doi:10.1086/628992

Veizer J, MacKenzie FT (2003) Evolution of sedimentary rocks. Treatise Geochem 7:369–407. doi:10.1016/B0-08-043751-6/07103-6

Verrecchia EP, Dumont JL (1996) A biogeochemical model for chalk alteration by fungi in semiarid environments. Biogeochemistry 35(3):447–470. doi:10.1007/BF02183036

Vivier ADD, Jacobson AD, Lehn GO, Selby D, Hurtgen MT, Sageman BB (2015) Ca isotope stratigraphy across the Cenomanian–Turonian {OAE} 2: links between volcanism, seawater geochemistry, and the carbonate fractionation factor. Earth Planet Sci Lett 416:121–131. doi:http://dx.doi.org/10.1016/j.epsl.2015.02.001

von Allmen K, Nagler TF, Pettke T, Hippler D, Griesshaber E, Logan A, Eisenhauer A, Samankassou E (2010) Stable isotope profiles (Ca, O, C) through modern brachiopod shells of T. septentrionalis and G. vitreus: implications for calcium isotope paleo-ocean chemistry. Chem Geol 269(3–4):210–219

Walker JCG, Hays PB, Kasting JF (1981) A negative feedback mechanism for the long-term stabilization of Earth's surface temperature. J Geophys Res 86(C10):9776–9782

Wang S, Yan W, Magalhaes H et al (2012) Calcium isotope fractionation and its controlling factors over authigenic carbonates in the cold seeps of the northern South China Sea. Chin Sci Bull 57:1325–1332

Wang Z, Hu P, Gaetani G, Liu C, Saenger C, Cohen A, Hart S (2013) Experimental calibration of mg isotope fractionation between aragonite and seawater. Geochim Cosmochim Acta 102:113–123. doi:http://dx.doi.org/10.1016/j.gca.2012.10.022

Wefer G (1980) Carbonate production by algae Halimeda, Penicillus and Padina. Nature 285(5763):323–324

West AJ, Galy A, Bickle M (2005) Tectonic and climatic controls on silicate weathering. Earth Planet Sci Lett 235(1–2):211–228

White AF, Schulz MS, Lowenstern JB, Vivit DV, Bullen TD (2005) The ubiquitous nature of accessory calcite in granitoid rocks: Implications for weathering, solute evolution, and petrogenesis. Geochim Cosmochim Act 69(6):1455–1471

White AF, Schulz MS, Vivit DV, Blum AE, Stonestrom DA, Anderson SP (2008) Chemical weathering of a marine terrace chronosequence, Santa Cruz,

California I: interpreting rates and controls based on soil concentration-depth profiles. Geochim Cosmochim Act 72(1):36–68. doi:10.1016/j.gca.2007.08.029

White AF, Schulz MS, Stonestrom DA, Vivit DV, Fitzpatrick JA, Bullen TD, Maher K, Blum AE, Anderson SP (2009) Chemical weathering of a marine terrace chronosequence, santa cruz, california ii: solute profiles, gradients and linear approximations of contemporary and long-term weathering rates. Geochim Cosmochim Act 72(1):36–68. doi:10.1016/j.gca.2007.08.029

White AF, Schulz MS, Vivit DV, Bullen TD, Fitzpatrick J (2012) The impact of biotic/abiotic interfaces in mineral nutrient cycling: a study of soils of the santa cruz chronosequence, California. Geochim Cosmochim Acta 77:62–85. doi:http://dx.doi.org/10.1016/j.gca.2011.10.029

White T, Brantley S, Banwart S, Chorover J, Dietrich W, Derry L, Lohse K, Anderson S, Aufdendkampe A, Bales R, Kumar P, Richter D, McDowell B, Giardino JR, Houser C (2015) The role of critical zone observatories in critical zone science. vol 19, Elsevier, Philadelphia, pp 15–78. doi:http://dx.doi.org/10.1016/B978-0-444-63369-9.00002-1

Wiegand BA, Schwendenmann L (2013) Determination of Sr and Ca sources in small tropical catchments (La Selva, Costa Rica)—a comparison of Sr and Ca isotopes. J Hydrol 488:110–117. doi:http://dx.doi.org/10.1016/j.jhydrol.2013.02.044

Wiegand BA, Chadwick OA, Vitousek PM, Wooden JL (2005) Ca cycling and isotopic fluxes in forested ecosystems in Hawaii. Geophys Res Lett 32(L11):404

Wieser ME, Buhl D, Bouman C, Schwieters J (2004) High precision calcium isotope ratio measurements using a magnetic sector multiple collector inductively coupled plasma mass spectrometer. J Anal Atom Spect 19:844–851. doi:10.1039/b403339f

Wilkinson BH, Algeo TJ (1989) Sedimentary carbonate record of calcium-magnesium cycling. Am J Sci 289:1158–1194

Wombacher F, Eisenhauer A, Heuser A, Weyer S (2009) Separation of Mg, Ca and Fe from geological reference materials for stable isotope ratio analyses by MC-ICP-MS and double-spike TIMS. J Anal Atom Spect 24:627–636. doi:10.1039/b820154d

Zekster I, Dzhamalov R (1988) Role of ground water in the hydrological cycle and in continental water balance. UNESCO Paris International Hydrological Programme IHP-II Project 2 3(41623):133

Zeller B, Martin F (1998) Contribution à l'étude de la decomposition d'une litiere de hetre, la liberation de l'azote, sa mineralisation et son prevement par le hetre (Fagus sylvatica L.) dans une hetraie de montagne du bassin versant du Strengbach (Haut-Rhin)

Zeng GM, Zhang G, Huang GH, Jiang YM, Liu HL (2005) Exchange of Ca^{2+}, Mg^{2+} and K^+ and uptake of H^+, for the subtropical forest canopies influenced by acid rain in Shaoshan forest located in Central South China. Plant Sci 168(1):259–266. doi:http://dx.doi.org/10.1016/j.plantsci.2004.08.004

Zhu P, Macdougall JD (1998) Calcium isotopes in the marine environment and the oceanic calcium cycle. Geochim Cosmochim Act 62(10):1691–1698

High Temperature Geochemistry and Cosmochemistry

Martin Schiller, Nikolaus Gussone and Frank Wombacher

Abstract

In this chapter, we summarize the evidence for high temperature Ca isotope fractionation in natural and experimental samples including minerals, igneous and metamorphic rocks and silicate melts and discuss the underlying isotope fractionation mechanisms. Furthermore, we outline the evidence for primordial nucleosynthetic variability of Ca isotopes in meteorites and their components, suggesting a diverse stellar origin of the Ca isotopes that make up our Solar System. We also present a brief synopsis of observed mass-dependent Ca stable isotope fractionation in meteoritic materials, which allow insights into the local conditions during the formative stage of our Solar System. Lastly, we provide an overview of the ^{40}K–^{40}Ca decay system and its use to track the evolution of Earth's mantle, continental and oceanic crust over time as well as its potential to date igneous and sedimentary rocks.

Keywords

High temperature isotope fractionation · Solar System · Meteorites · K-Ca dating · Radiogenic ^{40}Ca

M. Schiller
Natural History Museum of Denmark, University of Copenhagen, Copenhagen, Denmark
e-mail: schiller@snm.ku.dk

N. Gussone (✉)
Institut für Mineralogie, Universität Münster, Münster, Germany
e-mail: nikolaus.gussone@uni-muenster.de

F. Wombacher
Institut für Geologie und Mineralogie, Universität zu Köln, Köln, Germany
e-mail: fwombach@uni-koeln.de

1 High Temperature Ca Isotope Geochemistry of Terrestrial Silicate Rocks, Minerals and Melts

1.1 Calcium Isotope Fractionation Between Minerals Formed at High Temperatures

The comparatively few data on $\delta^{44/40}$Ca of minerals formed at high temperatures published so far suggests that significant isotopic differences

© Springer-Verlag Berlin Heidelberg 2016
N. Gussone et al., *Calcium Stable Isotope Geochemistry*,
Advances in Isotope Geochemistry, DOI 10.1007/978-3-540-68953-9_7

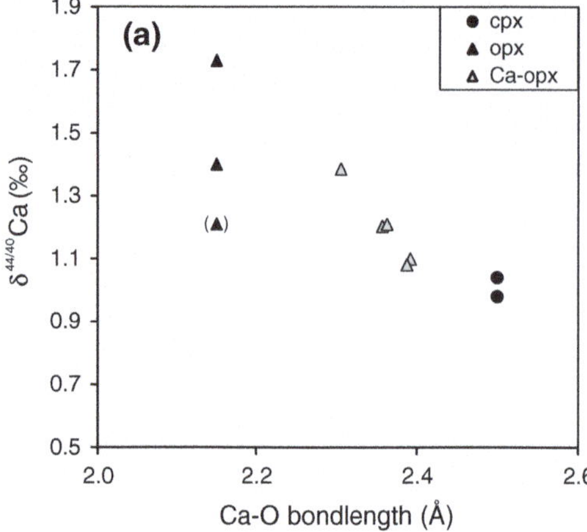

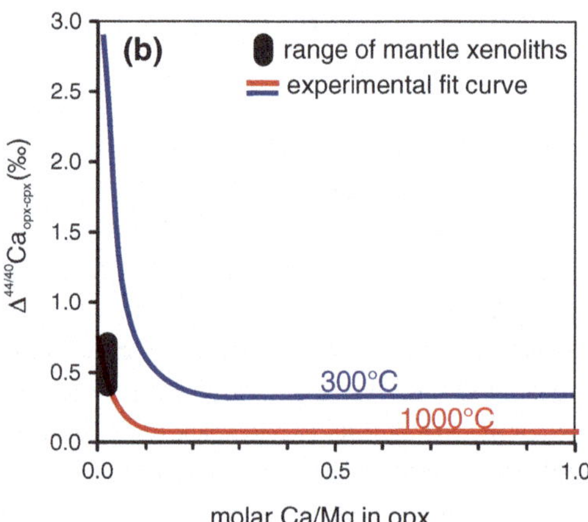

Fig. 1 Calcium isotope fractionation between ortho- and clinopyroxene. **a** $\delta^{44/40}$Ca of coexisting ortho- and clinopyroxene is offset by 0.4–0.7 ‰ in mantle xenoliths (filled symbols, Huang et al. 2010a). The orthopyroxene in brackets may be subject to contamination by Ca-rich mineral phases. $\delta^{44/40}$Ca of Ca-doped orthopyroxene depends on the Ca content and, thus, the Ca-O bond length (*open symbols*, Feng et al. 2014). **b** $\Delta^{44/40}$Ca$_{\text{opx-cpx}}$ decreases with increasing Ca content of the orthopyroxene and increasing temperature (redrawn from Feng et al. 2014, range of mantle xenolith data from Huang et al. 2010a)

between minerals are present in nature. In mantle xenoliths, $\delta^{44/40}$Ca of orthopyroxene (opx) ranges from about 1.2 to 1.7 ‰, but Huang et al. (2010a) suggested that the lower observed $\delta^{44/40}$Ca values ($\sim$1.2 ‰) may be contaminated by traces of Ca-rich phases such as carbonates. Therefore, they proposed $\delta^{44/40}$Ca$_{\text{opx}}$ between 1.4 and 1.7 ‰ for the upper Earth mantle. Coexisting clinopyroxene (cpx) from the same rocks cluster around 1.0 ‰

(Huang et al. 2010a, b; Simon and DePaolo 2010) (Fig. 1a). Consequently, the average $\Delta^{44/40}$Ca$_{\text{opx}}$$_{-\text{cpx}}$ in mantle xenoliths is about 0.55 ± 0.15 ‰. In addition to mineral separates from natural rocks, Feng et al. (2014) derived equilibrium isotope fractionation factors between coexisting clino- and orthopyroxene from first-principles calculation and melt experiments. Their results agree with the data of Huang et al. (2010a) demonstrating well resolvable Ca isotope fractionation between both minerals at temperatures between 300 and 1000 ° C, showing decreasing Ca isotope fractionation with increasing temperature. In addition, the experiments of Feng et al. (2014) reveal an influence of Ca content and thus Ca–O bond length in the orthopyroxene on the $\Delta^{44/40}$Ca$_{\text{opx-cpx}}$, namely decreasing fractionation with increasing Ca content (Ca–O bond length) of the orthopyroxene (Fig. 1b).

Strong evidence for high temperature fractionation of Ca isotopes is also provided by $\delta^{44/40}$Ca and $\delta^{44/42}$Ca of mineral separates from the Boulder Creek Granodiorite (Ryu et al. 2011) and orthogneisses from Greenland (Hindshaw et al. 2014). Calcium-rich and K-poor minerals from the Boulder Creek Granodiorite plot on a mass-dependent fractionation line, whereas K-rich minerals appear to deviate from this trend due to enrichment in ^{40}Ca resulting from decay of ^{40}K (Fig. 2). The $\delta^{44/40}$Ca compositions of the

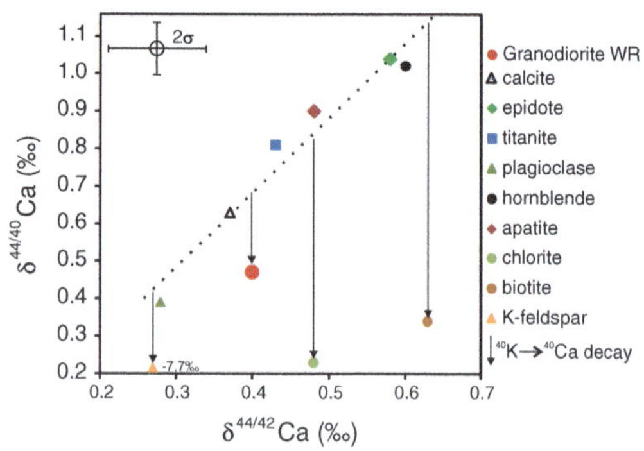

Fig. 2 $\delta^{44/40}$Ca and $\delta^{44/42}$Ca of minerals from the Boulder Creek Granodiorite. Calcium-rich minerals plot on the mass-dependent fractionation line within uncertainty, while the granodiorite whole rock and K-rich minerals biotite, K-feldspar and chlorite are enriched in radiogenic ^{40}Ca (Ryu et al. 2011)

K-poor minerals range from 0.4 to 1.0 ‰. Plagioclase is the lightest mineral phase with a typical $\delta^{44/40}$Ca value of 0.4 ‰, whereas calcite (0.6 ‰), titanite (0.8 ‰), apatite (0.9 ‰), hornblende (1.0 ‰) and epidote (1.0 ‰) become increasingly heavy. Ignoring the overprinted $\delta^{44/40}$Ca ratio, K-feldspar appears to be an isotopically light phase similar to plagioclase, whereas biotite is the heaviest phase in this rock and chlorite has an intermediate composition. Although it is not yet fully resolved if the reported apparent Ca isotope fractionation between all mineral phases of this rock reflects high temperature equilibrium fractionation or if some compositions are biased by later events, the data suggests possible Ca isotope fractionation during igneous differentiation processes.

1.2 Igneous Rocks

The overall range of $\delta^{44/40}$Ca observed in igneous rocks is 0.6–1.6 ‰. Volcanic rocks exhibit $\delta^{44/40}$Ca between 0.6 and 1.3 ‰. The presently available data neither show a significant difference between different volcanic rocks (average values in brackets) such as basalts (0.90 ‰), alkali basalts (0.91 ‰), tholeiitic basalts (0.88 ‰), tholeiites (0.75 ‰), olivine tholeiites (0.80 ‰), andesites and basaltic andesites (0.77 ‰), and rhyolites (0.84 ‰), nor a difference between basalts from different settings, e.g. ocean island basalts (0.89 ‰) and mid ocean ridge basalts (0.81 or 0.92 ‰) (Fig. 3). The observed scatter within a rock group may partly be related to analytical bias, but for most rock types no apparent offset between different studies exists. Currently it is unclear why $\delta^{44/40}$Ca of mid ocean ridge basalt reported by Zhu and Macdougall (1998) is significantly higher compared to other studies (Fig. 3).

Of the silicate rocks, dunites have the highest so far reported $\delta^{44/40}$Ca values (1.5–1.6 ‰). For peridotites, which range from 1.0 to 1.2 ‰, and serpentinites (0.8–1.1 ‰), increasing alteration appears to lower the $\delta^{44/40}$Ca (Sect. 1.3.3). Granites, granodiorites and diorites have $\delta^{44/40}$Ca between 0.6 and 1.1 ‰ (Fig. 3). Komatiites

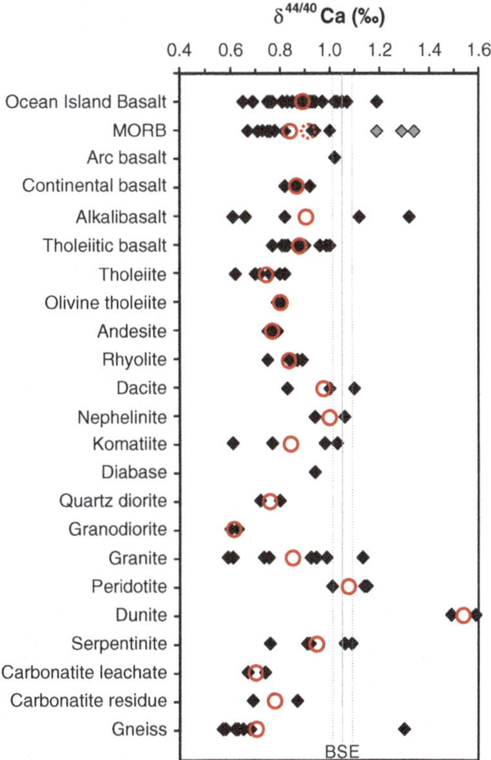

Fig. 3 Calcium isotope composition ($\delta^{44/40}$Ca) of different igneous and metamorphic rock types. Compiled from Richter et al. (2003), DePaolo (2004), Wombacher et al. (2009), Amini et al. (2009), Huang et al. (2010a, 2011a), Simon and DePaolo (2010), John et al. (2012), Ryu et al. (2011), Valdes et al. (2014), Jacobson et al. (2015) and Zhu and Macdougall (1998). *Red dots* show average value of a rock type. *Grey symbols* for mid ocean ridge basalt indicate data of Zhu and Macdougall (1998), which are significantly higher compared to other data. The *red dot* marks the average MORB excluding, the *dashed red dot* the average including the data of Zhu and Macdougall (1998). The *grey dashed vertical line* indicates the estimated $\delta^{44/40}$Ca of bulk silicate Earth (Huang et al. 2010a)

range from 0.6 to 1.0 ‰, and initial results suggest that Mg-rich, Precambrian komatiites have higher $\delta^{44/40}$Ca than Mg-poor Cretaceous komatiites (Amini et al. 2008). Carbonatites (leachates) range from 0.65 to 0.75 ‰, whereas the residues have $\delta^{44/40}$Ca between about 0.7 and 0.9 ‰. Most gneisses and schists fall into a range from 0.5 to 0.7 ‰, but one reported value is rather high (1.3 ‰) (Tipper et al. 2006; Moore et al. 2013).

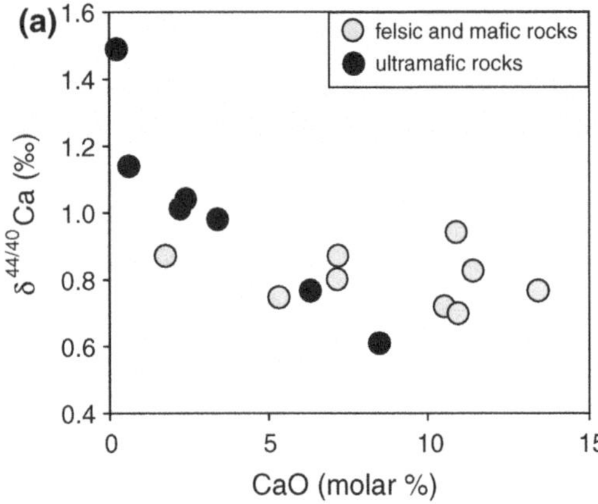
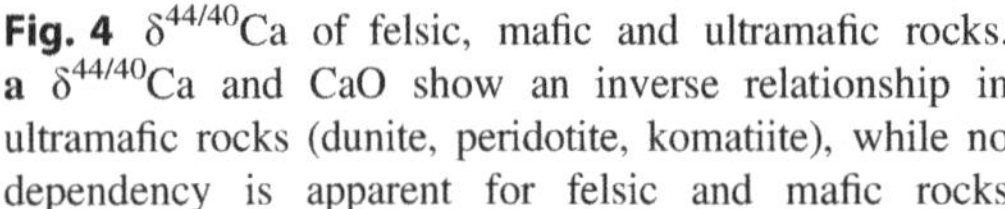
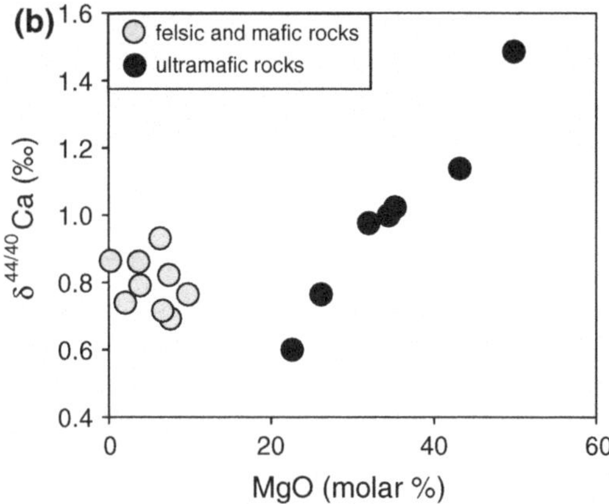

Fig. 4 $\delta^{44/40}$Ca of felsic, mafic and ultramafic rocks. **a** $\delta^{44/40}$Ca and CaO show an inverse relationship in ultramafic rocks (dunite, peridotite, komatiite), while no dependency is apparent for felsic and mafic rocks

(rhyolite, andesite, trachybasalt, diorite and basalt). **b** Similarly, $\delta^{44/40}$Ca and MgO are positively correlated in ultramafic rocks, whereas there is no such trend for mafic and felsic rocks (Amini et al. 2009)

There is a general relationship of $\delta^{44/40}$Ca and the CaO and MgO content of ultramafic rocks, such as dunite, peridotite and komatiite. This dependence is likely related to mineral specific Ca isotope fractionation and differentiation processes (Fig. 4). In contrast to ultramafic rocks, no such relationship is apparent for mafic and felsic rocks, such as rhyolite, andesite, trachybasalt, diorite and basalt (Amini et al. 2009).

1.3 The Earth's Silicate Reservoirs and Global Tectonics

1.3.1 Ca Isotope Composition of the Upper Mantle and Bulk Silicate Earth

Based on the $\delta^{44/40}$Ca of clinopyroxene (about 1.0 ‰) and orthopyroxene (1.4–1.7 ‰) from mantle xenoliths and modal abundances of orthopyroxene (27–32 %) and clinopyroxene (13–18 %) as well as average CaO contents of orthopyroxene (0.75–1.0 %) and clinopyroxene (about 21 %) in the upper Earth mantle, Huang et al. (2010a) estimated the Ca isotope composition of Bulk Silicate Earth ($\delta^{44/40}$Ca$_{BSE}$) to be 1.05 ± 0.04 ‰ relative to SRM 915a (Fig. 5). This value is identical within uncertainty to the

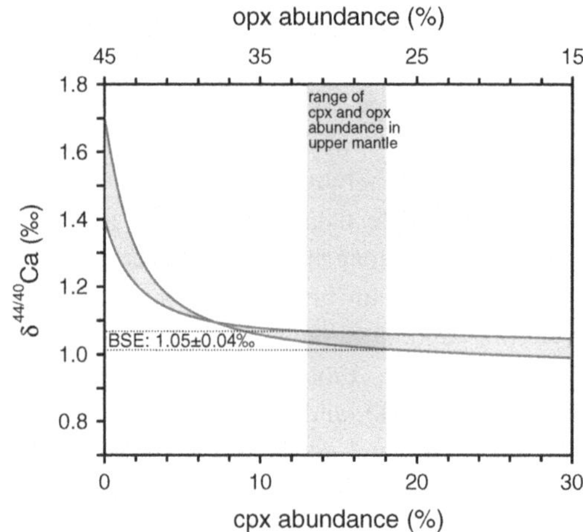

Fig. 5 Determination of upper mantle/Bulk Silicate Earth Ca isotope composition. Solid black curves show mixing models of two clino- and orthopyroxene pairs from upper mantle xenoliths. Based on CaO contents, modal abundance and $\delta^{44/40}$Ca of clino- and orthopyroxene, Huang et al. (2010a) determined the composition of the upper mantle. *Grey box* indicates range of clino- and orthopyroxene abundance in the upper mantle. *Dotted black lines* mark the intersection of *grey box* with mixing lines of clino- and orthopyroxene pairs, resulting in the estimated range of $\delta^{44/40}$Ca of the upper mantle, representative for Bulk Silicate Earth (BSE). Modified from Huang et al. (2010a)

estimate for bulk $\delta^{44/40}$Ca composition of Mars of 1.04 ± 0.09 ‰ (Magna et al. 2015).

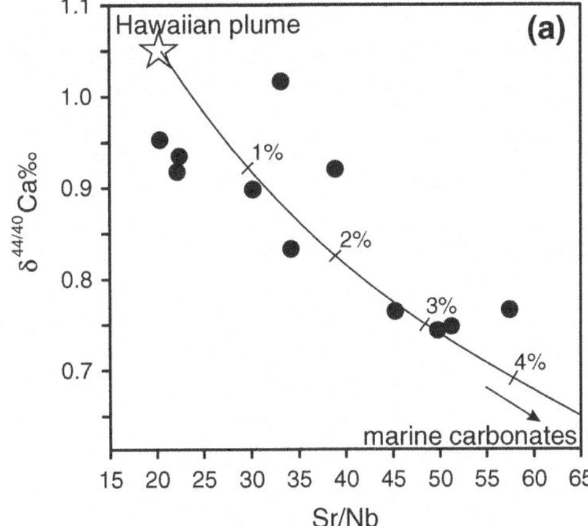

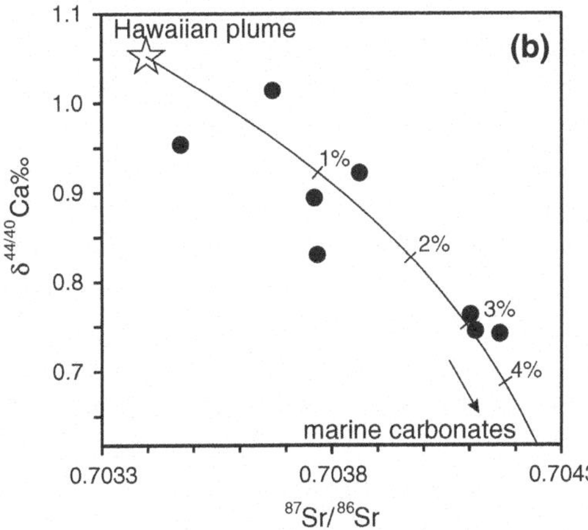

Fig. 6 $\delta^{44/40}$Ca of Hawaiian volcanic rocks. $\delta^{44/40}$Ca of Hawaiian volcanic rocks show a correlation with Sr/Nb (**a**) and ^{87}Sr/^{86}Sr (**b**) indicating a contribution of subducted ancient carbonate sediment to Hawaiian hotspot melts. Redrawn from Huang et al. (2011a)

1.3.2   Traces of Subducted Sedimentary Carbonate in Ocean Island Basalts

Huang et al. (2011a) suggested that the ^{40}Ca enrichment of ocean island basalts (OIB) relative to BSE relates to an admixture of an ancient subducted sedimentary component to hotspot magmas. This hypothesis is based on negative correlations of $\delta^{44/40}$Ca with Sr/Nb and ^{87}Sr/^{86}Sr ratios of Hawaiian lavas, suggesting decreasing $\delta^{44/40}$Ca with increasing sedimentary Ca contribution (Fig. 6). While Huang et al. (2011a) found no correlation between the degree of melting and $\delta^{44/40}$Ca, Valdes et al. (2014)

pointed out that the current database is not sufficient to exclude potential Ca isotope fractionation during magmatic processes. This reasoning is also in line with the mineral specific Ca isotope fractionation found in plutonic and metamorphic rocks (Ryu et al. 2011; Hindshaw et al. 2014) and also agrees with the variability in $\delta^{44/40}$Ca reported for Martian meteorites, which are not related to subduction processes (Magna et al. 2015).

1.3.3   Metasomatism

Subduction can cause Ca isotope variability in rocks not only by admixing of sedimentary carbonates to hotspot magma sources, but can also result in Ca isotope fractionation during high temperature and high pressure metamorphism. Calcium isotope analyses demonstrate significant ^{44}Ca enrichment in a blueschist facies quartz-carbonate vein and surrounding eclogitic reaction selvage relative to the unaltered blueschist. Rb/Sr systematics of this vein and reaction selvage from the Chinese Tianshan Mountains confirm vein formation at or near peak metamorphic conditions (510 °C and 2.1 GPa) and suggest that the system was not disturbed later (John et al. 2012). The $\delta^{44/40}$Ca of the silicate fraction increases from the blueschist (0.8 ‰) towards the vein (1.2 ‰), while the offset between silicate and carbonate fraction of the rocks ($\Delta^{44/40}$Ca$_{\text{silicate-carbonate}}$) is about −0.6 ‰ within the vein and the blueschist (Fig. 7).

The increase in $\delta^{44/40}$Ca towards the quartz-carbonate vein is accompanied by an increase in CaO content from 6 to 15 %, indicating the influx of a ^{44}Ca-enriched fluid during metamorphism. John et al. (2012) suggested that this could be achieved by a fluid with a primary heavy $\delta^{44/40}$Ca$_{\text{fluid}}$ or, alternatively, an initially light fluid that evolved towards a heavier isotopic composition during fluid-rock interaction along the fluid flow path (Fig. 8).

An isotopically heavy Ca source (Fig. 8a) could be present in partly altered lithospheric mantle with about 1–2 wt% water or ∼10–15 % of serpentinisation. This value is obtained from the negative correlation between the Ca isotope composition of serpentinised slab mantle rocks and the loss on ignition (LOI), demonstrating decreasing $\delta^{44/40}Ca$ with increasing degree of serpentinisation (supplementary information of John et al. (2012)). While these rocks exhibit high $\delta^{44/40}Ca$ values, their CaO content is

relatively low; therefore, it is unclear if dehydration of slab-mantle serpentinite can generate fluids with Ca concentrations that are sufficiently high for the observed metasomatism. Seawater has a high $\delta^{44/40}Ca$ and Ca concentration, but can be excluded as a significant component in the metamorphic fluid because of its very high $^{87}Sr/^{86}Sr$, which is incompatible with that of the vein and reaction selvage.

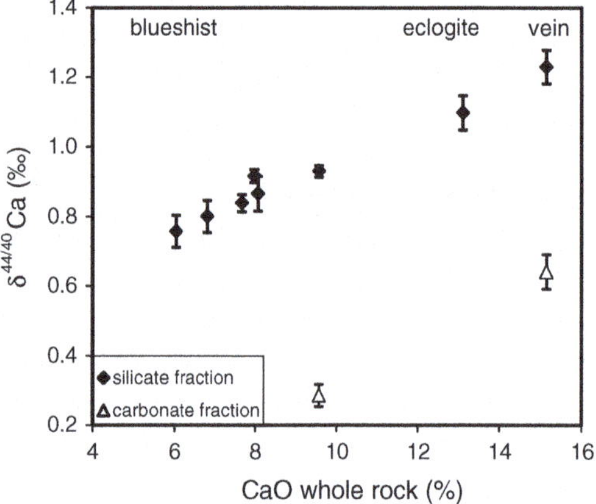

Fig. 7 $\delta^{44/40}Ca$ of a blueschist facies quartz-carbonate vein and adjacent rocks from the Tianshan, eclogitic reaction selvage and blueschist host rock. $\delta^{44/40}Ca$ of the silicate fraction increases from the blueschist host rock towards the quartz-carbonate vein. The $\delta^{44/40}Ca$ of blueschist host rock is in the range of altered oceanic lithosphere, while the high $\delta^{44/40}Ca$ values of the vein indicate influx of heavy Ca via fluids. Modified from John et al. (2012)

Another process that can provide sufficient amounts of Ca to subduction zone fluids is dewatering of altered oceanic crust (AOC), but $\delta^{44/40}Ca$ values are significantly lower than that of the proposed metasomatising fluid. However, the $\delta^{44/40}Ca$ of an initially AOC-like fluid may evolve to higher values during Ca-carbonate precipitation as the fluid is transported along the fluid pathway over large distances and channelized into the vein structures. As carbonate precipitation preferentially removes light Ca isotopes from the fluid, the $\delta^{44/40}Ca$ of the residual fluid is shifted towards heavier values (Fig. 8b). Considering a $\Delta^{44/40}Ca_{carb\text{-}fluid}$ of -0.6 ‰, about 40–70 % of the dissolved Ca are required to precipitate to explain the shift in $\delta^{44/40}Ca_{fluid}$ of 0.7 ‰ from the AOC value to that of the reacting fluid (John et al. 2012).

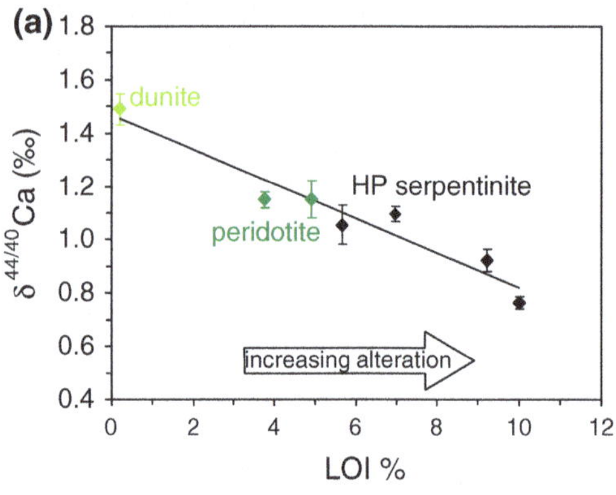

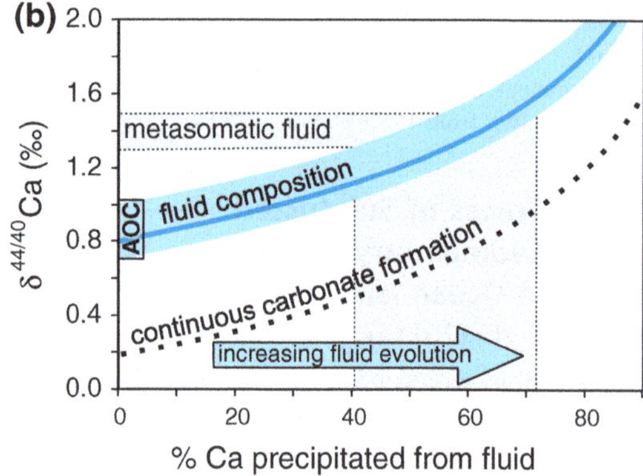

Fig. 8 a Calcium isotope composition of serpentinised slab mantle rocks. $\delta^{44/40}Ca$ is inversely correlated with the loss on ignition (LOI), indicating decreasing $\delta^{44/40}Ca$ with increasing degree of serpentinisation (supplementary information of John et al. (2012)). **b** Evolution of subduction zone fluid during ongoing $CaCO_3$ precipitation. $\delta^{44/40}Ca$ evolution of a fluid with initial altered oceanic crust (AOC)-like composition, during progressive fluid-rock interaction. The *blue arrow* indicates the evolution of the fluid by continuous Ca-carbonate precipitation (*dotted line*). During ongoing $CaCO_3$ precipitation $\delta^{44/40}Ca_{fluid}$ increases, as ^{40}Ca is preferentially incorporated into the solid carbonate. The intersection of $\delta^{44/40}Ca$ of the metasomatic fluid with the fluid evolution *curve* indicates a Ca consumption of 40–70 % during its passage. Redrawn from John et al. (2012)

1.4   Diffusion in Silicate Melts

Stable isotope fractionation during diffusion in liquids is commonly expressed by the ratio of the diffusion coefficients (D_i) of the isotopes with masses (m_i).

$$D_2/D_1 = (m_1/m_2)^{\beta} \qquad (1)$$

In liquid systems the exponent β is typically <0.5, with 0.5 being the value predicted by the kinetic theory for free evaporation (e.g. Richter et al. 2003). The expression implies that diffusional fractionation in gases and liquids, including melts, is governed by a power law relationship between the mass of the isotopes and their diffusion coefficients. In Eq. 1, the isotope masses are equal to those of the diffusing species, which is not necessarily given in all cases. The reduction of $\beta < 0.5$ is interpreted to result from the correlation of the movement of isotopes of interest to that of matrix atoms. A more detailed introduction into kinetic isotope fractionation theory is given in Chapter "Introduction".

1.4.1   Chemical and Self-diffusion

Elements in silicate melts diffuse along chemical gradients, with a net-migration from higher to lower concentrations. Experiments studying diffusion between different melt compositions reveal that this chemical diffusion is associated with a significant mass-dependent isotope fractionation, with light isotopes diffusing faster than heavy ones.

Diffusion experiments where mafic melt (CaO-rich) was placed next to a felsic melt (CaO-poor) demonstrate that Ca diffusion in silicate melts is associated with an isotope fractionation of several ‰ (Fig. 9, Richter et al. 2003; Watkins et al. 2009). Besides CaO gradients, Ca diffusion in silicate melts also depends on the properties of the melt-matrix. Watkins et al. (2009) showed that the speed of Ca diffusion and magnitude of Ca isotope fractionation are largely controlled by the composition of the melt and gradients of matrix elements such as SiO_2, Na_2O, Al_2O_3 (cf. Watkins et al. 2009, 2011, 2014). Calcium diffusion between rhyolite

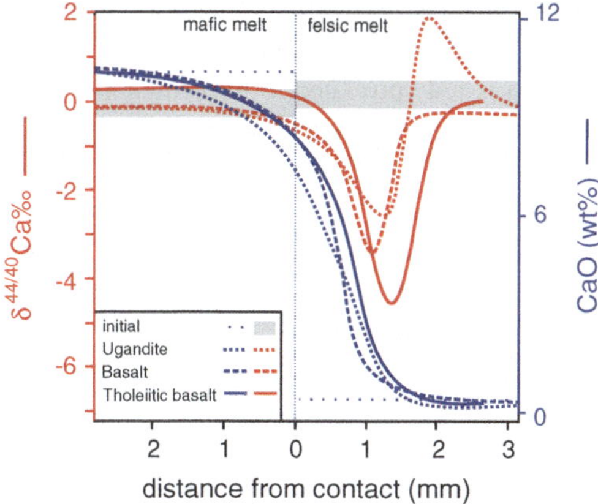

Fig. 9 Calcium isotope fractionation during chemical diffusion along chemical gradients of three diffusion couples of felsic and mafic melts. CaO (*blue*) and $\delta^{44/40}$Ca profiles (*red*) significantly differ between tholeiitic basalt-rhyolite (*solid line*), basalt-rhyolite (*dashed line*), ugandite-rhyolite (*dotted line*) diffusion couples, indicating a dependence of diffusion velocities and Ca isotope fractionation on the melt compositions of the diffusion couples. Modified from Richter et al. (2003) and Watkins et al. (2009)

and different mafic rocks show considerable differences in diffusion velocities and resulting Ca isotope fractionation in tholeiitic basalt, basalt and ugandite (Richter et al. 2003; Watkins et al. 2009). Tholeiitic basalt shows the largest depletion of ^{44}Ca of −5 to −6 ‰ relative to the $\delta^{44/40}$Ca of the initial rock (Richter et al. 2003). Calcium isotopes in the basalt experienced less pronounced depletion in ^{44}Ca of −3 ‰ relative to the initial composition. The ugandite, which is relatively low in SiO_2 and Al_2O_3 and therefore less polymerized, is characterized by the least depletion in ^{44}Ca relative to the initial composition (−2 ‰). The corresponding rhyolite to the ugandite is enriched in ^{44}Ca up to +2 ‰, which can be explained by diffusive coupling between CaO and Al_2O_3. Based on these results, Watkins et al. (2009) concluded that the magnitude of the observed Ca isotope fractionation relates to the difference in Al_2O_3 of the mafic and felsic member of the diffusion couple.

Based on different Ca diffusion characteristics with respect to the silicate concentration of the melt, Watkins et al. (2009) suggested that in SiO_2-rich melts Ca diffuses by site hopping through a quasi-stationary Al_2O_3–SiO_2 matrix, resulting in large isotope fractionation, where Ca diffusion is not correlated to movements of other melt components. On the other hand, in SiO_2-poor melts Ca diffusion is correlated to the movement of other melt components and mass discrimination for Ca isotopes is smaller (less isotope fractionation). As mass discrimination of Ca depends on the ratio of the diffusivity of Ca (D_{Ca}) and that of the SiO_2–Al_2O_3 network ($D_{SiO2-Al2O3}$), Ca isotope fractionation is enhanced in more polymerised melts, because higher degrees of polymerization reduce the diffusivity of the SiO_2–Al_2O_3–network (Watkins et al. 2009).

The studies on natural rock diffusion couples were succeeded by experiments using diffusion couples with less complex compositions, in order to clarify the effects of matrix elements on diffusion coefficients and Ca isotope fractionation (Watkins et al. 2011).

These experiments confirm that the diffusion coefficient of the element relative to that of the solvent (D_{Ca}/D_{Si}) determines the diffusion velocity and magnitude of isotope fractionation (Table 1, Watkins et al. 2011). In the albite-anorthite ($NaAlSi_3O_8$–$CaAl_2Si_2O_8$) diffusion couple, Ca diffuses 20 times faster than Si ($D_{Ca}/D_{Si} = 20$) and Ca isotope fractionation reaches a maximum of 9 ‰ ($\beta = 0.21$). In the albite-diopsite ($NaAlSi_3O_8$–$CaMgSi_2O_6$) diffusion couple, D_{Ca}/D_{Si} is about 6 and Ca isotope fractionation about 5.5 ‰ ($\beta = 0.16$). In the natural rock diffusion couple (basalt–rhyolite),

D_{Ca}/D_{Si} is only about 1 and β ranges from 0.035 to 0.075 (Watkins et al. 2011).

Diffusion experiment with different CaO–Na_2O–SiO_2 systems demonstrate that diffusion is relatively fast and Ca isotope fractionation is relatively large (2.5 ‰) in systems with CaO and Na_2O gradients and minor SiO_2 gradients, while diffusion is relatively slow and Ca isotope fractionation small (1.3 ‰) in a system with CaO–SiO_2 counter diffusion and homogeneous Na_2O distribution (Watkins et al. 2014).

The systematic link of diffusive isotope separation to the solvent normalized diffusivity D_{Ca}/D_{Si} is not only limited to magmatic melts. In aqueous fluids a similar relationship exists between isotope separation efficiency and D_{Ca}/D_{H2O}, but the efficiencies are smaller in the aqueous fluids (Watkins et al. 2011).

While the afore discussed experiments, using concentration gradients between different artificial and natural rock-melts, reveal significant Ca isotope fractionation induced by chemical diffusion, such Ca isotope fractionation effects have not yet been reported in natural melt systems. Their existence is, however, feasible given that Mg isotope fractionation on the order of 2–3 ‰, caused by chemical diffusion, is apparent from natural contacts of felsic (granitic) and mafic (basaltic) rocks (Chopra et al. 2012).

Diffusional Ca isotope fractionation may not be limited to systems with CaO gradients, as diffusion of Ca in a system with initially homogeneous CaO and $\delta^{44/40}Ca$ distribution may

Table 1 Diffusion coefficients of Ca and solvents in different silicate and aqueous liquids

Diffusing medium	Data type	Diffusion experiment	D_{Ca} (m^2/s)	$D_{solvent}$ (m^2/s)	β^a	$\pm$	Reference
Silicate melt	Experiment	basalt-rhyolite	3.6×10^{-12}	2.3×10^{-12}	0.035	0.005	a
Silicate melt	Experiment	basalt-rhyolite	1.3×10^{-11}	5.9×10^{-12}	0.075	0.005	b
Silicate melt	Experiment	albite-anorthite	4.5×10^{-11}	2.0×10^{-12}	0.21	0.015	c
Silicate melt	Experiment	albite-diopsite	2.5×10^{-11}	4.0×10^{-12}	0.16	0.01	c
Water	MD simulation	Liquid water (75 °C)	1.5×10^{-9}	5.5×10^{-9}	0	0.011	d
Water	Experiment	Liquid water (75 °C)	–	–	0.0045	0.0005	d

[a]Ca isotope discrimination expressed as β see Eq. 1, a: Watkins et al. (2009), b: Richter et al. (2003), c: Watkins et al. (2011), d: Bourg et al. (2010)

create CaO gradients and Ca isotope fractionation. Based on model results, Watkins et al. (2014) predicted an establishment of CaO gradients and diffusive Ca isotope fractionation on the order of 1 ‰ by counter diffusion of Ca induced by chemical gradients of the matrix compounds Na_2O and SiO_2.

Based on self-diffusion experiments (at 1250 °C and 7 kbar) with a large gradient in $^{44}Ca/^{40}Ca$, but homogeneous composition of the melt with respect to CaO and bulk chemical composition (Na_2O and SiO_2), Watkins et al. (2014) determined a Ca diffusivity (D_{Ca}^k) of 80 ± 20 $\mu m^2/s$, as weighted average of Ca isotopic diffusivities. They also concluded that self-diffusion can be faster or slower compared to chemical diffusion, depending on the properties of the system.

1.4.2 Thermal Diffusion

Elements driven by temperature gradients migrate in melts preferentially to either hotter (Si, Na, K) or colder areas (Ca, Mg). This so called Soret diffusion (or thermal diffusion) can be associated with substantial mass-dependent isotope fractionation. It has been shown in several experimental studies as well as molecular dynamic modelling, to be a mechanism that can generate isotope fractionation of alkaline earth elements (Ca, Mg and Sr) in the order of several ‰/amu (cf. Richter et al. 2008; Huang et al. 2010b, 2011b; Lacks et al. 2012).

The migration of the alkaline earth elements along the thermal gradient is associated with an enrichment of heavy isotopes at the cold site and lighter isotopes at the hot end (Fig. 10, Richter et al. 2008; Huang et al. 2010b). This effect is explained by a faster movement of the heavy isotope towards the cold end, and can be described by a 'persistence fraction' of the travel direction of heavy and light isotopes. The outcome of the model predicts that upon collision at a sufficiently large speed, a heavy isotope will keep its initial direction more likely than a light isotope. A further prediction of this model reveals different fractionation effects for network formers (Si, O) and network modifiers (e.g. Ca, Mg, Sr) (Huang et al. 2010b, 2011b).

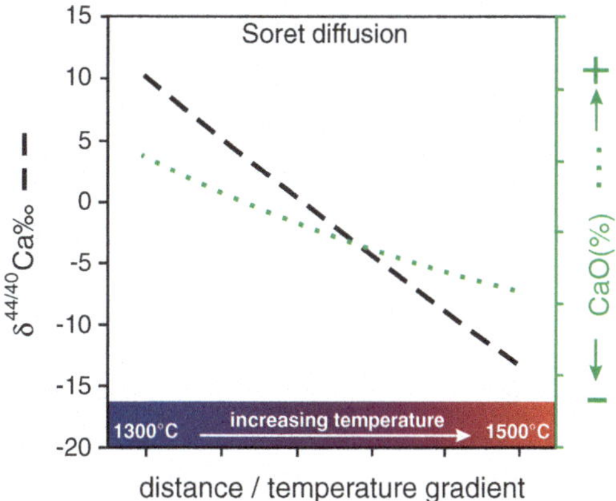

Fig. 10 Principle of Ca isotope fractionation caused by Soret diffusion in melt experiments. Based on Richter et al. (2009), and Huang et al. (2010b)

Mass-dependent isotope fractionation by Soret diffusion is not only apparent under controlled experimental conditions, but also relevant in natural systems. It has been suggested that isotope fractionation due to Soret diffusion may be responsible for the observed increase of heavy Mg and Fe isotopes during magmatic differentiation (Schüssler et al. 2009; Huang et al. 2009; Teng et al. 2011). In contrast to Mg and Fe isotopes, Ca isotope fractionation caused by Soret diffusion in natural melts has not yet been reported.

2 Extraterrestrial Materials

2.1 Scope and Framework of Ca Isotope Cosmochemistry

Our solar system most likely formed as part of a dense cluster of stars through the gravitational collapse of a part of a giant molecular cloud (Kennicutt and Evans 2012). After the initial collapse, a central star, our Sun, formed and through preservation of angular momentum the ongoing collapse led to the formation of a rotating disk of gas and dust around this central object. The planets and asteroids populating our Solar System today formed through accretion of this disk material shortly after the formation of our Sun. Although most of the initial dust and gas was thermally processed in some way during the initial phases of solar system formation, some pre-solar materials escaped this fate and are still present as individual grains incorporated in meteorites found on Earth or in interplanetary dust particles. Although the protoplanetary disk has long since dissipated, information about the initial conditions in our Solar System can be derived from meteorites and their components that were formed prior to or during this phase.

Meteorites are divided into primitive, undifferentiated meteorites (chondrites) and differentiated meteorites (achondrites). Chondritic meteorites did not experience sufficiently high temperatures for melt generation and are effectively cosmic sedimentary rocks, albeit most experienced variable degrees of thermal or aqueous metamorphism and/or shocks. The main components of chondrites are chondrules, which are round spherically molten objects, calcium-aluminum-rich inclusions (CAIs) and a fine-grained matrix. The matrix also contains small presolar grains, such as diamonds, graphite, carbides, oxides and silicates. Achondrites, on the other hand, are products of igneous processes and the group encompasses crustal-, mantle- and core-type rocks of early-formed and differentiated asteroids as well as matter ejected by impacts from Mars and the Moon.

Calcium is a refractory lithophile element and, thus, is prevalent in many of the minerals and rocks that formed during the earliest stages of our Solar System. This has made it one of the ideal target elements to study original components, processes and conditions that controlled the formation of our Solar System. The following section summarizes the current level of insight gained from Ca isotopes applied to the study of the stellar environment in which our Solar System formed and its initial conditions using data from differentiated and primitive meteorites and their components including presolar grains.

2.2   Nucleosynthesis of Ca Isotopes

Following Big Bang nucleosynthesis, which formed only the lightest nuclides (mainly H and He), the majority of nuclides are the result of nuclear reactions in stellar cores and supernovae. As part of galactic chemical evolution, where old stars die and new ones are formed, the average metallicity (the proportion of elements heavier than hydrogen and helium) in the galaxy increases. The interstellar dust present in the molecular cloud core that collapsed to form our Solar System contained signatures of different stellar sources in the form of its isotope compositions and, together with the gas phase, presents a snapshot of the galactic chemical evolution. The majority of matter contributing to our Solar System was isotopically homogenized as part of the dust destruction and reformation during its residence in the interstellar medium. However, a minor component of the original dust inventory, mainly in the form of refractory or newly formed dust, still preserves the nucleosynthetic signatures reflecting their stellar source. In this context, the isotopic composition of calcium present in solar system matter and, in particular, presolar grains can provide clues to the birth environment of our Solar System.

There are six stable or extremely long lived isotopes of Ca with masses 40, 42, 43, 44, 46 and 48, and one cosmochemically-relevant, short-lived isotope, ^{41}Ca, with a half-live of 103,000 years. The natural abundance of the stable isotopes of Ca differs by several orders of magnitude from the most to the least abundant isotope. Whereas ^{40}Ca is the by far most abundant isotope making up 96.94 % of the total Ca, ^{46}Ca with only 0.004 %, is the least abundant isotope (see Chapter "Introduction" for details). The differences in isotope abundances are primarily due to their nucleosynthetic origin. ^{40}Ca is a primary nucleus produced through oxygen and silicon burning and is a "double magic" isotope with 20 protons and 20 neutrons that grants it extra stability. With the exception of ^{48}Ca, the second "double magic" Ca isotope, other Ca isotopes (including the short-lived ^{41}Ca) are produced from ^{40}Ca through slow neutron capture in asymptotic giant branch (AGB) stars that are characterized by He shell burning around a carbon oxygen core. However, ^{44}Ca is overabundant with respect to the other minor isotopes of Ca. It owes its overabundance to the fact that it is a decay product of the short-lived ^{44}Ti (half-life of 60 days), which is synthesized along with ^{40}Ca during silicon and oxygen burning. The heaviest calcium isotope, ^{48}Ca, is 47 times more abundant than ^{46}Ca, the next heavy stable isotope. This can be explained through its "double magic" structure of 20 protons and 28 neutrons and points to its nucleosynthesis in thermal equilibrium such as the nuclear statistical equilibrium (NSE) or quasi-nuclear equilibrium (QSE) from a neutron-rich environment. For a detailed discussion of the environment required to produce significant amounts of ^{48}Ca, the reader is referred to Meyer et al. (1996).

Although ^{48}Ca can be formed under the right conditions in stars, the stellar source that is responsible for producing and ejecting the significant amounts of ^{48}Ca is still elusive. Hartmann et al. (1985) suggested that the innermost ejecta of a core-collapse supernova (CCSNe) could be the site of ^{48}Ca nucleosynthesis. However, this option was later rejected by Meyer et al. (1996) who favored rare, dense type Ia supernova explosions as sites of the origin of ^{48}Ca (Woosley 1997). More recently, Wanajo et al. (2013) returned to focus on CCSNe, suggesting that electron-capture supernovae, a sub-class of CCSNe arising from collapsing O–Ne–Mg cores of progenitor stars with a mass of 8–10 solar masses, may also be capable of ejecting substantial amounts of ^{48}Ca. Overall, it appears that ^{48}Ca follows an inherently different path in its nucleosynthesis compared to the other stable isotopes of Ca, which makes it a good tracer to search for different sources of presolar dust contribution to our Solar System.

2.3 Nucleosynthetic Ca Isotope Signatures in Presolar Grains

Presolar grains condense from outflows of giant stars and matter ejected from stellar explosions. The extrasolar origin of these grains makes them powerful samples to directly study the stellar sources that contributed dust to our Solar System. Although it is a relatively young field of study,

much work has been done to find and identify presolar grains in the matrix of primitive chondrites (and in interplanetary dust particles; see Zinner (2007) for a detailed review). The potential scientific value also sparked the Stardust sample return mission to a comet (McKeegan et al. 2006). Most initial work on presolar grains applied harsh chemical methods for grain extraction leading to a general bias to more refractory minerals (Nittler et al. 2008). More recent in situ techniques now allow also for detection of the more fragile presolar silicates, which are one of the dominant phases in the interstellar dust.

The identification of the source of nucleosynthetic signatures present in presolar grains is not trivial because the nucleosynthetic products of a star depend on variables such as the initial mass and metallicity. To achieve the best possible understanding of the stellar source of presolar grains, they are typically analysed for as many elements/isotopes as is feasible. In this context, ^{44}Ca is a good tracer isotope because it is the daughter product of the short-lived radionuclide ^{44}Ti, which in turn is primarily produced during He-burning in explosive SNe. Presolar grains, such as low-density graphite grains, often contain large ^{44}Ca excesses providing evidence for their SNe origin (Travaglio et al. 1999), while the ^{44}Ca signature is often absent in high-density graphite grains, which supports an origin of these grains from asymptotic giant branch (AGB) stars (Zinner and Jadhav (2013) and references therein). Large excesses in ^{42}Ca, ^{43}Ca and ^{44}Ca in high density graphite grains are incompatible with condensation from a SNe or AGB winds but rather reflect products of a very late thermal pulse of an AGB star (Jadhav et al. 2013).

Presolar hibonite grains are one of the widest studied groups of presolar grains. Large positive and negative ^{48}Ca anomalies have been found for individual grains and in some cases these are correlated with ^{50}Ti, another neutron-rich isotope (Zinner et al. 1986; Fahey et al. 1987; Ireland 1990). These large isotope effects for neutron-rich isotopes in hibonites are thought to signify a preservation of interstellar dust signatures within these mineral grains (Zinner et al. 1986).

Some hibonite grains, interpreted to have condensed in the winds of AGB stars, show inferred levels of ^{41}Ca ($t_{1/2}$ = 0.1 Myr) based on ^{41}K excesses that are in agreement with nucleosynthetic predictions for ^{41}Ca production in an AGB environment (Choi et al. 1999; Nittler et al. 2008). Further support for this interpretation could come from large excesses in ^{46}Ca in these grains. However, measurement of this isotope is difficult by in situ methods in hibonite grains due to the large ^{46}Ti interference on this mass and precise data is still lacking.

2.4 Nucleosynthetic Anomalies in Meteorites and Calcium-Aluminum-Rich Inclusions (CAIs)

When investigating the Ca isotope composition of meteorites and their components, it is important to realize that either mass-independent or mass-dependent processes can cause isotopic variability. Typically mass-independent Ca isotope effects in meteorites are linked to the admixture of nucleosynthetic contributions from distinct presolar sources, i.e. variable contributions of presolar grains. Variable Ca isotope abundances in bulk meteorites or meteoritic components result from mixing relationships. Another mass-independent effect is radiogenic ingrowth. In particular, over time ^{40}Ca is enriched in K-bearing minerals and rocks through the decay of ^{40}K and this decay system is discussed later in this chapter. In meteorites and their components, mass-dependent Ca isotope fractionation is typically the result of condensation and evaporation processes that are discussed further below.

Although mass-independent isotope effects are not directly related to mass-dependent isotope fractionation, for samples that contain both, mass-dependent and -independent effects, such as CAIs, the measured mass-independent isotope effects can be significantly affected by the choice of the mass fractionation law applied to account for the mass-dependent fractionation. For example, Schiller et al. (2015) showed that the choice of fractionation law for two CAIs with rare Fractionated and Unknown Nuclear (FUN) effects (Wasserburg et al. 1977) creates a spread in the calculated mass-independent ^{48}Ca component in each of these inclusions of 300–400 ppm when applying either the equilibrium or the power law for mass fractionation correction (see Chapters "Introduction" and "Analytical Methods" for a brief discussion of mass fractionation laws in nature and mass spectrometry).

Due to its refractory nature, Ca is one of the earliest major elements to condense from a gas of solar compositions (Albarède 2009) and consequently it is also a major component in the oldest solids that formed in our Solar System, CAIs. Nucleosynthetic effects in Ca isotopes (e.g. ^{40}Ca and ^{48}Ca) were first observed in rare FUN-type and "normal" CAIs (Lee et al. 1979; Jungck et al. 1984; Niederer and Papanastassiou 1984), all of which used thermal ionization mass spectrometry. Since then, the enriched ^{48}Ca signature of "normal" CAIs has been confirmed in several studies (Moynier et al. 2010; Huang et al. 2012; Dauphas et al. 2014; Schiller et al. 2015; Chen et al. 2015).

The early focus on the refractory inclusions can mainly be ascribed to the fact that, next to much smaller individual mineral grains, these ∼mm-sized inclusions contain the largest magnitude of the nucleosynthetic effects in Solar System materials. However, a result of the limited data set was that the nucleosynthetic effects found in CAIs were initially ascribed to incomplete mixing in the early solar system. Subsequent work that applied TIMS with larger ion beams shifted the focus from CAIs to other bulk solar system reservoirs. The resulting improved precision allowed Chen et al. (2011) to resolve ^{48}Ca variations in differentiated meteorites that correlate with variability in the other neutron-rich isotopes such as ^{54}Cr and ^{50}Ti in the same type of meteorites (e.g. Fig. 11a, b). However, they could not identify isotopic variability in the isotopes ^{43}Ca and ^{46}Ca when normalizing to $^{42}Ca/^{44}Ca$. The data by Chen et al. (2011) was, albeit at a slightly lower precision, expanded by Dauphas et al. (2014), who also could only identify nucleosynthetic effects on ^{48}Ca. The good

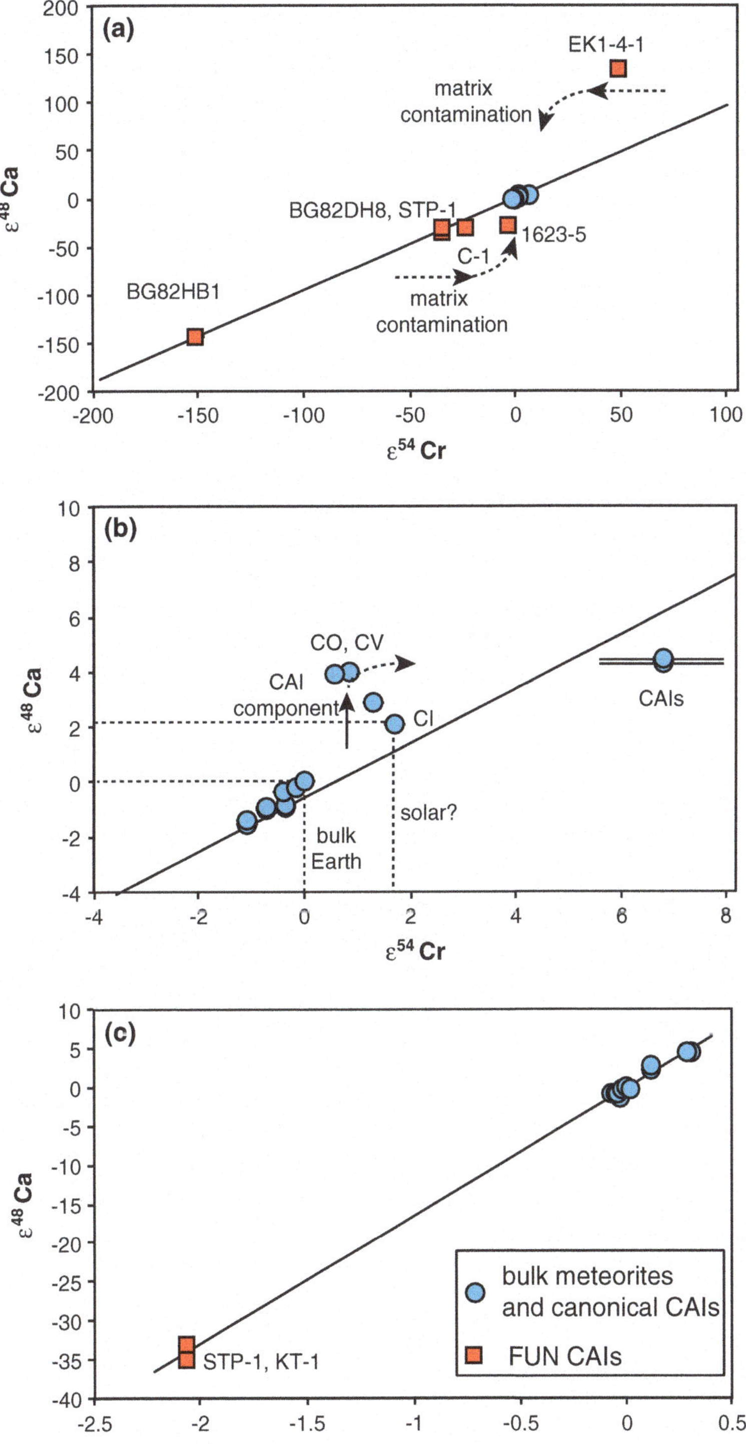
(a)
EK1-4-1
matrix contamination
BG82DH8, STP-1
1623-5
C-1
matrix contamination
BG82HB1
ε⁴⁸Ca
ε⁵⁴Cr

(b)
CO, CV
CAI component
CI
CAIs
bulk Earth
solar?
ε⁴⁸Ca
ε⁵⁴Cr

(c)
STP-1, KT-1
bulk meteorites and canonical CAIs
FUN CAIs
ε⁴⁸Ca
ε⁴³Ca

Fig. 11 Correlation of the mass-independent component of ^{48}Ca and ^{54}Cr in bulk meteorites and refractory inclusions including FUN CAIs (**a**) and a zoomed in view of the same data excluding FUN CAIs (**b**), correlation of ^{48}Ca and ^{43}Ca isotopic variability in bulk meteorites and refractory inclusions (**c**) (Lee et al. 1978; Papanastassiou 1986; Papanastassiou and Brigham 1989; Loss et al. 1994; Schiller et al. 2015; Dauphas et al. 2014; Trinquier et al. 2007; Holst et al. 2013; Larsen et al. 2011). The line shown in (**a**) and (**b**) is a regression through the FUN CAIs BG82HB1, BG82BH8, STP-1, canonical CAIs and non-carbonaceous meteorites and is close to a 1:1

correlation. Carbonaceous chondrites (and the achondrite NWA 2976) were excluded from the regression because of the potential effect of CAI-like material contribution to the bulk composition. Due to the different elemental abundances of Ca and Cr in CAIs and bulk solar material, small addition of CAIs can significantly increase the bulk ^{48}Ca excess with only a minute collateral effect on the bulk ^{54}Cr composition. The effect of remobilized matrix during parent body alteration, which, in part, may explain the lack of correlated nucleosynthetic effects in some FUN-CAIs between ^{48}Ca and ^{54}Cr, is indicated in (**b**). The regression in (**c**) includes all data shown

correlation of ^{48}Ca with other neutron-rich isotopes, which are all thought to be co-produced in type Ia SN, led Chen et al. (2011) to interpret their data as evidence for a late injection of type Ia SN material into the solar nebula or, alternatively, inefficient homogenization during the planetary formation process. A contribution from a nearby type Ia to our forming solar system, however, implicitly argues for an unusual setting for the formation of our Solar System, as their progenitor stars are not typically related to star forming regions. Identifying the origin of the ^{48}Ca enrichment is not trivial and the complications of this approach were discussed in depth by Dauphas et al. (2014). The same authors also argued that variable contributions of isotopically highly anomalous hibonite grains to different solar reservoirs cannot be the cause for the systematic nucleosynthetic variability of neutron-rich isotopes observed in bulk meteorites, which, in essence, rules out the incomplete mixing model. Dauphas et al. (2014) conclude that identification of the presolar carrier(s) of the isotopic anomalies is required before more detailed conclusions about their source can be made.

Unlike all previous work on nucleosynthetic Ca isotope effects in meteorites and their components that used TIMS, Schiller et al. (2015) took a high signal intensity MC-ICPMS approach to further improve the analytical precision. The work by these authors resolved nucleosynthetic isotopic variability of Ca in solar system materials in ^{48}Ca as well as ^{43}Ca and ^{46}Ca (when normalizing to a constant ^{42}Ca/^{44}Ca) confirming early detections of anomalous ^{43}Ca signatures in some meteorites (Simon et al. 2009; Moynier et al. 2010). The intriguing observation

by Schiller et al. (2015) was that the isotopic effects on ^{43}Ca and ^{46}Ca are positively correlated with ^{48}Ca and this correlation includes normal CAIs with enrichments in ^{43}Ca, ^{46}Ca and ^{48}Ca and also two FUN CAIs with large depletions in these three isotopes (Fig. 11c). The different nucleosynthetic origin of these three Ca isotopes led Schiller et al. (2015) to rule out incomplete mixing as the cause for these effects, in favor of thermal processing (Trinquier et al. 2009). In this model, enhancement or depletion of the nucleosynthetic signatures can be understood in the concept of an unmixing of a physically well mixed reservoir that contains isotopically homogenized old dust that has been processed in the interstellar medium and a relatively young component of freshly synthesized stellar dust with variable nucleosynthetic signatures (Schiller et al. 2015). During thermal processing the more labile component is preferentially destroyed, effectively unmixing the former homogenous reservoir and creating a spread in nucleosynthetic signatures that is related to the degree of thermal processing (Trinquier et al. 2009). Support for this model comes from sequential acid dissolution procedures of primitive meteorites (Moynier et al. 2010; Schiller et al. 2015) that identified signatures of carriers with variable and independent nucleosynthetic Ca isotope pattern, none of which have a signature that could create the observed correlation in bulk meteorites by themselves. Furthermore, it is unlikely that random sampling of these individual carriers could create a positive correlation between three different Ca isotopes in bulk solar system reservoirs (Schiller et al. 2015). However, to reiterate, an important task for the future is to identify and isolate these carrier grains to further understand their origin and the mechanism(s) that created correlated isotope effects within our solar system.

The former presence of live ^{41}Ca in our Solar System was initially demonstrated by Srinivasan et al. (1994). The initial ^{41}Ca abundance was later revised by Liu et al. (2012) through analysis of the same two CAIs from Efremovka in which the original discovery was made. The current estimate of the initial ^{41}Ca/^{40}Ca at the time of formation of CAIs is $\sim 4.2 \times 10^{-9}$ (Liu et al. 2012). This abundance is 4–16 times lower than the steady-state value from galactic nucleosynthesis and equivalent to 0.2–0.4 Myr of decay. This timescale is broadly consistent with the free-fall timescale of a collapsing molecular cloud to form a star but also requires the abundance ^{41}Ca/^{40}Ca in CAIs to be representative of the bulk solar system. In addition, there are alternative sources of ^{41}Ca, such as a last-minute input from a stellar source and early solar system irradiation, that currently cannot be ruled out (Liu et al. 2012). As of now, the limited amount of data is a key limiting factor in our understanding of the source of ^{41}Ca formerly present in CAIs.

2.5 Mass-Dependent Variations in CAIs and Related Experiments

CAIs are thought to be condensates from a gas phase of approximately solar composition (Grossman 1972). Some CAIs subsequently experienced melting and evaporation. Consequently, CAIs and their minerals show variable and large mass-dependent Ca isotope fractionation of several ‰/amu (Lee et al. 1979; Clayton et al. 1988; Ireland 1990; Ireland et al. 1992; Huang et al. 2012; Schiller et al. 2015; Niederer and Papanastassiou 1984; Hinton et al. 1988) (Fig. 12). The enrichment of light Ca isotopes in the early condensates is thought to be caused by disequilibrium condensation (Simon and DePaolo 2010). Heavy isotope signatures in CAIs are interpreted to signify evaporative loss (Grossman et al. 2000; Zhang et al. 2014). In principle, the magnitude of Ca isotope fractionation in CAIs, in

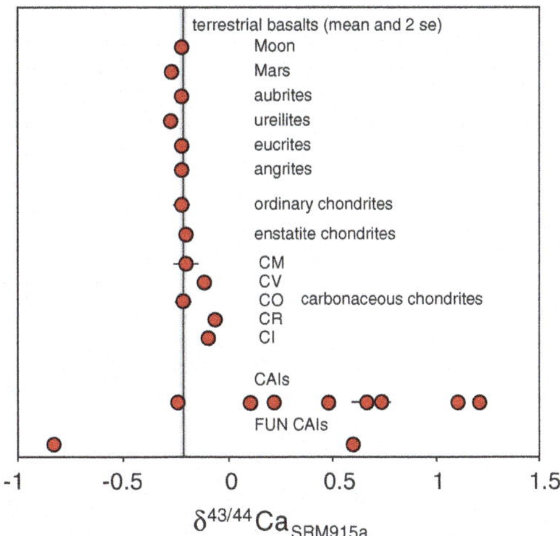

Fig. 12 Mass-dependent Ca isotope variability in chondrites, achondrites as well as in CAIs. Data are from Valdes et al. (2014), Huang et al. (2012), Schiller et al. (2015), and Magna et al. (2015). Note that light $\delta^{43/44}$Ca values are represented by positive values

particular in combination with isotope fractionation of other elements such as O, Si and Mg, provides clues to the environmental conditions under which these inclusions formed and fractionated (Davis and Richter 2014). However, in detail, the size of mass-dependent fractionation is dependent on numerous variables such as the initial chemical composition, oxygen fugacity, the total and partial pressure and temperature (Davis and Richter 2014). In an effort to better understand the fractionation law controlling evaporative loss of Ca from CAIs, Zhang et al. (2014) measured the Ca isotope composition of evaporation experiments of $CaTiO_3$ in a vacuum furnace. These authors concluded that Rayleigh distillation best explains the fractionation behavior of their sample (Fig. 13) with a mass dependence that is best described by the Rayleigh law, rather than the commonly used exponential law (see Fig. 4 in Introduction). So far, only Schiller et al. (2015) reported mass-dependent and mass-independent Ca isotope data for two canonical and FUN-type CAIs precise enough to distinguish between different fractionation laws. Based on the assumption that no or little variability in the mass-independent Ca isotope effects of canonical CAIs should exist, these authors

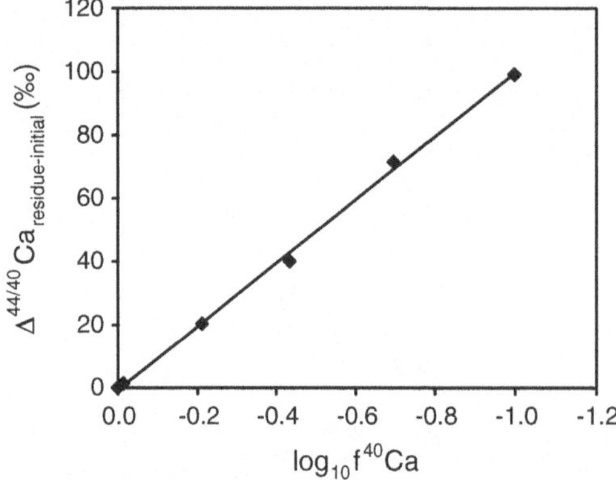

Fig. 13 Calcium isotope fractionation during evaporation. Calcium isotope composition of the residue relative to that of the initial material ($\Delta^{44/40}$Ca$_{\text{residue–initial}}$) shown as a function of the ^{40}Ca fraction (f^{40}Ca) remaining in the condensed phase. The experiments of Zhang et al. (2014) reveal a fractionation factor (α) of 0.9560 ± 0.0040, which is close to the theoretical value of 0.9535 for Rayleigh distillation, calculated using the inverse square root of the masses of the evaporating monatomic species $(m^{40}\text{Ca}/m^{44}\text{Ca})^{0.5}$

concluded that mass-dependent Ca isotope fractionation present in these inclusions followed an exponential law. However, due to the currently limited amount of available data, no final conclusions can yet be made.

2.6 Mass-Dependent Variations in Meteorites

In addition to the mass-independent and nucleosynthetic isotope effects, the mass-dependent fractionation provides another tracer for mass transport and accretion of planetary building blocks in our Solar System. Recent work by Simon and DePaolo (2010) suggests that enstatite chondrites are enriched in heavy Ca isotopes by 0.05–0.10 ‰/amu when compared to bulk Earth, while carbonaceous chondrites are depleted by a similar amount. If these results are correct, the difference in stable isotope composition of enstatite chondrites from that of the Earth potentially poses a significant complication

to the theory that enstatite chondrite-like matter is the primary source from which the Earth accreted (Javoy et al. 2010). Because of the importance of this issue, Valdes et al. (2014) investigated the mass-dependent Ca isotope composition of a large number of chondrites. These authors used a sample-standard bracketing approach by MC-ICPMS compared to the traditionally approach of double spike TIMS, and could not identify a difference in the mass-dependent Ca isotope composition of enstatite chondrites and Earth, but confirmed the light Ca isotope signature of carbonaceous chondrites (Fig. 12). Both, Simon and DePaolo (2010) and Valdes et al. (2014) did not find any mass-dependent Ca isotope variability in differentiated meteorites and, thus, it appears that, with the exception of carbonaceous chondrites, little mass-dependent Ca isotope fractionation is present on the scale of bulk asteroids. Recent investigation of mass-dependent Ca isotope variability in Martian meteorites that are on average indistinguishable from that of the Earth further suggests that this lack of mass-dependent Ca isotope variations can also be extended to the bulk composition of terrestrial planets (Magna et al. 2015). Although CAIs appear to be predominantly isotopically light, Valdes et al. (2014) pointed out that their presence in carbonaceous chondrites does not appear to sufficiently explain the variability of the mass-dependent Ca isotope composition of distinct carbonaceous chondrite groups. In particular, CI chondrites do not contain CAIs but are isotopically light (Fig. 12). Thus, it appears that another mechanism is required to create the isotopic offset between carbonaceous chondrites and other meteorites. A hypothesis to explain this disparity is that the enrichment of light isotopes in carbonaceous chondrites reflects material that condensed prior to complete condensation of the solar nebula (Simon and DePaolo 2010). Overall, it appears that some mass-dependent isotope variability is present between carbonaceous chondrites and the other asteroidal and planetary bodies, but its origin is not yet fully understood.

2.7   Lunar Samples

Currently, only little Ca isotope data for lunar samples exists. Russel et al. (1977) performed leaching experiments on lunar soils and found only very limited isotope fractionation between the leachate and residue, and samples have a mass-dependent isotopic composition similar to that of the Earth. Heavy Ca isotope signatures obtained in leachates were interpreted by Russel et al. (1977) to be associated with the grain surfaces and possibly result from solar wind ion sputtering.

3   The ^{40}K–^{40}Ca Decay System

In addition to mass-dependent fractionation processes in or between geological reservoirs, Ca isotope ratios that include ^{40}Ca can also be affected by the β-decay of ^{40}K to ^{40}Ca. Methological aspects related to radiogenic ^{40}Ca, including notation, decay constant and branching ratio of the ^{40}K decay are described in Chapter "Analytical Methods". Primarily, the radiogenic ingrowth of ^{40}Ca is a function of the K/Ca ratio and age of the respective sample or reservoir. However, systematics of mass-dependent Ca isotope fractionation may not only be disturbed by supported ^{40}Ca (formed by ^{40}K decay since rock formation), but also by inherited ^{40}Ca originating from already ^{40}Ca-enriched preexisting sources.

Consequently, consideration of the effect of radiogenic ^{40}Ca (expressed as ε_{Ca}) on the accuracy of measured mass-dependent Ca isotope ratios that utilize ^{40}Ca as either numerator or denominator are required for each sample. Besides potential pitfalls in the accurate interpretation of mass-dependent Ca isotope fractionation when utilizing Ca isotope ratios that include ^{40}Ca, the ^{40}K–^{40}Ca decay is a useful tool to investigate geochronological and petrogenetic questions. These aspects are beyond the primary scope of this book and therefore only briefly summarized.

3.1   Evolution of Earth's Reservoirs

3.1.1   Earth's Mantle

Since Earth's formation, the radiogenic ^{40}Ca content of the Earth mantle increased up to now by only about 0.01 ‰ ($\varepsilon_{Ca} \sim 0.1$), due to its low K/Ca ratio (~ 0.01) and, thus, can be considered as virtually constant over geological times (Fig. 14, Marshall and DePaolo 1989).

3.1.2   Earth's Crust

Due to the fact that Ca is less incompatible than K, the Earth's crust is enriched in K/Ca relative to the Earth's mantle and, thus, the Ca isotope composition of the crust is affected to a larger degree by ^{40}K decay. The level of radiogenic ^{40}Ca enrichment in different crustal rocks is a function of the K/Ca ratio of the rock, its age and, in some cases, especially in continental crust, unsupported ^{40}Ca inherited from older crustal sources. In general, oceanic crust experiences less radiogenic ^{40}Ca ingrowth than continental crust (Fig. 14)

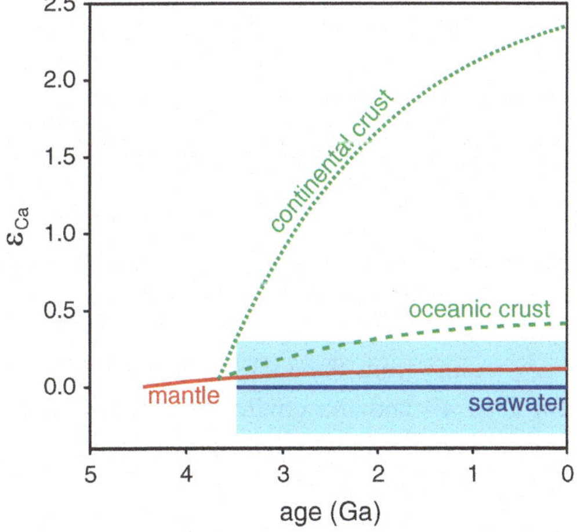

Fig. 14 Evolution of ε_{Ca} in different terrestrial reservoirs throughout geological times. Due to low K/Ca of the mantle (0.01), ε_{Ca} increased by only 0.1 since Earth formed (e.g. Marshall and DePaolo 1982). Evolution curves of oceanic crust and continental crust consider K/Ca of 0.05 and 0.35, respectively (Kreissig and Elliot 2005). ε_{Ca} of seawater has remained constant within ±0.35, as shown by analyses of marine carbonates throughout Earth history (Caro et al. 2010)

resulting from its generally lower K/Ca (0.05) when compared to average continental crust (0.35) (Rudnik and Fountain 1995). Significantly larger enrichment of radiogenic ^{40}Ca than the average continental crust composition, as indicated in Fig. 14, is possible in lithologies that are strongly enriched in K/Ca.

Young mid ocean ridge basalts (MORB) and ocean island basalts (OIB) are indistinguishable from the Earth's mantle within the quoted uncertainties that range from 0.5 to 2 ε_{Ca} (Marshall and DePaolo 1989; Nelson and McCulloch 1989; DePaolo 2004; Caro et al. 2010). For Island arc basalts, Marshall and DePaolo (1989) report slightly increased ε_{Ca} of 1–2, caused by the presence of crustal Ca in the magma sources. This range is consistent with the observation of Nelson and McCulloch (1989), who found ε_{Ca} of 0 ± 2 for samples of two island arcs. Also, ε_{Ca} reported for amphibolite (Hindshaw et al. 2014) and carbonatites (Nelson and McCulloch 1989) are within uncertainty identical to the mantle value.

Radiogenic ^{40}Ca enrichment in igneous rocks from the continental crust is more evident than in mantle-sourced basalts, whose ε_{Ca} is close to Earth mantle. ε_{Ca} observed in K-rich rocks including granites and gneisses are commonly in the order of several ε_{Ca} (Ryu et al. 2011; Hindshaw et al. 2014; Farkas et al. 2011; Nelson and McCulloch 1989; Marshall and DePaolo 1989), but occasionally old K-rich rocks and particularly K-rich minerals can exhibit significantly larger enrichment of ^{40}Ca. For instance, whole rock ε_{Ca} of pegmatites from the Jack Hills region range from 7 to 40 and muscovite mineral separates from these rocks show ε_{Ca} between about 2000 and 19,000 (Fletcher et al. 1997).

Granitic rocks demonstrate not only K-supported ^{40}Ca enrichment, but can also exhibit elevated initial ε_{Ca} values. This inherited radiogenic ^{40}Ca provides constraints, such as the K/Ca ratio, of deep crustal magma sources (Marshall and DePaolo 1989). Unsupported ^{40}Ca has also been found in kimberlitic and lamproitic intrusions (Nelson and McCulloch 1989). The ^{40}Ca-enrichment in these rocks is not associated

by elevated ^{87}Sr and, thus, Nelson and McCulloch (1989) interpreted this as a decoupling of major and trace element sources. In detail, these authors propose that the major elements may originate from the subcontinental lithosphere, whereas the trace elements are mantle derived. Thus, this example highlights the potential for additional information that can be derived from the K–Ca decay when combined with other radiogenic systems.

While the isotopic abundance of ^{40}Ca of bulk Earth increased only by a small amount since its formation, the isotope abundance of ^{40}K decreased by a factor of ~ 12 since 4.55 Ga, due to its half-life of 1.25 Ga (cf. DePaolo 2004). Consequently, major enrichments in ^{40}Ca are more likely found in old crustal reservoirs.

In an effort to identify a Ca isotope signature of a preexisting Hadean crust, Kreissig and Elliott (2005) searched for an inherited ^{40}Ca-signature in early Archean crustal rocks from Zimbabwe and Greenland, utilizing K-poor minerals (plagioclase) that faithfully record the initial $^{40}Ca/^{44}Ca$ at the time of their formation. If a K-enriched crust was present during the Hadean, radiogenic ingrowth of ^{40}Ca may have significantly altered the $^{40}Ca/^{44}Ca$ of the crust. Results of the rocks from Zimbabwe indicate that at 3.7 Ga, ε_{Ca} in the analysed minerals are indistinguishable from that of the mantle, providing no clear evidence for a precursor crustal source enriched in ^{40}Ca. On the other hand, mafic and felsic rocks from Greenland analysed as part of the same study exhibit ε_{Ca} (calculated at 3.7 Ga) values between 0 and 3 and, thus, contain potential evidence for an older crustal component. However, this interpretation is complicated by the fact that the reported ε_{Ca} of some of the plagioclases is too radiogenic to be compatible with crustal evolution models and, therefore, may rather be explained by secondary exchange of Ca between minerals during later thermal events (Kreissig and Elliott 2005). To further complicate matters, Kreissig and Elliott (2005) also identified a K-metasomatism event that occurred 3.6 Ga ago in some of the analysed mafic rocks. Consequently, Kreissig and Elliott (2005) concluded that Ca isotopes are not robust tracers to study the crustal evolution when active metamorphic and tectonic activities are involved. Nevertheless, the results of their study demonstrate the presence of slightly enriched ε_{Ca} in South West Greenland rocks and a lack of a clear signature of ^{40}Ca inherited from earlier (Hadean) crust.

3.1.3   Seawater and Exogenic Ca Cycling

Although weathering of K-rich continental crust provides radiogenic ^{40}Ca to the ocean, no radiogenic ^{40}Ca excesses appear to be present in the marine carbonate record ranging from recent to 3.5 Ga old sediments, suggesting that ε_{Ca} of seawater remained constant within a range of 0 ± 0.35 throughout Earth's history (Caro et al. 2010). Consequently, it appears that ε_{Ca} of seawater is controlled by hydrothermal input, whereas the Ca flux resulting from silicate weathering is one order of magnitude smaller (Caro et al. 2010). Water and sediments of rivers draining old crustal rocks such as the Himalayas or Greenland, on the other hand, exhibit ^{40}Ca-enrichment (ε_{Ca} up to 4) (Caro et al. 2010; Hindshaw et al. 2014).

Significant enrichment in radiogenic ^{40}Ca is also described for Archean pelagic sediments (ε_{Ca}: 3–26) and continental gypsum sediments deposited on top of Archean to early Proterozoic terranes in Australia (ε_{Ca}: 2–6), carrying the inherited ε_{Ca} signature of the weathered old K-rich rocks (Nelson and McCulloch 1989). Elevated ε_{Ca} are also present in soils developing on top of ancient K-rich rocks. Farkas et al. (2011) showed radiogenic ^{40}Ca-enrichment in soil profiles, originating from biotite weathering, which contribute up to 25 % of the Ca in the exchangeable soil pool. However, the radiogenic signature of the biotite is not found in the vegetation that growth from this soil (Farkas et al. 2011). Overall, even in relatively young reservoirs with low K contents, high ε_{Ca}, inherited from old or K-rich sources are reported and need to be considered for the interpretation of Ca isotope data.

3.2   Dating

3.2.1   Igneous Rocks

The decay of ^{40}K to ^{40}Ca in minerals of igneous rocks has been utilized by a number of studies to obtain crystallization ages (cf. Coleman 1971; Marshall and DePaolo 1982; Shih et al. 1994; Fletcher et al. 1997). Both, terrestrial and lunar granites have been successfully dated using the K/Ca method. For instance, Marshall and DePaolo (1982) obtained an age of 1041 $\pm$ 32 Ma from a K–Ca mineral isochron of Pikes Peak Granite, which is concordant with K–Ar, Rb–Sr, and U–Pb ages (Fig. 15).

A good agreement of K–Ca, Rb–Sr and K–Ar ages has also been reported for several lunar granites, however, with slightly older ages than corresponding U–Pb zircon ages (Shih et al. 1993). Good overall agreement between K–Ca ages and other chronometers is also evident from Proterozoic phlogopites that exhibit concordant Rb–Sr and K–Ca ages (Goppalan and Kumar 2008) and K–Ca ages of K-feldspar, which are in agreement to regional U/Pb ages (Heumann et al. 1979). Nägler and Villa (2000) observed good agreement between K–Ca and Ar–Ar ages of gem-quality muscovite, but a studied sanidine showed considerably younger Ar–Ar ages, likely

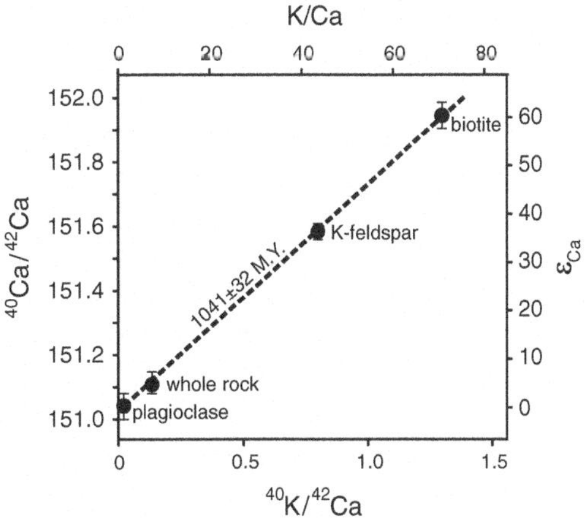

Fig. 15 K-Ca isochron. Mineral isochron of plagioclase, whole rock, biotite and K-feldspar from Pikes peak granite reveals an age of 1041 $\pm$ 32 Ma, in agreement to other dating techniques such as K-Ar and Rb-Sr. Redrawn from Marshall and DePaolo (1982)

related to Ar loss. Coleman (1971) also found a good general agreement between K–Ca and Rb–Sr ages for some mica samples, but also found some inconsistencies that are likely related to post crystallization histories. In this context, Shih et al. (1994) suggested, based on disturbed Lunar K–Ca and Rb–Sr isochrones, that the K–Ca system may be less susceptible to thermal metamorphism than the Rb–Sr system. For micas, Fletcher et al. (1997) concluded that closing temperatures of the K–Ca system are significantly lower compared to those of the Rb–Sr system, and that the Ca diffusion coefficient in muscovite is about one order of magnitude larger than that of Sr. This reasoning is based on combined Rb–Sr and K–Ca ages of Archean muscovite from pegmatites of the Jack Hills region (Australia), which revealed systematic differences between different mica phenotypes and between the Rb–Sr and K–Ca decay systems (Fletcher et al. 1997).

Additional insights into complex histories of heating events and dissolution-reprecipitation at lower temperatures have been gained through spatially resolved measurements of Ca and K isotopes by secondary ion micro probe analyses. Harrison et al. (2010) analysed doubly charged ions to suppress the K intensities relative to Ca. With this technique the authors were able to unravel complex genetic relationships and, thus, reconciled K–Ar and K–Ca systematics in alkali feldspars from the Klokken syenite (1166 Ma) in southern Greenland.

3.2.2 Authigenic Sedimentary Minerals

Similar to igneous rocks, radiogenic increase in ^{40}Ca is observed in K-rich authigenic minerals formed in sediments. Dating sedimentary rocks by the K/Ca method has been tested using authigenic sanidine and glauconite, but revealed severe complications. Nevertheless, although current results demonstrate large deviations between radiometric ages determined by the K/Ca method and stratigraphic depositional ages, ^{40}K-^{40}Ca systematics in K-rich authigenic minerals may provide valuable information about post-depositional histories of sediments.

Authigenic sanidine can demonstrate significant enrichment in ^{40}Ca. Reported ε_{Ca} values are between 50 and 200 in Ordovician sandstone (Marshall et al. 1986) and 4–5 in Cretaceous/Tertiary boundary sediments (DePaolo et al. 1983). Calculated K–Ca ages of the Ordovician sanidine are younger than depositional ages (20–50 Ma) and, thus, not representative for stratigraphic ages.

Similar complications are also apparent in Ordovician and Cambrian glauconite (Gopalan 2008; Cecil and Ducea 2011). Gopalan (2008) studied K–Ca and Rb–Sr systematics in glauconite from Ordovician sediments. In this case, both isotope systems have identical ages and are about 30–40 Ma younger than the depositional ages. For Cambrian glauconites Cecil and Ducea (2011) found highly variable apparent K–Ca ages, all younger than the stratigraphic age. Although K–Ca ages are not stratigraphic ages, they provide information about the thermal and hydrologic histories of sediment basins. Cecil and Ducea (2011), determined K–Ca closing temperatures in glauconite of about 60–90 °C for slow cooling rates of 1 °C/My. Considering that the mobility of Ca in glauconite is a function of temperature, the authors suggested that the K–Ca system may record thermochronological information on the post-depositional history of sediments such as diagenetic heating events and uplift related cooling.

3.2.3 Evaporites

Significant radiogenic enrichment in ^{40}Ca is also present in K-rich minerals from evaporites, such as sylvine, carnallite and langbeinite (Inghram et al. 1950; Polevaya et al. 1958; Wilhelm and Ackermann 1972; Heumann et al. 1979; Baardsgaard 1987). For example, Baardsgaard (1987) reported ε_{Ca} between 3 and 700 in evaporites. Calculated K–Ca ages based on evaporitic minerals are variable and in most cases significantly younger than the depositional ages. This is caused by salt metamorphism and recrystallization events. In such cases, the measured ε_{Ca} is significantly smaller than expected from the mineral K/Ca ratio and stratigraphic age.

References

Albarède F (2009) Volatile accretion history of the terrestrial planets and dynamic implications. Nature 461:1227–1233

Amini M, Eisenhauer A, Böhm F et al (2008) Calcium isotope ($\delta^{44/40}$Ca) fractionation along hydrothermal pathways, Logatchev field (Mid-Atlantic Ridge, 14°45′N). Geochim Cosmochim Acta 72:4107–4122

Amini M, Eisenhauer A, Boehm F et al (2009) Calcium isotopes (delta Ca-44/40) in MPI-DING reference glasses, USGS rock powders and various rocks: evidence for ca isotope fractionation in terrestrial silicates. Geostand Geoanalytical Res 33:231–247

Baadsgaard H (1987) Rb-Sr and K-Ca isotope systematics in minerals from potassium horizons in the prairie evaporite formation, Saskatchewan, Canada. Chem Geol 66:1–15

Bourg I, Richter F, Christensen J et al (2010) Isotopic mass dependence of metal cation diffusion coefficients in liquid water. Geochim Cosmochim Acta 74:2249–2256

Caro G, Papanastassiou DA, Wasserburg GJ (2010) ^{40}K–^{40}Ca isotopic constraints on the oceanic calcium cycle. Earth Planet Sci Lett 296:124–132

Cecil MR, Ducea MN (2011) K-Ca ages of authigenic sediments: examples from Paleozoic glauconite and applications to low-temperature thermochronometry. Int J Earth Sci 100:1783–1790

Chen H-W, Lee T, Lee D-C et al (2011) ^{48}Ca heterogeneity in differentiated meteorites. Astrophys J Lett 743:23

Chen H, Lee T, Lee D-C, Chen J-C (2015). Correlation of ^{48}Ca, ^{50}Ti, and ^{138}La heterogeneity in the allende refractory inclusions. Astrophysical J Lett 806:L21

Choi B-G, Wasserburg GJ, Huss GR (1999) Circumstellar hibonite and corundum and nucleosynthesis in asymptotic giant branch stars. Astrophys J 522:133–136

Chopra R, Richter F, Watson BE et al (2012) Magnesium isotope fractionation by chemical diffusion in natural settings and in laboratory analogues. Geochim Cosmochim Acta 88:1–18

Clayton R, Hinton RW, Davis A (1988) Isotopic variations in the rock-forming elements in meteorites. Philos Transac Royal Soc Lond Ser A-Math Phys Eng Sci 325:483–501

Coleman ML (1971) Potassium-Calcium Dates from Pegmatitic Micas. Earth Planet Sci Lett 12:399–405

Dauphas N, Chen JH, Zhang J et al (2014) Calcium-48 isotopic anomalies in bulk chondrites and achondrites: Evidence for a uniform isotopic reservoir in the inner protoplanetary disk. Earth Planet Sci Lett 407:96–108

Davis AM, Richter FM (2014) Condensation and evaporation of solar system materials. In: Holland HD, Turekian KK (eds) Treatise on geochemistry (2nd edn), pp 335–360

DePaolo DJ (2004) Calcium isotopic variations produced by biological, kinetic, radiogenic and nucleosynthetic processes. In: Johnson CM, Beard BL, Albarède F (eds) Geochemistry of non-traditional stable isotopes. Reviews in mineralogy and geochemistry. Mineralogical Society of America, Washington, pp 255–288

DePaolo DJ, Kyte FT, Marshall BD et al (1983) Rb-Sr, Sm-Nd, K-Ca, O, and H isotopic study of cretaceous-tertiary boundary sediments, Caravaca, Spain: evidence for an oceanic impact site. Earth Planet Sci Lett 64:356–373

Fahey AJ, Goswami JN, McKeegan KD et al (1987) ^{16}O excesses in Murchison and Murray hibonites: a case against a late supernova injection origin of isotopic anomalies in O, Mg, Ca, and Ti. Astrophys J 323:91–95

Farkaš J, Déjeant A, Novák M et al (2011) Calcium isotope constraints on the uptake and sources of Ca^{2+} in a base-poor forest: a new concept of combining stable ($\delta^{44/42}$Ca) and radiogenic (ε_{Ca}) signals. Geochim Cosmochim Acta 75:7031–7046

Feng C, Qin T, Huang S et al (2014) First-principles investigations of equilibrium calcium isotope fractionation between clinopyroxene and Ca-doped orthopyroxene. Geochim Cosmochim Acta 143:132–142

Fletcher IR, McNaughton NJ, Pidgeon RT et al (1997) Sequential closure of K-Ca and Rb-Sr isotopic systems in Archean micas. Chem Geol 138:289–301

Gopalan K (2008) Conjunctive K-Ca and Rb–Sr dating of glauconies. Chem Geol 247:119–123

Gopalan K, Kumar A (2008) Phlogopite K-Ca dating of Narayanpet kimberlites, south India: implications to the discordance between their Rb–Sr and Ar/Ar ages. Precambr Res 167:377–382

Grossman L (1972) Condensation in the primitive solar nebula. Geochim Cosmochim Acta 36:597–619

Grossman L, Ebel DS, Simon SB et al (2000) Major element chemical and isotopic compositions of refractory inclusions in C3 chondrites: the seperate roles of condensation and evaporation. Geochim Cosmochim Acta 64:2879–2894

Harrison TM, Heizler MT, McKeegan KD et al (2010) In situ ^{40}K–^{40}Ca "double-plus" SIMS dating resolves Klokken feldspar ^{40}K–^{40}Ar paradox. Earth Planet Sci Lett 299:426–433

Hartmann D, Woosley SE, El Eid MF (1985) Nucleosynthesis in neutron-rich supernova ejecta. Astrophys J 297:837–845

Heumann KH, Kubassek E, Schwabenbauer W et al (1979) Analytisches Verfahren zur K/Ca-Altersbestimmung geologischer Proben. Fresen Z Anal Chem 297:35–43

Hindshaw RS, Rickli J, Leuthold J et al (2014) Identifying weathering sources and processes in an outlet glacier of the Greenland Ice Sheet using Ca and Sr isotope ratios. Geochim Cosmochim Acta 145:50–71

Hinton RW, Davis AM, Scatena-Wachel DE et al (1988) A chemical and isotopic study of hibonite-rich refractory inclusions in primitive meteorites. Geochim Cosmochim Acta 52:2573–2598

Holst JC, Olsen MB, Paton C et al (2013) ^{182}Hf–^{182}W age dating of a ^{26}Al-poor inclusion and implications for the origin of short-lived radioisotopes in the early Solar System. Proc Natl Acad Sci 110:8819–8823

Huang F, Lundstrom CC, Glessner J et al (2009) Chemical and isotopic fractionation of wet andesite in a temperature gradient: experiments and models suggesting a new mechanism of magma differentiation. Geochim Cosmochim Acta 73:729–749

Huang S, Farkas J, Jacobsen SB (2010a) Calcium isotopic fractionation between clinopyroxene and orthopyroxene from mantle peridotites. Earth Planet Sci Lett 292:337–344

Huang F, Chakraborty P, Lundstrom CC et al (2010b) Isotope fractionation in silicate melts by thermal diffusion. Nature 464:396–401

Huang S, Farkaš J, Jacobsen SB (2011a) Stable calcium isotopic compositions of Hawaiian shield lavas: evidence for recycling of ancient marine carbonates into the mantle. Geochim Cosmochim Acta 75:4987–4997

Huang F, Chakraborty P, Lundstrom CC et al (2011b) Isotope fractionation in silicate melts by thermal diffusion reply. Nature 472. http://dx.doi.org/10.1038/nature09955

Huang S, Farkaš J, Yu G et al (2012) Calcium isotopic ratios and rare earth element abundances in refractory inclusions from the Allende CV3 chondrite. Geochim Cosmochim Acta 77:252–265

Inghram MG, Brown H, Patterson C et al (1950) The branching ratio of K 40 radioactive decay. Phys Rev 80:916

Ireland TR (1990) Presolar isotopic and chemical signatures in hibonite-bearing refractory inclusions from the Murchison carbonaceous chondrite. Geochim Cosmochim Acta 54:3219–3237

Ireland TR, Zinner EK, Fahey AJ et al (1992) Evidence for distillation in the formation of HAL and related hibonite inclusions. Geochim Cosmochim Acta 56:2503–2520

Jacobson AD, Andrews MG, Lehn GO et al (2015) Silicate versus carbonate weathering in Iceland: New insights from Ca isotopes, Earth Planet Sci Lett 416:132–142

Jadhav M, Pignatari M, Herwig F et al (2013) Relics of ancient post-AGB Stars in a primitive meteorite. Astrophys J Lett 777:27

Javoy M, Kaminski E, Guyot F et al (2010) The chemical composition of the earth: enstatite chondrite models. Earth Planet Sci Lett 293:259–268

John T, Gussone N, Podladchikov YY et al (2012) Volcanic arcs fed by rapid pulsed fluid flow through subducting slabs. Nat Geosci 5:489–492

Jungck MHA, Shimamura T, Lugmair G (1984) Ca isotope variations in Allende. Geochim Cosmochim Acta 48:2651–2658

Kennicutt RC, Evans NJ (2012) Star formation in the milky way and nearby galaxies. Ann Rev Astron Astrophys 50:531–608

Kreissig K, Elliott T (2005) Ca isotope fingerprints of early crust-mantle evolution. Geochim Cosmochim Acta 69:165–176

Lacks DJ, Van Orman JA, Lesher CE (2012) Isotope fractionation in silicate melts. Nature 482. http://dx.doi.org/10.1038/nature10764

Larsen KK et al (2011) Evidence for magnesium isotope heterogeneity in the solar protoplanetary disk. Astrophys J Lett 735:37

Lee T, Papanastassiou DA, Wasserburg GJ (1978) Calcium isotopic anomalies in the Allende meteorite. Astrophys J 220:21–25

Lee T, Russell W, Wasserburg G (1979) Calcium isotopic anomalies and the lack of Al-26 in an unusual Allende inclusion. Astrophys J 228:93–98

Liu M-C, Chaussidon M, Srinivasan G et al (2012) A lower initial abundance of short-lived ^{41}Ca in the early solar system and its implications for solar system formation. Astrophys J 761:137

Loss RD, Lugmair GW, Davis AM et al (1994) Isotopically distinct reservoirs in the solar nebula: Isotope anomalies in Vigarano meteorite inclusions. Astrophys J 436:193–196

Magna T, Gussone N, Mezger K (2015) The calcium isotope systematics of Mars. Earth Planetary Sci Lett 430:86–94

Marshall BD, DePaolo DJ (1982) Precise age determinations and petrogenetic studies using the K-Ca method. Geochim Cosmochim Acta 46:2537–2545

Marshall BD, Woodard HH, Krueger HW et al (1986) K-Ca-Ar systematics of authigenic sanidine from Wakau, Wisconsin, and the diffusivity of argon. Geology 14:936–938

Marshall BD, DePaolo DJ (1989) Calcium isotopes in igneous rocks and the origin of granite. Geochim Cosmochim Acta 53:917–922

McKeegan KD, Aléon J, Bradley J et al (2006) Isotopic compositions of cometary matter returned by Stardust. Science 314:1724–1728

Meyer BS, Krishnan TD, Clayton DD (1996) ^{48}Ca production in matter expanding from high temperature and density. Astrophys J 462:825

Moore J, Jacobson AD, Holmden C et al (2013) Tracking the relationship between mountain uplift, silicate weathering, and long-term CO_2 consumption with Ca isotopes: Southern Alps, New Zealand. Chem Geol 341:110–127

Moynier F, Simon JI, Podosek FA et al (2010) Ca isotope effects in Orgueil leachates and the Implications for the carrier phases of ^{54}Cr anomalies. Astrophys J Lett 718:7–13

Nägler TF, Villa IM (2000) In pursuit of the ^{40}K branching ratios: K-Ca and ^{39}Ar–^{40}Ar dating of gem silicates. Chem Geol 169:5–16

Nelson DR, McCulloch MT (1989) Petrogenetic applications of the ^{40}K-^{40}Ca radiogenic decay scheme—a reconnaissance study. Chem Geol 79:275–293

Niederer FR, Papanastassiou D (1984) Ca isotopes in refractory inclusions. Geochim Cosmochim Acta 48:1279–1293

Nittler LR, Alexander CMO'D, Gallino R et al (2008) Aluminum-, calcium- and titanium-rich oxide stardust in ordinary chondrite meteorites. Astrophys J 682:1450–1478

Papanastassiou DA (1986) Chromium isotopic anomalies in the Allende meteorite. Astrophys J 308:27

Papanastassiou DA, Brigham CA (1989) The identification of meteorite inclusions with isotope anomalies. Astrophys J 338:37–40

Polevaya NI, Titov NE, Elyaev VS et al (1958) Application of the Ca-method in the absolute age determination of sylvites, Geochem 8:897ff

Richter F, Davis AM, DePaolo DJ et al (2003) Isotope fractionation by chemical diffusion between molten basalt and rhyolite. Geochim Cosmochim Acta 67:3905–3923

Richter FM, Watson EB, Mendybaev RA et al (2008) Magnesium isotope fractionation in silicate melts by chemical and thermal diffusion. Geochim Cosmochim Acta 72:206–220

Richter FM, Watson EB, Mendybaev RA et al (2009) Isotopic fractionation of the major elements of molten basalt by chemical and thermal diffusion. Geochim Cosmochim Acta 73:4250–4263

Rudnick RL, Fountain DM (1995) Nature and composition of the continental crust: a lower crustal perspective. Rev Geophys 33:267–309

Russel WA, Papanastassiou DA, Tombrello T et al (1977) Ca isotope fractionation on the Moon. In: Proceedings of the 8th lunar science conference, 3791–3805

Ryu J-S, Jacobson AD, Holmden C et al (2011) The major ion, $\delta^{44/40}$Ca, $\delta^{44/42}$Ca, and $\delta^{26/24}$Mg geochemistry of granite weathering at pH = 1 and T = 25 °C: power-law processes and the relative reactivity of minerals. Geochim Cosmochim Acta 75:6004–6026

Schiller M, Paton C, Bizzaro M (2015) Evidence for nucleosynthetic enrichment of the protosolar molecular cloud core by multiple supernova events. Geochim Cosmochim Acta 149:88 102

Schuessler JA, Schoenberg R, Sigmarsson O (2009) Iron and lithium isotope systematics of the Hekla volcano, Iceland—evidence for Fe isotope fractionation during magma differentiation. Chem Geol 258:78–91

Shih C-Y, Nyquist LE, Wiesmann H (1993) K-Ca chronology of lunar granites. Geochim Cosmochim Acta 57:4827–4841

Shih C-Y, Nyquist LE, Bogard DD et al (1994) K-Ca and Rb-Sr dating of two lunar granites: relative chronometer resetting. Geochim Cosmochim Acta 58:3101–3116

Simon JI, DePaolo DJ (2010) Stable calcium isotopic composition of meteorites and rocky planets. Earth Planet Sci Lett 289:457–466

Simon JI, Depaolo DJ, Moynier F (2009) Calcium isotope composition of meteorites, Earth, and Mars. Astrophys J 702:707–715

Srinivasan G, Ulyanov A, Goswami J (1994) ^{41}Ca in the early solar system. Astrophys J 431:67–70

Teng FZ, Dauphas N, Helz RT et al (2011) Diffusion-driven magnesium and iron isotope fractionation in Hawaiian olivine. Earth Planet Sci Lett 308:317–324

Tipper ET, Galy A, Bickle MJ (2006) Riverine evidence for a fractionated reservoir of Ca and Mg on the continents: implications for the oceanic Ca cycle. Earth Planet Sci Lett 247:267–279

Travaglio C, Gallino R, Amari S et al (1999) Low-density graphite grains and mixing in type II supernovae. Astrophys J 510:325

Trinquier A, Elliott T, Ulfbeck D et al (2009) Origin of nucleosynthetic isotope heterogeneity in the solar protoplanetary disk. Science 324:374–376

Trinquier A, Birck J-L, Allegre C (2007) Widespread Cr-54 heterogeneity in the inner solar system. Astrophys J 655:1179–1185

Valdes MC, Moreira M, Foriel J et al (2014) The nature of Earth's building blocks as revealed by calcium isotopes. Earth Planet Sci Lett 394:135–145

Wanajo S, Janka H-T, Müller B (2013) Electron-capture supernovae as origin of ^{48}Ca. Astrophys J 767:26

Wasserburg G, Lee T, Papanastassiou D (1977) Correlated O and Mg isotopic anomalies in Allende inclusions II-magnesium. Geophysical Research Letters 4:299–302

Watkins J, DePaolo D, Huber C et al (2009) Liquid composition-dependence of calcium isotope fractionation during diffusion in molten silicates. Geochim Cosmochim Acta 73:7341–7359

Watkins J, DePaolo D, Ryerson F et al (2011) Influence of liquid structure on diffusive isotope separation in molten silicates and aqueous solutions. Geochim Cosmochim Acta 75:3103–3118

Watkins JM, Liang Y, Richter F et al (2014) Diffusion of multi-isotopic chemical species in molten silicates. Geochim Cosmochim Acta 139:313–326

Wilhelm HG, Ackermann W (1972) Altersbestimmung nach der K-Ca-Methode an Sylvin des Oberen Zechsteines des Werragebietes. Z Naturforsch 27:1256–1259

Woosley S (1997) Neutron-rich nucleosynthesis in carbon deflagration supernovae. Astrophys J 476:801–810

Wombacher F, Eisenhauer A, Heuser A et al (2009) Separation of Mg, Ca and Fe from geological reference materials for stable isotope ratio analyses by MC-ICP-MS and double-spike TIMS. J Anal Spectrom 24:627–636

Zhang J, Huang S, Davis AM et al (2014) Calcium and titanium isotopic fractionations during evaporation. Geochim Cosmochim Acta 140:365–380

Zinner E (2007) Presolar grains. In Holland HD, Turekian KK (eds) Treatise on geochemistry. Elsevier, USA, pp 1–33

Zinner E, Jadhav M (2013) Isochrons in presolar graphite grains from Orgueil. Astrophys J 768:100

Zinner E, Fahey A, McKeegan K (1986) Large Ca-48 anomalies are associated with Ti-50 anomalies in Murchison and Murray hibonites. Astrophys J 311:103–107

Zhu P, Macdougall D (1998) Calcium isotopes in the marine environment and the oceanic calcium cycle. Geochim Cosmochim Acta 62:1691–1698

Biomedical Application of Ca Stable Isotopes

Alexander Heuser

Abstract

This chapter summarizes the current knowledge on the Ca isotope composition of different tissues and Ca isotope cycling in the human and vertebrate body. The known sites of Ca isotope fractionation and the calcium isotopic composition of different compartments and body fluids are discussed. The first part of this chapter follows the journey of Ca in the body from the input (diet) to the output (urine, feces). Finally, current biomedical applications of Ca isotopes for the detection of bone loss during bed rest studies and during bone cancer are presented.

Keywords

Ca metabolism · Diet · Urine · Blood · Bed rest studies · Bone loss · Ca isotope transport model

1 Introduction

Initially, research on non-traditional stable isotopes focused on geosciences, but became increasingly important for life-sciences and biomedical applications during the past years. It was found that biological fractionation may serve as a potential diagnostic tools in biomedicine (e.g. Bullen and Walczyk 2009; Albarede 2015). Several studies investigated element metabolism from the isotope view and showed that Cu, Zn, Fe and Ca isotope systematics in humans can be used to detect diseases (e.g. Walczyk and von Blankenburg 2002; Walczyk and von Blankenburg 2005; Hotz et al. 2011; Aramendíaet al. 2013; Balter et al. 2013; Larner et al. 2015). Although much progress has been achieved, the potential use of biological fractionation of stable isotopes is not widely recognized in the medical community yet.

Calcium is one of the most important elements in human physiology. It is the fifth most abundant element in the human body. The Ca content of adults is about 1000–1200 g with about 99 % stored in the skeleton. Calcium plays an important role in the biology of most organisms. In physiology, Ca is used in biominerals for the formation and maintenance of bones and teeth (e.g. Pasteris et al. 2008). Furthermore, Ca is a

A. Heuser (✉)
Steinmann-Institut für Geologie, Mineralogie und Paläontologie, Universität Bonn, Bonn, Germany
e-mail: aheuser@uni-bonn.de

© Springer-Verlag Berlin Heidelberg 2016
N. Gussone et al., *Calcium Stable Isotope Geochemistry*,
Advances in Isotope Geochemistry, DOI 10.1007/978-3-540-68953-9_8

major participant in processes like muscle contraction, signal transmission, cell apoptosis, cell reproduction, and coagulation of blood (e.g. Bootman et al. 2001; Campbell 1990; Peacock 2010). In larger concentrations Ca is known to be cell poisoning and, hence, its concentration has to be held constant in very close limits (Ca homeostasis). Mainly, three organs control Ca homeostasis in vertebrates: the gastrointestinal tract where Ca is absorbed from the diet, the skeleton which is the main reversible Ca reservoir and the kidneys controlling the excretion and recycling of calcium from the blood. Disease related organ malfunction, physiological ageing processes and other diseases are known to be associated with disturbances of the calcium homeostasis. Studies on the Ca isotopic composition of mineralized tissues (bones, teeth), soft tissues (liver, kidney, muscles), body fluids (urine, blood) and feces have been carried out during the past years in order to determine isotope signatures for a healthy Ca metabolism and to detect Ca metabolism related diseases.

2 Ca Isotope Transport Model

Skulan and DePaolo (1999) reported the first systematic study on the fractionation of Ca isotopes during Ca transport in vertebrates and developed a simple transport model for Ca isotopes (Fig. 1a) which was later extended by Heuser and Eisenhauer (2010) and Heuser et al. (2016) (Fig. 1b).

In the following sections the transport and transport related fractionation of Ca in vertebrates will be discussed.

2.1 Ca Isotopic Composition of the Diet

Calcium from the diet is the only Ca source for the body. Thus the Ca isotopic composition of food defines the basis for the Ca isotopic composition of mineralized and soft tissues as well as body fluids. Chu et al. (2006) investigated human diet and found that dairy products are

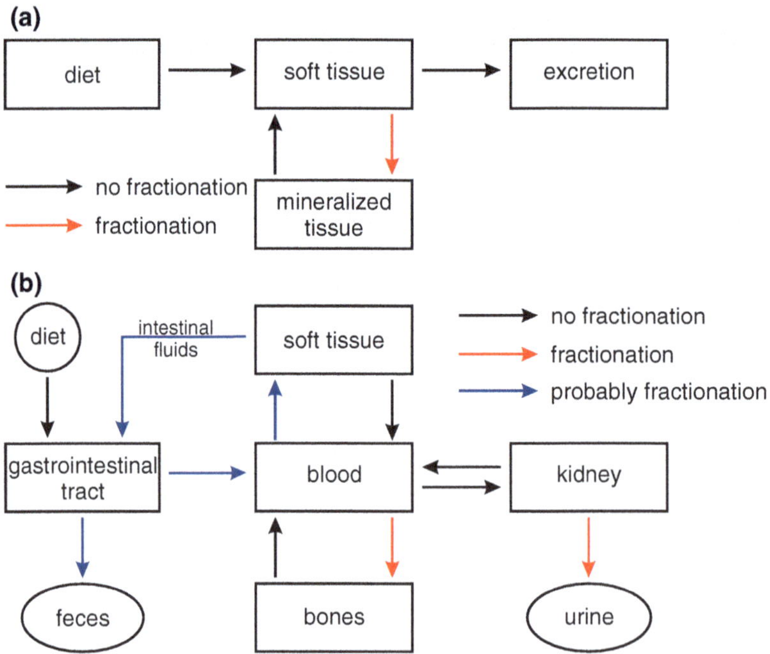

Fig. 1 **a** Simple Ca isotope transport model (after Skulan and DePaolo 1999). **b** Enhanced Ca isotope transport model (Heuser et al. 2016). *Red arrows* show fractionation of Ca isotopes between compartments, while *black arrows* represent Ca isotope transport without larger extent of fractionation. *Blue arrows* denote the fractionation which has not yet been studied but can be predicted based on pattern of already detected fractionations in the body

Table 1 Average Ca concentration, relative contribution of the type of meal to a typical Western diet, and the average assumed Ca isotopic composition of each group (Chu et al. 2006; Heuser and Eisenhauer 2010)

Group	Average Ca concentration (mg/100 g)	Fraction (%) of a Western diet	$\delta^{44/40}$Ca (‰ SRM 915a)
Dairy products	500	14	−1.20
Vegetables	100	21	−0.68
Fruit	20	14	−0.68
Cereal products	35	24	−0.56
Meat	10	6	−0.08
Fats	7	2	−1.00
Water	5	20	0.68

outstanding with regards to their importance for a normal western diet and their Ca isotopic composition (cf. Chapter "Biominerals and Biomaterial", Sect. 4.2).

From Table 1 it can be seen that the dietary Ca isotopic composition is dominated by dairy products due to their high Ca concentrations. Water and meat although having the heaviest Ca isotopic compositions do not have much influence on the dietary Ca due to their low Ca content. The $\delta^{44/40}$Ca$_{\text{diet}}$ for an average Western diet containing about 14 % of dairy products is about −1 ‰ (Fig. 2, red square). A simple modeling (Fig. 2) shows that a vegetarian diet and a 'normal' diet have nearly the same $\delta^{44/40}$Ca as the Ca concentration of meat is too low to cause significant changes in $\delta^{44/40}$Ca$_{\text{diet}}$. A vegan diet (Fig. 2, green triangle) which contains no milk and no meat is isotopically heavier by about 0.32 ‰ than a typical western diet. Currently, no systematic data on the Ca isotopic composition of different types of food exists and nothing is known about geographical variation of $\delta^{44/40}$Ca of human food. In order to properly interpret Ca isotopic variations in soft tissues, mineralized tissues, body fluids and feces the $\delta^{44/40}$Ca$_{\text{diet}}$ needs at least to be roughly estimated. During the reported bed rest studies all subjects were served the same diet to exclude that variations in $\delta^{44/40}$Ca$_{\text{urine}}$ or $\delta^{44/40}$Ca$_{\text{blood}}$ between subjects were caused by the diet.

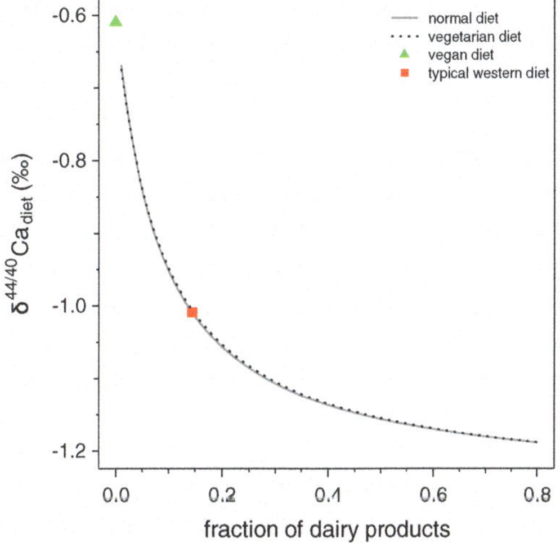

Fig. 2 Estimation of $\delta^{44/40}$Ca$_{\text{diet}}$ for a normal western diet (*grey line*), a vegetarian diet (*dot line*) and a vegan diet (*green triangle*). A typical western diet (14 % dairy products) results in a $\delta^{44/40}$Ca$_{\text{diet}}$ of about −1 ‰. Estimates were calculated using the data of Table 1 (see also Chapter "Biominerals and Biomaterial", Sect. 4.2)

2.2 From Food to Blood

Calcium from the diet enters the body by intestinal absorption into blood. During this transport of Ca the simple transport model (Fig. 1a) assumes no Ca isotope fractionation as $\delta^{44/40}$Ca$_{\text{soft tissue}}$ is in the same order of magnitude compared to $\delta^{44/40}$Ca$_{\text{diet}}$ (Skulan and DePaolo 1999). In contrast, the enhanced Ca isotope transport model (Fig. 1b) predicts that Ca isotopes are probably fractionated during intestinal absorption. Currently there is no direct experimental evidence that Ca fractionation is occurring during transcellular transport and there is no data proving directly the size or existence of a Ca isotope fractionation during intestinal absorption. Nevertheless, several observations suggest that isotope fractionation during Ca uptake is at least likely.

Tacail et al. (2014) investigated the Ca isotopic composition of blood (serum and red blood cells) and diet of sheep (n = 4). Their data show an offset between $\delta^{44/40}Ca_{diet}$ of 0.52 ± 0.15 ‰, ($\pm 1SD$) and $\delta^{44/40}Ca_{serum}$ of -0.18 ‰ ± 0.06 ($\pm 1SD$). This offset is likely caused by fractionation during intestinal Ca absorption, although other explanations cannot be completely ruled out.

An indirect evidence for a fractionation during intestinal Ca absorption is based on physiological considerations. On a cellular level intestinal Ca absorption and renal Ca reabsorption are similar. Renal Ca absorption is a process where Ca isotope fractionation is evident (cf. 2.4). Thus, a lack of fractionation during intestinal Ca absorption seems unlikely. It appears to be unlikely that the same types of Ca channels involved in intestinal Ca absorption show no preferential transport of the light Ca isotopes. In addition, transmembrane Ca transport was also suggested to be associated with significant isotope fractionation in protista (cf. Gussone et al. 2006, Chap. 4), metazoa (cf. Böhm et al. 2006, Chap. 4) and higher plants (Wiegand et al. 2005; Schmitt et al. 2013, Chap. 5).

An apparently convenient way to detect fractionation during intestinal Ca absorption is the analyses of the Ca isotopic composition of feces ($\delta^{44/40}Ca_{feces}$). If light Ca isotopes are preferentially absorbed in the intestine feces should display a depletion in light Ca isotopes compared to Ca nutritive source, i.e. showing a relative enrichment of heavy Ca isotopes. Surprisingly, existing data on $\delta^{44/40}Ca_{feces}$ (Tacail et al. 2014; Heuser et al. 2016) are not significantly higher than $\delta^{44/40}Ca_{diet}$. Rather $\delta^{44/40}Ca_{feces}$ show an enrichment of light Ca isotopes resulting in $\delta^{44/40}Ca_{feces}$ values being lower than

corresponding $\delta^{44/40}Ca_{diet}$. To solve this discrepancy Heuser et al. (2016) suggested that intestinal fluids have to be taken into account. Compartment models of the human body (e.g. Freeman et al. 1997; Mundy and Guise 1999) show that there is also a Ca transport into the intestine by intestinal fluids. This additional Ca source may change the isotopic composition of total intestinal Ca towards a lighter Ca isotopic composition. From this isotopically lighter diet/intestinal fluids mixture the light Ca isotopes are preferentially absorbed in the intestine. This scenario could then result in $\delta^{44/40}Ca_{feces}$ being close to $\delta^{44/40}Ca_{diet}$ and could be interpreted as an apparent lack of fractionation during intestinal Ca absorption.

The Ca concentration in blood is regulated to be within narrow limits (Ca homeostasis). Calcium uptake through the diet is fundamental for Ca homeostasis as it is the only source of Ca for the body. Figure 3 shows the intestinal part of a numbered compartment model for Ca homeostasis in the human body based on Ca fluxes determined/ discussed by Mundy and Guise (1999). Using this compartment model the expected $\delta^{44/40}Ca_{feces}$ can be calculated depending on the proportion of dietary Ca and Ca from intestinal fluids.

The total Ca input into the gastrointestinal tract is about 1150 mg/d and about 87 % of the total Ca is originating from the diet ($x_{diet} = 0.87$, $x_{if} = 1 - x_{diet} = 0.13$) (Fig. 3). About 74 % of the Ca in the intestine is excreted ($f_{ex} = 0.74$). It is possible to model the Ca isotopic composition of feces using Rayleigh fractionation:

$$R_{feces} = R_{intestine} \times f_{ex}^{(\alpha-1)} \qquad (1)$$

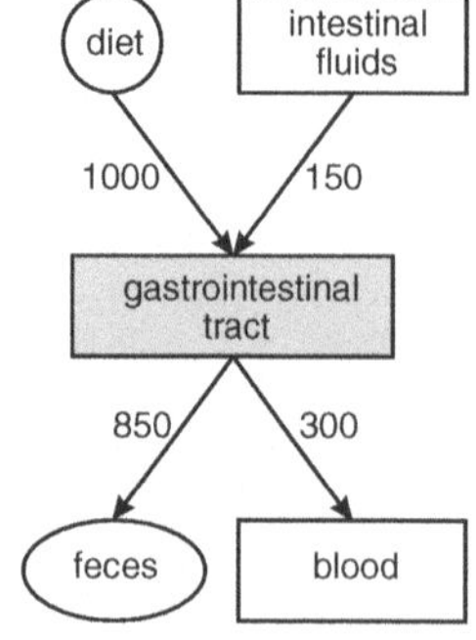

Fig. 3 Intestinal part of the enhanced Ca isotope transport model with Ca fluxes (mg Ca/d). Ca fluxes are taken from Mundy and Guise (1999) for a healthy adult i.e. under Ca homeostatic conditions

where R_{feces} and $R_{intestine}$ are the $^{44}Ca/^{40}Ca$ of feces and total intestinal Ca, respectively, f_{ex} is the fraction of excreted Ca and α is the fractionation factor between diet and feces. $R_{intestine}$ can be calculated by:

$$R_{intestine} = x_{diet} \times R_{diet} + x_{if} \times R_{if} \quad (2)$$

where $R_{intestine}$, R_{diet}, R_{if} are the $^{44}Ca/^{40}Ca$ of total intestinal Ca, Ca diet and Ca intestinal fluids. X_{diet} and x_{if} are the fractions of dietary Ca and Ca from intestinal fluids of the total intestinal Ca, respectively.

Combining Eqs. 1 and 2 leads to:

$$R_{feces} = (x_{diet} \times R_{diet} + x_{if} \times R_{if}) \times f_{ex}^{(\alpha-1)} \quad (3)$$

The fractionation factor α as well as the Ca isotopic composition of the intestinal fluids (R_{if}) are unknown, and thus have to be approximated. As reported by Heuser et al. (2016) it is reasonable to assume that intestinal fluids have a light isotopic composition in the order of 1–2 ‰ lower than the diet (cf. Sect. 2.4). These values are based on the observation that the Ca isotopic composition of milk is light compared to the diet and that both, milk and intestinal fluids, are excreted by glands. The fractionation factor α can be assumed to either represent a fractionation during intestinal absorption (α < 1) or no fractionation during intestinal absorption (α = 1).

Figure 4 shows the results of modeling the $\delta^{44/40}Ca_{feces}$ as a function of the x_{diet}. Two α values have been uses to show the difference between small fractionation (red curves, α = 0.9995) and no fractionation (blue curves, α = 1.0000) and two $\delta^{44/40}Ca_{if}$ values (−1.5 and −0.5 ‰; relative to the diet). Under homeostatic conditions (x_{diet} = 0.87) feces should be isotopically lighter than the diet if no fractionation during intestinal Ca absorption occurs. On the other hand, if Ca isotopes are becoming fractionated during intestinal absorption (light isotopes are preferentially absorbed) $\delta^{44/40}Ca_{feces}$ is close to $\delta^{44/40}Ca_{diet}$, which seemingly indicates an apparent lack of Ca isotope fractionation. If the amount of dietary Ca is lower than 1000 mg (i.e. x_{diet} < 0.87), $\delta^{44/40}Ca_{feces}$ is lower than

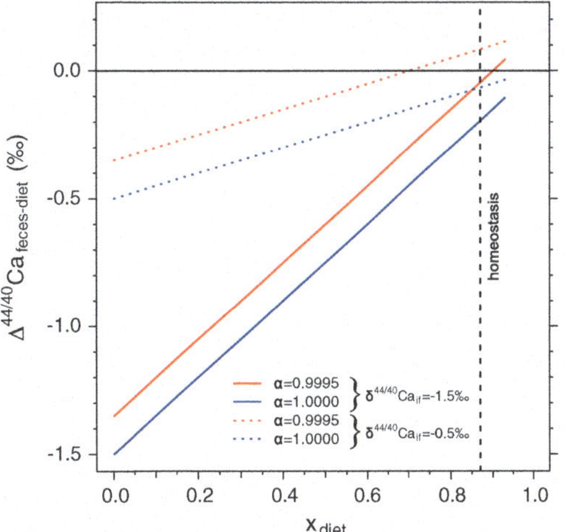

Fig. 4 Modelled difference between $\delta^{44/40}Ca_{feces}$ and $\delta^{44/40}Ca_{diet}$ ($\Delta^{44/40}Ca_{feces\text{-}diet}$) depending on changes of the Ca fraction originating from the diet (x_{diet}) using either no fractionation (*blue curve*, α = 1) or small fractionation (*red curve*, α = 0.9995) with an assumed $\delta^{44/40}Ca_{if}$ of −1.5 ‰ (relative to diet). The *dotted blue* and *light red curves* show the modeled $\delta^{44/40}Ca_{feces}$ assuming $\delta^{44/40}Ca_{if}$ of −0.5 ‰

$\delta^{44/40}Ca_{diet}$ which could be wrongly interpreted as a preferential absorption of the heavy isotopes.

2.3 Fractionation Between Soft Tissue and Mineralized Tissue

During the generation of mineralized tissues like bones and teeth Ca is fractionated probably during the transport to the sites of mineralization. This transport also favours the light Ca isotopes resulting in an enrichment of light isotopes in the mineralized tissue. The magnitude of fractionation between soft tissue and mineralized tissue was reported to be −1.3 ‰ (Skulan and DePaolo 1999), but this value may overestimate the actual fractionation as the fractionation during mineralization should be determined by comparing $\delta^{44/40}Ca_{mineralized\ tissue}$ with the Ca isotopic composition of blood (see Chapter "Biominerals and Biomaterial"). A critical look at the data of the Skulan and DePaolo (1999) shows that the fractionation between blood and bone is on average about −0.68 ± 0.23 ‰ (horse: −0.53 ‰, fur seal: -0.95 ‰, chicken B: -0.56 ‰). The fractionation between soft tissue and mineralized tissue of -1.3 ‰ reported by Skulan and DePaolo was calculated by comparing the Ca isotopic composition of the diet and bones assuming that no Ca fractionation occurs during absorption of dietary Ca (cf. Sect. 2.2). From a physiological point of view blood and soft tissue are different compartments. Blood is the main transporter of Ca to and into the different soft tissues. Currently, only few Ca isotope data of different soft tissues exist in vertebrates. The data of Tacail et al. (2014) give some more insights into the Ca isotopic composition of several soft tissues of sheep (Fig. 5).

Compared to $\delta^{44/40}Ca_{serum}$ all soft tissues (liver, kidney and muscle) are enriched in heavy Ca isotopes, whereas mineralized tissue (bone,

enamel) is isotopically lighter than the corresponding serum (Table 2).

Skulan and DePaolo (1999) state that the Ca isotopic composition of different soft tissues is influenced by the different residence times of Ca in the respective tissue. In blood for example the residence time of Ca is low so that the $\delta^{44/40}Ca_{blood}$ can change within hours and reflects changes from the last hours. On the other hand the residence time of Ca in muscles is higher thus the $\delta^{44/40}Ca_{muscle}$ and reflects changes of the Ca isotope metabolism over several days.

In contrast to bone formation, bone resorption is proposed to be a non-fractionating process (Skulan and DePaolo 1999), i.e. the Ca isotopic composition of bones is transferred one-to-one into blood. As a consequence, the $\delta^{44/40}Ca$ of blood is lowered during bone resorption as isotopically light Ca is released. Resorption of bone and formation of bone (ossification) are simultaneously occurring in the body and are lifelong processes and known as bone remodelling or bone metabolism. As a consequence of bone remodelling the Ca inventory of a bone or, more general, of the skeleton is exchanged during lifetime of the vertebrate organism. The cortical bone turnover is about 2–3 %/year and much higher for the trabecular bones (Clarke 2008). The consequences of bone remodelling in terms of Ca isotope fractionation is so far not well constrained. Skulan and DePaolo (1999) state that bone which is formed during periods of net bone loss should have low $\delta^{44/40}Ca$. This implies that remodelled bone tissue may have different $\delta^{44/40}Ca$ than un-remodelled (primary) bone tissue (Fig. 6) and that with increasing age bones should become isotopically lighter.

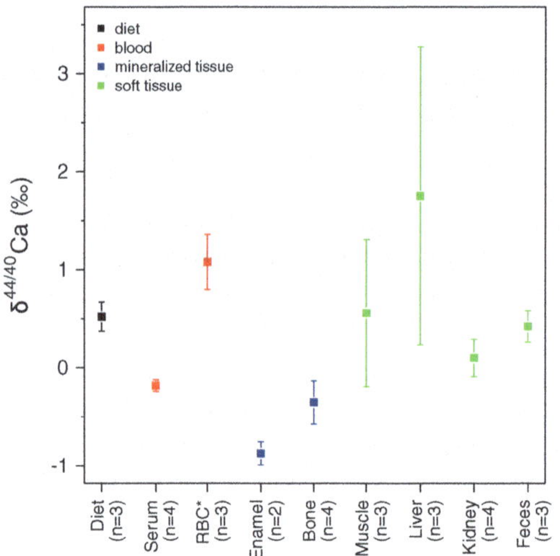

Fig. 5 $\delta^{44/40}Ca$ of soft tissues of sheep (data recalculated from Tacail et al. 2014). *RBC = red blood cells

Table 2 Differences in $\delta^{44/40}Ca$ (‰) of soft and mineralized tissues from corresponding $\delta^{44/40}Ca_{serum}$ presented by Tacail et al. (2014)

Sample	Enamel	Bone	Muscle	Liver	Kidney
A (9125)		-0.36	0.21	0.27	0.09
B (9169)	-0.84	-0.04		2.19	0.48
C (9351)		-0.29	0.48	3.34	0.34
D (9646)	-0.62	-0.01	1.58		0.18

Original data were given as $\delta^{44/42}Ca$ ‰ p. amu and have been converted to $\delta^{44/40}Ca$ by multiplying with 4.1

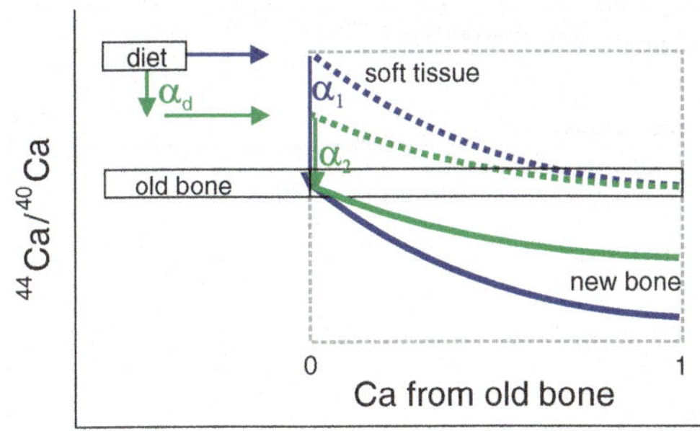

Fig. 6 Calcium isotope fractionation during bone remodelling, effect of net bone loss. *Green* and *blue* colours indicate Ca isotope fractionation and no isotope fractionation during digestion and Ca uptake, respectively

Indeed, Chu et al. (2006) reported small Ca isotopic effects of bone remodelling. They analysed $\delta^{44/40}Ca$ in four subsamples from a single jawbone and found a maximum difference of about 0.26 ‰. In combination with the uncertainty of their measurements (0.1 ‰) they state that intra-bone variations caused by bone remodelling are small and concluded that effects of bone remodelling on the Ca isotope composition are negligible. In a more recent study Heuser et al. (2011a) analysed $\delta^{44/40}Ca$ in three subsamples within the cortex of an *Apatosaurus* long bone and confirmed the general observation of Chu et al. (2006). In a polished section of a drill core across the bone cortex of an *Apatosaurus* long bone the areas of primary and remodelled bone tissues are well-defined by their different structures. This enabled sampling and direct comparison of primary and remodelled areas of the bone. Heuser et al. (2011a) report a difference of about 0.16 ‰ ($\delta^{44/40}Ca$) between primary fibrolamellar bone and remodelled bone. This result clearly shows that bone remodelling may have an effect on the Ca isotopic composition of bone tissue but at the same time these effects are small and may be even negligible compared to other factors influencing the Ca isotopic composition of bones.

2.4 Fractionation in the Kidneys

Several studies analyzed the Ca isotopic composition of urine in the human body (Skulan et al. 2007; Heuser and Eisenhauer 2010; Heuser et al. 2011b; Morgan et al. 2012). During the resorption of bone light Ca is released into blood without further fractionation (Skulan and DePaolo 1999). The lighter Ca isotopic composition of blood should also be transferred to urine. All three studies showed that urine is enriched in heavy Ca isotopes. Heuser and Eisenhauer (2010) suggested that this enrichment of urine in the heavy isotopes is the consequence of the preferential reabsorption of light Ca isotopes during the formation of urine. The formation of urine can be described as a two-step process (Fig. 7). The first step is the generation of primary urine from which in the second step mineral nutrients (e.g. Ca), organic molecules and water are reabsorbed. The residue of this second step is the secondary urine which then is collected in the bladder and finally is excreted. Hoenderop et al. (2005) report that about 98 % of the Ca entering the kidney is reabsorbed and thus only 2 % are excreted.

The reabsorption of Ca is achieved by a transcellular transport (Fig. 8), which can be described as a three step process: influx of Ca into the cytoplasm, transport of Ca through the cell and excretion of Ca out of the cell. The

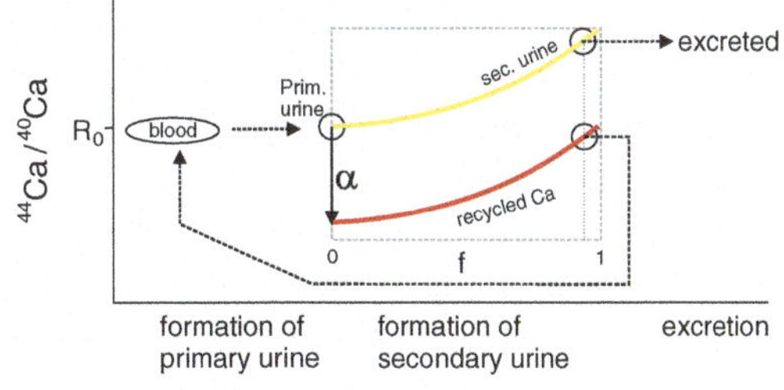

Fig. 7 Conceptual Ca isotope fractionation model during Ca recycling at urine formation

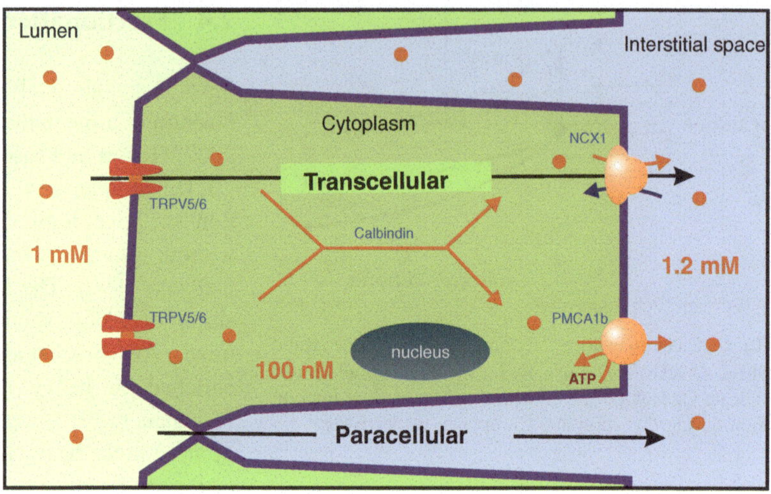

Fig. 8 Sketch of the transcellular and paracellular Ca transport (after Hoenderop et al. 2005). See text for details

influx of Ca into the cytoplasm is regulated/operated by so called Ca channels (Fig. 8, red structures). Calcium channels, especially from the Transient Receptor Potential Vanilloid (TRPV) family, are highly selective molecular structures which allow only ions of distinct charge density (ratio of charge to ionic radius) to pass. The transport through these channels is driven by the concentration gradient between the fluid adjacent to the cell (ca. 1 mM free Ca^{2+}) and the cytoplasm within the cell (ca. 100 nM free Ca^{2+}). Specialized proteins (e.g. Calbindin) within the cell transport Ca to the interstitial space. The excretion of Ca on the apical side of the cell is carried out by pumps against the concentration gradient and therefore requires energy. Prominent Ca pumps are sodium-calcium exchanger (NCX1) and plasma membrane Ca^{2+} ATPase (PMCA1b). The most prominent step where Ca isotopes may be fractionated is during the transport of Ca from the lumen through the Ca channels into the cytoplasma. Besides the transcellular pathway there is also a paracellular pathway. Both transport processes need to be subject of further investigation in order to identify their effectiveness of Ca isotope fractionation.

By the preferential reabsorption of light Ca from the primary urine the remaining secondary urine will become enriched in the heavy Ca isotopes. The magnitude of Ca isotope fractionation during the generation of urine depends on several variables and can be described by Rayleigh fractionation (Eq. 4):

$$R_{excreted} = R_{primary} \times f^{(\alpha-1)} \qquad (4)$$

with $R_{excreted}$ being the Ca isotopic composition of the secondary urine, $R_{primary}$ the Ca isotopic composition of primary urine, f the fraction of unreabsorbed Ca remaining in the secondary urine and α the fractionation between primary urine and blood. The Ca isotopic composition of excreted urine is depending on the Ca isotopic composition of the primary urine ($R_{primary}$), the fraction of Ca which is not reabsorbed from the primary urine (f) and the fractionation factor between primary urine and blood during reabsorption (α). Currently, there are no data on the fractionation factor α available and needs to be assumed. The Ca isotopic composition of primary urine may be assumed to be close to the Ca isotopic composition of blood. Hoenderop et al. (2005) report the fraction of Ca being not reabsorbed in the kidney to be in the order of 2 % (i.e. $f = 0.02$). Therefore, a change in the urinary Ca isotopic composition can be achieved by a change in the fraction of Ca (f) being reabsorbed, a change in the Ca isotopic composition of blood ($R_{primary}$) or a changing fractionation factor between primary urine and blood during reabsorption (α).

Intuitively, a change of the Ca isotopic composition of blood caused by net bone resorption (isotopically light Ca is released) or by a net bone growth (isotopically light Ca is utilized) will change the Ca isotopic composition of urine. Thus, the use of the urinary Ca isotopic composition as a diagnostic tool for diseases linked to disturbances of the Ca metabolism is currently under investigation. On the other hand, $\delta^{44/40}Ca_{urine}$ will also change if the fraction of Ca being reabsorbed is changing. This hinders the straightforward use of $\delta^{44/40}Ca_{urine}$ to determine body's net bone balance.

2.5 Calcium Isotope Fractionation During Milk Lactation

Milk is of special interest when investigating Ca isotope fractionation in mammals and especially in humans. Milk and dairy products belong to the most important Ca sources for humans. Chu et al. (2006) pointed out that the Ca isotopic composition of the dietary input to the body is mainly determined by the contribution of milk and dairy products to the consumed food. This is because of the high Ca content of milk and dairy products and the light Ca isotopic composition of milk. Chu et al. (2006) reported a difference of about -1 to -2 ‰ between diet of an animal and its milk.

Milk lactation is a very complex process and the Ca transport pathway inside the mammary gland into the milk is still under debate (e.g. VanHouten and Wysolmerski 2007; Hoenderop et al. 2005). As the calcium isotopic composition in milk is lighter than the diet, it becomes obvious that during Ca transport into milk a preferential transport of light Ca takes place. Chu et al. (2006) speculate that the preferential transport of light Ca isotopes is related to specialized Ca-binding proteins within the mammary gland. Currently, it is thought that most of the Ca in the mammary gland makes use of a transcellular pathway which includes similar Ca channels as have been described in the kidney and the intestine. These Ca channels may discriminate light and heavy Ca isotopes (cf. 2.4). This discrimination between light and heavy Ca isotopes taking place during transcellular Ca transport could easily explain the enrichment of light Ca isotopes in milk. Further detailed investigations on that topic are needed.

2.6 Calcium Use Index (CUI)

Skulan and DePaolo (1999) introduced the Calcium Use Index (CUI) which is defined as the isotopic difference between soft tissue (δ_s) and the diet (δ_d) divided by the difference between diet (δ_d) and mineralized tissue (δ_b). Based on their model (Fig. 1a) the CUI can also be expressed in terms of Ca fluxes (V_l: mineral loss, V_b: bone mineral gain, V_d: dietary input):

$$CUI = \frac{\delta_s - \delta_d}{\delta_d - \delta_b} \approx \frac{V_b - V_l}{V_d + V_l}$$

The CUI can be used to determine the net bone balance of a subject. A CUI of ~ 0 shows that the body is in steady state, i.e. bone resorption equals bone formation ($V_l = V_b$). If bone resorption prevails ($V_l > V_b$) the CUI becomes negative. If bone formation prevails ($V_b > V_l$) the CUI becomes positive.

A proper determination of the CUI with non-invasive methods is not possible as the Ca isotopic composition of bone and soft tissue is needed. Therefore, the CUI is currently of no practical importance. Additionally, the definition of the CUI is based on the simple Ca isotope transport model (Fig. 1a) which has been extended during the past years.

3 The Individuality of the Ca Metabolism

One of the most severe challenges using Ca isotopes for biomedical applications is the individuality of the (human) organism and its Ca metabolism. This individuality hampers the simple use of only one sample to reliably detect dysfunctions of the Ca metabolism related to diseases. Instead, studies investigating the isotopic Ca metabolism include a first phase in which a baseline value under healthy/normal conditions is determined. Bone loss during bed rest studies, changes in the dietary input (Ca amount or $\delta^{44/40}Ca_{diet}$) or the effectivity of drug therapies then were detected by changes in $\delta^{44/40}Ca$ (blood/urine) from the baseline. The individuality of the Ca metabolism is apparent from the observed Ca isotopic composition of urine and blood of subjects receiving a similar diet. This individuality has also been observed in archaeological bone samples (Reynard et al. 2010) although differences in the individual Ca intake and $\delta^{44/40}Ca_{diet}$ cannot be ruled out for these samples. The data of Skulan et al. (2007) show that $\delta^{44/40}Ca_{urine}$ values of different subjects span a range of about 1.9 ‰ which is comparable to the range of about 1.6 in $\delta^{44/40}Ca_{urine}$ between subjects at the beginning of a bed rest study (Heuser et al. 2011b). Channon et al. (2015) showed that large differences observed in $\delta^{44/40}Ca_{urine}$ are also found in $\delta^{44/40}Ca_{blood}$. Interestingly, the inter-individual differences in $\delta^{44/40}Ca_{blood}$ are in the order of about 1.1 ‰ while the differences in $\delta^{44/42}Ca_{urine}$ of the same study are about 1.5 times larger. This further illustrates the influence of renal Ca reabsorption and its effect on the Ca isotopic composition of urine and that the process of urine generation always has to be taken into account (Heuser et al. 2011b). In order to correct for the influence of the individual Ca metabolism more data from systematic studies under controlled conditions are needed and more information on the extent and places of Ca isotope fractionation in the body are needed.

4 Current Biomedical Application of Ca Isotopes

4.1 Bone Loss

The detection of bone loss is the most widespread biomedical application of Ca isotopes so far. Calcium isotopes may be a much more sensitive marker to assess current net bone mineral balances compared to conventional methods (e.g.

imaging techniques, serum biomarker). Two different scenarios of bone loss have been investigated using Ca isotopes: osteoporosis and bone loss during spaceflight simulated in bed rest studies. Osteoporosis is a severe disease resulting in huge amounts of fractures causing high costs in the order of several billion dollars per year (Blume and Curtis 2011) and represents a major public health problem. Osteoporosis is marked by an increased loss of bone resulting in a decrease of bone density resulting in an increase in bone fragility. Osteoporosis is defined as a bone density of 2.5 standard deviations below the bone density of a young adult (WHO 2003) and is typically diagnosed by measuring the bone density using DXA (dual-energy x-ray absorption). Besides DXA measurements several other methods exist to detect and diagnose osteoporosis (cf. WHO 2003 and references therein).

Astronauts are threatened by bone loss during manned space flight. Due to a mechanical unloading of bones under microgravity or zero-gravity conditions bone resorption is increased resulting in a decrease of bone density and thus in an increase in bone fragility. Bed rest studies are used to simulate conditions of spaceflight.

Both scenarios of bone loss have in common that bone loss prevails and Ca from the skeleton is released into blood. The Ca isotopic composition of urine should monitor the increased release of isotopically light Ca from bones (cf. Fig. 1). Thus, the Ca isotopic composition of urine should monitor the actual bone mineral balance of a body.

In a pilot study Heuser and Eisenhauer (2010) investigated the Ca isotopic composition of urine of a 4 year-old boy and 63 year-old woman with diagnosed osteoporosis. It was found that the average Ca isotopic composition of urine over the investigated period of time ($\sim$5 days) was significantly different between the two subjects ($\Delta^{44/40}Ca_{boy\text{-}woman} \sim 1.2$ ‰, Fig. 9). This difference in the Ca isotopic composition was mainly attributed to the different bone mineral balance (growth vs. osteoporosis) of the two subjects, but could also be due to the individuality of the Ca metabolism (see Sect. 3) or related to differences in the renal Ca reabsorption (see Sect. 2.4)

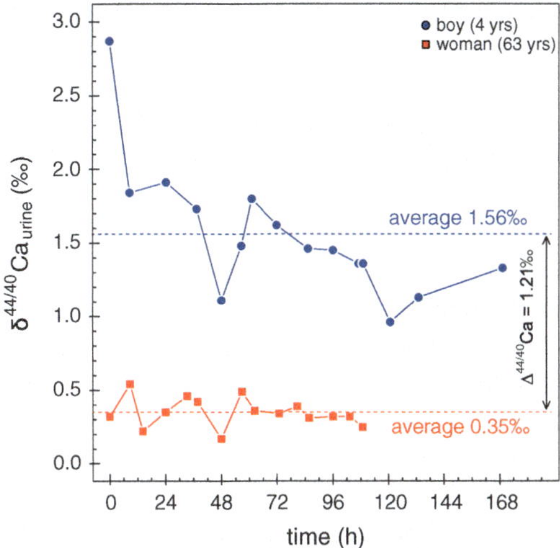

Fig. 9 Heuser and Eisenhauer (2010) showed that the average $\delta^{44/40}Ca_{urine}$ of a 4 year-old boy and 64 year-old woman, suffering from osteoporosis, are significantly different ($p < 0.001$). The average Ca isotopic composition of the woman is lower by 1.1 ‰ which was attributed to a net bone loss. This study confirmed the findings of Skulan et al. (2007). In a more recent bed rest study Morgan et al. (2012) further confirmed that the Ca isotopic composition of urine is sensitive to bone loss

Bone loss during spaceflight and possible interventions have been investigated to a somewhat bigger extend. The effect of spaceflight on the skeleton can be simulated by a prolonged bed rest. Possible countermeasures against bone loss (drugs, diet or exercises) have also been investigated during bed rest studies. Bed rest studies have the big advantage that prior to the bed rest phase the baseline Ca isotopic composition of the subject can be determined and that all subjects are fed on the same diet.

In the pioneering study of Skulan et al. (2007) the effect of a 17 weeks lasting bed rest was investigated by analyzing the Ca isotopic composition of urine. Three different groups were investigated: (1) a control group with no intervention, (2) a group doing daily exercises to prevent bone loss and (3) a group medicated with alendronate ($C_4H_{13}NO_7P_2$) to suppress bone loss. It was found that the Ca isotopic composition of urine shifted to isotopically lighter values for the control group and that this change was correlated to biomarkers typically linked to bone resorption (e.g. NTX). Both intervention groups showed

success i.e. no bone loss was detectable using biomarkers. The success of the therapies was also clearly visible in the urinary Ca isotopic composition. For all subjects from the intervention groups the urinary Ca isotopic composition remained nearly constant over the complete 17 weeks of bed rest which can be expected if no bone loss occurs (Fig. 10).

Heuser et al. (2011b) and Morgan et al. (2012) also analysed $\delta^{44/40}Ca_{urine}$ during a bed rest study including a bed rest phase of 3 and 4 weeks, respectively. Both studies show that at the end of the bed rest phase $\delta^{44/40}Ca_{urine}$ is lower than baseline $\delta^{44/40}Ca_{urine}$ (average $\delta^{44/40}Ca_{urine}$ measured before the onset of bed rest) indicating bone loss during bed rest (Fig. 10). Additionally, Morgan et al. (2012) used the change of $\delta^{44/40}Ca_{urine}$ from baseline to quantify bone loss. They reported that bone loss during bed rest was about 0.25 % of bone mass. Extrapolation of this value to a 90-days period resulted in a bone loss of about 1.0 % of skeletal mass which is in good agreement with bone mass losses in long-term bed rest studies determined by X-ray densiometry scans.

In order to neglect the influence of renal Ca reabsorption rates Channon et al. (2015) measured the Ca isotopic composition of blood.

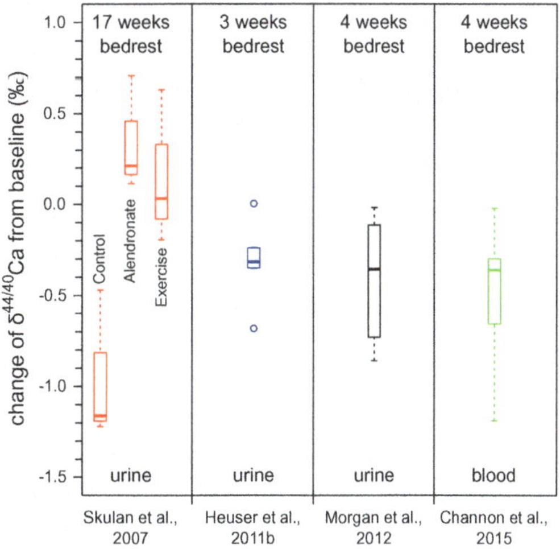

Fig. 10 Boxplot of the changes of $\delta^{44/40}Ca$ of urine and blood from baseline values during bed rest studies. All data were converted to $\delta^{44/40}Ca$ (‰). *Open circles* indicate outliers

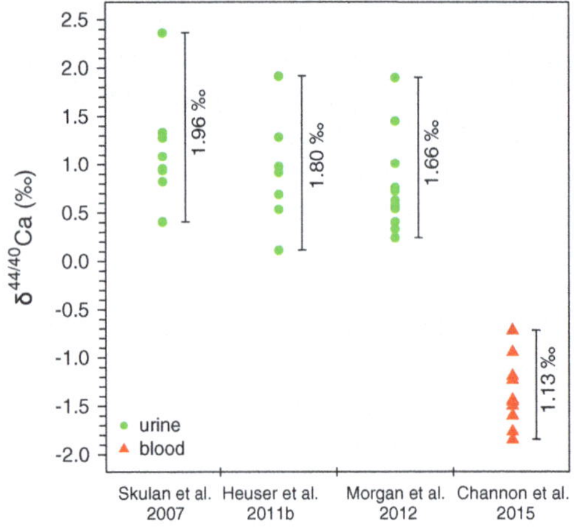

Fig. 11 Measured $\delta^{44/40}$Ca of blood and urine of the different subjects from bed rest studies before intervention (i.e. real bed rest phase) started. $\delta^{44/40}$Ca values show large inter-individual differences, which cannot be explained by different $\delta^{44/40}$Ca$_{diet}$ only

These samples are from the same subjects and the same bed rest study already reported by Morgan et al. (2012). As expected, changes in $\delta^{44/40}$Ca$_{blood}$ from baseline values also indicate bone loss and confirm the findings using $\delta^{44/40}$Ca$_{urine}$ data. When comparing the range in $\delta^{44/40}$Ca$_{urine}$ and range in $\delta^{44/40}$Ca$_{blood}$ before the bed rest phase it can be seen that the range in urine is larger (1.66 ‰) compared to blood (1.13 ‰) (Fig. 11). The larger range in $\delta^{44/40}$Ca$_{urine}$ values may indicate the effect of different renal Ca reabsorption rates.

4.2 Bone Cancer

Multiple myeloma (MM) is a severe disease where bone is destroyed. Osteoporosis, fractures and lytic lesions are typical markers of MM related bone destruction. Imaging methods used to detect and quantify this bone disease are insensitive. More than 30 % of trabecular bone needs be lost in order to clearly observe anomalies (Gordon et al. 2014). Serum biomakers which also may be used to detect bone loss or bone formation are normally used before or during therapy when a baseline value has been

determined. Serum biomarkers are not useful to estimate net bone mineral balance (Gordon et al. 2014). Gordon et al. (2014) explored if $\delta^{44/40}$Ca$_{blood}$ values are more sensitive to diagnose if MM is in an active or non-active state. They found that patients with active disease had statistically significant lower $\delta^{44/40}$Ca$_{blood}$ values than patients with non-active disease. A statistically significant difference of 0.26 ‰ in average $\delta^{44/40}$Ca$_{blood}$ between the active disease and the non-active disease group was found (recalculated from $\delta^{44/42}$Ca). The $\delta^{44/40}$Ca$_{blood}$ values of both groups span a large range (>2 ‰) which is probably linked to the individual diet and individuality of the Ca metabolism.

Gordon and coworker also state that the lower $\delta^{44/40}$Ca$_{values}$ are the strongest indicator for active disease compared to other statistical strong indicators. This is the first clinical study which systematically investigated the possibility to use Ca isotopes as an important sensitive diagnostic tool. The results demonstrate that analysing natural Ca isotope variations may be a useful tool in biomedicine.

5 Summary and Outlook

Calcium is an essential element for the vertebrate/human body and thus the Ca metabolism is well explored. But only little is known about fractionation of Ca isotopes in the human/vertebrate body. The data collected during the past years clearly show that bones are isotopically lighter than blood, that urine is enriched in heavy isotopes compared to blood and that milk is enriched in light Ca isotopes. More recent data also points to the likeliness that during intestinal absorption Ca becomes fractionated. The observed fractionation of Ca isotopes is closely linked to the transport of Ca into organs and body fluids but the mechanism causing the fractionation is not well constrained. Besides this, more data on the Ca isotopic composition of the diet, of other body fluids (e.g. intestinal fluids, sweat) and of more tissues/compartments (e.g. hair, teeth) are needed to complete Ca isotope metabolism. This is an important prerequisite to apply variations of Ca isotopes as a sensitive diagnostic tool in biomedicine.

Studies performed so far look promising to establishing the use of Ca isotopes as a sensitive diagnostic tool especially for bone loss. The detection of bone loss has been shown using invasive (blood) as well as non-invasive (urine) samples. Clinical studies which properly evaluate the sensitivity of Ca isotopes on a bigger set of patients/subjects are needed. However, the precise measurement of the Ca isotopic composition of samples is still challenging and time consuming which limits the operability of big clinical studies.

References

Albarède F (2015) Metal stable isotopes in the human body: a tribute of geochemistry to medicine. Elements 11:265–269

Aramendía M, Rello L, Resano M et al (2013) Isotopic analysis of Cu in serum samples for diagnosis of Wilson's disease: a pilot study. J Anal At Spectrom 28:675

Balter V, Lamboux A, Zazzo A et al (2013) Contrasting Cu, Fe, and Zn isotopic patterns in organs and body fluids of mice and sheep, with emphasis on cellular fractionation. Metallomics 5:1470–1482

Böhm F, Gussone N, Eisenhauer A et al (2006) Calcium isotope fractionation in modern scleractinian corals. Geochim Cosmochim Acta 70:4452–4462

Blume SW, Curtis JR (2011) Medical costs of osteoporosis in the elderly Medicare population. Osteoporos Int 22:1835–1844

Bootman MD, Collins TJ, Peppiatt CM et al (2001) Calcium signalling—an overview. Semin Cell Dev Biol 12:3–10

Bullen TD, Walczyk T (2009) Environmental and biomedical applications of natural metal stable isotope variations. Elements 5:381–385

Campbell AK (1990) Calcium as an intracellular regulator. Proc Nutr Soc 49:51–56

Channon MB, Gordon GW, Morgan JLL et al (2015) Using natural, stable calcium isotopes of human blood to detect and monitor changes in bone mineral balance. Bone 77:69–74

Chu N-C, Henderson GM, Belshaw NS et al (2006) Establishing the potential of Ca isotopes as proxy for consumption of dairy products. Appl Geochem 21:1656–1667

Clarke B (2008) Normal bone anatomy and physiology. Clin J Am Soc Nephrol 3(Suppl 3):S131–S139

Freeman SP, King JC, Vieira NE et al (1997) Human calcium metabolism including bone resorption measured with 41Ca tracer. Nucl Instrum Meth B 123:266–270

Gordon GW, Monge J, Channon MB et al (2014) Predicting multiple myeloma disease activity by analyzing natural calcium isotopic composition. Leukemia 28:2112–2115

Gussone N, Langer G, Thoms S et al (2006) Cellular calcium pathways and isotope fractionation in Emiliania huxleyi. Geology 34:625–628

Heuser A, Eisenhauer A (2010) A pilot study on the use of natural calcium isotope (^{44}Ca/^{40}Ca) fractionation in urine as a proxy for the human body calcium balance. Bone 46:889–896

Heuser A, Tütken T, Gussone N et al (2011a) Calcium isotopes in fossil bones and teeth—diagenetic versus biogenic origin. Geochim Cosmochim Acta 75:3419–3433

Heuser A, Frings-Meuthen P, Rittweger J et al (2011b) Calcium isotopes in human urine under simulated microgravity conditions. Miner Mag 75:1019

Heuser A, Eisenhauser A, Scholz-Ahrens KE et al (2016) Biological fractionation of stable Ca isotopes in göttingen minipigs as a physiological model for Ca homeostasis in humans. Isot Environ Health Stud doi:10.1080/10256016.2016.1151017

Hoenderop JGJ, Nilius B, Bindels RJM (2005) Calcium absorption across epithelia. Physiol Rev 85:373–422

Hotz K, Augsburger H, Walczyk T (2011) Isotopic signatures of iron in body tissues as a potential biomarker for iron metabolism. J Anal At Spec 26:1347–1353

Larner F, Woodley LN, Shousha S, Moyes A, Humphreys-Williams E, Strekopytov S, Halliday AN, Rehkämper M, Coombes RC (2015) Zinc isotopic compositions of breast cancer tissue. Metallomics: Integr Biometal Sci 7:112–117

Morgan JLL, Skulan JL, Gordon GW et al (2012) Rapidly assessing changes in bone mineral balance using natural stable calcium isotopes. Proc Nat Acad Sci 109:9989–9994

Mundy GR, Guise TA (1999) Hormonal control of calcium homeostasis. Clin Chem 45:1347–1352

Pasteris JD, Wopenka B, Valsami-Jones E (2008) Bone and tooth mineralization: why apatite? Elements 4:97–104

Peacock M (2010) Calcium metabolism in health and disease. Clin J Am Soc Nephrol 5(Suppl 1):S23–30

Reynard LM, Henderson GM, Hedges REM (2010) Calcium isotope ratios in animal and human bone. Geochim Cosmochim Acta 74:3735–3750

Schmitt AD, Cobert F, Bourgeade P et al (2013) Calcium isotope fractionation during plant growth under a limiting nutrient supply. Geochim Cosmochim Acta 110:70–83

Skulan JL, DePaolo DJ (1999) Calcium isotope fractionation between soft and mineralized tissues as a monitor of calcium use in vertebrates. Proc Nat Acad Sci 96:13709–13713

Skulan J, Bullen TD, Anbar AD et al (2007) Natural calcium isotopic composition of urine as a marker of bone mineral balance. Clin Chem 53:1155–1158

Tacail T, Albalat E, Télouk P et al (2014) A simplified protocol for measurement of Ca isotopes in biological samples. J Anal At Spectrom 29:529–535

VanHouten JN, Mysolmersky JJ (2007) Transcellular calcium transport in mammary epithelial cells. J Mammary Gland Biol 12:223–235

Walczyk T, von Blanckenburg F (2002) Natural iron isotope variations in human blood. Science 295:2065–2066

Walczyk T, von Blanckenburg F (2005) Deciphering the iron isotope message of the human body. Int J Mass Spectrom 242:117–134

WHO Scientific Group on the Prevention and Management of Osteoporosis (2003) Prevention and management of osteoporosis. World Health Organization, Geneva

Wiegand BA, Chadwick OA, Vitousek PM et al (2005) Ca cycling and isotopic fluxes in forested ecosystems in Hawaii. Geophys Res Lett 32:L11404. doi:10.1029/2005GL022746

The manufacturer's authorised representative in the EU is Springer
Nature Customer Service Centre GmbH, Europaplatz 3, 69115 Heidelberg,
Germany. If you have any concerns regarding our products, please
contact ProductSafety@springernature.com

Printed and bound by CPI Group (UK) Ltd, Croydon, CR0 4YY
08/01/2026
02031749-0005